INTRODUCTION TO PLANT PHYSIOLOGY

Bernard S. Meyer
The Ohio State University

Donald B. Anderson
University of North Carolina

Richard H. Bohning
The Ohio State University

Douglas G. Fratianne
The Ohio State University

D. Van Nostrand Company
New York/Cincinnati/Toronto/London/Melbourne

D. Van Nostrand Company Regional Offices:
New York   Cincinnati   Millbrae

D. Van Nostrand Company International Offices:
London   Toronto   Melbourne

Library of Congress Catalog Card Number: 72-6329

ISBN: 0-442-25328-1

Published by D. Van Nostrand Company
450 West 33rd Street, New York, N.Y. 10001

Published simultaneously in Canada by
Van Nostrand Reinhold Ltd.

10   9   8   7   6   5   4   3

PREFACE

This book is essentially a modernized and updated revision of Meyer, Anderson, and Böhning's *Introduction to Plant Physiology,* first published in 1960, which was in turn a condensed version of the older well-known *Plant Physiology* by Meyer and Anderson, last revised in 1952. Although the topical sequence and general philosophical approach of the first edition of this book have been retained in the main, the text has been largely rewritten in order to bring it abreast of recent advances in this field of knowledge. The order of presentation of the major topics appears to the authors to be both logical and psychological, but they can be used in other sequences if so preferred. A studied attempt has been made to integrate discussions from chapter to chapter in order to assist in the realization that a plant operates as a coordinated unit.

This text is designed primarily for use in a one-semester or a one- or two-quarter first course in plant physiology. Background training only in general botany or biology and general chemistry is assumed for the students. Although a background in organic or biochemistry would obviously be advantageous even to students in a beginning plant physiology course, it has not been assumed in writing this text, and basic principles of these disciplines are incorporated into the discussion as needed. With appropriately selected supplementary readings the book could be used as core reading material for an advanced course. The many monographs, advanced textbooks, reviews, and encyclopedic articles now available in the field make this kind of a reading program an especially feasible procedure.

The central presentation is concerned with the whole plant, with ecological undertones. The authors feel strongly that this is a proper first approach to an understanding of the physiology of plants. Cellular and biochemical levels of interpretation are not neglected, however, especially insofar as they contribute to an understanding of the operation of a plant as a coordinated unit. And we have attempted to inculcate some feeling for the wide diversity of physiological reactions—rooted in genetic differences—which exist from one kind of plant to another.

The authorship of an introductory book which does not run to too many pages is in many ways a more difficult task than the preparation of a more extended treatment of the same general subject. The knowledge explosion which has invaded all fields of science in recent years has not made the writing and organization of a compact textbook any easier. The burden of evaluation becomes formidable, as to which topics to cover, the proportionate coverage of the various topics, and the depth of interpretation which is appropriate for each of the topics.

As is true in most introductory textbooks, many interpretations must be more or less simplified or curtailed. The most important considerations regarding any simplified or curtailed treatment of any field of science is that it not only be internally consistent, but also that it be consistent with the more sophisticated interpretations that some students will encounter later as they move into more advanced studies in the field.

Since, in a short textbook, space limitations preclude the development of the subject in depth either historically or intellectually, such a book is unavoidably invested with an aura of authoritarianism which does not represent the true spirit of science. An important role of the teacher is to recognize this built-in shortcoming of a brief text and to attempt to dispel the concept of authoritarian science in the minds of his students.

We have written this book for students and hope that it will be judged by fellow plant physiologists in the light of this objective. Preliminary drafts of most of the chapters have been used as assigned readings in plant physiology classes for one or more years. Such use has resulted in a considerable constructive feedback from students, which has proved helpful and for which we are appreciative. We are indebted to Dr. Paul J. Kramer, of Duke University, for the opportunity of examining certain chapters of his book, *Plant and Soil Water Relationships: A Modern Synthesis,* while still in manuscript form; also to Dr. Carroll A. Swanson, of The Ohio State University, Dr. Francis H. Witham, of The Pennsylvania State University, and Dr. Carlos O. Miller, of Indiana University, for their critical reading of certain chapters. Figures taken from or adapted from other sources are properly credited in the legends.

BERNARD S. MEYER
DONALD B. ANDERSON
RICHARD H. BÖHNING
DOUGLAS G. FRATIANNE

CONTENTS

Osmotic Pressure.   Turgor Pressure.   Quantitative Aspects of Osmosis.   Factors Influencing the Osmotic Potential of Solutions. Imbibition.   Dynamics of Imbibition.   Osmotic Effects on Imbibition.   Quantitative Aspects of Imbibition.

Variations in The Water Potential of Plant Cells.   Permanent Wilting.
Internal Redistributions of Water in Plants.   Drought Resistance.

istry of Gibberellins. Roles of the Gibberellins. Synthesis and Translocation of the Gibberellins. The Cytokinins. Roles of the Cytokinins. Synthesis of Cytokinins. Abscisic Acid. Roles of Abscisic Acid. Vitamins. Ethylene. Effects of Ethylene. Other Hormone-like Substances in Plants. Mechanism of Hormone Action. Interactions among Hormones.

1 | INTRODUCTION

When the progenitor of this book was written more than thirty years ago, the world of mankind was a smaller place. Since then men have landed on the moon and found it a lifeless and barren globe. Our probes have surveyed the planet Mars, long considered a likely habitat for living organisms, and, as nearly as can be discerned, have found only a bigger moon. If the cosmos is as vast and timeless as we conceive it to be, it seems certain that life in some manifestations exists in other galaxies, but they are probably forever beyond our reach, except in the imagination. More than ever before it is becoming clear that life as we know it on this earth is a unique phenomenon, at least within our own solar system.

Green plants are the basic architects of this earth-bound heritage of life. They are the only significant bridge between the existence of a living world of which we are a conscious part and a dead one in which no living organism lives or breathes. With inconsequential exceptions, they are the only kind of organism which can utilize some of the simple inorganic substances present in their environment and construct from them the complex organic molecules out of which living organisms of all kinds are composed and from which they derive their energy. They alone among all kinds of living organisms have the capacity of converting the energy of sunlight into the chemical energy of such complex organic compounds. In short, green plants are the ultimate source of all food which is used, not only to sustain themselves, but also all other kinds of living organisms, including man himself.

The world of chlorophyllous (green) plants includes hundreds of thousands of species which are almost unbelievably diverse in size, configuration,

1

structure, and physiological behavior. Because of their diversity, such organisms are present in almost every conceivable kind of habitat. Green plants occupy almost every variety of land surface from marshlands to semi-arid deserts and exist under tropical, temperate, and subarctic temperature regimes. They are also present, often in great abundance, in most parts of oceans, lakes, and rivers. A large proportion of the green plants which live in aquatic habitats are free-floating, unicellular, or colonial organisms, which, as a group, go under the name of phytoplankton. Plants thus spread their green mantle over the surface of the earth. Only extremely dry or extremely cold regions are devoid of plant life or virtually so. And, as is true of all other living organisms, plants are self-perpetuating.

An understanding of plants in all of their biological versions is thus a most fundamental branch of human knowledge. In an empirical way man has studied plants ever since he first learned to poke seeds into the ground so that plants would grow where he wished them to, instead of in the wild. But in our modern complex civilization, the empirical knowledge of plants is not adequate. It is only by a disciplined and organized scientific approach to the study of plants that knowledge about them becomes available in such forms and of such validity that it can be used successfully in managing or controlling them to human advantage.

Interrelationships of Plant Physiology with Other Branches of Science. Botany is often defined as the science of plant life. As commonly understood, however, the term botany does not entirely embrace this field in a completely comprehensive manner. Many phases of several other well-recognized disciplines such as agronomy, horticulture, and forestry clearly come under the heading of the science of plant life. Workers in such fields have contributed substantially to the development of botany in the broad sense of the word.

Plants can be studied from various points of view, and the science of plant life is conventionally divided into a number of different branches, of which physiology, anatomy, genetics, ecology, morphology, pathology, and taxonomy are the most familiar. Such compartmentalizations of knowledge appeal to the orderly mind and are useful for emphasizing the major aspects of plant life. But nature abhors pigeonholes and exists as a continuum, every facet of which grades into every other facet. Hence any definite boundaries set to the field of plant physiology or to any other phase of botany are bound to be somewhat arbitrary. Each phase of the subject grades almost imperceptibly into others.

Basically, however, plant physiology is considered to be that branch of botany which comprises a knowledge of the processes which occur in plants. The generalized machinery of physiology is the same in all of the chlorophyllous plants. If this were not true, there could be no science of plant physiology. But the obvious diversity of plants in their outward configuration, internal anatomy, and behavioral pattern reflects a corresponding diversity of operation at the physiological level. The physiology of an oak tree differs from that of a tomato plant in the same sense that the physiology of a horse differs from that of a cat. Both of these animals possess many elements of physiology in common: yet there are obvious physiological differences between them. The same statement can be made regarding the two kinds of plants.

The diversity of plants in all of their aspects reflects a basic influence of their genetic constitution. The differences in physiological reactions between one species and another are as much a reflection of their genetic differences as are their usually more obvious differences in external morphology. Physiological manifestations of genetic differences are often more subtle than morphological ones. For example, some varieties of a given species of plant may be much better able than other varieties of the same species to tolerate low temperatures, yet the two varieties may be largely or entirely indistinguishable morphologically. Whenever the same process or plant response is studied comparatively for two different species, or even for two different varieties of the same species, the effect of their genetic makeup comes under implied if not explicit consideration.

However, as discussed later in this book, the control of metabolism does not rest with genetic constitution alone, but in the hand-in-glove interaction between genetic factors and environmental conditions. A study of the effects of various environmental factors on plant processes is therefore an integral phase of plant physiology. Investigations of such factors as light, temperature, humidity, and soil moisture on such processes as photosynthesis, respiration, transpiration, and growth are examples of studies which can be carried out either in the laboratory or in the field. This area of plant physiology overlaps the field of plant ecology in a borderline domain often called physiological ecology.

It has long been a basic tenet of plant physiology that the usually complex processes occurring in plants can be resolved into the relatively simpler processes of chemistry and physics. The field of plant physiology thus grades imperceptibly into the fields of plant biochemistry and plant biophysics. The use of chemical and physical techniques of experimentation has led to substantial progress in the elucidation of many of the physiological processes which occur in plants. Complete explanations of most of these processes in terms of chemical and physical principles are still lacking, however, not because the basic precept is erroneous but because many of the processes which occur in living organisms are infinitely more complex than those which occur in the inanimate world. Living systems have the baffling propensity of seeming, at least, to add up to more than the parts into which they can be resolved.

Plant processes do not occur in empty space but in structural units at the cellular and sub-cellular level. The relationship between structure and process is an intimate one and a reciprocal one. Growth, which is a complex of co-ordinated physiological processes, results in the development of cells, tissues, and organs. Once the organization of a cell or the cellular structure of a tissue has materialized, however, it has marked effects upon the physiological processes which proceed within it. Structures are forged by processes, and processes in turn are influenced by structures; thus process and structure are inseparable facets of plant growth and behavior.

Plant Physiology as a Science.   This book is an epitome of all that has gone before in plant physiology. As with all other sciences the body of knowledge which we call plant physiology has been built up, step by step, by generations of workers in this field. Both the authors of this book and the readers of this book owe a tremendous debt to the many men who have helped advance this field of knowledge. The contribution of many of these scientists and scholars is relatively small and for some consists of little more than a holding operation whereby accumulated knowledge is preserved and disseminated. The whole science of plant physiology is illuminated, from time to time, as are others, by the occasional worker who emerges as an intellectual leader in the field. The names of only a few such eminent plant physiologists can be mentioned in an introductory text; all others must remain anonymous, although their contributions, in the aggregate, have been enormous.

The starting point of all science is the observation of natural phenomena. Some sciences, such as geology and astronomy, must by their very nature be constructed principally on a foundation of observationally obtained data. Only nature's own experiments can be observed. In other fields of science, of which physics and chemistry are the prime examples, most basic facts are obtained as a result of deliberate experimentation by man himself. The biological sciences occupy an intermediate position between those sciences which are largely or entirely observational, and those which depend mostly on experimentation for their substructure of facts.

Plant physiology is principally an experimental branch of biological science, but the appropriate lines of the experimentation to be followed are often suggested as a result of perceptive observations of plant behavior. In the most critically performed experiments plants are grown under such conditions that all of the environmental factors known to affect their process and development are under control. The biological unit of study may be organelles, cells, tissues, organs, whole plants, or even groups of plants, depending upon the objectives of the experiment.

A less rigorous, but for some purposes very informative kind of experimentation with plants, is to measure the rates of processes or patterns of morphogenic development under field conditions, which means that the plants are exposed to the complex of factors constituting a natural environment. In investigations of this sort, if they are to be subject to meaningful interpretations, it is necessary to make parallel and more or less continuous measurements of the presumably effective environmental conditions over the duration of the experimentation.

The basic substructure of any science consists of the facts and data obtained by observation and experimentation. But facts alone are not enough to erect an edifice of scientific knowledge. A further step that is also necessary to this edifice is the formulation of generalizations. Such generalizations are sweeping statements which are valid across a wide spectrum of the science. They usually begin as postulated theories or hypotheses which are tested and retested by further experimentation, often by other scientists than the one who originally proposed the hypothesis. Continued experimentation may lead to a substantiation of the original theory, or to its modification, or to its rejection.

Experiments often raise more questions than they answer. New approaches to the problem under consideration as well as desirable new lines of inquiry are constantly opening up to the alert investigator. In this way experimentation leads to more experimentation, more facts accumulate, and more theories are proposed. Some of the suggested hypotheses are confirmed, others are rejected, and still others are modified. Most of them, sound or fallacious, in turn suggest further observations and experimentation. As a result of such endless and painstaking labors, there is slowly built up that vast, complex, and ever-changing body of knowledge which we refer to as a science.

The system of subjecting all hypothetical explanations of natural phenomena to experimentation is the essence of experimental science. Progressive modification of accepted concepts in the light of new experimental findings continually increases the soundness of scientific generalizations. Thus there are incorporated into any science theories and generalizations in various stages of acceptance. Some stand upon such a firm substructure of facts that they are accepted by all the authorities in the field. Others, less securely supported by experimental results, are subscribed to by some but rejected by other workers. Finally, in any science there are always some theories which are so dubious that they will find only a few advocates. Furthermore, some of the theories now widely held sooner or later will be discarded or modified as a result of new findings or as a consequence of the different interpretations of the facts already known.

However, not all scientists are always in agreement regarding the interpretation of the same set of facts. Although this state of affairs is entirely consistent with the spirit of scientific research, it is frequently puzzling to students and laymen. Differences of opinion regarding the hypotheses which suitably explain scientific phenomena are most likely to arise when experimental data are inadequate. Disagreements regarding the interpretations of experiments and observations are often inevitable steps in the clarification of scientific generalizations. Controversies usually focus attention upon gaps in our factual information. Frequently, therefore they are stimulating to research and often lead to a further enrichment of human knowledge.

Non-botanists are often perplexed by the fact that much experimental work on plants is done with species which are of very little use or interest to man. A plant physiologist interested in the basic aspects of the science often chooses as his experimental plants species which are especially well adapted for one reason or another to furnishing insight into the particular process or phenomenon he wishes to investigate. The unicellular alga *Chlorella,* for example, has been the most widely used organism in the experimental work done on photosynthesis. This alga, which grows satisfactorily on culture media, can be readily subjected to the wide variety of laboratory manipulations which are essential to a study of the mechanism of photosynthesis, without encountering some of the complications which the use of structurally more complex plants would entail. To cite another example, the cocklebur, a useless and uninspiring weed, has been widely used in experiments on photoperiodism. The photo-periodic reaction leading to flowering can be evoked in this plant by exposure to one light-dark cycle of the proper proportions. For this reason, cocklebur is especially well adapted to studies of this mechanism.

The results obtained with such "guinea pig" plants as *Chlorella* and cockle-bur can usually be generalized, at least to a substantial degree, to other species. However the point should be emphasized again that physiological processes as they occur in one plant do not take place in exactly the same manner in all others. Differences in physiology which occur from one species to another may be qualitative or quantitative. All green plants carry on photosynthesis, for example, but the metabolic pathways followed in accomplishing this process are not the same in all species. Quantitative differences in physiology are of even more widespread occurrence from one kind of plant to another than qualitative ones. To cite one of the many possible examples, the rate of tran-spiration from plants of different species will usually be quite different, even when all of them are exposed to identical environmental conditions.

Plant Physiology and Human Welfare.   The point has already been made that plants are the ultimate source of food for all living organisms including man. Human beings use many plants and plant products directly as foods. In addition, the animals and animal products used as foods by man all have their ultimate origin in plants. Even the fish and other aquatic organisms used as human food all derive their own food ultimately from the chlorophyllous phytoplankton. The food chains leading from the larger predatory fish and the other aquatic animals often include a number of organisms as intermediate steps, but all such food chains have a common origin in the phytoplankton.

With the burgeoning of the world population, increasing the supply of food available for mankind has become more and more a matter of concern. The pressure on agriculture, especially in undeveloped countries, to produce more food has been enormous. Marked increases in the yield of crops have been obtained in some countries, and an application of some of the principles of plant physiology has contributed to this greater abundance of food. There are limits, however, beyond which agriculture cannot be pushed. The ultimate solution to the food problem is not more plants but fewer people.

The plant products besides foods of importance to human welfare in the material sense include lumber, fibers (cotton, flax, sisal, and hemp), vegetable oils, rubber, and drugs. Important animal products, of which leather and wool are the best examples, are also indirectly of plant origin.

Until comparatively recent times the vast majority of the people of the world spent most of their waking hours in agricultural pursuits, and in the majority of countries of the world today agriculture still remains the predominant occupation. In most of the more highly developed industrial nations, however, agriculture has become so efficient that the proportion of persons engaged in this occupation is relatively small and is still shrinking. More and more of the people live in towns and cities, and fewer of them depend directly upon plants for a livelihood. For many such urban populations the aesthetic and recreational aspects of plants are of great interest. For millions of persons house plants, ornamental shrubbery, flowers, vegetable crops, fruit trees, and shade trees are the kinds of plants with which they have the most direct contact and in which they have the greatest interest. Such plants are sources of avocational occupation, aesthetic pleasure, and often of physical exercise for these urban dwellers. Fulfilling in a somewhat different manner the recreational needs of many people are the numerous municipal, state, provincial, and national parks, forests, and game preserves, whose attractiveness for such purposes depends largely upon the presence of natural vegetation.

Man, therefore, can be expected to be vitally concerned with plants as long as he continues to inhabit the earth. A better understanding of, and more control over the growth of plants, has been and will continue to be the significant contribution of plant physiologists to the peoples of the world.

GENERAL REFERENCES

Each of the references listed here covers a wide range of topics in plant physiology or a closely related field. References listed at the end of subsequent chapters pertain primarily to the topics discussed in that chapter.

Annual Reviews of Plant Physiology. Vol. 1– . Annual Reviews, Inc., Stanford, 1950–

Baum, S. J., *Introduction to Organic and Biological Chemistry.* The Macmillan Company, New York, 1970.

Bonner, J., and J. E. Varner. *Plant Biochemistry.* Academic Press, Inc., New York, 1965.

The Botanical Review. Vol. 1– . The Botanical Review, New York, 1935–

Devlin, R. M. *Plant Physiology, 2nd Edition.* Van Nostrand Reinhold Company, New York, 1969.

Eastin, J. D., F. A. Haskins, C. Y. Sullivan, and C. M. H. Van Bavel. *Physiological Aspects of Crop Yield.* American Scociety of Agronomy, Madison, Wisconsin, 1969.

Encyclopedia of Plant Physiology, 18 vols. Springer-Verlag, Heidelberg, 1955-1967.

Janick, J. *Horticultural Science, 2nd Edition.* W. H. Freeman and Company, San Francisco, 1972.

Kozlowski, T. T. *Growth and Development of Trees*. 2 vols. Academic Press, Inc., New York, 1971.

Lehninger, A. L. *Biochemistry*. Worth Publishers, Inc., New York, 1970.

Luckwill, L. C., and C. W. Cutting, Editors. *Physiology of Tree Crops*. Academic Press, Inc., New York, 1970.

Salisbury, F. C., and C. Ross. *Plant Physiology*. Wadsworth Publishing Company, Belmont, Calif., 1969.

Spector, W. S., Editor. *Handbook of Biological Data*. W. B. Saunders Company, Philadelphia, 1956.

Steward, F. C. Editor. *Plant Physiology: A Treatise*. 6 vols. Academic Press, Inc., New York, 1959-1972.

Zimmerman, M. H., and C. L. Brown. *Trees: Structure and Function*. Springer-Verlag, New York, 1971.

2 | SOLUTIONS AND COLLOIDAL SYSTEMS

The dynamics of living systems can be largely interpreted in terms of the physio-chemical properties of solutions and colloidal systems, one component of which is water. In physiologically active plant cells, water is the most abundant compound present. Because of its great solvent power, water in a liquid state in plant cells is never pure but contains other substances dissolved in it. In addition it usually contains dispersed particles not in true solution. When the dispersed particles are within a certain range of sizes, this system of water plus particles is referred to as a colloidal system.

Similarly, in the natural environment of living organisms, water invariably contains dispersed particles as well as substances in solution. This is true of the water of streams, lakes, or oceans as well as of the water in the soil. Even raindrops, the products of natural distillation, contain gases and other substances dissolved in them from the atmosphere.

General Nature of Solutions. Simple solutions contain at least two components in which one (the solute) is dispersed throughout the other (the solvent) in the form of molecules or ions. In most familiar solutions the solvent is a liquid. Naturally occurring solutions usually contain a number of different solutes and are often very complex. Since the commonest solvent in both the inorganic and the organic worlds is water, the subsequent discussion will be principally in terms of aqueous solutions.

Methods of Expressing the Composition of Solutions. If the molecular weight in grams (1 mole) of a substance is dissolved in enough water to make

9

exactly 1 liter of solution at 20°C, the result is a *volume molar* solution, usually called a *molar solution*. Since the gram molecular weights of all substances contain the same number of molecules (6.02 × 10²³), equal volumes of all solutions of the same molarity will contain the same number of solute molecules. If a given volume of a 1 molar solution is diluted with an equal volume of water, the result is a 0.5 *M* solution. A 0.5 *M* solution can also be obtained by dissolving one-half of the gram molecular weight of a substance in enough water to make 1 liter of solution at 20°C.

If a mole of any substance is completely dissolved in 1000 g of water, the result is a *weight molar* or *molal solution*. Such solutions are used principally in experimental work upon various osmotic phenomena. The addition of a mole of most solids to a liter of water increases the volume of the resulting solution to more than 1 liter. This increase in volume is called the *solution volume* of the solute. The solution volume of many substances is very small, and for a few it is even negative, *i.e.,* there is a shrinkage in volume when the solute is added to the solvent. On the other hand, the solution volume of some compounds, especially the sugars, is considerable. When a mole of sucrose is added to 1000 g of water, the resulting solution will have a volume of 1207 ml at 0°C. Hence the solution volume of a mole of sucrose is 207 ml. The solution volume of a mole of sodium chloride, on the other hand, is only about 18 ml. Since every solute has a different solution volume, it follows that equal volumes of molal solutions of different substances do not contain the same number of either solvent or solute molecules. Dilution of a given volume of a 1 molal solution with an equal volume of water does not result in a 0.5 molal solution, because such solutions must be diluted in terms of the volume of solvent present, not in terms of the total volume of the solution.

In physiological work it is often convenient to make up solutions on a percentage basis. The simplest procedure to follow is to make all such solutions strictly on the basis of percentage by weight. A 10 per cent solution of sodium chloride, on this basis, is made by dissolving 10 g of sodium chloride in 90 g of water. Similarly, a 20 per cent solution of acetone is made up mixing 20 g of acetone with 80 g of water. Such solutions are designated as per cent w/w (weight/weight) solutions. When a solid is dissolved in a liquid, the designation per cent w/v (weight in grams/volume in milliliters) is often used. If water is the solvent in which a solid is dissolved, a given per cent w/v solution can equally and appropriately be designated as a per cent w/w solution because 1 milliliter of water weighs 1 gram (at 4°C). Solutions of liquids in liquids are sometimes made up on a per cent v/v (volume/volume) basis. For example, the solutions used in chromatographic work are usually prepared in terms of proportions by volume. A 3:3:2 solution of butanol—acetic acid—water would be made from three volumes each of butanol and acetic acid and two volumes of water. Since the total volume of the liquids after being mixed is sometimes less than the sum of their separate volumes before mixing, the volume of each liquid should be measured out individually before mixing.

Another system of expressing concentration is that of parts per million (ppm). This system is based upon the fact that a liter of water contains 1000 ml, each milliliter of which weighs 1000 mg at 4°C; there are thus one million (1,000,000) mg per liter. If, for example, 5 mg of a substance is dissolved in water to 1 liter, the result is a 5 ppm solution. Parts per million of an aqueous solution are equivalent to milligrams per liter (mg/l).

Electrolytes and Nonelectrolytes.   Some aqueous solutions conduct an electrical current; others do not. The former are called electrolytes; the latter nonelectrolytes. Electrolytes may be further, but not rigidly, reclassified into strong electrolytes and weak electrolytes. The solutions of many acids, bases, and salts are strong electrolytes. The solutions of organic acids, an important kind of compound in plant metabolism, are weak electrolytes. Examples of nonelectrolytes are the solutions of sugars, alcohols, and ketones.

The conduction of an electrical current by a solution is dependent upon the presence in solution of charged particles called *ions*. Positively charged ions are called cations; negatively charged ions, anions. A single cation may carry from one to several positive charges; a single anion may carry from one to several negative charges. At any given moment the total number of positive charges on all the cations in a solution is equal to the total number of negative charges on all of the anions.

The ions formed by different kinds of electrolytes in solution originate in different ways. Many electrolytes, such as NaCl and similar salts, have an ionic structure even in crystal form. The solution of such compounds results in breaking down the structural crystal lattice, thus releasing preexisting cations and anions into the solution in electrostatically equivalent amounts. Compounds of this sort are invariably strong electrolytes.

Some substances which are not ionic in the pure state become so when put in solution. A good example of this is HCl which is a strong electrolyte when dissolved in water.

Weak electrolytes, of which acetic acid (CH_3COOH) is an example, belong in still a different category. Such compounds remain largely in the molecular state in solution, but dissociate slightly, forming equivalent amounts of cations and anions in low concentrations:

$$CH_3COOH \rightleftharpoons CH_3COO^- + H^+$$

A dynamic equilibrium is considered to exist between the ions and the undissociated molecules. A 1 M solution of acetic acid is only about 0.4 per cent dissociated. With increasing dilution, the proportion of dissociated molecules increases. A 0.0001 M solution of acetic acid, for example, has about 15 per cent of the molecules present in the dissociated state. In extremely dilute solutions dissociation is virtually complete even with weak electrolytes.

It is generally considered that strong electrolytes such as NaCl are 100 per cent dissociated when in solution. The effects of such a solute on certain properties of a solution such as its osmotic potential (Chapter 4) are not quantitatively what would be expected in terms of its complete dissociation. According to the Debye-Hückel theory, the fact that strong electrolytes do not exert efforts consistent with their complete ionization is accounted for by interionic attractions. The anions and cations of NaCl, for example, attract each other because of their opposite charges with the result that some of them operate as ion pairs, rather than as individual ions. An ion pair has only half the effect on a colligative property such as the osmotic potential that the two ions would have if acting individually. Those ions which are associated with each other as pairs are considered to be in dynamic equilibrium with the free ions in the solution.

Acids, Bases, and Salts.   An acid may be defined as a substance which forms hydrogen ions (H^+) when dissolved in water. Since a hydrogen ion is, in effect, a proton, acids are sometimes defined as compounds that can *donate* a proton to other compounds. For various reasons it is considered unlikely that free hydrogen ions (protons) wander at large in a solution, and there is evidence that each hydrogen ion becomes loosely attached to a water molecule, becoming a *hydronium* ion (H_3O^+). From this point of view the ionization of HCl represents a reaction with water:

$$HCl + H_2O \rightleftharpoons H_3O^+ + Cl^-$$

Since the hydronium ion is, in effect, a hydrated proton, and the water of hydration is often omitted from chemical equations, the previous equation is usually simplified to read:

$$HCl \rightleftharpoons H^+ + Cl^-$$

The "strength" of an acid or base depends upon its degree of dissociation at a given concentration. A base may be defined as a substance which produces hydroxyl ions (OH^-) when in solution. Bases are also sometimes defined as substances which can *accept* protons.

Salts are formed when an acid and a base are brought together in a solution, as a result of a chemical union between the hydrogen ion(s) of the acid and the hydroxyl ion(s) of the base, forming water. This reaction is called *neutralization*. The following are several examples:

$$HCl + NaOH \rightleftharpoons NaCl + H_2O$$
$$H_2SO_4 + 2KOH \rightleftharpoons K_2SO_4 + 2H_2O$$
$$2HCl + Ca(OH)_2 \rightleftharpoons CaCl_2 + 2H_2O$$

A salt may therefore be defined as a compound resulting from the union of the anion(s) of an acid with the cation(s) of a base. The reverse reaction, indicated in the previous equations by the arrows pointing to the left, is called *hydrolysis*.

Normal Solutions.   The concentrations of acids and bases are most commonly expressed in terms of normal solutions. A 1 N solution contains a gram-equivalent weight (1.008 g) of ionizable hydrogen or its equivalent per liter of solution at 20°C. Thus a 1 N solution of an acid contains 1.008 g of ionizable hydrogen per liter; a 1 N solution of a base contains 17.008 g (1 gram-equivalent) of ionizable hydroxyl per liter. The dilution of a normal solution results in a reduction in its normality in proportion to the magnitude of the dilution. The normality of an acid solution is a measure of its *total acidity, i.e.* of its concentration in terms of ionizable hydrogen ions. Similarly, the normality of a base solution is a measure of its *total basicity.* Since 1.008 g of ionizable hydrogen represents the same number of ions as 17.008 g of ionizable hydroxyl, it is evident that equal volumes of acid and base solutions of equal normality will exactly neutralize each other.

Hydrogen Ion Concentration.   Some of the most fundamental of the physiological phenomena are markedly influenced by the concentration of hydrogen ions in the medium in which they occur. For many purposes, therefore, it is more important to have an index of the concentration of hydrogen ions than of the total acidity of the solution.

Total acidity, as already noted, is customarily expressed in terms of normality. It is also entirely possible to use the normal system for expressing the concentration of hydrogen ions. A 1 N solution of acetic acid is only about 0.42 per cent dissociated. Such a solution is 1 N in terms of the ionizable (titratable) hydrogen present, but only 0.0042 N in terms of the dissociated hydrogen ions.

Since no acid ever behaves as if it is completely dissociated, a normal solution of any acid will always be less than normal when its concentration is expressed in terms of the hydrogen ions present. In order to prepare a normal solution of hydrogen ions it is necessary to make up a solution which is more than normal in terms of total acidity. Such a solution must be of the precise strength and degree of dissociation that exactly 1.008 g of the ionizable hydrogen present are actually in the dissociated form—as ions—per liter of the solution.

Although hydrogen ion concentration can be readily expressed in terms of normalities, in actual practice this system is not generally used because it is cumbersome, especially when it is necessary to refer to the very small concentrations of hydrogen ions usually dealt with in biological problems. The hydrogen ion concentration of a solution is generally defined in terms of its pH value. The pH value bears a simple mathematical relation to the hydrogen ion concentration of a solution in terms of its normality. Because of the practically universal acceptance of this system, it is necessary to understand the significance of the term pH and its relation to the hydrogen ion concentration expressed in terms of normality.

The relation between pH and hydrogen ion concentration in terms of normality is a logarithmic one (Table 2.1). The pH value for any solution is the

TABLE 2.1 THE RELATION OF pH TO HYDROGEN-ION CONCENTRATION
EXPRESSED IN TERMS OF NORMALITY

	OH ion concentration in terms of normality ⟶		pOH	pH	⟵ H ion concentration in terms of normality	
	.00000000000001	10^{-14}	14	0	10^{0}	1
	.0000000000001	10^{-13}	13	1	10^{-1}	.1
	.000000000001	10^{-12}	12	2	10^{-2}	.01
Acid	.00000000001	10^{-11}	11	3	10^{-3}	.001
Range	.0000000001	10^{-10}	10	4	10^{-4}	.0001
	.000000001	10^{-9}	9	5	10^{-5}	.00001
	.00000001	10^{-8}	8	6	10^{-6}	.000001
	.... .0000001	10^{-7}	7	7	10^{-7}	.0000001 ....
	.000001	10^{-6}	6	8	10^{-8}	.00000001
	.00001	10^{-5}	5	9	10^{-9}	.000000001
Alkaline	.0001	10^{-4}	4	10	10^{-10}	.0000000001
Range	.001	10^{-3}	3	11	10^{-11}	.00000000001
	.01	10^{-2}	2	12	10^{-12}	.000000000001
	.1	10^{-1}	1	13	10^{-13}	.0000000000001
	1	10^{0}	0	14	10^{-14}	.00000000000001

negative of the logarithm of the hydrogen ion concentration in terms of normality. Zero is the logarithm of 1, -1 is the logarithm of 0.1, -2 is the logarithm of 0.01 and so forth. The pH of a 1 N solution is therefore 0, of a 0.1 N solution 1, of a 0.01 N solution 2, and so forth.

All aqueous solutions as well as pure water contain hydrogen ions in some concentration. The mathematical product of the concentration of hydroxyl ions and the concentration of hydrogen ions in a solution is a constant. This may be expressed as follows:

$$(H^+) \times (OH^-) = K \qquad K = 10^{-14} \text{ at } 22°C$$

Corresponding to the concentration of hydrogen ions, therefore, there is always a certain definite concentration of hydroxyl ions. As the pH of a solution is increased, the pOH decreases and vice versa. For example, if the pH increases from 5 to 6, the pOH decreases from 9 to 8. Furthermore, only at pH 7, which is also pOH 7, can the concentration of the hydrogen ions equal the concentration of the hydroxyl ions. This is, therefore, the neutral point on the pH scale (at 22°C), and it corresponds to the dissociation of pure water. This pH value represents a dissociation of only one water molecule in approximately every 555,000,000.

Values below 7 on the pH scale represent the acid range, those above 7 the basic or alkaline range of the scale. An "acid" solution is one with a larger concentration of hydrogen ions than hydroxyl ions, while in an "alkaline" solution the reverse is true. The lower the pH value, the greater the hydrogen ion concentration of a solution. A pH value of 5 represents ten times the hydrogen ion concentration of a solution with a pH of 6 and one hundred times the hydrogen ion concentration of one with a pH value of 7. This is a consequence of the fact, previously mentioned, that the numbers on the pH scale are related to each other as logarithms and not as ordinary arithmetic numbers.

The hydroxyl ion concentration of a solution could be expressed in pOH units instead of in pH units. For alkaline solutions especially, this would seem to be a logical practice. But because of the definite mathematical relationship between the pH and pOH, the pH value alone also defines the pOH value. Hence both the acidity of a solution in terms of $H+$ ions and its alkalinity in terms of OH^- ions may be, and customarily are, expressed in terms of pH units.

Buffer Action. The addition of 1 ml of a 1 N hydrochloric acid solution to 10 ml of a 1 N sodium acetate solution will cause the pH of the solution to change from approximately 8.8 to only about 6. If, however, 1 ml of a 1 N hydrochloric acid solution is added to 10 ml of a 1 N sodium chloride solution with an initial pH of about 7, the pH of the resulting mixture drops to a value close to 1 (Fig. 2.1).

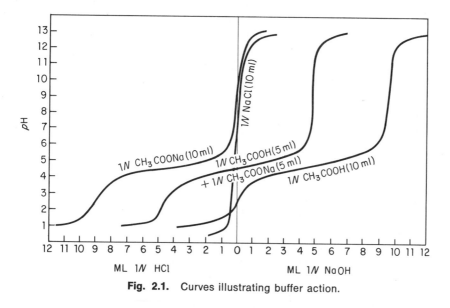

Fig. 2.1. Curves illustrating buffer action.

It is evident that the sodium acetate solution retards in some way the changes in the pH value upon the addition of an acid, while a solution of sodium chloride does not. Solutions of such compounds as sodium acetate which are relatively resistant to changes in the pH resulting from the addition or loss of hydrogen or hydroxyl ions are known as *buffer solutions,* and this property of solutions is called *buffer action.* Solutions such as sodium chloride which show no buffer effect are called *unbuffered solutions.*

If 1 ml of a 1 N sodium hydroxide solution is added to 10 ml of a 1 N solution of sodium chloride, a marked increase in the pH will occur. If 1 ml of the same solution is added to 10 ml of a 1 N solution of sodium acetate, its pH will increase markedly. In other words, the sodium acetate solution is buffered against the addition of an acid, but not against the addition of a base. If, on the other hand, 10 ml of a 1 N solution of acetic acid is substituted for the sodium acetate, it will be found that this solution is strongly buffered against the introduction of hydroxyl ions into the solution. A mixture of equal volumes of molar sodium acetate solution and molar acetic acid solution will exhibit buffer action against both the acids and the bases over a considerable range of the pH scale.

Different buffer solutions vary greatly in their effectiveness in maintaining pH stability. Some are strongly buffered against acids and weakly buffered against bases; of others the converse is true. The commonest buffer systems are those composed of a weak acid plus one or more of its salts. The sodium acetate-acetic acid buffer systems already described is such a system. Practically all of the buffer solutions of importance in living organisms belong to this group.

Buffer action consists essentially in the tying up of free hydrogen or hydroxyl ions nearly as rapidly as they are introduced into the solution by the formation of compounds which are only slightly dissociated. The ensuing change of the pH is therefore relatively small in proportion to the volume of acid or base added. As an illustration let us consider once more a solution consisting of both sodium acetate and acetic acid dissolved in water.

These two compounds dissociate as follows:

$$CH_3COONa \rightleftharpoons CH_3COO^- + Na^+$$
$$CH_3COOH \rightleftharpoons CH_3COO^- + H^+$$

The sodium acetate is strongly dissociated, but the acetic acid will dissociate only slightly, being a weak acid.

Suppose now that a little HCl be added to this solution. This is equivalent to adding H^+ and Cl^- ions and HCl molecules; the latter, however, will dissociate, forming additional ions as rapidly as the H^+ and Cl^- ions already present are bound up in the chemical combination. The H^+ and CH_3COO^- ions cannot exist side by side in the same solution in appreciable concentrations, since CH_3COOH is a poorly dissociated compound. Hence the added H^+ ions are almost all tied up in the formation of CH_3COOH. The Cl^- ions form NaCl with the Na^+ ions, which dissociates in the usual way. The result is that there is only a slight increase in the concentration of hydrogen ions in the solution, and hence only a very slight reduction in the pH value.

Suppose now that instead of HCl, a little NaOH is added to this solution. This is equivalent to adding Na^+ and OH^- ions and NaOH molecules; the latter, however, will produce additional ions by dissociating as rapidly as the Na^+ and OH^- ions already present are tied up in the chemical combination. But OH^- and H^+ ions cannot exist side by side in the same solution in appreciable concentrations, since H_2O is only slightly dissociated. Most of the added OH^- ions therefore combine with the H^+ ions produced by the CH_3COOH and form H_2O. More of the CH_3COOH dissociates producing more H^+ ions, which in turn unite with more of the OH^- ions. This continues until practically all of the added OH^- ions are tied up. The Na^+ ions form CH_3COONa with the CH_3OOO^- ions which dissociates in the usual way. The final result is that there is only a very slight decrease in the concentration of H^+ ions in the system, and hence only a very slight increase in its pH value.

General Nature of Colloidal Systems.   Colloidal systems, like solutions, are two-phased systems. Unlike solutions, however, the particles of the dispersed phase are not in the molecular or ionic condition, but, with certain exceptions to be noted shortly, are molecular aggregates. One colloidal particle is often composed of hundreds or even thousands of molecules lumped together. The molecular aggregates must not be so large, however, that the particles settle out of the system, as stability is one of the essential attributes of colloidal systems. In general, if the dispersed particles fall within the range of 0.001–0.1 μ in diameter, the system is considered a colloidal system; if larger than this, a suspension; and if smaller, a solution. The individual molecules of some substances (certain dyes and some proteins) are so large as to bring them within the colloidal range of dimensions. Hence molecular dispersions of such substances are simultaneously both solutions and colloidal systems. The limits generally accepted for the range of sizes of colloidal particles have been somewhat arbitrarily set and actually there is no sharp boundary between colloidal systems and suspensions on the one hand, or between colloidal systems and solutions on the other. There is a perfect gradation in properties from one type of system to the next. The properties of suspensions in which the suspended particles are relatively small in dimensions approach those of colloidal systems, while the smaller the dispersed particles in a colloidal system, the more closely it approaches a solution in its properties.

Particles of suspensions can ordinarily be seen under a microscope, but those of colloidal systems cannot. Colloidal particles can be detected, however, under the electron microscope or under the ultramicroscope.

Interfacial Area of Colloidal Systems.   The tremendous surface area produced by the subdivision of particles to colloidal size can be appreciated by the following example. The exposed surface of a cube of solid matter 1 cm on an edge, if cut up into cubes 0.001 μ on an edge (the approximate size of the smallest colloidal particles), will be 6000 square meters. This represents an increase of 10,000,000 times over the exposed surface area of a cube 1 cm on an edge. The contact surfaces between a solid and a liquid or between immiscible liquids are known as *interfaces.* This term also applies to the contact surfaces between colloidal particles and the liquid in which they are dispersed. The interfacial area of colloidal systems is enormous in proportion to the actual mass of material which is dispersed. Some of the most important properties or colloidal systems are a consequence of their enormous interfacial area.

Adsorption.   Interfaces are characteristically the seat of the phenomenon called adsorption. A simple example of adsorption can be seen by stirring about 1 g of activated charcoal with about 50 ml of 0.025 per cent methylene blue solution; upon filtering, the filtrate is found to be colorless; the methylene blue has been adsorbed by the carbon black.

Adsorption consists of an interfacial concentrating of molecules such as occurs to those of methylene blue at a carbon-water interface and is a phenomenon of very general occurrence, taking place at all kinds of interfaces. The adsorption of gas molecules at solid-gas interfaces is a common phenomenon. Solutes may be adsorbed at solid-liquid, or liquid-gas interfaces. Solvent molecules, such as those of water, also become adsorbed at certain kinds of interfaces. Adsorption is probably of universal occurrence at the interfaces of colloidal systems. In some such systems solute molecules are adsorbed, in others solvent molecules, and in some both solute and solvent molecules. Adsorption, as the charcoal-methylene blue example shows, also occurs at the interfaces of non-colloidal dimensions. There are many such interfaces in plant cells.

The Nomenclature of Colloidal Systems.   Colloidal systems, as has already become evident, are composed of two phases—a continuous phase, and a discontinuous phase—the latter is composed of discrete particles, each entirely separated from its fellows by the intervening continuous phase. The continuous phase is commonly called the *dispersion medium,* and the discontinuous phase the *disperse phase.* According to another terminology, applicable however only when the dispersion medium is a liquid, each individual dispersed particle is called a *micelle,* while the continuous phase of the system is called the *intermicellar liquid.*

Some colloidal systems possess the property of fluidity and can be poured more or less readily from one vessel to another. To the unaided eye they often appear to be true solutions, but an examination by means of an ultramicroscope reveals their colloidal nature. Such a system is called a *sol.* Non-fluid colloidal systems, called *gels,* are discussed later in the chapter.

Sols may be classified into hydrophobic (Greek "water-fearing") and hydrophilic (Greek "water-loving") types. In the latter an appreciable affinity exists between the dispersed phase and the water; in the former no such affinity is present. The affinity between the two phases of a hydrophilic sol manifests itself by the hydration of the micelles. Hydration is the association of one or more molecules of water with an ion, molecule, or micelle.

There is no general agreement regarding the exact relationship between the micelle of a hydrophilic colloid and its water of hydration, but only two conceptions are widely held. One relates this hydration to an actual solution of some of the water in the substance of the micelle. The other holds that no actual solution occurs but that water molecules are oriented around each dispersed particle, forming a "shell" many layers of molecules in thickness. The fact that micelles of hydrophilic colloids may, under certain conditions, lose their water of hydration very rapidly would seem to favor the latter theory. It is possible, of course, that in some hydrophilic systems the water is actually dissolved in the micelles, that in others it is present only as a shell of oriented molecules, while in still others both forms of hydration may exist.

Most colloidal systems composed of metallic substances dispersed in water are examples of hydrophobic sols. Gelatin, agar, starch, and gum accacia sols are examples of hydrophilic systems. Protein sols also belong in this group. Actually all possible graduations exist from highly hydrophilic sols to highly hydrophobic sols.

Brownian Movement. In 1828 the botanist Robert Brown observed through a microscope that pollen grains suspended in water showed a rapid vibratory motion. Brown at first was inclined to attribute this motion to the fact that the pollen grains were alive, but examination of preparations of dead pollen grains and spores showed that they likewise exhibited such a motion. It became evident, therefore, that this movement was in no way connected with living processes. We now know that any particle up to about 4 or 5μ in diameter will exhibit this movement when suspended in a liquid. This phenomenon is termed *Brownian movement,* after its discoverer.

Brownian movement is caused by the kinetic activity of the molecules of the solvent. Even the smallest particles in which Brownian movement can be observed are very large in proportion to the size of the solvent molecules which impinge upon them. A particle suspended in a liquid, such as water, undergoes a continual bombardment by the molecules of the liquid. At any given moment the sum total effect of the blows which the particle sustains on one side may be greater than the effect of the blows sustained on any other side, causing the particle to move. The next moment a greater impetus may be given to the particle from some other direction, and the course of its movement is changed. In this way the highly erratic movements of suspended particles, known as Brownian movement, originate. Increase in temperature increases the rate of Brownian movement because of an increase in the kinetic energy of the solvent molecules. This phenomenon is the nearest approach we have to actual visible evidence of the validity of the kinetic theory of matter. It almost brings before our eyes the veritable "dance of the molecules."

Electrical Properties. The micelles of most colloids possess electrical charges. By their movement in an electrical field it has been found that some micelles are negatively charged while others are positively charged. The origin of the charge on the colloidal particle is generally thought to result either from the adsorption of a particular kind of ion or to the ionization of the molecular aggregate of which the micelle is constructed. Ordinarily all the dispersed particles in any one system bear a charge of the same sign. Since like charges repel each other, aggregation is prevented, a fact which, together with Brownian movement, accounts in part for the stability of colloidal systems.

Although the micelles are charged, the system is electrically neutral, because in effect the dispersion medium is also charged. When the micelles are negatively charged, the dispersion medium is positively charged and vice versa. Because of electrostatic attraction, each particle is therefore surrounded by a "shell" of ions of the opposite charge. This arrangement of charges at the surface of a micelle is called an *electric double layer* (Fig. 2.2).

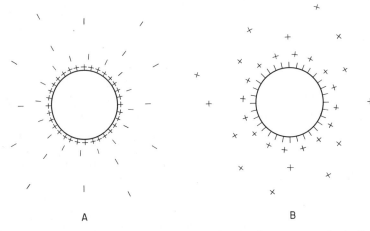

A   B

Fig. 2.2. Diagrammatic representation of the electric charges around micelles: (A) Positively charged; and (B) negatively charged.

Flocculation. As was mentioned previously, a colloidal system is usually stable because the particles bear like changes. When the charge on the particles is neutralized by the addition of ions or micelles of the opposite charge (Fig. 2.3), Brownian movement causes a collision of the micelles. Agglomeration of the individual particles into larger particles, which rapidly settle out, occurs. This phenomenon is called *flocculation, precipitation,* or *coagulation.* The point at which there is no difference of electrical potential across the double layer is known as the *isoelectric point* of the sol. Hydrophobic sols will flocculate at the isoelectric point. Hydrophilic sols, on the other hand, possess an additional stabilizing factor, that of water of hydration, which "cushions" the particles in collision and prevents agglomeration even though their charge may have been neutralized. In order to cause flocculation of a hydrophilic sol, the dehydration of the micelles must be accomplished as well as the neutralization of the

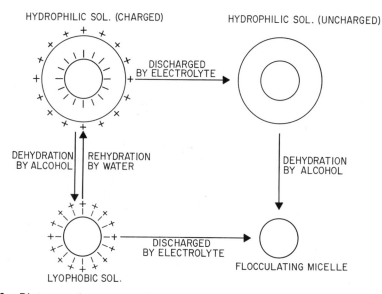

Fig. 2.3. Diagrammatic representation of the flocculation of a micelle of a hydrophilic sol.

charges (Fig. 2.3). Dehydration may be brought about by the addition of alcohol or relatively large amounts of certain salts. The latter method will also cause the neutralization of the charges and flocculation will occur. When this happens the phenomenon is referred to as *salting out*.

The addition of a small amount of hydrophilic sol, such as gelatin or gum arabic sol, to a hydrophobic sol makes flocculation of the latter by electrolytes or micelles of the opposite charge difficult or impossible. This effect is termed *protective action* and apparently results from the adsorption of hydrophilic micelles around the micelles of the hydrophobic sol, imparting to the entire system the properties of a hydrophilic sol.

Properties of Gels.   Under certain conditions most hydrophilic sols change into gels. Any gelatin sol which is not too dilute, for example, will "set" upon standing and form a gel. Everyone is familiar with such gelatin gels, which often appear on the table as desserts. Other familiar gels are the agar gels widely used as a medium for the culturing of bacteria, fungi, and algae, ordinary household jellies which are basically pectin gels, and starch gels. Some hydrophilic sols, however, do not ordinarily form gels. This is true of sols of gum acacia and of some protein sols.

The gel-forming capacity of some substances is very remarkable. A gelatin sol containing as low a proportion as one part of gelatin to 100 parts of water will usually gelate. Agar gels containing only 0.15 per cent of agar can be prepared. In such a gel one part of agar has the property of binding nearly 700 parts of water.

Certain gels are heat reversible. When heated such gels are converted into sols *(solation),* and when cooled such sols resume their gel condition *(gelation).* Usually this reversal of state can occur a number of times to a colloidal system without greatly affecting its physical properties when in either the sol or gel condition.

The most generally favored hypothesis of a gel structure is that both the solid and liquid phases of a gel are continuous. The solid phase is most usually visualized as a meshwork of long, tangled fibrillae of submicroscopic dimensions, the spaces within this interwoven mesh being occupied by the liquid phase. This theory is often called the "brushpile" theory in allusion to the supposed jumble of the intermeshing threads of the solid phase. The fact that diffusion, electrical conductivity, and the velocities of chemical reactions occur at approximately the same rate in a gel medium as in a water medium lends credibility to this theory. Since elastic gels can be readily transformed into hydrophilic sols, and hydrophilic sols into gels, it seems probable that many hydrophilic sols also possess a fibrillar structure.

SUGGESTED FOR COLLATERAL READING

Christensen, H. N. *pH and Dissociation.* W. B. Saunders Company, Philadelphia, 1964.

Daniels, F., and R. A. Alberty. *Physical Chemistry.* John Wiley & Sons, Inc., New York, 1955.

Giese, A. C. *Cell Physiology, 3rd Edition.* W. B. Saunders Company, Philadelphia, 1968.

West, E. S., W. R. Todd, H. S. Mason, and J. T. Van Bruggen. *Textbook of Biochemistry, 4th Edition.* The Macmillan Company, New York, 1966.

3 | PLANT CELLS

Advances in knowledge of the structure of plant and animal cells have been inextricably bound up with improvements in the techniques of microscopy. Cells were first described by Robert Hooke, one of the first microscopists who, in 1665, observed them in cork tissue. Actually he observed only the cell walls since cork cells are dead and have lost their living components. With further improvements in the quality of the microscopes and in the technique of microscopy, other cell components were gradually discovered and studied through the years. By the early twentieth century most of the major parts of cells had been identified.

However, the techniques of light microscopy are subject to severe limitations, principally because of the nature of light itself. The theoretically obtainable maximum magnification under a light microscope is about 2000 times, but in most work the practical upper limit of magnification is about 1000–1200 times. Beginning about 1950 the techniques of electron microscopy have been successfully employed in the study of the structure of cells. The electron microscope operates on the same general principle as the light microscope, except that a beam of electrons is used as the resolving agent instead of a beam of light. Magnifications of up to 200,000 times are theoretically possible with this instrument, although most observations with it are done at lower magnifications. In other words an electron microscope yields magnifications approximately 100 times as great as a light microscope. Furthermore, at equal magnifications, say 1000 ×, the electron microscope yields a much sharper image than the light microscope. The employment of the techniques of electron microscopy has led to spectacular advances in knowledge of the fine structures of cells.

Hitherto unsuspected cell components have been discovered, and the more detailed structure of some of the larger components has been revealed. The living cell has been unveiled as a fascinating miniature world which is enormously complex in structure.

The Origin and Development of Cells. The principle that cells arise only from preexisting cells was not generally accepted until about the middle of the nineteenth century. The vast majority of cells arise by division, either mitotic or meiotic, of preexisting cells, a process often called *cytokinesis*. In higher plants mitotic cell divisions occur chiefly in certain restricted regions called meristems (Chapters 19 and 21). The principal meristems are the apical meristems located at the tips of stems and roots, the secondary meristems of perennial monocots, the vascular cambium of gymnosperms and dicots, the cork cambium of dicots, and the intercalary meristems of some species. One other way in which new cells form from preexisting ones is by fusion, which in the higher plants occurs in the processes of fertilization and triple fusion (Chapter 21). While the number of new cells formed in these ways is vastly smaller than the number formed by cell divisions, these processes, and especially fertilization, are pivotal ones in the life history of the plant. The development of a new cell, whether it arises by division or by fusion, involves its subsequent enlargement and maturation (Chapter 19).

The Structure of Plant Cells. Practically every cell in the tissues of the higher plants is a tiny, many-sided compartment enclosed by a water-impregnated wall otherwise composed principally of cellulose and other compounds closely related to carbohydrates (Fig. 3.1). Plant cells vary greatly in both size and shape, and a number of different kinds are discussed and appear in the illustrations in the following chapters. Some kinds of plant cells are isodiametric, that is, no one dimension greatly exceeds any other. Cells of this type are characteristic of many parenchymatous tissues. Other kinds of cells, especially those in the conductive tissues, are elongate. Examples of these are tracheids, vessel elements, sieve-tube elements, phloem and wood fibers, and phloem and xylem ray cells (Chapters 8 and 17). Cells also show a great range of variability in size. Relatively few of the isodiametric cells of vascular plants have diameters which are less than 10 μ or more than 200 μ. In some elongate kinds of cells one of the dimensions may greatly exceed this range. A cotton fiber is a single cell, and in some varieties the fibers exceed 4 cm in length. An even more extreme example is the phloem fiber cells of the ramie plant *(Boehmeria nivea)* which are known to reach a length of 55 cm.

The bulk of the interior of almost all mature living plant cells is occupied by a single large *vacuole*, which is filled with *cell sap*. The cell sap is composed of water in which a great variety of substances are dissolved or colloidally dispersed.

The metabolic process of the cell are largely restricted to the *cytoplasm* and to the particulate structures located therein. In young, meristematic plant cells the cytoplasm occupies virtually all of the cell interior, but in mature plant cells it is present as a relatively thin layer, lining the cell wall, and bounded on its interior margin by the vacuole.

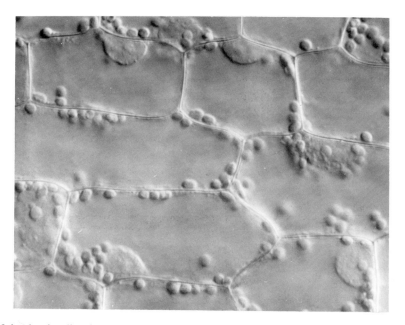

Fig. 3.1. Leaf cells of waterweed *(Anacharis canadensis)* showing nuclei, chloroplasts, and cell walls. Photographed by Nomarski interference contrast. × 825. Photograph by H. P. Hostetter.

The bulk of the cytoplasm consists of the undifferentiated ground substance which is transparent or nearly so as viewed under the light microscope. In active cells water is by far the largest single constituent of the cytoplasmic ground substance as well as of the various particulate structures contained therein. The nonaqueous portion of the ground substance is largely proteinaceous and in a colloidal condition, most commonly a sol, but sometimes a gel. A substantial proportion of it undoubtedly consists of enzymes, of which there are many different kinds (Chapter 9).

Within the ground substance matrix occur a number of kinds of structural units called *organelles*. Most organelles had been observed under the light microscope, but the structural complexity of many of them was not appreciated until they were subjected to the greater magnifications of the electron microscope. A few of the smaller organelles were not even discovered until the electron microscope revealed their presence (Fig. 3.2).

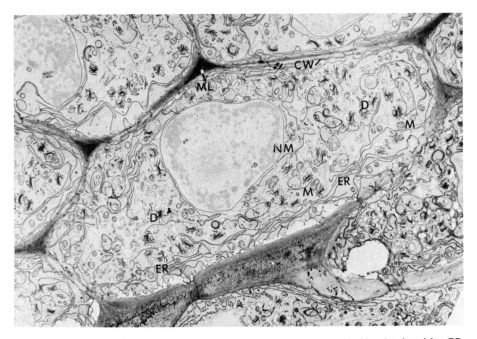

Fig. 3.2. Electron micrograph of a root cap cell of maize. × 2,100. *M*, mitochondria; *ER*, endoplasmic reticulum; *D*, dichtyosome; *NM*, nuclear membrane; *CW*, cell wall; and *ML*, middle lamella. Photograph from H. H. Mollenhauer.

The cytoplasm in many, but by no means all, plant cells is in a fluid state. This is evidenced by the phenomenon of *streaming* (also called *cyclosis*), which can be observed in a number of kinds of plant cells. The simplest manifestation of streaming consists of a rotation of the cytoplasm round the inner surface of the cell wall. Where cytoplasmic strands extend through the vacuole, as in the cells of *Tradescantia* stamen hairs, the circulation of the cytoplasm may become very complex. The plastids and other visible organelles are carried passively around the cell by the moving cytoplasm. The manner in which some organelles, such as the endoplasmic reticulum (see later), are influenced by cyclosis is not known. The dynamics of cyclosis are unclear, but it is obvious that an expenditure of energy is required to keep such a process in operation. It is accelerated by increases in temperature up to the point where injury occurs and is checked by a low temperature, ceasing at temperatures a little above the freezing point. Cyclosis stops in the absence of oxygen, implying that the process of aerobic respiration (Chapter 13) is involved.

While plant and animal cells are basically similar in their structural components, it should be recognized that three features are distinctive to plant cells in general in contrast to animal cells in general. These are the possession by each plant cell of a firm wall, the presence in mature plant cells of a large

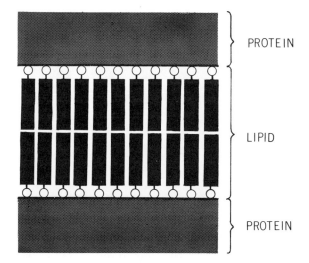

Fig. 3.3. Diagrammatic representation of a unit membrane.

central vacuole, and the incidence in the cytoplasm of plastids, of which the chloroplasts are the most conspicuous examples.

Cell Membranes.   Surrounding the cytoplasm on the side adjacent to the wall is the *plasma membrane* or *plasmalemma.* Correspondingly, the cytoplasm lying adjacent to the vacuole is bounded by the *vacuolar membrane* or *tonoplast* (Chapter 5). The plasma membrane is only about 7–8 nm[1] in thickness and is not visible under the light microscope. It is a flexible structure which, aside from water, is composed largely of proteins and lipids. There is considerable evidence that these two kinds of compounds are arranged in the plasma and vacuolar membranes as shown diagramatically in Fig. 3.3. Under very high magnifications with the electron microscope, the membrane appears as two dark lines each about 2–3 nm wide with a lighter area, also about 2–3 nm wide, between. This arrangement is called a *unit membrane* (Fig. 3.4).

Most, if not all, of the organelles are surrounded by membranes, many of which seem to be of the unit membrane type. As discussed later there are interrelations between the plasma and vacuolar membranes and the membranes of some of the organelles, especially the nucleus and the endoplasmic reticulum.

The plasma and vacuolar membranes both possess the property of *differential permeability* (Chapter 5). Metabolically generated energy is sometimes required to transport solutes across these membranes, often in opposition to

[1]A nanometer (nm) is the metric unit which is equal to 10^{-9} meters, 10^{-6}mm, or $10^{-3}\mu$. The term millimicron (mμ) was formerly widely used for this unit. One nanometer equals 10 Å (Angstroms), another unit frequently used in discussing wavelengths of light and the dimensions of cell organelles.

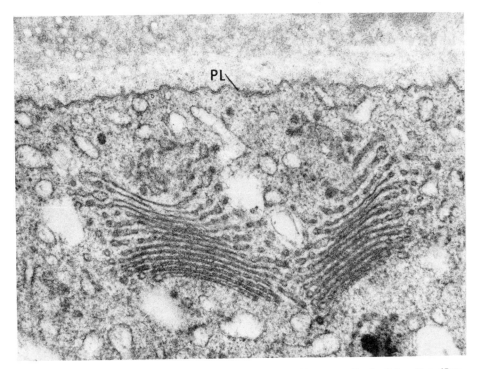

Fig. 3.4. Electron micrograph of an apical secretory rhizome cell of wild coffee *(Psychotria bacteriophila)* showing unit membrane (plasmalemma, *PL*), and dictyosome. × 47,000. Photograph from H. T. Horner, Jr.

a concentration gradient (Chapters 15 and 17). The membranes surrounding the organelles doubtless have similar differential permeability characteristics.

Organelles. *The nucleus.* This is the most prominent organelle in most plant cells. It is a more or less spherical body the diameter of which commonly falls within a range of 5 μ to 25 μ. In the vascular plants there is usually only one nucleus to a cell, although in certain types of cells several may be present. Mature sieve tube cells (Chapter 17), are the only well-known examples of living cells in the higher plants in which an organized nucleus is absent, although a nucleus is present in these cells when they are first formed. The nucleus is surrounded by a unit membrane in which pores are present (Fig. 3.5), which range from about 50–100 nm in diameter. These pores undoubtedly facilitate the passage of materials into and out of the main body of the nucleus. Extensions of the nuclear membrane occur into the cytoplasm where they appear to be continuous with the endoplasmic reticulum as discussed later.

The nucleus is filled with *nuclear sap,* in which there are one or more bodies called *nucleoli* (singular: *nucleolus*). These are denser than the surrounding sap and usually more or less spherical in shape. The nucleoli appear to be

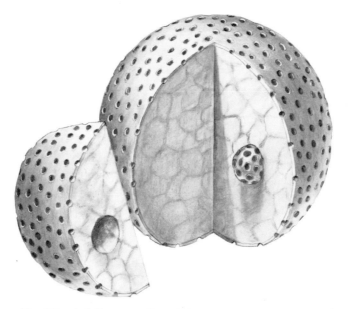

Fig. 3.5. Artist's conception of the structure of a nucleus.

centers of ribosome construction and have a high RNA content. Ribosomes are also present in the nuclear sap and are involved in protein synthesis, as are the cytoplasmic ribosomes to be discussed later. *Chromatin,* the actual genetic material, containing DNA, also occurs in the nuclear sap. In resting cells the chromatin may not be distinct or may appear as a network. Just prior to nuclear division in mitosis or meiosis, the chromatin in higher plants becomes organized into chromosomes.

Evidence that the nucleus is the control center of the cell is overwhelming. Operating through a mechanism in which DNA, RNA, and ribosomes play key roles, genetic information is translated into the synthesis of proteins and enzymes which largely control the metabolic destiny and structural development of the cell (Chapter 16).

The Plastids. These well differentiated bodies are conspicuous in the cytoplasm of most if not all plant cells. Plastids are usually centers of certain kinds of metabolic activity. In addition to the well-known chloroplasts, which contain chlorophyll and carotenoid pigments (Chapter 10), there are present in the cytoplasm of plant cells *chromoplasts* which contain other pigments other than chlorophyll, and the colorless *leucoplasts.* The chloroplasts (Fig. 10.5) are centers of photosynthesis and starch synthesis and are described more fully in Chapter 10. The chromoplasts contain red or yellow carotenoid pigments. The red and yellow colors of some kinds of fruits and flowers result from the presence of chromoplasts. Among the leucoplasts are the *amyloplasts* which, like chloroplasts, are centers of starch synthesis (Chapter 14). Other leucoplasts appear to operate as centers of the accumulation of proteins and still others as centers of the accumulation of oils.

Mitochondria.   The existence of these bodies (also called *chondriosomes*) in cells has been recognized for many years. As seen under the light microscope they appear to be spherical, rod-like, or filamentous in shape. Because of their small size (0.5 to 1.0 μ in diameter and 1 to 2 μ in length), however, nothing was ascertained regarding their internal structure until they could be observed under the electron microscope (Fig. 3.2). The structure of the mitochondria has proved to be quite complex and is shown diagrammatically in Fig. 13.7. Each mitochondrion is bounded by an inner and an outer membrane. The inner membrane invaginates into the interior of the mitochondrion in the form of projections called *cristae* (singular:*crista*). Such projections may be shelf-like or tubular and frequently interconnect. Each of the membranes appears to be a unit membrane of the type previously described. Mitochondria apparently can multiply within at least some kinds of cells by a process of division. The basic role of the mitochondria in respiration is discussed in Chapter 13. They also carry out a limited amount of protein synthesis (Chapter 16).

Endoplasmic Reticulum.   The existence of this cell organelle was not even suspected until observations of cellular structure were made under the electron microscope. The endoplasmic reticulum is a branching system of membranes which ramifies throughout the cytoplasm. In some cells the endoplasmic reticulum is in the shape of narrow tubes, but more commonly it consists of long, folded sheets of membranes (Fig. 3.6). As observed in sections of cells under the electron microscope, it appears as a series of parallel pairs of unit membranes (Fig. 3.2). Extensions of the nuclear membrane interconnect with the endoplasmic reticulum constituting a single membrane system throughout the cell. There is some evidence of similar interconnections with the plasma and vacuolar membranes, but this is not certain. It is possible that the endoplasmic reticulum constitutes a translocative system within the cell, but this also is by no means a certainty.

Ribosomes.   These are relatively small, approximately spherical particles about 20 nm in diameter which are associated with the endoplasmic reticulum in many but not all kinds of cells. They are located only on the outside of the endoplasmic reticulum (Fig. 3.2). When the latter structure has ribosomes adhering to it, it is classed as rough; if the endoplasmic reticulum is free of ribosomes, it is classed as smooth. Ribosomes also occur free in the cytoplasm as well as in nuclei and chloroplasts. As mentioned earlier, ribosomes are fabricated in the nucleoli, although whether or not this is their sole origin is not clear. Apart from water they are composed almost entirely of proteins and ribonucleic acid (RNA). The basic roles of the ribosomes in protein synthesis are discussed in Chapter 16.

A now outmoded but nearly synynomous term for a ribosome was the term *microsome*. A microsome was the name applied to ribosomes of rough

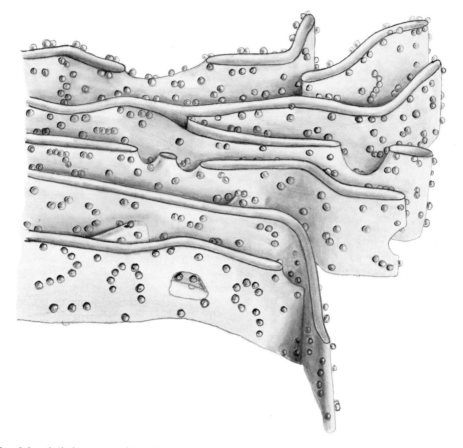

Fig. 3.6. Artist's conception of the structure of the endoplasmic reticulum, showing ribosomes.

endoplasmic reticulum, as these ribosomes with attached fragments of endoplasmic reticulum were collected by centrifugation.

Golgi Apparatus.   The very existence of this kind of organelle was a matter of controversy for a long time after their existence was first claimed by Golgi in 1903. With the advent of the electron microscope, however, their presence in both plant and animal cells was confirmed, and their structure revealed in considerable detail. Compressed sac-like membranes called *cisternae* are found in stacks within organelles known as *dictyosomes* (Fig. 3.2, 3.4, and 3.7). The peripheral portions of the membranes making up the dictyosome form an interconnecting meshwork of tubules. Some tubules terminate in pouch-like protuberances known as *vesicles*. The dictyosome vesicles appear to be important in the deposition of cell wall materials other than cellulose at the cell wall/plasmalemma interface. In the area of this interface, the vesicles are thought to become pinched off from the dictyosome proper and to migrate to the plas-

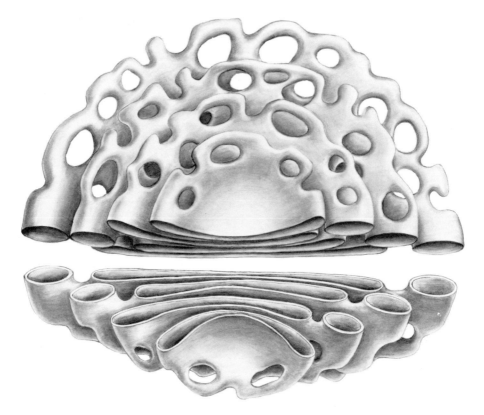

Fig. 3.7. Artist's conception of the structure of a dictyosome.

malemma, eventually possible uniting with this membrane and depositing their contents to the outside of the plasmalemma. While dictyosomes may sometimes appear in photomicrographs as separate organelles, there is evidence to suggest that a given dictyosome may have membrane connections with a number of other dictyosomes in the cell. A complex of several dictyosomes is referred to as a Golgi apparatus.

Microtubules. These organelles are extremely small bodies (about 25 nm in diameter) which have been found in both plant and animal cells. Several functions have been ascribed to them, especially in animal cells. In plants they appear to play a role in cell wall formation as mentioned later.

Peroxisomes and Glyoxysomes. These two organelles are superficially very similar, both lying within a size range of 1 to 2 μ, but functionally they are very different. Peroxisomes play an important role in the process of photo-respiration (Chapter 12), while glyoxysomes are involved in certain phases of fat metabolism (Chapter 14).

The Vacuole. One of the most distinctive features of most mature plant cells is the presence of a large central vacuole filled with cell sap and entirely surrounded by a layer of cytoplasm. In such mature plant cells the vacuole may occupy as much as 90 per cent of the total volume of the cell. Cambium initials and meristematic cells in the tips of stems and roots usually contain numerous small vacuoles scattered throughout the cytoplasm. Mature cells, however, whether they arise from the primary root or stem meristems or from cambium cells, typically contain one large central vacuole which originates, as the cell enlarges, by the coalescence of the numerous smaller vacuoles present in the meristematic cells.

Among the various substances present as solutes in the vacuoles are sugars, mineral salts, organic acids, amino acids, fatty acids, amides, glycosides, and water soluble pigments such as anthocyanins. Fats and related compounds may occur in a finely emulsified form. Proteins, tannins, mucilages, and lipids are among the substances which are often present in the colloidal state. Crystals of calcium oxalate also occur in the vacuoles of some mature cells.

The Cell Wall.   The presence of a relatively rigid, although often elastic, cell wall, composed predominantly of water-impregnated carbohydrate or carbohydrate-like compounds is one of the distinctive features of plant cells. Fabrication of the wall of a plant cell starts during the process of cell division and continues through subsequent stages in the growth of the cell.

In the vascular plants the division of a plant cell (cytokinesis) is preceded by the division of the nucleus. Nuclear division (mitosis) is a complicated process involving organization of the nuclear reticulum into unit chromosomes, duplication of those chromosomes, separation of the duplicates, and finally the reconstitution of a daughter nucleus from each set of daughter chromosomes. At the anaphase stage of mitosis a new membrane begins to develop in the center of the cell. The margins of this membrane gradually extend, eventually making contact on all sides with the existing cell wall, thus forming a septum called the *cell plate,* between the two newly formed cells (Fig. 3.8). The first membrane, the *middle lamella,* is composed largely, if not entirely, of amorphous pectic compounds (Chapter 14). The middle lamella shows no structural organization above the molecular or colloidal level.

As the cells begin to enlarge, a thin *primary wall* composed of cellulose, hemicelluloses, and pectic compounds is constructed on each of the two sides of the middle lamella. These additional wall layers become joined to the existing walls at the sides of the cells in such a way that the protoplast of each daughter cell becomes encased within a wall which is continuous, except for certain interconnections between the cells, to be noted later. The primary wall is relatively thin, ranging between 1–3 μ in thickness in most cells.

The structure of the primary wall is complex (Fig. 3.9) and consists basically of *microfibrils* of cellulose, embedded in a dense and amorphous hydrogel of pectic compounds and hemicelluloses (Chapter 14). The cellulose microfibrils provide the basic framework of the cell wall. Proportions of these several kinds of compounds present in cell walls may differ considerably from one kind of cell wall to another.

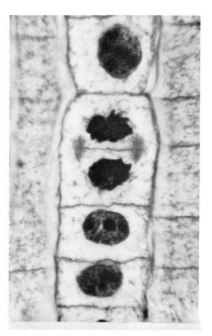

Fig. 3.8. Cells in an onion root tip illustrating stages in the formation of cell walls during cell division. The top cell has not divided recently. The middle cell shows a cell plate between two daughter nuclei. The lower two cells represent a stage in which a new cross wall completely separates two daughter cells. Photomicrograph by Tillman Johnson.

The cellulose microfibrils are built up of cellulose "molecules" which are long and chain-like. Each such molecule consists of many D-glucose residues in β 1–4 linkage (Chapter 14). The number of glucose residues present in one cellulose "molecule" is variable but is probably seldom less than a thousand and often many times greater. A single microfibril is of the order of 20 nm in diameter and up to about 4 μ in length. In at least portions of each microfibril the cellulose molecules are arranged in a crystalline fashion, that is, the molecules are oriented along their long axis in a regular arrangement relative to each other. In addition, amorphous portions of each microfibril with non-oriented cellulose chains also exist. In cross walls and in all the walls of isodiametric cells the microfibrils are usually arranged in a random fashion.

Many kinds of cells, however, especially those in conductive tissues, elongate much more along one axis than others. In such walls the first microfibrils to be laid down in the side walls are oriented in parallel fashion and at right angles to the direction of the cell elongation. As the elongation of the cell proceeds, more microfibrils are constructed towards the interior of the cell with the same basic arrangement, but the earlier developed ones are pulled more and more nearly into a vertical position. Thus the elongation of a primary wall results in a graded arrangement of microfibrils from a nearly transverse

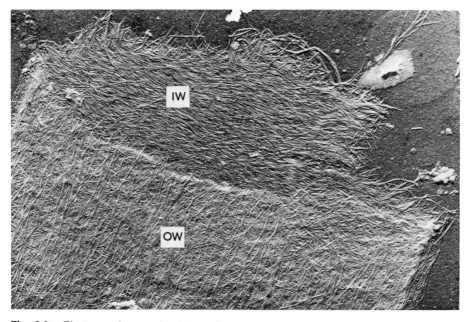

Fig. 3.9. Electron micrograph of the primary cell wall of *Juncus,* showing microfibrils. The microfibrils of the inner part of the wall (IW) are oriented perpendicular to the long axis of the cell; those of the outer wall (OW) are oriented parallel to the long axis of the cell, which is the axis along which elongation takes place. × 29,000. Photograph by P. A. Roelefsen. From Jensen and Park, *Cell Ulrastructure.* Copyright 1967, Wadsworth Publishing Company, Inc.

orientation in the innermost wall layer to a nearly vertical one in the layer near the middle lamella (Fig. 3.9).

At about the time the enlargement of a cell ceases, further deposition of cellulose begins on the inner surface of the primary wall of many kinds of cells, a process which may continue until the wall becomes conspicuously thickened. These added layers constitute what is known as the *secondary wall* which is commonly from 5–10 μ thick. In some kinds of cells the secondary wall may increase in thickness until the wall occupies most of the cell interior. This is true, for example, of a cotton fiber, which is actually a single elongate seed epidermal cell. As in the primary wall, the cellulose in secondary walls occurs in the form of microfibrils. Each layer of the secondary cell wall is made up of more or less parallel bundles of microfibrils generally oriented in the same direction. Each new ply is usually oriented at a different angle relative to the layers next to it (Fig. 3.10).

Many secondary walls, especially those of xylem tissues, contain a non-saccharidic constituent known as *lignin* (Chapter 14). Next to cellulose, this is the most abundant compound present in the secondary cell walls. It is not incorporated into the fibrillar structure of the wall, but is an ingredient of the

Fig. 3.10. Electron micrograph of secondary cell wall of *Valonia* (an alga) showing different orientation of microfibrils in different layers. × 11,300. Photograph by K. Muhlenthaler. From Bonner and Varner, *Plant Biochemistry,* p. 160. Copyright, 1965, Academic Press, Inc.

matrix surrounding the cellulosic microfibrils. Much of the rigidity and strength of woody tissues results from the presence of this compound in the secondary walls of constitutent cells. Small quantities of lignin are also often present in the middle lamella and primary wall of some mature cells.

Two other non-carbohydrate constituents of plant cell walls of considerable importance are *cutin* and *suberin*. The former is the name applied to the mixture of waxlike materials found on the outer surface of the epidermal cell walls of leaves, stems, fruits, and other organs. It is the principal constituent of the cuticle. Cutinized cell walls are relatively impermeable to water. The presence of cutin in the outer walls of the epidermal cells greatly reduces the evaporation of water from such cells. Suberin is an important constituent of cork cell walls and, like cutin, is highly impermeable to water. Nearly all perennial stems and roots sooner or later become encased in layers of cork cells. As a result, water loss from the older parts of such organs is extremely low. Neither cutin nor suberin have been clearly characterized chemically.

All cells have a middle lamella and a primary wall, but secondary walls are present only in certain kinds of cells. The cambium, parenchyma, and collenchyma are examples of tissues in which secondary cell walls do not develop. Phloem fibers, stone cells, tracheids, vessel elements, and wood fibers are examples of cells with prominent secondary cell walls. Many kinds of such cells, especially those in the xylem, retain a functional role in the plant long after their protoplasts have disappeared.

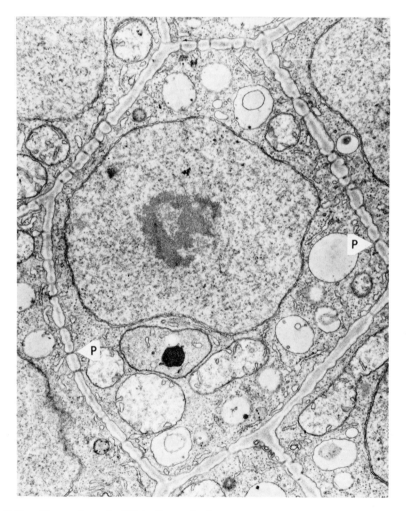

Fig. 3.11. Plasmodesmata *(P)* in the wall of a sporogenous cell of *Selaginella lepido-phylla.* × 19,300. Electron micrograph from H. T. Horner, Jr.

Primary cell walls are flexible and elastic within limits, retain a capacity for increasing in area, and often exhibit reversible changes in thickness. Secondary walls are nonelastic and often quite rigid, have little capacity for expanding in area, and do not as a rule show reversible changes in thickness.

Although superficially each protoplast appears to be a unit imprisoned within its own cell walls, sufficiently refined microscopical studies show that interconnections exist from one cell to its neighbors. Small thin areas occur in primary cell walls which are called *primary pits*. In some tissues fine threads of cytoplasm pass through the intervening cell wall layers, connecting the cytoplasm of one cell with that of an adjoining one (Fig. 3.11 and Fig. 3.12). These *plasmodesmata* (singular: *plasmodesma*), as they are called, are often concentrated within the pit areas, as many as ten of these cytoplasmic threads passing through one such area. *Bordered pits* are another type of cell structure which facilitates at least some kinds of intercellular traffic. They are found only in secondary walls and are especially prominent features of various cells and elements of the xylem (Figs. 8.6, 8.7).

Very little is known about the dynamics of cell wall fabrication except that it results principally and perhaps entirely from the activity of cytoplasmic constituents. Several different organelles have been at least tentatively implicated

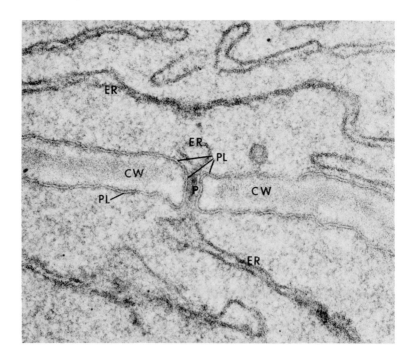

Fig. 3.12. A plasmodesma *(P)* through the cell wall of a cortical cell of maize root tip. *CW*, cell wall; *PL*, plasmalemma; and *ER*, endoplasmic reticulum. × 103,000. Electron micrograph by H. H. Mollenhauer.

in this process by various investigators. Among these are the microtubules, the Golgi apparatus, the ribosomes, and the endoplasmic reticulum.

SUGGESTED FOR COLLATERAL READING

Clowes, F. A. L., and B. E. Juniper *Plant Cells.* Blackwell Scientific Publications, Oxford, 1968.

Esau, Katherine. *Plant Anatomy, 2nd Edition.* John Wiley & Sons, New York, Inc., 1965.

Frey-Wyssling, A., and K. Muhlenthaler. *Ultrastructural Plant Cytology.* Elsevier Publishing Company, Amsterdam, 1965.

Gibbs, M., Editor. *Structure and Function of Chloroplasts.* Springer-Verlag, New York, 1971.

Jensen, W. A. *The Plant Cell, 2nd Edition.* Wadsworth Publishing Company, Belmont, California, 1970.

Jensen, W. A., and R. B. Park. *Cell Ultrastructure.* Wadsworth Publishing Company, Belmont, California, 1967.

Kirk, J. T. O., and R. A. E. Tilney-Bassett. *The Plastids, Their Chemistry, Structure, Growth and Inheritance.* W. H. Freeman and Company, San Francisco, 1967.

Ledbetter, M. C. and K. R. Porter. *Introduction to the Fine Structure of Plant Cells.* Springer-Verlag, New York, 1970.

Nobel, P. S. *Plant Cell Physiology.* W. H. Freeman and Company, San Francisco, 1970.

Pridham, J. B., Editor. *Plant Cell Organelles.* Academic Press, Inc., New York, 1968.

Siegel, S. M. *The Plant Cell Wall.* The Macmillan Company, New York, 1962.

4 | DIFFUSION, OSMOSIS, AND IMBIBITION

The body of a plant is composed of some of the commonest chemical elements found on the earth's surface. These elements occur in the environment of plants as relatively simple inorganic compounds. It is principally in the form of these compounds or in the form of their constituent cations or anions that these elements enter plants. Some of these substances enter mainly through the aerial organs and others primarily through the roots. Carbon dioxide and oxygen enter principally by way of the stomates. Water and the cations and anions of inorganic salts usually enter from the soil by way of the roots.

Not only are substances continually entering plants from the environment, but substances also are constantly escaping from plants. Large quantities of water vapor pass out of leaves and other aerial organs into the atmosphere. Under certain conditions carbon dioxide escapes from the plant, and under other conditions oxygen passes from the plant to the atmosphere. Certain volatile compounds also escape from the aerial organs of many plants. Carbon dioxide and other substances may also pass out of the roots into the soil.

A substantial proportion of the movement of substances into and out of plants is accomplished by the process called *diffusion*. Some of the movement of substances from place to place within a plant is also brought about by diffusion. However the translocation of substances from one plant organ to another, sometimes over relatively long distances, is largely accomplished by more complex mechanisms (Chapters 8 and 17).

Diffusion of Gases. Many examples of the diffusion of gases might be cited. If a bottle of ammonia, ether, peppermint oil, perfume, or any other readily volatile substance with a characteristic odor is opened indoors, within a very short time the distinctive odor can be detected in all parts of the room. Such a dispersal of gas molecules is in part the result of diffusion, although air currents often assist in such a distribution of molecules.

If a vial of a readily visible gas such as bromine is broken under a bell jar previously evacuated of air, in which there are no air currents, the entire jar quickly fills with the brownish vapor of bromine. The distribution of bromine gas throughout the bell jar has been accomplished by the kinetic activity of the bromine molecules and is a simple example of diffusion. If the vial of bromine is broken under a bell jar which has not been evacuated of air, the time required for the bromine gas to completely occupy the bell jar will be longer than when diffusion of the gas occurs in a vacuum. Under such conditions the freedom of movement of the bromine molecules is impeded by the molecules of the gases of air, and hence the rate of diffusion is retarded. If the pressure of the air within the bell jar is increased, the rate of diffusion of bromine gas through the air will be further decreased. Regardless of the air pressure, the jar will eventually fill with bromine gas. Thus only the rate and not the direction of diffusion of molecules of a gas is influenced by the presence of the molecules of another gas.

Because of their kinetic energy, molecules are constantly in motion. The direction of movement of a particular molecule at any specific time cannot be predicted. The molecule moves in a straight line until deflected from this path by collision with some other molecule or object. Following a collision, the path of motion is again a straight line but in a different direction. Such collisions occur with great frequency, and the route of a molecule therefore becomes highly erratic. A given molecule may thus move great distances from its starting place or be deflected back into the region whence it came.

For example, if a bottle of ammonia gas is opened in a room, molecules of ammonia will, in time, enter the bottle as well as escape from it. Because initially there is a greater number of ammonia molecules per unit volume (greater concentration) inside than outside the bottle, there will be a greater chance of ammonia molecules moving out of it in a given time period than of ammonia molecules moving into it. As long as the concentration of ammonia molecules inside the bottle remains greater than the concentration of ammonia molecules outside it, the number of molecules moving out of the bottle per unit time will be greater than the number of molecules entering it per unit time. When the concentration of ammonia molecules inside the bottle becomes equal to the concentration of ammonia molecules outside the bottle, the number of molecules moving into and out of the bottle per unit time will be equal.

Since molecular motion is still occurring and equal numbers of molecules are moving per unit time into and out of the bottle, this situation is termed a *dynamic equilibrium*. We will not use the term diffusion to refer to the movement of molecules under such conditions but will restrict it to characterize situations in which there is a net movement, *i.e.*, a gain or loss of molecules of a certain kind in one part of a system at the expense of other parts. According

to this concept, diffusion can occur only when the concentrations of the diffusing substance are not uniform throughout the system, and the process can
continue only as long as differences of concentration are maintained.

Since, temperature remaining constant, the pressure exerted by any gas
is directly proportional to its concentration (number of molecules per unit
volume), it follows that the diffusion of gases may be interpreted in terms of
the differences in *partial pressure* exerted by that gas in the different parts of a
system. Diffusion of a gas is the net movement of molecules from a region of
their greater partial pressure to a region of their lesser partial pressure.

Diffusion phenomena should be clearly distinguished from *mass movements*
in which an assemblage of molecules is moving more or less as a unit because
of some external force rather than, as in diffusion, as individual molecules
because of their own kinetic energy. Winds and air currents representative of
the physical process called *convection* are examples of mass movements. Mass
movements of gases and liquids can be caused in many ways, and numerous
examples of such movements are common knowledge.

The *direction* in which any gas diffuses is controlled entirely by its own
difference in partial pressures and is not influenced by either the direction or
the rate of diffusion of other gases in the system. Hence in a given system one
or more gases may be diffusing in one direction, while one or more other
gases are diffusing in the opposite direction. Each gas diffuses in the direction
determined by its own difference in partial pressures, and at a speed determined
by the factors which influence the rate of diffusion of that particular gas.

Factors Influencing the Rate of Diffusion of Gases.   *Density of the Gas.*
Different gases diffuse at different rates even when influenced by the same set
of environmental factors. Hydrogen, for example, diffuses more rapidly than
any other gas. The relative speeds of the diffusion of different gases are inversely proportional to the square roots of their relative densities. By relative
density is meant the weight of a given volume of gas as compared with the
weight of the same volume of hydrogen. Representing the density of hydrogen
by 1, the relative density of oxygen is 16. Hence the rate of diffusion of hydrogen is proportional to $1/\sqrt{1}$ while that of oxygen is proportional to $1/\sqrt{16}$.
Hydrogen gas will therefore diffuse four times as rapidly as oxygen gas under
the same conditions of temperature and pressure. The relative density of carbon
dioxide is 22; hence hydrogen gas will diffuse nearly five times as rapidly as
carbon dioxide gas.

Temperature.   An increase in temperature increases the speed of diffusion.
This results, at least in part, from the correlated incease in the kinetic activity
of the molecules of the diffusing gas. Actual measurements of the Q_{10}[1] of diffusion generally yield values between 1.2 and 1.3. Such values are characteristic of purely physical processes.

---

[1]The Q_{10} of any process is defined as the number of times the rate of that process
increases with a 10°C rise in temperature. If the rate of the process is doubled, the Q_{10} is
2, etc.

Diffusion Gradient.   The fact that the *direction* of the diffusion of a gas is from a region of its greater partial pressure to its lesser partial pressure has already been emphasized. The *speed* of diffusion is also influenced by differences in partial pressures. In general the greater the difference in partial pressures between two regions, the more rapidly diffusion will occur, but the distance through which the diffusing molecules must travel between the two regions is also a factor. Both are components of the *diffusion gradient.* The rate of diffusion is directly proportional to the difference in partial pressures and inversely proportional to the length of the diffusion path. The diffusion gradient is equivalent to the difference in partial pressures between the delivering and receiving ends of the system divided by the length of the distance between them (Fig. 4.1). The greater or steeper this gradient the more rapidly diffusion occurs.

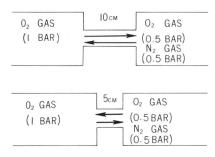

Fig. 4.1. Diagram illustrating that the length of the diffusion gradient is a factor influencing its steepness.

Concentration of the Medium Through Which Diffusion Occurs.   In general the more concentrated the medium, *i.e.,* the more molecules per unit volume in the medium through which the diffusing molecules must pass, the slower the rate of diffusion. Bromine gas, as pointed out earlier in this chapter, diffuses more rapidly through a vacuum than through air.

Diffusion of Solutes.   The molecules or ions of a solute possess sufficient kinetic energy to move from place to place within the limits of a solution. The simplest method of demonstrating the diffusion of a solute is to introduce a crystal of potassium permanganate, or some other compound which is colored when in solution, into the bottom of a tall glass cylinder filled with water. The cylinder should then be placed in an environment of constant temperature where it will be free from disturbance. The diffusion of the molecules or the ions which pass into solution in the water can be followed by the slow change in color of the water. One of the most striking facts illustrated in such experiments is the extremely slow rate of diffusion of solutes through water. This results partly from the fact that in such an experiment the steepness of the diffusion gradient decreases with time, but principally from the fact that the densely packed molecules of the liquid enormously impede the diffusion of the dissolved molecules or ions. Such experiments, however, can lead to misconceptions. Diffusion may be very rapid if the diffusion gradient is short. Because of the short distances involved, usually measured in terms of microns, diffusion within cells and from one adjacent cell to another may occur very rapidly.

The *direction* of the diffusion of any solute occurs in accordance with its own differences in concentration, regardless of the rate or direction of diffusion of other solutes in the same system. The *rate* of diffusion of solute particles, like gases, is influenced by temperature and the steepness of the diffusion gradient, but by virtue of being in solution the rate of diffusion of solutes may also be influenced by the following additional factors.

Size and Mass of the Diffusing Particles.   Small molecules or ions diffuse more rapidly than large ones. A hydrogen ion, for example, diffuses many times more rapidly than a glucose molecule. Similarly, highly hydrated ions diffuse more slowly than those which have fewer water molecules bound to them, since the association of the water of hydration with a molecule or ion in effect increases its size. The mass of the particle is also a factor influencing the speed of its diffusion. As between two particles of the same size but different masses, the heavier particle will diffuse more slowly.

Solubility.   In general, the more soluble a substance is in a liquid, the more rapidly it will diffuse through that liquid. This influence of solubility upon diffusion rates can be interpreted principally in terms of its effect upon the diffusion gradient, since steeper gradients can be built up if the solute is very soluble in the liquid than if it is only slightly soluble.

Osmosis.   An understanding of the dynamics of this process and of the significance of the physical quantity termed osmotic pressure is essential to an interpretation of the water relations of plant cells and tissues.

A simple demonstration of osmosis can be prepared as diagramed in Fig. 4.2. A sac-like membrane is completely filled with a strong sucrose solution and immersed in a beaker of water. The top of the sac is tightly plugged with a rubber stopper. The membrane is prepared of a substance which is slightly elastic and which is permeable to water, but impermeable or practically so to sucrose.

After a short time the originally limp sac becomes rigidly distended. This is a result of a net movement of water through the membrane into the interior of the sac. Such a net movement of water is an example of *osmosis.* The pressure developed as a result of the entrance of water prevails throughout the solution and is also exerted against the inside of the membrane. If the membrane is virtually impermeable to the solute, eventually the entire system will come to a dynamic equilibrium, after which there will be no further increase in the volume of water inside of the membrane.

It should be emphasized that osmosis involves the movement of a *solvent,* not a solution, across a membrane. In living organisms water is the only important solvent which moves by osmosis, hence discussion of this process will be restricted to water. The *direction* in which water moves in osmosis is con-

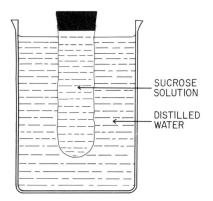

SUCROSE
SOLUTION

DISTILLED
WATER

Fig. 4.2. Apparatus for the demonstration of osmosis through a differentially permeable membrane.

trolled by differences in the quantity called the *water potential,* discussed later in greater detail.

The exact mechanism of osmosis is a matter of some uncertainty. It was long regarded as a special case of diffusion in which the solvent molecules are the diffusing particles. Various considerations have caused some authorities to consider that osmosis is a mass flow phenomenon through the membrane rather than the molecule by molecule type of movement inherent in diffusion. Quite possibly the process is predominantly or entirely a mass flow phenomenon through some kinds of membranes and predominantly or entirely a diffusional phenomenon through others. Regardless of the mechanism involved, however, the *direction* of the osmotic movement of water follows the same principles which govern the direction of diffusion.

Solutes, in molecular or ionic form, may also move through any membrane which is permeable to them. When such a movement of solutes occurs by diffusion, it takes place in accordance with the differences in concentration of each individual kind of solute particle. The diffusion of solutes across a membrane occurs independently of the movement of water by osmosis across the same membrane. Hence there is seldom any proportionality between the rate of water movement across a membrane and the rate of movement of any kind of a solute.

Membranes and Permeability.   The concept of permeability is inseparable from that of a membrane. Permeability is a property of the membrane, not of the substance which diffuses through it. Thin layers or sheets of many different kinds of substances such as rubber, parchment paper, collodion, gelatin and copper ferrocyanide may serve as membranes.

Some membranes are impermeable to all, or virtually all, substances; others allow all or most substances to diffuse through them with little impediment. Many biologically important membranes are of the type known as *differentially permeable.* Such membranes allow some substances to pass through them much more readily than others. They may be, and often are, impermeable or virtually so to some substances, while others diffuse through them quite freely. The membrane of the osmotic system described in the previous section is an example of a differentially permeable membrane.

The simplest kinds of membranes are essentially molecular sieves. The pores of a given membrane of this type are such that micelles, molecules, and ions below a certain size can diffuse through them, while those exceeding this size cannot pass through. Membranes of collodion, parchment paper, and copper ferrocyanide owe their differential permeability largely to their sieve-like properties.

Some membranes are differently permeable because certain substances are more soluble in them than others. An example of a membrane of this type in operation can be seen in an experiment set up as in Fig. 4.3. A thin layer of water is introduced on top of a layer of chloroform in a test tube; the tube is nearly filled with ether and stoppered. A second tube is prepared in the same way except that xylene is substituted for the ether. After several days it will be observed that the level of the water layer in the first tube has risen, while in the second tube it has fallen, although the distance through which the layer moves is not as great as in the second tube. In the first tube ether is diffusing through the water membrane more rapidly than the chloroform; hence the volume of liquid below the membrane is increasing and the layer of water rises. In the second tube chloroform diffuses through the water membrane more rapidly than

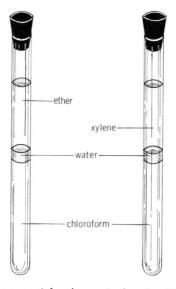

Fig. 4.3. Experimental arrangement for demonstrating the differential permeability of a water membrane.

the xylene; hence the water layer falls in the tube. Much less chloroform diffuses in a given time through the water layer in the second tube than does ether in the first tube. Of these three compounds, ether is the most soluble in water, chloroform next most soluble, and xylene least soluble. Permeability of the water membrane to these three compounds is clearly correlated with their solubility in water.

In this section the basic principles regarding permeability are discussed solely in terms of inorganic membranes. These principles also apply to both the boundary and the intracellular membranes of plant cells, discussed in more detail in other chapters, but do not explain much of their operation. Many such membranes not only exhibit permeability in the simple sense of the word, but also exert metabolic controls over the ingress and egress of solutes.

Free Energy and the Concept of Water Potential.   According to accepted thermodynamic principles, every component of a system possesses *free energy* which is capable of doing work under constant temperature conditions (Chapter 9). Ideally diffusion phenomena such as those discussed previously should be analyzed in terms of their differences in free energies, but for such relatively simple systems, analysis in terms of concentrations or partial pressures results in an acceptable interpretation. The movement of water, however, as in osmosis, cannot be accurately explained in terms of the differences in concentration in its usual sense of quantity per unit volume. The basic driving force in osmosis is the difference in the free energies of the water on the two sides of the membrane. Absolute values for the free energy of any substance cannot be readily ascertained. With reference to water the free energy per mole is termed the *water potential* and is arbitrarily given a value of zero at the prevailing temperature and under atmospheric pressure. Although the absolute water potential cannot be measured or calculated, departures from this zero value, plus or minus, can be ascertained. For quantitative purposes the water potential may be expressed either in energy terms such as ergs per gram, or in pressure units such as dynes per square centimeter or bars. One *bar* equals 10^6 dynes per square centimeter. A familiar pressure unit which has been widely used is the *atmosphere*. One bar is equal to 0.987 atmospheres. Pressure units will be used in this discussion, because they are more meaningful than energy units in analyzing many features of the water relations of plants. In osmotic systems and others of which it is a component, water moves from the regions of greater water potential (less negative) to the regions of lesser water potential (more negative).

If pressure is imposed on water, as by a piston in a closed system, or by the confining walls of a closed osmometer, or by a plant cell when the volume of the vacuolar sap is expanding as the result of osmosis, the water potential of the confined water increases by the amount of the imposed pressure. If the imposed pressure is 8 bars, for example, the water potential increases by 8 bars. Correspondingly, if water is subjected to a tension, which is physically equivalent to a negative pressure, its water potential decreases by the amount of the tension. The effect of any given pressure on the water potential is quantitatively the same, whether the pressure is imposed on pure water or whether it is imposed on a solution. The same statement is true for any given imposed tension.

In an aqueous solution the water potential is also influenced by the proportion of solute particles (ions plus molecules) to water molecules. The greater this proportion, the more negative the water potential, because the solute particles decrease the free energy of water. Within a wide range of concentrations of the solute molecules or ions, the diminution in the water potential is closely proportional to the ratio of the solute particles to the solvent molecules.

Matric forces, whereby water molecules are bound to molecules or micelles largely by adsorption, and thus are less free to diffuse, also operate to reduce the water potential in some systems. As discussed later such matric forces play a significant role in some kinds of plant cells and are the most important factor in determining the water potential in most soils (Chapter 7).

Osmotic Pressure.   This term designates the maximum pressure which develops in a solution enclosed within an osmometer under certain ideal conditions. An osmometer is an apparatus for measuring the magnitude of osmotic pressures. If a manometer or pressure guage were inserted through the stopper of an apparatus set up as in Fig. 4.2, the arrangement would serve as a crude osmometer. Most of the precise measurements of osmotic pressures have been made with osmometers constructed of cylindrically-shaped porous clay cups in the pores of which have been precipitated differentially permeable membranes of cupric ferrocyanide. The pressure developed is measured with a sensitive mercury manometer or in other ways. The necessary ideal conditions, only attained under rigorous experimental conditions, are that the membrane be permeable only to the solvent, that it be immersed in the pure solvent, and that the pressure equilibrium be attained without any appreciable dilution of the enclosed solution. The temperature of the experimental setup must also be controlled and recorded, as the osmotic pressure of a solution is in part a function of its temperature.

By a widely used extension of its primary meaning as the designation for an actual pressure, the term osmotic pressure has also been employed in the sense of a potential pressure rather than an actual one. According to this usage a 1 molal solution of sucrose standing in a beaker may be designated as having an osmotic pressure of 27 bars at 25°C. If this solution is allowed to come to equilibrium with pure water in an ideal osmotic system at 25°C, an osmotic pressure of 27 bars will develop. As will become clearer in the later discussion, the osmotic pressure is basically a quantitative index of the lowering of the water potential in a solution, insofar as this results from the presence of solutes. This component of the water potential is now generally called the *osmotic potential* rather than the osmotic pressure. The osmotic potential of the 1 molal sucrose solution is therefore −27 bars. It should be noted that the osmotic potential value carries a negative sign, while the numerically equal osmotic pressure value carries a postitive sign. In an open system, the osmotic potential value and the water potential value are numerically identical and have the same sign.

Turgor Pressure.   Turgor pressure is the *actual* pressure which develops in a closed osmometer or a plant cell as a result of osmosis or imbibition (see the following). The terms "turgor pressure" and "osmotic pressure" are not equivalent since the latter refers only to the maximum possible pressure which can develop under certain specified conditions as described previously, while turgor pressure represents that portion of this potential which has actually come into existence. Commonly the turgor pressure which can prevail in a solution ranges in magnitude between zero and its osmotic potential and has a positive sign. If a tension prevails in the solution, however, the turgor pressure carries a negative sign. The turgor pressure prevailing within a solution is also exerted against the confining walls of the system in which it is present.

Quantitative Aspects of Osmosis.   Let us suppose that a solution with an osmotic potential of −20 bars completely fills an essentially inelastic membrane which is permeable only to water and arranged like that shown in Fig. 4.2. Let us further assume that this membrane is immersed in pure water. The water potential in the solution is −20 bars; that of the water outside of the sac, zero. Water therefore moves across the membrane into the solution, *i.e.,* from the region of greater (zero water potential) to that of lesser (more negative) water potential. The entering water engenders a gradually increasing turgor pressure throughout the solution in the sac, because fewer molecules of water move out of the sac then enter it per unit of time. Subjection of the water in the solution to pressure increases its water potential, because under pressure the free energy of the water in the solution is increased (made less negative). If we assume that the dilution of the solution is so slight that it can be disregarded, a dynamic equilibrium will be attained when the turgor pressure of the solution reaches 20 bars. At this point the −20 bars water potential which was present at the start as a result of the presence of the solutes is offset by the 20 bars turgor pressure. The resulting water potential in the solution becomes zero, which is the same as that of the pure water outside of the membrane. When the water potential becomes the same on both sides of the membrane, osmosis ceases, and the number of water molecules moving through the membrane in the inward direction will be the same as the number moving in the outward direction per unit of time. In other words, a dynamic equilibrium has been established. Since the membrane is assumed to be inelastic or nearly so, the osmosis of only a very small amount of water into the solution is necessary to raise its turgor pressure to 20 bars. Since there is virtually no dilution of the solution, its final osmotic potential, for all practical purposes, is the same as its initial osmotic potential.

In living organisms water usually osmoses across a membrane from one solution to another solution rather than from pure water into a solution. Let us consider, therefore, a situation exactly like that described here previously except that there is a solution which maintains an osmotic potential of -12 bars on the outside of the membrane instead of pure water. Initially, before any turgor pressure develops, the water potential of the internal solution is -20 bars; that of the external solution, -12 bars. Water therefore osmoses across the membrane from the external into the internal solution, *i.e.,* from the region of the less negative to the region of the more negative water potential. As the turgor pressure of the internal solution increases, the water potential of that solution increases. When a turgor pressure of 8 bars has been attained, the water potential of the internal solution will have increased from -20 bars to -12 bars, the latter being the water potential of the external solution. Dynamic equilibrium is thus attained in this particular system when the turgor pressure prevailing in the internal solution reaches 8 bars.

In the foregoing discussion an inelastic membrane has been assumed. Many membranes, especially in living organisms, are elastic within limits. When a membrane is appreciably elastic, the further complication of shifts in the osmotic potential of an internal solution during osmosis is introduced. Let us assume an osmotic system identical with that described in the preceding paragraph, except that the membrane is elastic enough to permit a 25 per cent increase in the volume of the internal solution before a dynamic equilibrium is attained. As the internal solution increases in volume it becomes more dilute, *i.e.,* less concentrated. The osmotic potential of a solution becomes less negative (*i.e.,* approaches zero) with a diminution in its concentration. When the effects of ionization and hydration are disregarded (see later), the change in osmotic potential with a change in the volume of the solution can be calculated from the equation $P_1V_1 = P_2V_2$ (P_1 = original internal osmotic potential, V_1 = original internal volume, P_2 = resulting internal osmotic potential, and V_2 = resulting internal volume). Hence the final osmotic potential in this osmotic system will be -16 bars ($-20 \times 1 = P_2 \times 1.25$). A turgor pressure of only 4 bars would be required under these conditions to bring the water in the internal solution into dynamic equilibrium with that in the external solution, which maintains an osmotic potential of -12 bars. In such a system the turgor pressure of the internal solution is increasing (from 0 to 4 bars) while its osmotic potential is also increasing (from -20 to -16 bars). If the volume of the external solution is large relative to that of the internal solution, variations in the osmotic potential of the former are usually negligible. It should be evident from the foregoing discussion that the water potential of an aqueous solution is always equal to its osmotic potential plus the turgor pressure within it.

Expressed mathematically:

$$\psi_w = \psi_s + \psi_p \qquad (\psi = \text{Greek letter psi})$$

in which ψ_w is the water potential, ψ_s is the osmotic potential, *i.e.,* that component resulting from the presence of solutes, and ψ_p is the pressure potential, usually turgor pressure.

As applied to the initial water potential in the second example given here:

$$\psi_w = \psi_s + \psi_p$$

External solution $-12 = -12 + 0$
Internal solution $-20 = -20 + 0$

At dynamic equilibrium the water potential of the internal solution will become equal to that of the external solution, so its ψ_p *at equilibrium* can be calculated as follows:

$$-12 = -20 + \psi_p \qquad \psi_p = +8 \text{ bars}$$

The osmotic potential (ψ_s) is always negative in value except in pure water in which it is zero. The pressure potential (ψ_p) is usually positive in value, but becomes negative in a liquid which is under tension, and has a zero value in a liquid under atmospheric pressure. The water potential (ψ_w) may be positive, zero, or negative in value. In plants it is usually negative.

Factors Influencing the Osmotic Potential of Solutions.   *Concentration.*

An increase in the concentration of a solution invariably results in its osmotic potential becoming more negative. If the solute is a nonelectrolyte and its molecules do not acquire water of hydration, the osmotic potential is almost strictly proportional to the molal concentration, that is, to the proportion of solute to solvent molecules. The theoretical osmotic potential of a 1 molal solution of such a solute is -22.7 bars at $0°C$. In solutions in which the solute is dissociated or hydrated or both, an increase in the osmotic potential is not proportional with an increase in molality.

Ionization of the Solute Molecules. Osmotic potential is a colligative property, that is, one which depends on the number of solute particles, regardless of kind, in proportion to the number of solvent particles in a solution. The same number of ions or of solute molecules or of colloidal micelles in a given weight of water will result in a system with the same osmotic potential. Actually the concentrations of micelles in colloidal systems are usually so low that their influence on the osmotic potential is negligible or nearly so. If the solute dissociates, either in part or entirely, the osmotic potential will be more negative than otherwise. If an electrolyte such as NaCl were present at a 1 molal concentration completely in the form of free ions, the theoretical osmotic potential of the solution would be -45.4 bars (2×-22.7). The measured osmotic potential of a 1 molal solution of NaCl is -43.2 bars, which indicates that a condition of maximum ionic mobility is closely approached in such a solution.

Hydration of the Solute Molecules. We have seen (Chapter 2) that the micelles of all hydrophilic sols are highly hydrated. In a somewhat similar fashion water molecules adhere to many kinds of solute molecules and ions. Water thus associated with the particles of a solute is called water of hydration. Different species of molecules and ions have different numbers of water molecules associated with them. Water molecules which are bound to solute particles as water of hydration are no longer effective as part of the solvent. In effect a solution containing hydrated solute particles is more concentrated than its molality would indicate, and its osmotic potential is correspondingly more negative. A 1 molal solution of sucrose, for example, has an osmotic potential of -25.1 bars. The deviation from the theoretical value (-22.7 bars) for the osmotic potential of a sucrose solution is believed to result from the hydration of the sucrose molecules.

Temperature. The osmotic potential of a solution becomes more negative with an increase in temperature. For an ideal solution, the osmotic potential is proportional to the absolute temperature.

Imbibition. If a handful of pea or bean seeds is dropped into water, within a few hours they will have swollen visibly (Fig. 4.4). Seeds of any other species

Fig. 4.4. Photograph of the pea seeds, showing imbibition. Each cylinder contains the same number of seeds. Air-dry seeds are in the cylinder on the left. Seeds in the cylinder on the right were allowed to imbibe water for 24 hours before photographing. Photograph by Alan Heilman.

in which the coats are permeable to water behave in like fashion when brought into contact with water. Many other materials swell in a similar way when immersed in water. Among them are starch, cellulose, agar, gelatin, and kelp stipe. Some substances swell similarly when immersed in other liquids. All of these phenomena are examples of the process called *imbibition*. The amount of water which may enter substances in imbibition is often very great in proportion to the dry weight of the substance which swells. A piece of dried kelp stipe, for example, can absorb as much as fifteen times its own weight in water.

Water may be imbibed as a vapor as well as in the liquid state. The swelling of doors and woodwork during damp weather is a familiar example of this phenomenon. Plant structures, if sufficiently low in water content, also imbibe water vapor.

Dynamics of Imbibition.   As mentioned earlier the free energy of water and its corresponding water potential are reduced in magnitude by the existence in any system of adsorptive and microcapillary forces which result in the binding and immobilizing of water molecules. The molecules of imbibants have a great affinity for water molecules and as a result substantial negative water potentials prevail in such materials as long as they are relatively dry. Imbibition occurs whenever an imbibant is immersed in a medium with a greater (less negative) water potential than that of the imbibant. An equilibrium will be attained only when the water potential of the imbibant and that of the external medium have attained the same value.

Osmotic Effects on Imbibition.   Water moves by imbibition into a substance only when its water potential exceeds that of the imbibant. The introduction of a solute into water invariably has the effect, already discussed, of reducing the water potential of the water in the resulting solution as compared to pure water. Hence the magnitude of the osmotic potential of the solution in contact with the imbibant influences both the rate of imbibition and the equilibrium water content of an imbibant.

The lower (more negative) the osmotic potential of the solution in which an imbibant is immersed, the less the amount of water held per unit of dry weight at equilibrium (Table 4.1). Since at equilibrium the water potentials of the imbibing substance and the surrounding liquid must be equal, the basis for this osmotic effect upon the imbibition is evident. The relation between the osmotic potential and the amount of water imbibed is not, however, a strictly proportional one. At lower external solution concentrations, a decrease of a few bars in the osmotic potential causes a certain decrease in the amount of water imbibed; but at higher concentrations a decrease in the osmotic potential of several hundred bars is required for an equivalent change in the volume of the water imbibed. This is evidence that the first increments of water passing into any imbibant are held by forces tremendously greater than those which hold the increments which are imbibed subsequently.

TABLE 4.1 IMBIBITION OF WATER BY COCKLEBUR SEEDS IMMERSED IN SOLUTIONS OF DIFFERENT OSMOTIC POTENTIALS[*]

Volume molar concentration of solutions	Osmotic potential of solutions, atmos.	Water imbibed by seeds at equilibrium (48 hours). Per cent of air dry weight
H_2O	0.0	51.58
0.1 *M* NaCl	−3.8	46.33
0.2	−7.6	45.52
0.3	−11.4	42.05
0.4	−15.2	40.27
0.5	−19.0	38.98
0.6	−22.8	35.18
0.7	−26.6	32.85
0.8	−30.4	31.12
0.9	−34.2	29.79
1.0	−38.0	26.73
2.0	−72.0	18.55
4.0	−130.0	11.76
Sat. NaCl	−375.0	6.35
Sat. LiCl	−965.0	−0.29

[*]Data of Shull, *Bot. Gaz.* **62**, 1916: 86.

Pressures, sometimes of an enormous magnitude, are often exerted by a swelling imbibant if it is confined in some manner. Blocks of rock, for example, have been successfully quarried by drilling holes along the desired cleavage planes, driving tight-fitting wooden stakes into the holes, and pouring water on the stakes. The pressure developed upon imbibition of the water by the wood is sufficient to split the rock.

The actual pressures which develop as a result of imbibition, like those which develop as a result of osmosis, can logically be termed turgor pressures. No turgor pressure develops in an unconfined imbibant any more than it does in an unconfined solution. Only when complete swelling is prevented, as by enclosure within an inelastic wall, does a turgor pressure develop within an imbibant.

In an imbibant the component analogous to the osmotic potential (ψ_s) of a solution is the *matric potential* (ψ_m) which results primarily from adsorptive forces binding water molecules to micelles or molecules of the imbibant. In a solution, the greater the amount of water present in proportion to a given amount of solute, the less negative the osmotic potential. Likewise, in an imbibant, the greater the amount of water present in proportion to a given amount of imbibant, the less negative the matric potential. With reference to pure water, matric potentials like osmotic potentials, are always negative in value. Thoroughly dry imbibants, such as air dried seeds, have a matric potential of -1000 bars or even lower. When such seeds become partially swollen, their matric potential is much less negative; when they have reached a condition of maximum swelling in pure water under atmospheric pressure, their matric potential has risen to zero.

Quantitative Aspects of Imbibition.   The water potential of an imbibant is equal to its matric potential (always negative) plus any turgor or other pressure which may be imposed upon it:

$$\psi_w = \psi_m + \psi_p$$

Since, if the inbibant is unconfined, no pressure factor is involved, the equation for such systems simplifies to:

$$\psi_w = \psi_m$$

If an imbibant with a matric potential of -100 bars is immersed in pure water, its initial water potential is -100 bars, because no pressure component is involved. When equilibrium is attained between the imbibant and the water, both will have a water potential of zero. In other words water molecules now occupy all possible adsorption sites in the imbibant. If a relatively small amount of such an imbibant is immersed in a large volume of solution with an osmotic potential of -20 bars, when equilibrium has been attained, both the matric potential and the water potential will be -20 bars. Under these conditions not all the possible adsorption sites will be occupied by the water molecules. In an unconfined imbibant, equilibrium of its water potential with that of water in a bathing solution occurs by an adjustment of its matric potential.

Consider now the situation in which an imbibant with an initial matric potential of -100 bars is enclosed in an inelastic, water-permeable wall and is immersed in pure water. Equilibrium will be attained by the movement of an infinitesimal amount of water into the imbibant. The equilibrium water potential of the imbibant will be zero, not because of a decreased negativity of its matric potential but because a pressure potential (ψ_p) of 100 bars has developed. If such an imbibing system is immersed in a large volume of solution with a ψ_s of -20 bars rather than in pure water, the ψ_w of the imbibant at equilibrium will be -20 bars, again not because of a decreased negativity of its matric potential, but because it becomes subjected to a turgor pressure of 80 bars.

SUGGESTED FOR COLLATERAL READING

Kozlowski, T. T., Editor. *Water Deficits and Plant Growth,* 3 vols. Academic Press, New York, 1968, 1968, 1972.

Kramer, P. J. *Plant and Soil Water Relationships. A Modern Synthesis.* McGraw-Hill Book Company, New York, 1969.

Slatyer, R. O. *Plant-Water Relationships.* Academic Press, Inc., New York, 1967.

WATER RELATIONS
OF PLANT CELLS

5

The Plant Cell as an Osmotic System. For many years it has been the practice to view the vacuolate plant cell as a tiny closed osmotic system. Considering the intricate organization of a cell, much of which has been realized only in comparatively recent years, this is an oversimplified viewpoint, but it is still a serviceable one in analyzing the cell to cell movement of water in plants and the movement of water into or out of plant cells from a contiguous solution. Although other membranes are present in plant cells, the only ones through which water and solutes must pass in moving into or out of a plant cell are the cell wall and some or all of the cytoplasmic layers.

The cell wall usually consists of several layers each with a complex organization of its own (Chapter 3). In general, cell walls composed of cellulose, hemicelluloses, and pectic compounds, such as those of the cells in most parenchymatous tissues, are quite permeable to both water and solutes. Water and solutes probably penetrate such walls through the hydrophilic intermicellar material. Lignified cell walls are usually quite permeable to water and solutes. Cell walls in which suberin or cutin are present in appreciable quantities, on the other hand, are always relatively impermeable to water and solutes and often almost completely impermeable to these compounds. Plasmodesmata may contribute to the apparent permeability of many cell walls.

Lining the cell wall of most mature plant cells is a layer of cytoplasm, the bounding layers of which constitute distinct membranes, the plasma membrane or plasmalemma next to the cell wall, and the vacuolar membrane or tonoplast next to the vacuole (Chapter 3). Although the presence of these membranes is clearly revealed only by the electron microscope, their existence in living cells had long been recognized as the result of various kinds of physiological experimentation.

The cytoplasmic membranes are differently permeable and are responsible for the behavior of plant cells as osmotic systems. The term "cytoplasmic membranes" is used in a loose sense to refer to the plasmalemma plus the interior cytoplasm plus the tonoplast. It is virtually impossible to ascertain in which of these membranal components the various aspects of the property of differential permeability actually reside. The plasmalemma is often regarded as the most important unit in the cytoplasmic system of membranes, since many substances entering a plant cell are intercepted and utilized in the cytoplasm and similarly compounds synthesized in the cytoplasm may pass out of the cell without crossing the tonoplast.

The results of permeability studies of the cytoplasmic membranes present apparent anomalies. The permeability of the cytoplasmic membranes to most kinds of solutes of physiological importance, such as sugars, electrolytes, organic acids, and amino acids appears to be quite variable. It differs from one kind of cell to another and varies in the same cell from one time to another. The cytoplasmic membranes of many mature plant cells appear to be relatively impermeable to solutes in these categories. The cytoplasmic layers, however, must be regarded as more than passive membranes, since they are metabolically active. In many kinds of cells there is an active transport of ions or molecules of solutes across membranes, often in opposition to concentration gradients (Chapters 15, 17, and 19). Such active transport of molecules or ions across membranes is powered by the energy released in respiration (Chapter 13).

Plasmolysis. If a vacuolate plant cell is immersed in a solution with a more negative osmotic potential than its cell sap, and if its cytoplasmic membranes are relatively impermeable to the solute in the external solution; a characteristic series of changes takes place in the appearance of that cell. If the cell walls are sufficiently elastic that the cell is initially somewhat distended, the first detectable occurrence is a gradual shrinkage in the volume of the entire cell as a result of outward osmosis, which results in a reduction in the pressure exerted by the cell sap against the cytoplasm and the cell wall. This shrinkage in volume can be detected in many cells by measurement under a microscope, although in cells with relatively inelastic walls little or no change in volume takes place. Once the lower limit of elasticity of the cell wall is reached, no further decrease in the volume of the cell occurs.

Since the cell wall is quite freely permeable to the water and the solutes of the surrounding solution, and hence the external solution is in contact with the plasmalemma, the cell sap continues to lose water, just as if no cell wall were present. The cell sap therefore continues to shrink in volume after contraction of the cell wall has ceased. As a consequence, the cytoplasmic layer begins to recede from the cell wall at this stage and, if the external solution is concentrated enough, this separation of the cytoplasm from the cell wall becomes very pronounced (Fig. 5.1). This phenomenon is called *plasmolysis*. The pattern followed by the shrinkage of the cytoplasm in plasmolysis is more or less typical for each kind of cell, although it may be modified somewhat depending upon physiochemical conditions within the cytoplasm and the kind of solute used in the plasmolyzing solution. The space between the cell wall and the outer membrane of the cytoplasm will be filled, after separation of the latter from the wall, with the external solution.

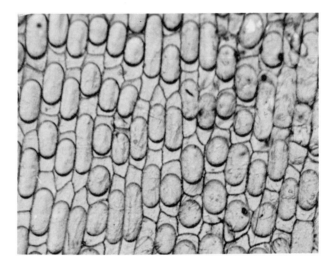

Fig. 5.1. ⋅ Plasmolysis in cells of onion bulb epidermis. Photo-micrograph by Eduard Stadelmann.

If a plasmolyzed cell is immersed in water, it will gradually recover and regain a turgid state as a result of the osmotic movement of water into the vacuole. Similarly, if such a cell is immersed in a solution with a less negative osmotic potential than the cell sap, recovery will also ensue, but the degree of turgidity attained will be less than in pure water.

Magnitude of the Osmotic Potentials in Plant Cells. As one of the main components of the water potential of plant cells, the magnitudes of their osmotic potentials are significant. Different organs or tissues of the same plant may differ widely from one to another in the osmotic potentials of their cells. Leaf cells, for example, almost invariably are more negative in their osmotic potential than the cells of younger roots on the same plant. Even similar organs on the same plants, such as leaves, may differ considerably among themselves in the average osmotic potential of their cells. Within any one organ differences in the osmotic potential from one tissue to another are frequent. The mesophyll cells of leaves, for example, usually have a more negative osmotic potential than the non-green epidermal cells on the same leaf.

In general, the osmotic potentials of the plant cells of mesic species lie within a range of −2 to −40 bars. The osmotic potential of a given plant cell or group of plant cells is not a fixed quantity and often fluctuates considerably. More or less regular diurnal variations occur in the osmotic potentials of many kinds of plant cells (Fig. 6.5). Seasonal variations of considerable magnitude also occur, especially in the leaf cells and other tissues of a temperate zone species (Fig. 5.2). Such variations in the osmotic potential have important effects on the water potential of the cells in which they occur.

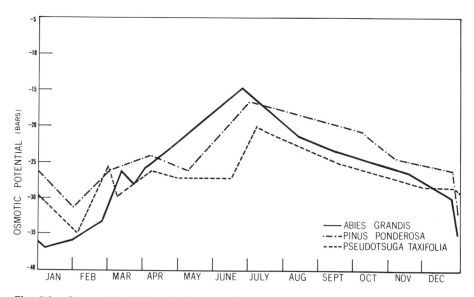

Fig. 5.2. Seasonal variations in the osmotic potential of the leaves of evergreens. Data of Gail, *Botan. Gaz.* **81**, 1926:437.

Factors Influencing the Osmotic Potential of Plant Cells. Any factor which influences either the water content of a plant cell or the solute content of its cell sap will have an effect on the magnitude of the osmotic potential of that plant cell. The water content of the plant as a whole, and hence of its constituent cells, is controlled principally by the relative rates of transpiration and the absorption of water (Chapter 8). The latter process is markedly influenced by the water content and other conditions prevailing in the soil. Individuals of the same species invariably have more negative osmotic potentials when growing under drought conditions than when provided with a favorable water supply (Table 5.1). This is at least partially a result of the relatively lower water content of the leaves which results when the available soil water supply becomes low. Other factors which are also probably involved are a decrease in the growth rate of the plant, which often permits an accumulation of mineral salts and soluble foods, and a shift of the insoluble carbohydrate (starch)/soluble carbohydrate (sugar) equilibrium toward the side of the soluble carbohydrates.

TABLE 5.1 THE EFFECTS OF DIFFERENT SOIL WATER CONTENTS
UPON THE OSMOTIC POTENTIAL OF MAIZE PLANTS[a]

Water content of soil, per cent dry weight	Osmotic potential of tops, bars	Osmotic potential of roots, bars
31	−22.29	−5.99
23	−23.38	−7.32
16	−24.68	−7.89
14	−25.37	−9.36
13	−25.80	−11.49
11	−26.82	−12.14

[a]Based on Hibbard and Harrington, *Physiol. Res.*, **1**, 1916: 452.

The solute content of the cell sap is controlled by the specific metabolic processes of the plant and by the absorption of mineral salts by the plant from its environment. The rate of photosynthesis is an important factor in influencing the osmotic potential of the plant cells, particularly those of the leaf tissues. Increases in sugar content resulting from photosynthesis result in more negative osmotic potentials within the cells. Inherent metabolic processes affect the kinds and concentrations of the various types of soluble organic compounds present in cells, such as simple carbohydrates, organic acids, and amino acids, and hence have important effects on the magnitude of cellular osmotic potentials. Metabolic conditions and their effects upon osmotic potentials may also be altered by environmental conditions. A well-known example of this is the difference in the osmotic potentials of sun and shade leaves on the same tree. The former almost invariably have the more negative osmotic potential, presumably at least in part because of their greater photosynthetic activity.

The mineral salts which contribute to the osmotic potentials of plant cells are absorbed from the soil or in aquatic species from the water in which part or all of the plant is immersed. Different species of plants vary greatly in their toleration of high concentrations of mineral salts in the soil. All species can become adjusted, within limits, to a change in the mineral salt content of the substratum. One aspect of this adjustment is a decrease in the osmotic potential of the plant with a decrease in the osmotic potential of the medium from which it obtains its mineral salts (Table 5.2).

Species indigenous to saline soils usually have relatively negative osmotic potentials. Such soils are rich in soluble salts, and the highly negative osmotic

TABLE 5.2   THE EFFECT OF THE OSMOTIC POTENTIAL OF THE SOIL
SOLUTION UPON THE OSMOTIC POTENTIAL OF THE
ROOTS OF MAIZE[a]

Osmotic potential of soil solution, bars	Osmotic potential of sap from root cells, bars
−1.23	−4.65
−2.02	−5.55
−3.42	−6.70
−5.02	−7.60
−7.31	−8.30

[a]Based on McCool and Millar, *Soil Sci.,* **3,** 1917: 129.

potentials found in species native to such soils result from the absorption of
relatively large quantities of mineral salts. In fact, the most negative osmotic
potential known for any species of plant, −205 bars, was found in the saltbush,
Atriplex confertifolia, a saline soil species. Salt marsh plants also generally
have markedly negative osmotic potentials. This is also a correlation with the
relatively high salt content of the substratum. Except in such halophytes, however, a larger proportion of the osmotic potential of most kinds of plant cells
results from the presence of organic solutes more than of mineral salts. Root
cells are more likely to prove exceptions to this last statement than cells in
the aerial organs.

The Osmotic Quantities of Plant Cells.   Let us consider three similar plant
cells, each with a zero turgor pressure, and each with a cell sap osmotic potential of −14 bars. Cell *A* is immersed in pure water, cell *B* is immersed in a
solution with an osmotic potential of −10 bars, and cell *C* is immersed in a
solution with an osmotic potential of −14 bars (Fig. 5.3). Since there is no
turgor pressure in each of the cells, their initial water potential will equal their
osmotic potential of −14 bars. The volume of the liquid in which each cell is
immersed is assumed to be very large in proportion to the volume of the cell.

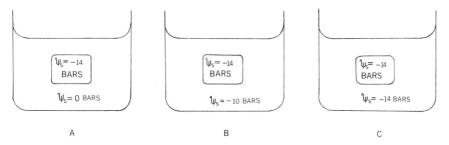

Fig. 5.3.   Similar cells immersed in solutions of different osmotic potentials.

Certain simplifying assumptions will be made regarding these cells, and unless stated to the contrary, regarding all other cells discussed in the remainder of this chapter: (1) that the cells and the surrounding solutions are all at the same temperature; (2) that the cytoplasmic membranes are permeable only to water while the cell walls are permeable to both water and solutes; (3) that the walls of all the cells are equally elastic; (4) that volume changes in the cells are so small that any resulting effects upon the osmotic potential of the cell sap are negligible and can be disregarded.

The osmosis of water into cell A begins immediately upon its immersion, because the water potential of the cell sap is -14 bars and that of the surrounding water is zero. The entrance of water into the cell results in a gradually increasing turgor pressure inside the cell, just as it does in an osmometer. This turgor pressure not only prevails within the cell sap, but is also exerted against the cytoplasmic layer and in turn against the cell wall. Unless the cell wall is totally inelastic, the increase in turgor pressure results in some distention of the cell. Imposition of a pressure of 14 bars on the cell sap will increase its water potential by 14 bars. Since the initial water potential of the cell sap was -14 bars, the imposition of a 14 bar turgor pressure increases its water potential to zero. This is also the water potential of the surrounding water; hence a dynamic equilibrium is attained when the turgor pressure in the cell sap reaches 14 bars.

The water potential of the cell sap of cell B is initially -14 bars; that of the surrounding solution -10 bars. Hence there will be a net movement of water into the cell. A turgor pressure thus develops within the cell, but at its maximum under these conditions it will attain a value of only 4 bars. Since the initial water potential of the cell sap was -14 bars, the imposition of a turgor pressure of 4 bars reduces this to -10 bars which is the water potential of the surrounding solution. Hence the attainment of a turgor pressure of only 4 bars results under these conditions in an osmotic equilibrium.

Cell C will remain, under the prescribed conditions, in a state of incipient plasmolysis. The water potential of the cell sap is equal to that of the surrounding solution since their osmotic potentials are equal, and neither is under any pressure, except, of course, atmospheric pressure. Hence a dynamic equilibrium is established as soon as the cell is immersed in the solution. Since there is no net movement of water into the cell, no turgor pressure will develop, a situation which will prevail in all plant cells in the state of incipient plasmolysis.

In cells A and B a dynamic equilibrium has been attained by the adjustment of the water potential of the cell sap until it is equal to the water potential of the surrounding medium. This adjustment was attained in each of these cells by an increase in the magnitude of the turgor pressure, which is one of the two main components determining the water potential of the cell sap, the other being the osmotic potential. In cell C no shift in the magnitude of the turgor pressure occurred, because the water potential was initially the same in both the cell sap and the solution. Unlike an unconfined solution the water potential of cell sap is not numerically equal to its osmotic potential, except when the turgor pressure prevailing in it is zero.

The interrelationships among the osmotic potential, turgor pressure, and water potential of a plant cell are shown in Fig. 5.4. This diagram also takes into account the influence of the volume changes in the cell upon these

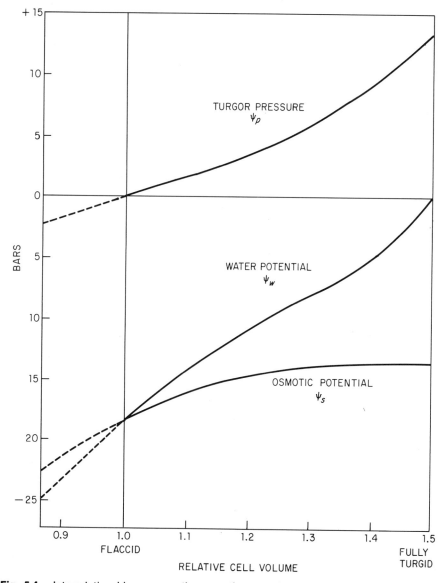

Fig. 5.4. Interrelationships among the osmotic potential, turgor pressure, water potential, and cell volume of a plant cell. Based on the data of Höfler, *Ber. Deut. Botan. Ges.,* **38,** 1920:288–298.

physical quantities. In the interests of simplicity the effects of such volume changes upon the osmotic quantities of cells have been disregarded in the discussion up to this point. When the cell is completely flaccid (relative volume = 1.0), its water potential is equal to its osmotic potential (−20 bars) while its turgor pressure is zero. As the volume of the cell increases, resulting from an influx of water, its osmotic potential becomes less negative because of the dilution of the cell sap. The pattern of the change in the magnitude of the turgor pressure with an increase in cell volume is not the same in all kinds of cells, but often follows a trend such as that shown in Fig. 5.4. The water potential of the cell increases progressively toward zero as its turgidity increases. When the cell attains a condition of maximum turgidity (relative volume = 1.5), its turgor pressure is equal, but opposite in sign, to its osmotic potential, and the water potential of the cell therefore has a zero value.

In many types of cells, volume changes are much less marked than indicated in Fig. 5.4, and hence can often be disregarded in generalized considerations of the water relations of plant cells. In some species many of the cells have walls which are virtually inelastic. This is especially true of many cells in the tissues of xerophytes.

The fundamental interrelations among the osmotic potential, turgor pressure, and water potential of a plant cell are the same as for a closed osmometer and are expressed by the same equation:

$$\psi_w = \psi_s + \psi_p$$

In plant tissues many of the cells are under a pressure imposed upon them by the surrounding cells. In addition to its own turgor pressure the water in such a cell is also subjected to this pressure of external origin. Such a pressure is just as much a factor in determining the water potential of a cell as its own turgor pressure. The true water potential of such a cell is therefore less negative than that of the cell considered as an individual unit by the amount of this added pressure. An example of such a situation exists in at least some kinds of leaves in which the epidermal layer of cells exerts a pressure on underlying mesophyll cells. This kind of pressure also exists in leaf blades, petioles, and stems, when they bend in the wind.

The water in the vessels and some kinds of plant cells frequently passes into a state of tension (Chapter 8). The imposition of a pressure from an external source raises the magnitude of the water potential. The imposition of a tension, which is in effect a negative pressure, has precisely the opposite effect. In cells and vessels this can happen only if the enclosed water shrinks in volume to such a point that the encompassing cytoplasm and/or cell walls are pulled inward as a result of the adhesion between them and water. The resulting counterpull exerted by the walls on the cell sap, results in throwing it into a state of tension. Under such conditions the turgor pressure is negative in value, and the tension (negative pressure) developed within the cell sap will be equal to the negative turgor pressure. See the dotted extension of the line for turgor pressure to the left in Fig. 5.4. The water potential of a cell in which the turgor pressure is negative is more negative than its osmotic potential.

A term which has been widely used in many books and articles evaluating the water relations of plants is the *diffusion pressure deficit* (DPD). This quantity is defined as the amount by which the diffusion pressure of the water in a solution or imbibant or soil is less than that of pure water at the same temperature and under atmospheric pressure. In basic concept the DPD is equivalent to the water potential except that the negativity is inherent in the term itself and is not indicated by giving the numerical values a negative sign. In other words a DPD of 10 bars is equal to a water potential of −10 bars.

Dynamics of the Osmotic Movement of Water in Plants. In order to simplify this part of the discussion, changes in the osmotic potential of the cell sap resulting from the volume changes in the cell, as shown in Fig. 5.4, will be disregarded, as usually these are not great enough to modify substantially any generalized picture of the water relations of the plant cells.

Let us imagine a certain cell X to have an osmotic potential of −14 bars and a turgor pressure of 8 bars, and a second cell Z to have an osmotic potential of −12 bars and a turgor pressure of 4 bars. Let us further suppose that these two cells can be brought into such intimate contact that the osmotic movement of water can occur from one to the other (Fig. 5.5).

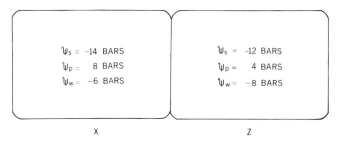

X   Z

Fig. 5.5. Diagram of two adjacent cells used in explanation of the mechanism of the cell-to-cell movement of water.

The water potential of the cell sap of X would be −14 bars were it under no pressure. The turgor pressure of 8 bars, however, raises the water potential of the cell sap to −6 bars. Similarly the water potential of the cell sap of Z would be −12 bars were it not also influenced by the turgor pressure of 4 bars. Hence the water potential of cell Z is −8 bars. The water potential of cell Z is therefore more negative than that of cell X, and water osmoses from X to Z. Water continues to show a net movement from X to Z until the water potentials of the two cells are equal. In the movement of water from cell to cell in plants, it is the water potentials and not the osmotic potentials which tend to equilibrate.

This is only a special aspect of the fundamental tendency of water potentials to attain a uniform value throughout any aqueous system. It is by no means impossible, therefore, as in this example, for water to move from a cell of more negative to one of less negative osmotic potential.

After a dynamic equilibrium has been established between two cells, their water potentials are seldom an exact average of their initial water potentials. The only general statement which can be made is that at equilibrium the water potentials of the two cells are equal, and this value lies somewhere between the two original values for the cells.

Whenever the water potentials of two adjacent cells are dissimilar, a water potential gradient exists between them. Net movement of water molecules from one cell to another can only occur when such a gradient exists. Other conditions being equal, the steeper this gradient, *i.e.,* the greater the difference in water potentials, the more rapidly one cell gains water from the other. The term "water potential gradient" can also be applied to a chain of cells in which the water potential increases serially from cell to cell. Several examples of such gradients are discussed later in subsequent chapters.

The Imbibitional Mechanism of the Movement of Water. In the preceding discussion plant cells have been considered purely as osmotic systems. There is no doubt that the osmotic mechanism of the cell to cell movement of water operates in most plant cells, at least at some stage of their development. Certain of the movements of water which occur in plants, however, result from the operation of an imbibitional mechanism rather than an osmotic mechanism. As discussed in the preceding chapter, such mechanisms result from the presence of a matric potential in the imbibant. Such matric potentials are always negative in value. Whenever the water potential in an imbibant is more negative than in a contiguous solution, water moves from the solution into the imbibant.

Not only does water pass into dry, mature seeds by imbibition as the first step in germination; but there is evidence that water moves by this process into ovules which are ripening into seeds in the ovulary. During the latter stages of the development of seeds in the cotton boll, for example, their water potentials are more negative than can be accounted for by their osmotic potentials. This indicates that an osmotic mechanism does not adequately account for the movement of water into the maturing seeds during this period and a likely explanation of these results is that the negative water potentials of maturing cotton seeds represent largely matric potentials. It seems probable also that imbibition plays a proportionately large role in the movement of water into the cells of many actively growing meristematic tissues in which vacuoles are lacking or undeveloped. In cells in which the interior is occupied by hydrophilic colloidal substances, imbibition also undoubtedly plays a large role in the movement of water. During the growth process imbibition accounts for much of the water that moves into the expanding and thickening cell walls.

Relation Between Osmotic and Matric Potentials. Up to this point the osmotic and matric potentials have been discussed as independent components of the water potentials of plant cells. In many cells both kinds of potentials are present. It has sometimes been considered that their effects are additive and that a summation of the two should be used in calculating the water potential of a cell as a unit. This interpretation, however, is not a valid one, at least for vacuolate cells. Let us visualize a cell in which the cell sap has a certain osmotic potential and in which hydrophilic colloids are also present. The latter will have a matric potential but it will be in equilibrium with the osmotic potential of the cell sap, just as the matric potential of seeds immersed in a solution will come to equilibrium with the osmotic potential of the solution. Two phases thus in equilibrium will act as a unit in influencing the water potential of a cell, and the magnitude of this effect will be the equilibrium value of the osmotic and matric potentials.

In intracellular movements of water a matric potential may predominate in one part of a cell and an osmotic potential in another. For example, the water potentials of the walls of a mesophyll cell, primarily a result of a matric potential, increase in negativity when an evaporational water loss occurs during transpiration. As soon as the water potential of the cell wall becomes more negative than that of the cell sap, of which the osmotic potential is a main component, water moves through the cytoplasm from the cell sap into the wall.

Possible Metabolically Activated Movement of Water in Plants. For many years it has been considered that, in addition to the osmotic and imbibitional mechanisms already considered, metabolic mechanisms of the movement of water also exist in plants. Such a mechanism would utilize the energy of respiration in causing the transport of water across the membranes. Water movement thus energized might be in opposition to a water potential gradient or might reinforce movement along such a gradient. Correlations between the rate of the water uptake or movement and respiration have been demonstrated in some plant tissues, a kind of finding which lends credence to the concept of a metabolically-activated water movement. It is highly probable, however, that such relationships are only indirect. Many other processes occurring in plants are powered by respiratory energy, and water movement is often tied to such processes rather than to respiration itself. Among such processes are the movement of solutes (Chapters 15 and 17), cyclosis, the fabrication of protoplasm, and the expansion of the cell walls in area, the latter two occurring during the growth process. Furthermore, cell membranes are so highly permeable to water that relatively enormous amounts of energy would be required to move water through them in opposition to a water potential gradient. There is no evidence for the expenditure of energy in such quantities in conjunction with the movement of water from cell to cell. In brief, the metabolic movement of water in plants appears to be either nonexistent or of negligible magnitude.

SUGGESTED FOR COLLATERAL READING

Kozlowski, T. T., Editor. *Water Deficits and Plant Growth,* 3 vols. Academic Press, New York, 1968, 1968, 1972.

Kramer, P. J. *Plant and Soil Water Relationships. A Modern Synthesis.* McGraw-Hill Book Company, New York, 1969.

Nobel, P. S. *Plant Cell Physiology.* W. H. Freeman and Company, San Franciso, 1970.

Slatyer, R. C. *Plant-Water Relationships.* Academic Press, Inc., New York, 1967.

6 LOSS OF WATER FROM PLANTS

It is commonplace knowledge that all plants require water for their existence and development and most plants require it in considerable quantities. It is not so generally recognized, however, that in most species of plants an overwhelmingly large proportion of the water absorbed from the soil is lost by the plant into the atmosphere and takes no permanent part in its development or in its metabolic processes. The lack of this general realization probably results from the fact that while water is supplied to and absorbed by plants in its familiar liquid form, by far the greater part of that lost escapes in the invisible form of water vapor.

The loss of water vapor from living plants is known as *transpiration*. Loss of water vapor may take place from any part of a plant which is exposed to the air. This applies even to roots in contact with the soil atmosphere. Generally speaking, however, the leaves are the principal organs of transpiration. Most of the transpiration from leaves occurs through the stomates; this is termed *stomatal transpiration*. Smaller amounts of water vapor are lost from leaves by direct evaporation from the epidermal cells through the cuticle; this is called *cuticular transpiration*. All aerial parts of plants lose some water by transpiration, although because of the presence on some organs of superficial layers almost impervious to water, the rate of loss from such organs is very low. Some of the transpiration from herbaceous stems, flower parts, and fruits is of the cuticular type, but is small in amount. Most herbaceous stems, fruits, and flower parts bear stomates which permit the occurrence of stomatal transpiration from such organs. The loss of water vapor also takes place through the lenticels of fruits and woody stems; this is called *lenticular transpiration*.

The Mechanics of Foliar Transpiration.   The subsequent discussion will deal almost entirely with transpiration from leaves, since in most plants the amount of water vapor lost from other organs is comparatively small. It should be recalled that the vacuoles of all of the living cells of a leaf are filled with water, which also saturates the protoplasm and the cell walls, this water being supplied to the leaf cells through the water-conducting tissues of the vascular bundles. Hence evaporation of water will occur from these wet cell walls into the internal atmosphere of the intercellular spaces just as it will occur from any wet surface into the surrounding air. The intercellular spaces constitute a connected system, ramifying throughout the leaf (Fig. 6.1). Under certain unusual conditions the intercellular spaces can become injected with liquid water, but under all normal conditions they are occupied by air.

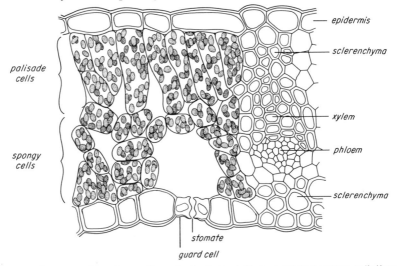

Fig. 6.1.   Cross section of a portion of a leaf of a tulip tree *(Liriodendron tulipifera)*.

If the stomates are closed, the only effect of evaporation from the mesophyll cell walls will be the saturation of the entire volume of the intercellular spaces with water vapor. When the stomates are open, however, diffusion of water vapor may occur through them into the outside atmosphere. Such outward diffusion will always take place unless the atmosphere has a vapor pressure equal to or greater than that of the intercellular spaces, a condition which does not commonly exist during the daylight hours. The rate of such diffusion will depend principally upon the excess of the vapor pressure in the leaf over that of the atmosphere, although the "diffusive capacity" of the stomates, discussed later in this chapter, is also an important factor.

   One side of every epidermal cell on a leaf is also exposed to the atmosphere. The evaporation of water occurs into the atmosphere directly from these cell surfaces. The surfaces of practically all aerial leaves are covered with a layer of waxlike substance known as *cutin*. This is not readily permeable to water and hence greatly reduces transpiration directly through the walls of the epidermal cells. The thickness of the cutin layer varies with the species of the plant and the environmental conditions under which the leaves have developed. The layer of cutin is usually thicker on leaves which have developed in bright sunlight, for example, than on leaves of the same species which have developed in the shade. Even in leaves which are heavily coated with cutin, some cuticular transpiration occurs, possibly largely through tiny rifts in the cutin layer. In most species of plants of the temperate zone less than 10 per cent of the foliar transpiration occurs through the cuticle, the remainder being stomatal transpiration.

Stomates.   Since 80 to 90 per cent of all the water vapor lost represents stomatal transpiration, let us consider the structure and operation of the stomates before continuing with the discussion.

The most important physiological fact about the stomates is that they are sometimes open and sometimes closed. When open, they serve as the principal pathways through which gaseous exchanges take place between the intercellular spaces of the leaf and the surrounding atmosphere. When closed, gaseous exchanges between a leaf and its environment are greatly retarded. The gases of the greatest physiological importance which enter or depart from a leaf principally through the stomates are oxygen, carbon dioxide, and water vapor. The movement of gases through the stomates in either direction is primarily a diffusion phenomenon, although under certain conditions, as will be discussed later, the mass movement of gases may occur through the stomates. Although the stomates are the principal portals through which the entry and escape of gases take place, the fact should not be overlooked that at least small quantities of gases pass directly through the epidermis and cuticular layers of all leaves. In submerged vascular aquatics all gaseous exchanges between the plant and its environment occur through the epidermis.

Strictly speaking, the *stomates* or *stomata* (singular *stomate* or *stoma*) are minute pores which occur in the epidermis of plants (Fig. 6.2). They are surrounded by two distinctive epidermal cells known as the *guard cells* (Fig. 6.3). Stomates may occur on any part of a plant except the roots, but in most species are most abundant upon the leaves. The size of the stomatal pore varies in most plants, depending upon the turgidity of the guard cells, and often, especially at night, the pore is entirely closed. The structure of the stomatal apparatus shows marked variations in detail in different species of plants, but the essential feature of a pore between two guard cells is common to all species of vascular plants. Guard cells which are roughly kidney- or bean-shaped are typical of more species of plants than any other kind. A perspective view of such a stomate and the surrounding guard cells is shown in Fig. 6.4. Unlike other epidermal cells, the guard cells contain chloroplasts. They also appear to contain a larger proportion of cytoplasm than the epidermal cells.

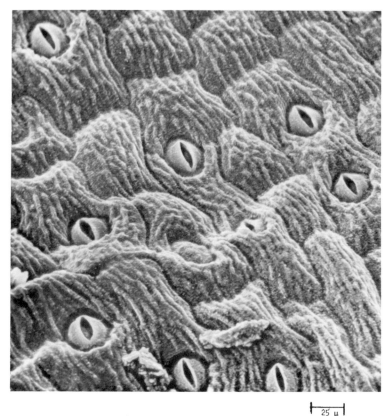

25 μ

Fig. 6.2. Surface of the leaf of a species of primrose (*Dianthus plumarius*), as shown under the scanning electron microscope. Photograph by Y. and J. Heslop-Harrison.

Size and Distribution of the Stomates. The size of the stomatal pore varies greatly according to the species of plant, and somewhat among the individual stomates on any one plant. The pores are always very minute, however, their dimensions being expressed in terms of microns (Table 6.1). Minute as these openings appear to be from a human scale of values, they are enormous when compared with the size of the gas molecules which diffuse through them. The calculated diameter of a water molecule is 0.000454 μ. More than 2000 water molecules would have to be placed side by side to measure a distance of 1 μ. The molecules of both carbon dioxide and oxygen are larger than water molecules. Since the stomatal diameters usually are considerably in excess of 1 μ, it is evident that the stomates afford relatively enormous portals to the gas molecules which diffuse through them.

In general, the number of stomates present in the epidermis of leaves may range from a few thousand to over a hundred thousand per square centimeter, the exact number depending upon the species and upon the environmental conditions under which the leaf has developed.

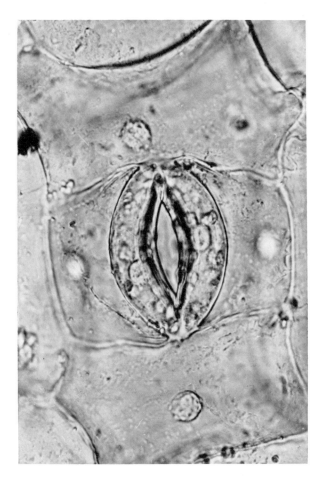

Fig. 6.3. Open stomate in the lower epidermis of *Zebrina pendula*. Photomicrograph by Tillman Johnson.

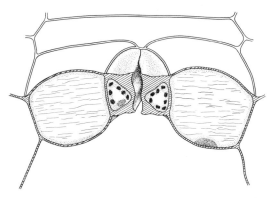

Fig. 6.4. Perspective view of a stomate and adjacent cells (semidiagrammatic).

TABLE 6.1 SIZE AND DISTRIBUTION OF THE STOMATES ON THE LEAVES OF VARIOUS SPECIES OF PLANTS[*]

Species	Average no. of stomates per cm²		Size (length × breadth) of pore when fully open (Lower epidermis)	Reference
	Upper epidermis	Lower epidermis		
Alfalfa				
(*Medicago sativa*)	16,900	13,800	—	M
Apple				
(*Pyrus malus*) var.	0	29,400	—	M
Barberry				
(*Berberis vulgaris*)	0	22,900	—	K
Bean				
(*Phaseolus vulgaris*)	4,000	28,100	7 × 3 μ	E
Begonia				
(*Begonia coccinea*)	0	4,000	21 × 8 μ	E
Black Oak				
(*Quercus velutina*)	0	58,000	—	Y
Black poplar				
(*Populus nigra*)	2,000	11,500	—	S
Black walnut				
(*Juglans nigra*)	0	46,100	—	K
Cabbage				
(*Brassica oleracea*)	14,100	22,600	—	M
Castor bean				
(*Ricinus communis*)	6,400	17,600	10 × 4 μ	E
Cherry				
(*Prunus cerasus*) var.	0	24,900	—	M
Coleus				
(*Coleus blumei*)	0	14,100	10 × 5 μ	E
English ivy				
(*Hedera helix*)	0	15,800	11 × 4 μ	E
English Oak				
(*Quercus robur*)	0	45,000	—	S
Geranium				
(*Pelargonium domesticum*)	1,900	5,900	24 × 9 μ	E
Holly				
(*Ilex opaca*)	0	17,000	12.5 × 6.5 μ	K
Jimson weed				
(*Datura stramonium*)	11,400	18,900	—	K
Lilac				
(*Syringa vulgaris*)	0	33,000	—	K
Linden				
(*Tilia vulgaris*)	0	13,000	—	S
Maize				
(*Zea mais*)	5,200	6,800	19 × 5 μ	E

TABLE 6.1 (Continued)

Species	Average no. of stomates per cm²		Size (length × breadth) of pore when fully open (Lower epidermis)	Reference
	Upper epidermis	Lower epidermis		
Mulberry				
(*Morus alba*)	0	48,000	—	K
Nasturtium				
(*Tropaeolum majus*)	0	13,000	12 × 6 μ	E
Nightshade				
(*Solanum dulcamara*)	6,000	26,300	—	K
Oat				
(*Avena sativa*)	2,500	2,300	38 × 8 μ	E
Pea				
(*Pisum sativum*)	10,100	21,600	—	K
Peach				
(*Prunus persica*) var.	0	22,500	—	M
Potato				
(*Solanum tuberosum*)	5,100	16,100	—	M
Red oak				
(*Quercus rubra*)	0	68,000	—	Y
Scarlet oak				
(*Quercus coccinea*)	0	103,800	—	Y
Scilla				
(*Scilla nutans*)	5,500	5,100	—	S
Sunflower				
(*Helianthus annuus*)	8,500	15,600	22 × 8 μ	E
Sycamore				
(*Platanus occidentalis*)	0	27,800	—	K
Tomato				
(*Lycopersicon esculentum*)	1,200	13,000	13 × 6 μ	E
Tree of heaven				
(*Ailanthus glandulosa*)	0	38,600	—	K
Wandering Jew				
(*Zebrina pendula*)	0	1,400	31 × 12 μ	E
Wheat				
(*Triticum sativum*)	3,300	1,400	38 × 7 μ	E
Willow oak				
(*Quercus phellos*)	0	72,300	—	Y
Wood sorrel				
(*Oxalis acetosella*)	0	4,500	—	S
Yew				
(*Taxus baccata*)	0	11,500	—	S

[a]Eckerson, *Bot. Gaz.*, **46**, 1908: 221; Salisbury, *Phil. Trans. Royal Soc. (London)*, **B 216**, 1927: 1; Kisser, *Tabulac Biologicae* (W. Junk, Berlin), **5**, 1929: 242; Miller, *Plant Physiology*, 2d ed., New York, 1938: 422; Yocum, *Plant Physiol.*, **10**, 1935: 795.

The average numbers of stomates which have been found per square centimeter on the leaves of a number of representative species are listed in Table 6.1. However, marked deviations from such average values are possible for any species, depending upon the environmental conditions under which the leaves have developed. The number of stomates per unit area of leaf surface may be quite different on leaves of two plants of the same species if one grew in a greenhouse and the other grew in the open, or upon the leaves of plants of the same species which have developed during different seasons.

As shown in Table 6.1, stomates occur in both the upper and lower epidermis of many species of plants. In numerous others, especially woody species, they are confined to the lower epidermis. Even in species in which stomates occur on both surfaces of the leaf, they are commonly, but not always, more abundant in the lower epidermis. In floating leaves, such as those of the water lily, stomates occur only in the upper epidermis. Species in which the stomates are relatively small usually have more per unit area than species in which stomates are relatively large. In general, no correlation has been found between transpiration rates and either the size or distribution of the stomates, other factors being much more important in determining the rate of loss of water vapor from the intercellular spaces.

Mechanism of the Opening and Closing of the Stomates.   The degree of the stomatal opening is influenced both by changes in the turgor of the guard cells and by changes in the turgor of the epidermal cells, especially those adjacent to the guard cells. A predominant role is usually played by changes in the turgor of the guard cells. In general, an increase in their turgor relative to that of the epidermal cells leads to a widening of the stomatal aperture and vice versa. The greater this turgor difference, the wider the stomatal aperture.

The mechanism of the effect of changes in the turgor of the guard cells upon the size of the stomatal aperture varies with the structure, form, and position of the stomates. In one type of guard cell, found in many species of plants, the cell wall is thicker on the side bordering the stomatal pore than on the side bordering the epidermal cells (Fig. 6.4). With an increase in turgor the thinner walls of the guard cells are stretched more than the thicker; this causes the thicker-walled sides to assume a concave shape and results in the appearance of a gap—the stomatal pore—between the two guard cells.

Unless other conditions, most commonly an internal water deficit or an unfavorable temperature, are limiting, the stomates of most species open upon exposure to light and remain open as long as illuminated. With the advent of darkness the stomates close and remain in this condition until they are exposed again to illumination. Usually, therefore, the stomates are open in the daytime and closed at night, although there are many exceptions to this statement.

The usual diurnal pattern in the opening and closing of the stomates indicates that the guard cells are more turgid during the daylight hours than at night. Measurements of the osmotic potentials of the guard cells of a number of species have shown that they are usually more negative during the daylight hours than during the night hours. The osmotic potential of the epidermal cells does not change appreciably during the course of the daylight period and approximates that of the guard cells at night (Fig. 6.5).

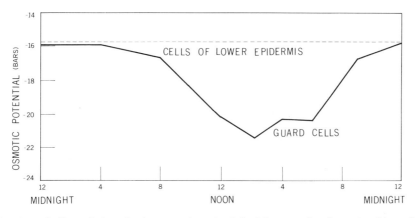

Fig. 6.5. Daily variations in the osmotic potential of the guard cells and epidermal cells of English Ivy *(Hedera helix)*. Data of Beck, *Plant Physiol.* **6**, 1931:320.

Increase in the negativity of the osmotic potential of the guard cells in the morning results in an increase in the negativity of their water potential relative to that of contiguous epidermal cells. Water therefore moves into the guard cells, increasing their turgor pressure, which results in a widening of the stomatal aperture. Movement of water into the guard cells from neighboring cells also results in a loss of turgor by the latter, which probably facilitates the opening of the stomate. Decrease in the negativity of the osmotic potential of the guard cells leads to a reduction in their turgor pressure and a narrowing of the stomatal aperture as a result of a reversal of the series of processes which leads to stomatal opening.

The processes whereby the osmotic potential of the guard cells becomes more negative in the light and less so in the dark are not entirely clear. Only an increase in the solute content of the cell sap can cause a decrease in its osmotic potential, because at the very time this increase is occurring the water content of the guard cells is usually increasing. Increase in the solute content of the cell sap of the guard cells can occur by synthesis of sugars or other soluble compounds within them, by conversion of insoluble carbohydrates such as starch into soluble ones, or by the transfer of solutes from adjacent cells into the guard cells. It seems probable that all three of these processes are involved, at least indirectly. Some photosynthesis occurs in the guard cells, but quantitatively the amount of carbohydrate synthesized seems inadequate to account for the quantities often found to be present. Some of the carbohydrates present in the guard cells undoubtedly move in from nearby mesophyll cells in which they are synthesized and, as the later discussion shows, potassium ions move into guard cells upon the exposure of the latter to light. The conversion of insoluble polysaccharides in the guard cells into soluble compounds also occurs, as discussed later. Contrariwise, the cessation of photosynthesis, the movement of solutes out of the guard cells, and the conversion of soluble compounds into insoluble ones within the guard cells could all result in their having a less negative osmotic potential.

A close correlation has been found between the carbon dioxide concentration of the intercellular spaces and the opening and closing of the stomates. In general a low carbon dioxide concentration leads to stomatal opening and a higher concentration to stomatal closing. A connecting link between light and the opening and closing of the stomates may lie in the fact that during photosynthesis the carbon dioxide concentration of the intercellular spaces decreases while in the absence of photosynthesis it increases as a result of respiration. Stomates of many species are closed in the dark if carbon dioxide is present in approximately the usual atmospheric concentration, but open if exposed to carbon dioxide free air. The concentration of dissolved carbon dioxide in the guard cells and other leaf cells fluctuates in accordance with its concentration as a gas in the intercellular spaces.

The relationship between the occurrence of photosynthesis and the carbon dioxide concentration of the guard cells is fairly clear, but the relation of this latter factor to changes in the osmotic potential and the turgor of the guard cells is much less so. A good possibility, however, is that changes in their carbon dioxide concentration so influence the *pH* of the guard cells that the activity of certain enzyme systems is affected (Chapter 9). An increase in carbon dioxide concentration lowers the *pH;* decrease in its concentration raises it.

The guard cells of many species contain starch which may have been synthesized from photosynthate made in the guard cells, from photosynthate translocated into them from the mesophyll cells, or from photosynthate of both origins. Changes in the guard cells from starch to sugars upon the advent of light and from sugars to starch upon the advent of darkness have been detected in some species. Such reversible transformations from insoluble to soluble carbohydrates are undoubtedly a factor, and perhaps a major one, in the fluctuations in the osmotic potential of the guard cells. Such carbohydrate transformations necessarily are mediated by enzyme systems whose activity would be influenced by the *pH* of the guard cells. An increase in the *pH* of the guard cells of some species has been detected during stomatal opening. Comprehensive knowledge of the enzymic complement of the guard cells and of the effects of *pH* variations upon their activity is lacking.

The preceding discussion can be summarized in the form of a plausible osmotic theory for the opening of stomates in response to the light factor, as follows: light → photosynthesis → reduction of carbon dioxide concentration throughout the leaf → increase in *pH* of guard cells → activation of enzyme systems which convert insoluble polysaccharrides into soluble carbohydrates → increase in negativity of the osmotic potential of the guard cell sap → increase in negativity of its water potential → movement of water into the guard cells from contiguous cells → increase in turgor of the guard cells → opening of the stomate.

Closure of the stomates could be induced by the cessation of photosynthesis, engendering a reversal of this sequence of processes.

This theory of stomatal opening and closing which, in essence, is of long standing, may need to be discarded or at least modified in view of more recent discoveries regarding a role of the potassium ion in this process. It has been shown in several species that, when the guard cells are exposed to light, potassium ions move into them from nearby epidermal cells. It is not certain what anions accompany such potassium ions, although it seems probable that they may be derived from organic acids. The quantities of such potassium salts which move are adequate to account for the decreased negativity in the osmotic potential of the guard cells when they are exposed to light. Such decreased negativity leads to inward osmosis and increased turgidity of the guard cells. This role of potassium appears to be a specific one since other similar ions found in plants, such as sodium, do not operate in this way relative to the guard cells. The movement of potassium and the accompanying anions into the guard cells is not simple diffusion but a metabolically energized process. There are indications that it may be powered by noncyclic photophosphorylation (Chapter 10). With the advent of darkness potassium ions move out of the guard cells, their osmotic potential becomes less negative, water osmoses out of them into the adjacent epidermal cells, and their turgidity decreases. Further studies with other kinds of plants will be necessary before it is known whether this is the prevalent or perhaps only mechanism of stomatal opening and closing or whether it operates in some kinds of plants and a mechanism depending upon changes in the soluble carbohydrate content of the guard cells operates in others.

The hydrodynamic status of the leaf also has an important influence on the operation of the stomatal mechanism. The development of an internal water deficit in plants during the daytime is of frequent occurrence, especially on clear, warm days (Chapter 8). A shrinkage in the total volume of water in a plant results in general in a diminution in the volume of water in each individual cell, although all cells will not necessarily be affected equally. Such a decrease in the water content of the leaf cells, not sufficient to induce visible wilting, is called *incipient wilting*. Under such conditions the guard cells usually decrease in turgor as a result of the osmotic movement of water into contiguous cells. Reduction in the turgor pressure of the guard cells as a result of a diminution of the volume of water within them will bring about a partial to complete closure of the stomates. Because of the development of internal water deficits, stomates may often be closed even during periods when light and temperature conditions are favorable to their being open. It seems probable that on bright, warm days the degree of stomatal opening is determined by the degree of balance achieved between the light-induced opening effect and water deficit-induced closing effect.

Temperature also influences the opening and closing of the stomates. For example, under favorable light and internal water relations, stomatal opening in cotton (Fig. 6.6) and tobacco increases with a rise in temperature up to 25–30°C. The range of temperatures most favorable to stomatal opening undoubtedly differs considerably from one species of plant to another.

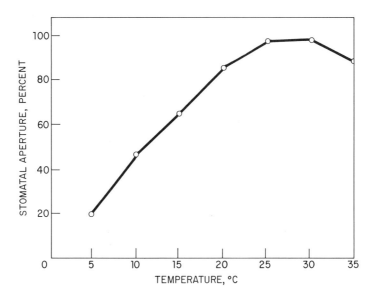

Fig. 6.6. Relation between temperature and stomatal aperture in cotton. Data of Wilson, *Plant Physiol.* **23,** 1948:25.

Diffusive Capacity of the Stomates.   Since gaseous exchanges between the intercellular spaces and the atmosphere take place principally through the stomates, their *diffusive capacity* is a significant consideration. Although the aggregate area of the fully open stomates of a leaf is probably never more than 3 per cent of the stomate-bearing surface and is often more nearly 1 per cent, the rate of the water vapor loss from leaves per unit area may be 50 per cent or even more of the evaporation from an exposed water surface of the same area. Stomatal transpiration from a leaf of birch *(Betula pubescens),* for example, has been found to be as much as 65 per cent of that from the same area of an open evaporating surface. Leaves of this species bear stomates only on the lower surface. Assuming the aggregate area of the open stomates to be 1 per cent of the leaf area, it is evident that water vapor often diffuses through them at rates ranging up to at least fifty times greater than it diffuses away from an equal area of open evaporating surface.

The relatively high rate of water vapor loss from a leaf in spite of the small aggregate area of the open stomates results because relatively steep vapor pressure gradients are generated through small pores such as stomates. If water vapor diffuses from an open water surface into the atmosphere the space over the entire surface will be thickly populated with water vapor molecules which will gradually thin out with increasing distance from the surface. As a result the vapor pressure gradient will be a relatively gradual one. Water vapor molecules diffusing through a small pore, however, spaced at some distance from neighboring pores as are the stomates, diffuse out in a hemispherical pattern in all directions as they emerge through the pore. This results in lower vapor pressures in the atmosphere above the pore than is the situation when diffusion is occurring from an open surface and thus in a steeper vapor pressure gradient. Hence diffusion through the pore occurs more rapidly than it does from the equivalent area of an open surface. When diffusion occurs simultaneously through many pores the aggregate effect is a relatively high rate of water vapor loss relative to the total area of all the pores involved.

Not all of the movement of gases through the stomates occurs by diffusion. In a wind, for example, the back and forth bending of the leaves causes the alternate compression and expansion of the intercellular spaces with a corresponding outward or inward mass flow of gases. A mass flow of gases also undoubtedly occurs when leaf temperatures (see later in this chapter) are rapidly fluctuating. This is commonly the situation when wind velocities are variable on a bright day or when leaves are alternately exposed to direct sunlight and shade, as on a day with scattered clouds in the sky. With each increase in temperature there is an increase in the volume of gases in the intercellular spaces and a resultant outflow of gases through the stomates; with each decrease in temperature, there is a resultant inflow of gases.

The Magnitude of Transpiration.   Transpiration may be computed per unit of leaf area, per unit of fresh or dry weight, per plant, or per unit area of field or forest. Rates may be calculated for hourly, daily, seasonal, or yearly periods. Transpiration rates show an enormous variation from one kind of plant to another, and for the same kind of plant under different environmental conditions.

Transpiration rates in broad-leafed plants of temperate regions range up to about 5 g per dm^2 of leaf area per hour. The phrase "leaf area" should be interpreted literally; the exposed *surface* of any leaf of the broad-leafed variety is twice its area. The usual rates, under conditions favorable for stomatal transpiration, fall within a range of 0.5 to 2.5 g per dm^2 per hour. At night, or during periods when a dry soil, a low temperature, or other conditions unfavorable to stomatal transpiration prevail, the rate may fall to 0.1 g per dm^2 per hour, or even less. Under favorable conditions many herbaceous plants transpire several times their own volume of water in a single day.

The transpiration rates of large plants such as mature trees obviously cannot be measured directly, but can only be estimated from the data on the leaf population of the tree and the known rates of transpiration of some of the leaves. To arrive at an accurate estimate of the leaf population of a large tree is in itself a laborious procedure. Few such data are available; one which may be cited is a 47 foot open grown silver maple *(Acer saccharinum)* which bore 177,000 leaves with an aggregate leaf area of 68,000 square decimeters. Likewise, calculations of transpirational water loss per acre of cropped land or of natural vegetation are necessarily only approximations. Such calculations indicate that sufficient water may transpire from maize plants during the course of a growing season to cover the field to a depth of nearly 40 centimeters. Transpiration of a deciduous forest, largely oak, in the southern Appalachian mountains of the United States has been estimated to be equivalent to 40–55 cm of rainfall per year.

Factors Affecting Transpiration.   The rate of transpiration of a plant or any leaf on a plant varies from day to day, from hour to hour, and frequently from minute to minute. Variations in the rapidity with which water vapor is lost by plants result from the effects of environmental factors upon the physiological conditions within the plant. The more important environmental factors influencing the rate of transpiration are (1) solar radiation, (2) humidity, (3) temperature, (4) wind, and (5) availability of soil water.

Solar Radiation.   This term refers to the visible light and other forms of radiant energy (infrared and ultraviolet radiations) reaching the earth from the sun (Chapter 10). The principal effects of solar radiation upon transpiration result from the influence of light upon the opening and closing of the stomates. In most of the species of plants which have been studied the stomates are usually closed in the absence of light, thus causing a virtually complete cessation of stomatal transpiration during the hours of darkness. Since none of the other environmental factors can have influence upon stomatal transpiration except when the stomates are open, light occupies a position of prime importance among the environmental conditions influencing transpiration.

A second important effect of solar radiation upon transpiration is its influence on leaf temperatures. This effect will be analyzed later in this chapter.

Humidity.   The basic unit for expressing humidity values is the *vapor pressure* which represents the partial pressure of water vapor. For any given temperature of an atmosphere there is a maximum vapor pressure which can prevail called the *saturation vapor pressure* (Table 6.2). A given atmosphere may have any vapor pressure in the range from zero to its saturation vapor pressure.

It has long been customary to give quantitative expression to vapor pressures in terms of their equivalent pressures in millimeters of mercury. The metric system unit for expressing such values, however, is the *millibar* which is one-thousandth of a bar (Chapter 4). One millimeter of mercury is equivalent to 1.32 millibars. A familiar unit for expressing humidity values is the *relative humidity* which is the percentage saturation of an atmosphere or, otherwise

TABLE 6.2   SATURATION VAPOR PRESSURES
AT DIFFERENT TEMPERATURES

| Temperature | Saturation vapor pressure | |
°C	mm/Hg	millibars
15	12.79	16.9
20	17.54	23.2
25	23.76	31.4
30	31.82	42.0
35	42.18	55.7
40	55.32	73.0

stated, the ratio of the actual vapor pressure to the saturation vapor pressure at the same temperature multiplied by one hundred. Given the relative humidity and the temperature of an atmosphere, its actual vapor pressure can be calculated by multiplying the saturation vapor pressure at that temperature by the relative humidity. Relative humidity is usually not a satisfactory unit in which to express humidity values in physiological work because it is a function of both the vapor pressure and the temperature.

In general, the greater the vapor pressure of the atmosphere, other factors remaining unchanged, the slower the rate of transpiration. Whenever the stomates are substantially open, the rate of diffusion of water vapor out of a leaf is largely conditioned by the difference between the vapor pressure in the intercellular spaces and the vapor pressure in the outside atmosphere.

Temperature. The effects of temperature upon the rate of stomatal transpiration can be most clearly analyzed in terms of its effect upon the difference in vapor pressures between the intercellular spaces and the outside atmosphere. Suppose that the temperatures of a leaf with open stomates and the surrounding atmosphere both increase from 20°C to 30°C. Unless the leaf is markedly deficient in water this will result in an increase in the vapor pressure of the intercellular spaces from approximately 23.2 mbars to approximately 42.0 mbars, these being the values for a saturated atmosphere at 20°C and 30°C, respectively. The atmosphere of the leaf intercellular spaces is in direct contact with the relatively extensive evaporating surface of the mesophyll cells. In order to simplify this part of the discussion, it will be assumed that the atmosphere of the leaf intercellular spaces maintains essentially a saturation vapor pressure for the prevailing leaf temperature. In the surrounding atmosphere, however, vapor pressure conditions are very different. On clear days, that is, on the very type of day upon which the highest rates of transpiration occur, there is frequently little change in the vapor pressure of the atmosphere during the course of a single day over land surfaces where the free movement of the air is occurring (Fig. 6.7). Significant changes in vapor pressure may occur, however, in the vicinity of rapidly transpiring plants when air movement is impeded. If we as-

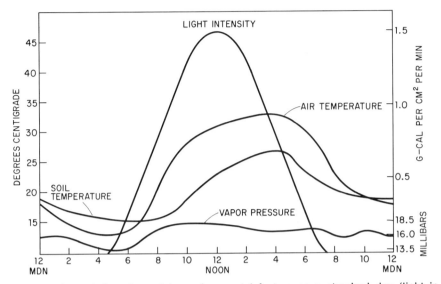

Fig. 6.7. Daily variations in certain environmental factors on a standard day (light intensity measured at horizontal incidence).

sume, for the purpose of our specific example, that the vapor pressure of the atmosphere at 20°C is half that of a saturated atmosphere at that temperature, 11.6 mbars (23.2 ÷ 2), then the difference in vapor pressures at 20°C is 11.6 mbars (23.2 − 11.6). At 30°C, however, the vapor pressure of the inter-cellular spaces would have increased to about 42.0 mbars, while the increase in the vapor pressure of the surrounding atmosphere would in most situations be so small that it can be disregarded in analyzing the effect of temperature upon transpiration. The difference in vapor pressures is now 30.4 mbars (42.0 − 11.6) which will result in the diffusion of water vapor out of the leaf at a rate nearly three times as fast as at 20°C. The effect of a rise in the temperature therefore is principally an increase in the steepness of the vapor pressure gradient through the stomates, and hence an increase in the rate of transpiration.

In the preceding discussion it was assumed that the leaf and the surrounding atmosphere are at the same temperature. As the later discussion shows, the temperature of leaves exposed to direct sunlight is usually higher than that of the ambient atmosphere. If the temperature of a leaf is increased above that of the surrounding atmosphere by the absorption of solar radiation, a usual effect is an increase in the magnitude of the excess vapor pressure of the intercellular spaces over that of the outside atmosphere. At 30°C, under the conditions stated here, the vapor pressure differences between the intercellular spaces and the atmosphere is 30.4 mbars. If, however, as a result of the absorption of solar radiation, the temperature of the leaf rises to 35°C, water would evaporate from the walls of the mesophyll cells until the vapor pressure of the intercellular spaces approximated that of a saturated atmosphere at that temperature (55.7 mbars). Assuming that the air temperature remains at 30°C, the vapor pressure gradient between the intercellular spaces and the atmosphere will be 44.1 mbars (55.7 − 11.6) which would result in a greater rate of transpiration than would occur if both leaf and air were at 30°C.

Taking into account the fact that a leaf is usually slightly warmer on a sunny day than the air that surrounds it, it should be clear from the previous discussion that it is possible for transpiration to occur from a leaf, even if the surrounding atmosphere is at its saturation vapor pressure (100 per cent relative humidity). For example, with the air saturated at 30°C (vapor pressure 42.0 mbars) and a leaf with saturated intercellular spaces at 35°C (vapor pressure 55.7 mbars), a vapor pressure difference of 13.7 mbars would exist.

The rate of diffusion of water vapor out of a leaf through the stomates is also influenced by temperature. In general, the higher the temperature for a given vapor pressure gradient, the higher the rate of diffusion. Because of the low Q_{10} of diffusion (Chapter 4), however, this effect is not a very marked one and is decidedly secondary in significance as compared with the effect of temperature on the steepness of the vapor pressure gradient through the stomates.

Wind. The effects of wind upon the rate of transpiration are complex and depend in part upon prevailing environmental conditions. Winds of relatively low velocity have two principal and often opposing effects upon the rate of transpiration.

In a very quiet atmosphere water vapor often accumlates in the vicinity of transpiring leaves. A breeze, even of very moderate velocity, largely or entirely disperses such an accumulation of water vapor molecules, resulting in an increase in the steepness of the vapor pressure gradient through the stomates, and a consequent increase in the rate of loss of water vapor. Such an effect of wind on transpiration will obviously be less pronounced when the stomates are largely or entirely closed, or if the atmosphere is at or near a saturation vapor pressure.

Wind also has an effect on the temperature of a leaf, which may, in turn, influence its rate of transpiration. Commonly exposure to wind results in a lower leaf temperature than that which would prevail in a quiet atmosphere. Such a cooling effect is most likely to prevail under conditions of intense solar radiation. Under less intense solar radiation wind has little effect on leaf temperature or may result in increasing it.

The overall effect of wind on transpiration at any one time will depend on its relative influences on the steepness of the vapor pressure gradient through the stomates and on the leaf temperature.

The swaying of branches and shoots and the bending and fluttering of leaf blades in the wind all contribute to the complexity of the effects of this factor on transpiration. Winds of very high velocity may result in a marked retardation in the rate of transpiration which may result from a loss of turgor by the guard cells as result of a high rate of evaporation from them under such conditions.

Availability of Water. Although transpiration can continue for short periods at rates considerably in excess of the rate of the absorption of water (Chapter 8), in general if soil conditions are such that the absorption of water is appreciably retarded, the rate of transpiration will soon show a corresponding retardation. The availability of soil water to the plant is therefore an important and in fact often the limiting factor in transpiration. The principal soil factors which affect the rate of absorption of water by plants are (1) the available soil water, (2) the soil temperature, (3) the aeration of the soil, and (4) the concentration of solutes in the soil solution (Chapter 7). All of these factors indirectly influence the rate of transpiration.

Effects of Structural Features of Plants on the Rate of Transpiration. Different kinds of plants, even when growing side by side under virtually identical environmental conditions, may have very different rates of transpiration. In part, such differences in transpiration rate from one kind to another result from species differences in internal processes and conditions such as osmotic potentials of the leaf cells, imbibitional capacities of the protoplasm and cell walls, and the behavior of the stomates. Structural differences, particularly in the leaves, also account in part for unlike rates of water vapor loss from plants of different species growing in the same environment. Structural features, such as (1) the total leaf surface, (2) the distribution and gross morphology of the root system, (3) the thickness of the cutin layer, (4) the presence of epidermal hairs, (5) the size, spacing, distribution, and structural peculiarities of the stomates, and (6) the ratio of the internal exposed surface area of mesophyll cells to the external surface area of the leaf, may influence in one way or another the rate of transpiration. It has proved impossible, however, to draw correct inferences regarding the relative rate of transpiration of a plant on the basis of its observed anatomical peculiarities. Many species which, on the basis of their distinctive structural features, have been judged to have a low rate of transpiration, proved upon experimentation to transpire very rapidly when environmental conditions were favorable.

The Daily Periodicity of Transpiration. No kind of living organisms is more inescapably at the mercy of its environment than a rooted plant. A number of the factors in the environment of plants exhibit more or less regular daily periodicities. The physiologically most important factors of which this is true are solar radiation and temperature, both soil and air. Other usually less significant factors in which daily cyclical variations may occur, at least in some habitats, include wind velocity, atmospheric vapor pressure, and the carbon dioxide content of the atmosphere. Since the rates of plant processes are conditioned in part by the environmental factors to which the plant is exposed, they also exhibit more or less regular daily periodicities.

Daily cycles of environmental conditions vary greatly from one plant habitat to another. Furthermore, in any given habitat such cycles vary from

season to season and during any one season with day-to-day variations in me-
terological conditions. Hence the daily periodicity of any plant process may vary
considerably in pattern from one day to another depending upon the prevailing
cycle of environmental conditions. It is desirable, therefore, to choose a definite
type of diurnal cycle of environmental factors as a reference standard in terms
of which to discuss daily variations in the rates of various plant processes. For
this purpose we shall select a representative summer day as our standard. We
shall consider this hypothetical day to be characterized by a relatively cloudless
sky, a soil water content at approximately the field capacity (Chapter 7), and
a maximum temperature in the range of 30–35°C. We will also assume that the
cyclical daily variations in solar radiation, air and soil temperatures, wind velocity, and atmospheric humidity will be representative of a day upon which the
conditions as defined here prevail. For convenience we shall refer to these as
"standard day conditions" (Fig. 6.7). Such environmental conditions will actually be approximated on many summer days in moist temperate and tropical regions.

The process of transpiration characteristically exhibits a daily periodicity
which varies to some extent with the kind of plant and markedly with the daily
pattern of environmental conditions. The daily periodicity of transpiration in
alfalfa, as exhibited under approximately "standard day conditions," is illustrated in Fig. 6.8. Similar daily periodicities of transpiration are shown by many
other species under comparable conditions.

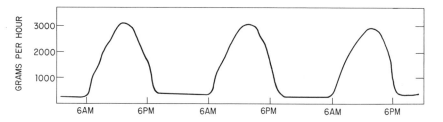

Fig. 6.8. Daily periodicity of the transpiration of alfalfa on three successive days under
approximately standard day conditions. Transpiration expressed as grams per hour per
plot of alfalfa about 2 meters square (average of two plots). Data of Thomas and Hill,
Plant Physiol., **12**, 1937:304.

During the hours of darkness the transpiration rate is generally low, and
in most species water vapor loss during this period may be regarded as almost
wholly cuticular. The periodicity aspect of the process resides almost entirely in
stomatal transpiration. The following interpretation of the dynamics of the daily
periodicity of stomatal transpiration is partly theoretical but in accordance with
known facts. This phenomenon can be analyzed in terms of the two factors:
(1) the diffusive capacity of the stomates and (2) the steepness of the vapor
pressure gradient through them.

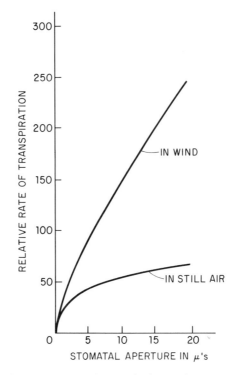

Fig. 6.9. Relation between stomatal transpiration and aperture in still air and in wind. Data of Bange, *Acta Botan. Neerl.* **2,** 1953:273.

For a given steepness of the vapor pressure gradient, the diffusive capacity of a stomate increases progressively, although not proportionately, with an increase in the stomatal aperture (Fig. 6.9). This term could be given quantitative expression as the amount of water vapor diffusing through a stomate per unit of time at a given temperature under a standard steepness of the vapor pressure gradient. In analyzing the dynamics of transpiration, however, it is necessary to consider a large population of stomates. The significant quantity is the aggregate diffusive capacity of all the stomates on the plant or plant part under consideration. Since not all stomates are equally open at one time, however, it is virtually impossible to give any precise expression to such a quantity as the aggregate diffusive capacity.

The rate of water vapor loss does not necessarily parallel the diffusive capacity of the stomates, because the steepness of the vapor pressure gradient may intervene as a limiting factor. If the ambient air is moving, the vapor pressure gradient usually remains approximately constant and the water vapor loss is closely proportional to the diffusive capacity of the stomates. This is because the vapor pressure of the atmosphere adjacent to the leaf surface remains essentially constant under such conditions. Unless there are appreciable changes in temperature this is also generally true of the vapor pressure of the intercellular spaces, although there may be some exceptions to this statement as mentioned later.

In very still air, however, the situation is different (Fig. 6.9). The rate of water vapor loss does not keep pace with an increase in the diffusive capacity of the stomate, measured in terms of the stomatal aperture. In very quiet air the steepness of the vapor pressure gradient through the stomates cannot be long maintained because transpiration itself results in building up atmospheric layers of high vapor pressure close to the leaf surface. As discussed previously, one of the main reasons that wind often results in increased rates of transpiration is that it removes accumulations of water vapor from the vicinity of the leaf surfaces.

The initial rise in the transpiration rate in the morning is brought about by the opening of the stomates resulting in a gradual increase in their aggregate diffusive capacity. As each stomate opens, a vapor pressure gradient is established through it between the atmosphere of the intercellular spaces and the outside atmosphere. During the hours of darkness the leaf cells increase in turgidity and the intercellular spaces usually become saturated with water vapor. Thus, when the stomates open in the morning, the vapor pressure of the intercellular spaces is at the maximum possible for the prevailing leaf temperature. On clear days this is almost always in excess of the vapor pressure of the atmosphere when the stomates open, and often considerably so. Hence the outward diffusion of water vapor through the stomates usually starts as soon as they begin to open.

The rate of transpiration continues to increase, however, for some time after the stomates, in the aggregate, have attained their maximum diffusive capacity. This is because of a gradual increase in the steepness of the vapor pressure gradient through the stomates. As the day progresses the temperatures of both the atmosphere and the leaf increase; if the latter is in direct sunlight its temperature is usually somewhat in excess of that of the atmosphere. On clear days, as previously described, the vapor pressure gradient between the internal atmosphere of the leaf and the external atmosphere therefore increases progressively during the earlier part of the day, and this is the important factor accounting for the rise in the transpiration rate, once the stomates have attained approximately their maximum diffusive capacity.

Almost from the moment when the stomates begin to open in the morning a train of events is set in operation in the plant which ultimately causes a reduction in the rate of transpiration. During approximately the first half of the daylight period, however, these factors are more than offset by the factors resulting in an increase in the rate of transpiration. In most plants, while transpiration is occurring rapidly, the rate of the absorption of water does not keep pace with the rate at which water vapor is lost from the leaves (Chapter 8). This results in a reduction in the water content of the entire plant and especially of the leaves. This diminution in leaf water content results in a gradual closing of the stomates. Some of the stomates on a plant probably begin to close even before the peak transpiration rate is attained. In all likelihood stomates near the tip or margin of a leaf begin to close before those in the middle, since the effects of a deficiency of water usually appear first in the marginal regions of a leaf. With increasing leaf water deficits, gradual closure is induced in more and more of the stomates. As a result there is a gradual reduction in the aggressive diffusive capacity of the stomatal population of the plant during the afternoon hours.

The diminished water content of the leaves may also affect the steepness of the vapor pressure gradient through the stomates. A decrease in the turgor of the mesophyll cells results in a greater negativity of the water potential of the cell wall and the cell sap. There are indications, although this is not certain, that such an increased negativity of water in the cell walls retards the rate of evaporation from them. Conceivably this could prevent the maintenance of a saturation vapor pressure throughout the intercellular spaces. As a result the vapor pressure gradient through the stomates would become less steep and the rate of the escape of water vapor would become diminished.

By late afternoon the air temperature and the intensity of the solar radiation begin to diminish appreciably thus engendering a decrease in the temperature of the leaf. This lowering of leaf temperature decreases the vapor pressure of the intercellular spaces, and therefore decreases the steepness of the vapor pressure gradient, since changes in temperature have very little effect on the vapor pressure of the outside atmosphere on a clear day. The result of a reduction in the steepness of the vapor pressure gradient through the stomates in decreasing the rate of transpiration continues with augmented effect as the hours of darkness approach. Also, as night approaches, stomatal closure is accelerated by the steady diminution in the intensity of the solar radiation and the decrease in the magnitude of the leaf water deficit. The decreasing effects on stomatal transpiration of a diminishing steepness of the vapor-pressure gradient through the stomates and of a gradual stomatal closure overlap during most of the afternoon hours. By nightfall complete closure of virtually all the stomates has occurred and stomatal transpiration is terminated. During the ensuing hours of darkness water vapor loss from the plant is usually confined to cuticular transpiration.

Variations in temperature, light intensity, humidity, and soil water supply may all markedly influence the trend of the transpiration periodicity as well as the magnitude of the daily water vapor loss. The effect of one of these, soil water supply, is clearly shown in Fig. 6.10.

Factors Influencing Leaf Temperatures; The Role of Transpiration.   The temperature of a leaf, in general, does not deviate greatly from that of the circumambient atmosphere, but there are numerous exceptions to this statement. Theoretically the temperature of a leaf may be regarded as conditioned by four different influences: (A) thermal emission, (B) thermal absorption, (C) internal endothermic (energy-using) processes such as photosynthesis and transpiration, and (D) internal exothermic (energy-releasing) processes such as respiration.

Thermal emission refers to the loss of heat from a leaf by the physical

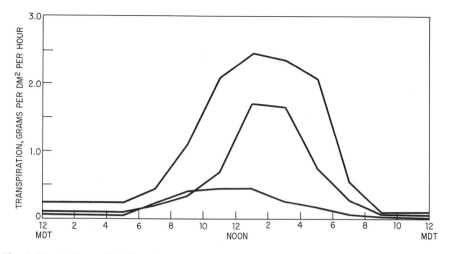

Fig. 6.10. Daily periodicity of the transpiration of bean (*Phaseolus vulgaris*) for three successive days during a period when the soil was gradually becoming drier. Data of Chung, Ph.D. Dissertation, The Ohio State University, 1935.

processes of conduction (which also involves convection) and radiation. *Thermal absorption* refers to the gain of energy by a leaf as a result of these same physical processes.

Heat transmission which is brought about by intermolecular contacts is known as *thermal conduction*. The greater the temperature difference between a leaf and its environment, the more rapidly conduction of heat occurs from the leaf to the gases of the atmosphere, if its temperature is the higher, or in the opposite direction if the temperature of the atmosphere is the higher. Whenever loss of heat is occurring by conduction from a leaf to gas molecules in the atmosphere, convection currents are set up in the vicinity of the leaf. Cooler gas will displace the gas in proximity to the leaf surfaces which has been warmed by thermal conduction from the leaf. This accelerates the rate at which conduction can occur from the leaf. Wind or other air movement also accelerates heat loss by conduction for the same reason.

The other component of thermal emission from leaves is long wavelength (infrared) radiation. In accordance with a general principle this occurs only when the leaf is at a higher temperature than other bodies to which such radiation can be transmitted.

Thermal absorption involves gain of heat by conduction as indicated here and also involves the absorption of radiant energy. The best known kinds of radiant energy are light, infrared radiations, and ultraviolet radiations (Chapter 10). Such radiations are absorbed by leaves from the sunlight which impinges on them either directly or as a result of reflection. Thermal absorption of long wave radiation by leaves from other sources may also occur but only if the radiation source is at a higher temperature than the leaf.

The only endothermic processes which we need mention are photosynthesis and transpiration. In spite of its importance as the basic energy-storing process of the biological world, the energy consumption of photosynthesis is small relative to the impinging solar radiation and ordinarily has only a negligible effect on the temperature of the leaves in which it occurs. Transpiration, however, is often a major factor in influencing leaf temperatures.

The main internal exothermic process which influences leaf temperatures is respiration, but its effect upon leaf temperatures is usually very small (see Chapter 12 for some exceptions to this statement).

Using the letter designations given previously in the first paragraph of this section it should be evident that if $A + C = B + D$ that the leaf will be in thermal equilibrium. If, however, $B + D$ exceeds $A + C$, heat is accumulating in the leaf and its temperature rises. If, on the other hand, $A + C$ exceeds $B + D$ the leaf is becoming deficient in heat, and its temperature decreases.

For most evaluations of the energy relations of leaves only A, B, and the transpiration component of C are of significance. The relative roles of these different quantities will first be considered under approximately standard day conditions as defined earlier in this chapter. On such a day solar radiation is the most important component of B (thermal absorption). In direct noonday summer sunlight the rate of receipt of solar energy is often as much as 1.3 g-cal per cm² of leaf area per minute (Fig. 10.2) and it can be even greater. This figure includes both solar energy impinging directly on the leaf plus that which is indirectly reflected from the sky. As with any kind of radiant energy impinging upon a leaf, some is reflected, some is transmitted, and some is absorbed. Only that fraction of the incident solar energy which is absorbed can influence the temperature of the leaf. The proportion of such absorbed energy varies with the kind of leaf, but 50 per cent is probably a representative value. Hence in bright sunlight about 0.65 g-cal of solar radiant energy are absorbed per square centimeter of leaf area per minute. Some absorption of reflected radiant energy from other sites than the sky may occur as well as some absorption of reradiated infrared radiations, but these radiation sources are relatively small.

Such a rate of receipt of solar energy is sufficient to raise the temperature of 1 g of water (specific heat = 1.00 g-cal), 0.65°C per minute. However, the specific heat of leaf material is less than that of water. It has been determined to be 0.88 g-cal for a sunflower leaf which is probably a representative value. Furthermore the mass corresponding to 1 cm² of leaf area is much less than 1 gram. This will vary in accordance with the thickness of the leaf, but 0.02 ± 0.005 g represents a range of values which covers a number of different kinds of leaves. The rise in temperature of a representative leaf fully exposed to sunlight of this intensity would be: $\dfrac{0.65}{0.02 \times 0.88}$ or about 37°C per minute unless the absorbed energy is emitted from the leaf or unless it is utilized in some energy-using process.

Actual measurements show that the temperatures of leaves exposed to direct sunlight are usually at least a few degrees higher than the surrounding atmosphere and sometimes substantially so (up to about 15°C), but there is no evidence for increases in their temperature of the magnitude indicated in the preceding calculation. Obviously efficient energy-dissipating mechanisms are at work which prevent excessive increases in leaf temperatures.

One of these mechanisms is thermal emission, previously discussed. A second heat-dissipating mechanism is transpiration in which radiant energy is converted into kinetic energy as a result of the vaporization of water. It requires 580 g-cal at 30°C to convert 1 g of liquid water into water vapor. Although higher rates probably sometimes occur, in general a transpirational water vapor loss of 3 g per dm² per hour is a very high rate for most plants. Such a rate would result in the utilization of 0.29 g cal per cm² per minute $\left(\dfrac{580 \times 3}{100 \times 60} \right)$ which should be compared with the 0.65 g-cal per cm² per minute receipt of the solar radiation.

Under strong sunlight, therefore, with conditions favorable to a high rate of transpiration, it appears that from one-third to one-half of the absorbed radiant energy is utilized in transpiration.

If the rates of both the absorption of radiant energy and the transpiration are high, a substantial part of the absorbed energy is dissipated in the evaporation of water as described earlier, although it seems unlikely that transpiration can ever dispose of all the absorbed energy under such conditions. If, on the other hand, the rate of absorption of the radiant energy is high, but because of an internal water deficiency or other reasons, the rate of the transpiration is low, most of the absorbed energy is disposed of by thermal emission.

When the rates of absorption of the radiant energy are relatively low, but conditions are still favorable to transpiration, this process often becomes a major factor in the dissipation of the absorbed solar radiation. High rates of transpiration coupled with low rates of thermal absorption can sometimes result in leaves being cooler than the ambient atmosphere.

The usually prevailing low transpiration rates at night naturally have very little effect upon leaf temperatures during the hours of darkness. Thermal exchange between a leaf and its environment, however, does not cease during the hours of darkness. At night leaves may gain energy by absorbing infrared radiations from a soil or rock surface which is warmer than they are. Simultaneously such leaves may be losing energy by long wavelength (infrared) radiation to the relatively cold gases of the atmosphere. If the rate of radiant energy emission by such leaves exceeds their rate of radiant energy absorption, they will become cooler than the surrounding atmosphere. Some interchange of heat between a leaf and its environment may also occur at night by conduction, but this is usually small.

The temperatures of other aerial organs of plants are controlled by mechanisms similar to those which control the leaf temperatures.

The Loss of Water in Liquid Form from Plants. If a pot of young oat plants is copiously watered and then enclosed in a bell jar, in a relatively short time a slow exudation of sap begins at the tip of each leaf. The drops which form at the leaf apexes gradually enlarge and eventually may run down the side of the leaf or fall off. This process of the escape of liquid from uninjured plants is called guttation (Fig. 6.11). It is of very general occurrence, having been recorded in plants of more than 300 genera, although there are many species in which it has not been observed. Guttation occurs most frequently and abundantly under conditions which favor the rapid absorption of water by the roots, but which result in a reduced rate of transpiration. In most temperate regions such conditions occur most often during the late spring when there is often an alternation between relatively cool nights and relatively warm days. Guttation is frequently observed at that season, usually taking place at night or during the early hours of the morning. The drops of guttation water which form at the tips of grass blades and the tips or edges of the leaves of other plants are often erroneously considered to be dew.

The process of guttation occurs from porelike structures known as hydathodes located at the margins of the leaves. These structures are also sometimes referred to as water stomates or water pores. The exudation of water through the hydathodes is considered to result from a pressure which develops in the sap of the xylem elements, and not from any locally developed pressure in the hydathode itself. This pressure is generally believed to be identical with the so-called "root pressure" (Chapter 8). It is supposed that the water is forced from the xylem

Fig. 6.11. Guttation from tomato leaves. Photograph from J. H. Gourley.

vessels through the intercellular spaces and out of the plant through the pore of the hydathode. This water is not pure but contains at least traces of solutes, including sugars, amino acids, and mineral salts. While the volume of water exuded by most plants of the temperate regions in guttation is usually small, some tropical species lose large quantities in this process. A young leaf of *Colocasia nymphaefolia,* a native of India, has been observed to lose as much as 100 ml of liquid water in a single night by guttation.

Glands are found on leaves, flower parts, and other organs of the plant. Certain types of glands secrete water or, more accurately, a dilute solution. The exudation of water or a dilute solution by glands is apparently caused by a physiological mechanism within the gland itself, and not by a pressure developed in the sap of the xylem conduits as appears to be true of the hydathodes. Sugars (as in nectar) and certain mineral salts are secreted by some glands. Non-water-soluble substances, such as resins and volatile oils, are secreted by the glands on some kinds of plants.

SUGGESTED FOR COLLATERAL READING

Esau, Katherine. *Plant Anatomy, 2nd Edition.* John Wiley & Sons, Inc., New York, 1965.

Gates, D. M. and L. E. Papian. *Atlas of Energy Budgets of Plant Leaves.* Academic Press, Inc., New York, 1971.

Kozlowski, T. T., Editor. *Water Deficits and Plant Growth,* 3 vols. Academic Press, New York, 1968, 1968, 1972.

Kramer, P. J. *Plant and Soil Water Relationships. A Modern Synthesis.* McGraw-Hill Book Company, New York, 1969.

Ledbetter, M. C., and K. R. Porter. *Introduction to the Fine Structure of Plant Cells.* Springer-Verlag, New York, 1970.

Meidner, H. and T. A. Mansfield, *Physiology of Stomata.* McGraw-Hill Book Company, New York, 1969.

7 | ABSORPTION OF WATER

Although water may be absorbed through almost any surface of a plant, only a negligible quantity is absorbed through organs other than roots in terrestrial plants. From observations of plants in their native habitats it is evident that the absorption of both water and mineral salts is primarily from the soil by way of the roots. Any thorough consideration of the absorption of water, therefore, requires an understanding of the properties of soils, particularly as they affect the movement of water from the soil into the root.

Components of the Soil. In general, five different components of the soil complex are distinguished:

(1) *The Mineral Matter of the Soil.* The parent substance of practically all soils is rock, of which there are numerous varieties. By various weathering processes rock strata are reduced to fragments of diverse sizes, and these compose the bulk of most soils. The rock particles in soils vary in size from stones and gravel down to submicroscopic particles of colloidal clay. The mineral portion of the soil is customarily classified into several fractions, depending upon the particle size. Such classifications usually disregard the very large particles of the soil (rock fragments, pebbles, etc.). The classification now generally used is as follows:

COARSE SAND	FINE SAND	SILT	CLAY
←——————→	←——————→	←————→	←———→
2.0 mm	0.2 mm	0.02 mm	0.002 mm

The proportions of these fractions are very different in various kinds of soils.

Fertile soils usually contain a large amount of clay, and many of the properties of soils which are of the greatest significance to plant growth result from the presence of this fraction. The particles in the clay fraction are flat and plate-like. Most of them are of colloidal dimensions and exhibit the characteristic properties of colloidal systems. The particles of the clay complex, like those of many other colloidal systems, retain water within the structure of the particle. In contrast, the sand and silt particles of the soil retain water only on their surfaces. The presence of a considerable proportion of clay in a soil therefore endows it with a high water-retaining capacity.

The micelles of colloidal clay usually bear a negative charge when in contact with water. The presence of these charged particles makes possible the phenomenon of *cation exchange* which will be discussed later in connection with the absorption of mineral salts (Chapter 15). In the presence of calcium ions the individual clay particles are more or less completely flocculated into compound particles. Much of the colloidal clay in most soils exists as enveloping films around the larger soil particles and is also closely associated with the organic material in the soil. The flocculation of the clay particles therefore usually results in the formation of compound granules including sand and silt particles and organic matter in addition to the clay. These are called soil crumbs. For agricultural purposes a well-developed granular structure of the soil is highly desirable, because such a structure favors both a high moisture-retaining capacity and a good aeration of the soil.

(2) *The Organic Matter of the Soil.*   Most soils contain organic matter which has been derived principally from the partial decomposition of plant residues. Small quantities may also originate from animal residues and excretions. The proportion of organic matter present may vary from almost none, as in some sand deposits, to 95 per cent or better in some peat soils. In ordinary agricultural soils the amount present seldom exceeds 15 per cent.

The organic matter is the seat of most of the microbiological processes occurring in the soil. One of the most important of these is the oxidation of the organic matter, a process resulting largely from the metabolic activities of bacteria and fungi, although a limited amount of purely chemical decomposition probably also occurs. Under conditions that are exceptionally favorable for the activities of microorganisms, the organic matter of the soil is oxidized completely and disappears. For this reason the organic matter of soils in tropical regions, particularly when under cultivation, is very low. Even in more temperate regions cultivation of the soils generally results in a rapid reduction in organic matter content, principally as a result of the better aeration induced by tillage. For this reason the regular addition of organic matter to soils is a general practice in good soil management.

As a result of the decay process there is present in most soils organic matter in various stages of decomposition. A large proportion of the organic material which is added to some soils survives in the form of a dark-colored amorphous substance called *humus*. Humus is composed principally of the degradation products of the cellulose and lignin derived from plant remains.

The accumulation of humus in soils is furthered by conditions unfavorable to the oxidative decomposition of organic matter.

(3) *Soil Water and the Soil Solution.*   Water is universally a component of soils, although the amount present may vary from the merest trace to a quantity sufficient to saturate the soil, *i.e.,* completely fill all of the spaces between the soil particles. Dissolved in the soil water are varying quantities of numerous chemical compounds. These originate principally from the dissolution or chemical weathering of the rock particles, from the decomposition of the organic matter, from the activities of the microorganisms, and from the reactions between the roots of plants and the soil constituents. It is thus more accurate to speak of the *soil solution* than of the soil water, although in discussions of the water relations of soils it is a common practice to disregard the presence of solutes in the soil water.

(4) *The Soil Atmosphere.*   The irregularity of the soil particles in size, shape, and arrangement insures the existence of a certain amount of space among them, even in the most tightly packed soils. This is termed the *pore space* of a soil. A soil in which the pore space is about equally divided between large and small pores is a more favorable environment for the roots of most kinds of plants than one in which most of the pores are large or in which most of the pores are small. In such a soil the large pores permit adequate drainage and aeration, while the small pores permit considerable capillary retention of water. The pore space of any given soil depends upon the physical and chemical conditions to which the soil is subjected. Conditions favoring a crumb structure of a soil, for example, usually result in an increase in its pore space. The interstitial spaces of a soil may be occupied entirely by gases, as in desiccated soils, entirely by water, as in saturated soils, or, as is usually true, partly by water and partly by gases. The relative proportions of water and gases present in any given soil vary, depending upon the water content of the soil.

In well-aerated soils the *soil atmosphere* often does not deviate greatly from the atmosphere proper in the proportion of gases present. Contrariwise in poorly aerated soils, marked variations in composition of the soil atmosphere may occur (see later in this chapter).

(5) *Soil Organisms.*   The soil flora includes bacteria, fungi, and algae. The bacteria are generally the most abundant of all the living organisms present in any soil. Among them are the nitrifying, sulfofying, nitrogen-fixing, ammonifying, and cellulose-decomposing bacteria. The bacteria accomplishing the oxidative decomposition of cellulose and similar compounds are the most important agents in the formation of humus. The numbers of bacteria present vary greatly from soil to soil, and in any one soil vary with seasonal and other fluctuations in soil conditions. Most soil contain between two million and two hundred million bacteria per gram of soil. The number of bacteria decreases rapidly with increasing depth, subsoils being sterile or practically so. In general an abundant representation of most species of bacteria is favored in soils by warm temperatures (35–45°C), good aeration, and a good but not superabundant water

supply. Some species of bacteria, on the other hand, are anaerobic and thrive when the aeration of the soil is deficient. The denitrifying bacteria and certain of the nitrogen-fixing bacteria (*Clostridium spp*) are examples of anaerobes.

Fungi are, in general, most abundant in acidic soils. In such soils they largely replace bacteria as agents of the decomposition of organic matter.

The soil fauna includes protozoa, nematodes, earthworms, insects, insect larvae, and the burrowing species of the higher animals. The earthworms are generally credited with having the most important effects on the soil structure, at least in many soils. Their activities result principally in a general loosening of the soil, which facilitates both aeration and distribution of water. Many of the other soil animals have similar effects on the structural organization of the soil.

Soil-Water Relations Under Field Conditions.   In many regions there exists below the surface of the soil a level at which the soil is completely saturated with water. Water will stand at this level in any hole bored into the ground to this depth. This point in the soil at which the complete saturation with water is called the *water table*. The depth of the water table in any locality usually fluctuates, sometimes very markedly, with seasonal periodic changes in the relative rates of rainfall, evaporation, transpiration, and other factors. Relatively impermeable soil layers sometimes impede the downward percolation of water sufficiently to cause the development of temporary water tables which may be far above the level of the true water table. Such temporary water tables have essentially the same effects on soil-water relations as true water tables except that their influence is often only a transient one.

For many years an important role was ascribed to the capillary rise of the water from the water table into the soil above in maintaining the moisture conditions within that soil, but it is now clear that the importance of this source of water in soils has been overemphasized. Experiments indicate that water seldom rises through the soil by capillarity from a water table at an appreciable rate for heights of more than several meters. In a typical loam soil the absolute maximum capillary rise of water does not exceed 3 meters. Not only is the height to which water moves in a soil by capillarity much less than it was once thought to be, but the actual rate of movement in most soils is very slow.

In river valleys or in close proximity to large bodies of water the water table will usually be reached at a depth not exceeding a few meters below the soil surface. In many soils, however, the water table lies so far below the soil surface that it has little or no effect on the soil moisture conditions in the soil layers which are penetrated by the roots of most plants. In many arid regions there is no water table at all.

The more significant source of soil water is that part of the precipitation— usually rain, although often melting snow or ice—which percolates downward into the soil. In dry regions this is the only source of water naturally available to plants. Even in more humid regions this is the principal or only source of available soil water in most soils during the dry season of the year. In many dry regions irrigation is an important way in which water is supplied to crop plants. The percolation behavior of irrigation water is essentially similar to that of the water from natural precipitation.

Let us now consider how percolating water becomes distributed in a soil. We will assume that the soil under consideration is relatively dry, *i.e.,* that its water content is substantially below the field capacity (see the following), and that it is essentially homogeneous in structure to a depth of a number of meters. Furthermore, we will assume that the water table is so far below the soil surface that it has no effect on soil-water conditions in the upper layers of the soil.

Following a rain or irrigation the water which enters the soil through its surface layer increases its water content sufficiently for capillary distribution of water within the soil to begin. The force of gravity also has an influence, but under the conditions postulated it is so small in comparison with the downward capillary forces at work that it may be disregarded. As the water gets distributed through a soil, the water films are gradually attenuated until eventually the downward capillary movement of water becomes negligible. When this attenuation of the soil water has reached its limit, water is believed to occupy only the smaller interstices in the soil.

The depth of this moist blanket of surface soil will depend upon the limit to which water can be transferred in the downward direction by capillary movement, being greatest in sandy soils and least in clay soils for a given amount of water. Furthermore, the water content of this moist layer of soil will be approximately uniform, and the boundary between it and the zone of drier soil below will be fairly sharp.

The water content of the moist layer of soil after the capillary movement of water has become negligible is termed its *field capacity.* Most soils reach their field capacity within a period of two days or less after a rain or irrigation, unless a water table is present within a few meters of the surface. It is important to realize that a soil at its field capacity is far from being saturated. Field capacities range from about 5 per cent of the dry weight of the soil in the very sandy soils to about 45 per cent of the dry weight in the clays.

Let us now assume that another equal quantity of water is added to this same soil before any appreciable amount is lost from its upper layers by evaporation or transpiration. After a new approximate capillary equilibrium has been attained, the upper portion of soil will not have increased in water content, but an additional layer of the soil below this will now be moistened up to its field capacity. In other words, the addition of a second increment of water equal in volume to the first does not increase the water content of the already moist layer of the soil above its field capacity, but results, because of further capillary movement, in raising to its field capacity a second layer of soil, lying directly under the zone which had been moistened by the addition of the first increment of rain, and approximately equal to it in thickness. Although the exact mechanics of the distribution of water under these conditions is probably more complicated, the effect produced is as if the second increment of water simply flows through the moist blanket of soil under the pull of gravity, after which it is distributed to the dry soil layer underneath by capillarity.

Successive irrigations or rainfalls would thus continue to deepen the layer of soil which has been moistened up to its field capacity. If the water table is not too deep and if sufficient rainfall or irrigation water penetrates the soil, not too much of which is utilized by plants growing thereon, eventually the entire soil from its surface down to the water table will be moistened to its field capacity. If additional water is applied to the soil, after the entire soil mass down to the water table has attained its field capacity, this water percolates down through the soil under the influence of gravity and becomes a part of the ground water. Such conditions prevail in many of the soils of the more humid regions at least during the wetter seasons of the year provided that the water table is not located at too great a depth. This downward percolation of water to the water table is an important factor in influencing its depth below the soil surface, which usually fluctuates considerably from season to season.

In the preceding discussion only the simplest possible situation, that of a homogeneous soil, has been considered. However, most soils consist of a vertical succession of several distinct horizons, each with more or less distinctive physical and chemical properties. Even after the long continued disturbance of a soil by plowing or other cultural activities, at least some semblance of the original soil stratification usually persists. In the horizons below those reached by the plow the original structural organization of the soil is usually maintained practically intact. Although the individual horizons of a soil are often fairly homogeneous within themselves, each such horizon in a given soil may have a distinct field capacity of its own. The water contents of the different horizons, even after an equilibrium in the capillary distribution of water has been attained, may therefore be very different.

Permanent Wilting Percentage.   This quantity (also called *wilting percentage, wilting point,* or *wilting coefficient*) is defined as the percentage of water content, on a soil dry weight basis, to which a soil is reduced when the plant or plants growing in it have just reached a condition of permanent wilting. A permanently wilted plant is one that will not recover its turgidity unless water is supplied to the soil. The permanent wilting percentage is a physiological index of the soil water relations.

Prior to extensive measurements of permanent wilting percentages it was supposed that plants differed markedly in their capacity to reduce the water content of a soil. It was assumed, for example, that species which could endure drought conditions could deplete the moisture content of a soil to a lower percentage before showing permanent wilting than could those species which were soon injured or killed when subjected to drought. Experiments have shown, however, that most kinds of plants reduce the water content of a given type of soil to approximately the same value before permanent wilting sets in. Some xeric species may prove to be exceptions to this statement.

Although the permanent wilting percentage for a given soil shows no appreciable variation when measured by most kinds of plants growing in it, its value varies greatly with the type of soil. The percentage of water remaining in a soil when the permanent wilting of plants growing in it occurs ranges from about 1 per cent of the dry weight of the soil in some coarse sands to approximately 25 per cent in heavy clays.

The significance of the permanent wilting percentage lies in the fact that it is the percentage of soil moisture below which no growth of a plant occurs. Wilted plants continue to reduce the water content of a soil, but at such a slow rate that the restoration of turgidity is impossible, because the rate of transpiration, largely cuticular, even from a wilted plant exceeds the rate of absorption of water from a soil at its permanent wilting percentage or any lower soil water content.

The Soil-Water Potential.   In preceding discussions of the water relations of plant cells and tissues it has been shown that the most significant unit for the expression of the dynamics of such water relations is the water potential. Since all problems of the absorption of water by plants involve a consideration of the relation of the water in the soil to the water in the plants, it is desirable to extend the concept of water potential to the soil if the absorption of water is to be interpreted in terms of dynamic units (Fig. 7.1).

The potential of the soil water is always negative and has two major components. One is the osmotic potential of the soil solution which in most kinds of soils is seldom more negative than a few tenths of a bar, although saline and alkali soils are marked exceptions to this statement. The other component is the attractive forces between the soil particles and water molecules. Such matric forces, which are also involved in imbibition (Chapter 4), may attain a very considerable magnitude, especially in dry soils. In moist soils, those at a field capacity or a higher water content, the osmotic potential component is principally responsible for the negative soil water potential. In drier soils the matric component is almost solely responsible for the still more negative soil water potentials which prevail. In the majority of soils the water potential is only slightly less than zero at saturation, not more than a few tenths of a bar below zero at the field capacity, and is in the vicinity of -15 bars at the permanent wilting percentage. In the range of soil water contents below the wilting percentage the negativity of the water potential escalates so rapidly that very little water can be absorbed by a plant. In the range between the field capacity and soil saturation, the absorption of water is not impeded because of a low soil water potential, but in many kinds of plants is retarded in rate because of deficient soil aeration.

In most soils the water readily available to plants is that in the range between the field capacity and the permanent wilting percentage. In general, this range is narrow in sandy soils, wider in loam soils, and widest in clay soils.

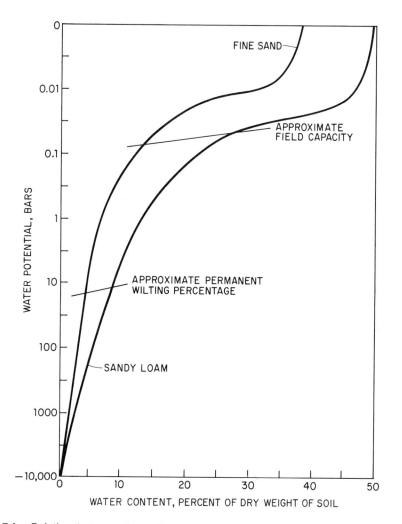

Fig. 7.1. Relation between the soil water potential and the soil water content in two soils over the entire soil water range. Note that the vertical scale in this figure is a logarithmic one. Data of Russell, *Proc. Soil Sci. Amer.* **4,** 1939:53.

Hence, at the field capacity, clay soils contain the greatest quantities of both available and unavailable water, loam soils the next, and sandy soils the least.

Roots and Root Systems. The root system of a plant is often as distinctive in form and structure as its aerial portions (Fig. 7.2). Prevailing soil conditions, however, exert a pronounced effect upon the distribution of the root systems. Although the depth to which roots penetrate is in part a species characteristic, this can be modified by various soil factors. A plant may develop a deep, pro-

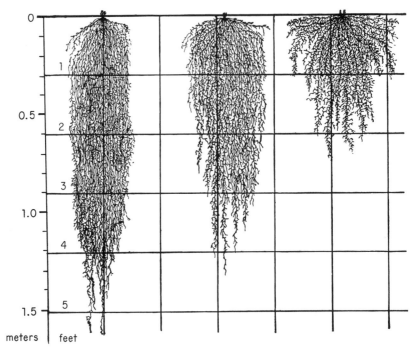

Fig. 7.2. Average root development of winter wheat in fertile silt loam under a precipitation of 26 to 32 inches (left), 21 to 24 inches (center), and 16 to 19 inches (right). From Weaver and Clements, *Plant Ecology, 2nd Edition*, McGraw-Hill Book Co., New York, 1938, p. 299.

fusely branched root system in a well-drained soil, for example, while another individual of the same species will produce a shallow root system of an entirely different configuration in a soil which is waterlogged to within a decimeter or two of its surface. The presence of rock strata or extremely tight layers of soil such as a hardpan may also limit the depth of the penetration of roots.

In well-drained soils the bulk of the roots of most crop plants is located in a zone between the surface and a depth of 1 to 2 meters. Some individual roots penetrate to greater distances; in many crop plants a few roots reach a depth of 3 meters. The roots of trees do not penetrate the soil to the depths often presumed. As a rule most of the root system of the vast majority of trees will be found in the upper 1 or 2 meters of the soil. Growth of tree roots to a depth of more than 4 meters beneath the soil surface is uncommon.

When a seed germinates, the first root which appears is called the *primary root*. This develops from an apical growing region which is already differentiated in the embryo. The primary root, which may be considered a downward extension of the main axis of the plant, gradually elongates, grows in diameter, and produces lateral branches. The branches and subbranches of the primary root are called *secondary roots*.

The primary root and its branches considered collectively are called the *primary root system.* In seed plants primary root systems develop only from embryos. In many species the primary root system remains the only, or at least the conspicuous root system throughout the life of the plant. In perennial plants, especially certain tree species, such primary root systems may attain an enormous size.

All other roots, regardless of the organ of the plant on which they develop, are termed *adventitious roots.* The roots which develop from bulbs, tubers, corms, rhizomes, and cuttings are classed in this category. Adventitious roots may even arise from the leaves of some species such as begonia, bryophyllum, and the walking fern. Such roots also develop from the lower nodes of the vertical stems of many species, especially monocots. In some species, maize for example, they may arise from the nodes above the soil surface, becoming the so-called prop roots. When they develop from the stems, adventitious roots most commonly arise at the nodes.

Two very generalized types of root systems which are often distinguished are *tap root systems* and *fibrous root systems.* Practically all adventitious root systems belong in the latter category. In the former the primary root system is predominant, the primary root itself often being conspicuous.

The Absorbing Region of Roots.   Most of the absorption of water and mineral salts occurs in the terminal portions of the roots. Because of the extensive branching of the roots there are often millions of root tips on the root system of a mature plant. From a physiological point of view the number of root tips borne by a root system is probably the most important index of its effectiveness in obtaining water and mineral salts from the soil.

The external morphology of a root tip can be observed most easily in roots which have developed in moist air. Upon the close examination of a root tip, four distinct but intergrading regions can usually be discerned with no greater magnification than that afforded by a hand lens or, in many species, even with the naked eye. At the apex of the root is an extremely short region, white in color, which is known as the *root cap.* Just above the root cap, and partly covered by it, is the *meristematic region,* the zone of the maximum cell division, which is seldom more than a millimeter in length and is usually distinguishable by a yellowish color. Next above the meristematic region is the *region of cell enlargement,* usually not more than a few millimeters in length, in which most increase in the length of the root occurs. Above this region is usually present the *root-hair zone,* which bears the slender hairlike outgrowths of the epidermal cells known as the root hairs. The root-hair zone varies in length, depending on the species and the conditions to which the root is subjected during its development.

Most evidence indicates that the region of the root tip in which the maximum absorption of water occurs corresponds to the root-hair zone or the zone in which the root hairs would be present if they had developed. The rate of absorption of water in the zone of cell enlargement is also high, but relatively little movement of water into the root occurs through the root cap and meristematic region.

Fairly uniform amounts of mineral salts are absorbed throughout the root tip from the root hair zone to the meristematic region. The meristematic region of the root tip also possesses a high capacity for the accumulation and retention of mineral salts (Chapter 15).

Measured from the apex, the length of the absorbing region of a root varies greatly with the species, the age of the root, and the conditions under which it has developed. Above the root-hair zone one or more of the outer cell layers of the root become suberized or lignified or both, and in older roots cork cambiums develop (see later in this chapter).

It has often been assumed that little absorption of water occurs through the older suberized roots. There is experimental evidence, however, that substantial amounts of water can be absorbed through such portions of the roots of many woody species. It is probable that more water is absorbed through the suberized portions of the roots of forest trees than through the unsuberized terminal portions of such roots.

Anatomy of Roots.   If a longitudinal median section is cut through a root tip such as has just been described and examined under a microscope, the anatomy of the several regions of the root can be observed (Fig. 7.3). The root cap is a more or less thimble-shaped assemblage of cells covering the distal end of the meristematic region. The growth of a root tip through the soil gradually tears off the outer terminal cells of the root cap, but these are replaced by new root cap cells which are formed by cell divisions occurring in the peripheral layers of the cells of the meristematic regions.

The meristematic tissue of a root tip is composed of small, thin-walled cells with prominent nuclei. As new cells are formed by cell division they begin to enlarge, principally in the direction of the long axis of the root. The division of the meristematic cells and their subsequent elongation result in projecting the growing region and the root cap forward through the soil and account for the growth in the length of the roots. The region of the cell elongation is seldom more than a few millimeters in length. Only a small part of the root tip—a few millimeters in length at most—is pushed through the soil. Since, in a growing root tip, the elongation of the cells ensues as soon as they are formed by cell division, a cell which is one day in the region of cell division is the next day in the region of cell elongation, subsequent divisions in the meristematic tissue having produced additional layers of cells beyond it (Chapter 19).

The anatomy of a representative root, as shown in cross section through the root-hair zone, is illustrated in Fig. 7.4. The structure of a young root shows a number of distinctive features. The cortex is relatively thick; in fleshy roots, for example, it is often many times thicker than the radius of the stele (central vascular core). Intercellular spaces in the root cortex are also very prominent.

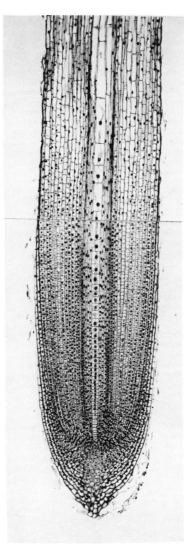

Fig. 7.3. Median longitudinal section (approximately **3 mm** in length) of onion roof tip. Photomicrograph by Tillman Johnson.

An endodermis is almost invariably present in roots; this is generally considered to represent the innermost layer of the cortex. Just within the endodermis is present a narrow zone of parenchymatous pericyclic tissue. Usually this is continuous, but in some species, as described later, it may be discontinuous.

In roots the primary xylem and the primary phloem are present in a radial pattern. The primary xylem as seen in cross section appears as a number

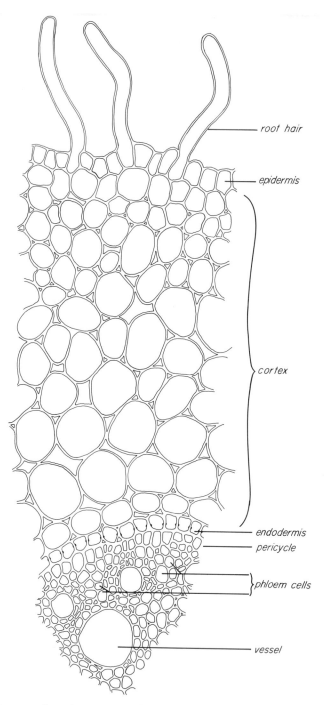

root hair

epidermis

cortex

endodermis
pericycle

phloem cells

vessel

Fig. 7.4. Cross section of a segment of a young maize root through the root hair zone.

(usually 2 to 5, although sometimes as many as 20) of radially situated strands. In many plants the central portion of the root is composed of xylem; in some, especially monocots, it is composed of pith. Usually the xylem strands terminate radially in contact with the pericycle, but in some species they abut directly on the endodermis, breaking up the ring of the pericycle, as seen in cross section, into a discontinuous series of arcs. The primary phloem of the roots occurs as patches of tissue (as seen in cross section) which alternate with the strands of the xylem (Fig. 7.5).

The structure of the individual types of cells occurring in the root tissues is essentially similar to the structure of the corresponding types of cells occurring in the stem (Chapter 8).

With few exceptions the roots of all perennials and many annuals grow in diameter as they increase in age by means of a cambium layer. The cambium layer is initiated in young roots in such a way that it lies inside of the strands of phloem tissues, and outside of the xylem strands. In cross section the original cambium layer appears as a wavy band, passing inside of each phloem strand and outside of each xylem strand. Once differentiated, this cambium layer produces secondary xylem on its inside face and secondary phloem on its outside face. The initial formation of secondary tissues by a root cambium is usually most rapid in the segments of the cambium internal to the primary phloem strands. Because of this differential growth rate, the cambium of a root rapidly attains a circular aspect in cross section, after which the division rate of cambial cells is approximately equal around the entire circumference of the cambium.

Most perennial roots sooner or later become encased in layers of cork cells. The initial cork cambium often originates in the pericycle. As layers of cork cells are produced by the cork cambium, the cortex of the root, including the endodermis, is ruptured and the cells of these tissues die and decay away. Older roots therefore have a characteristic smooth brownish, corky covering that is pierced only by lenticels. With increasing age, secondary cork cambiums may arise progressively more and more deeply in the phloem tissues. This results in the gradual loss of the pericycle and the older phloem tissues. The bark of older roots, therefore, is essentially similar to that of the trunks or larger branches of trees (Chapter 19).

In the roots of species in which no secondary thickening occurs, as in many monocots, the epidermal layer of cells may persist intact, usually becoming suberized. In other such species the epidermis may die and decay, but when this occurs an underlying layer of the cortex cells in turn becomes suberized.

The lateral branches of the primary root system usually originate in the pericycle, most of them being formed in the region just above the root-hair zone. Usually the locus of origin of a lateral root is opposite one of the primary xylem strands. The first step in the formation of a lateral root in most species is the development of a group of meristematic cells by the division of several pericyclic cells in the layer just inside of the endodermis (Fig. 7.6). By successive divisions these cells rapidly form an apical root meristem with its char-

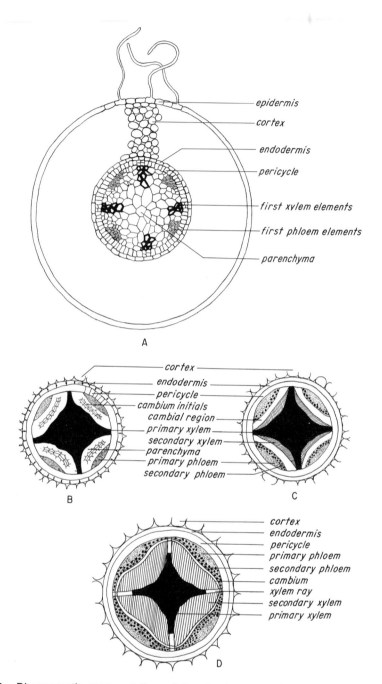

Fig. 7.5. Diagrammatic representation of the development of secondary phloem and xylem in a dicotyledonous root. (A) Cross section showing first xylem and phloem elements. (B) Central region of the root showing the development of primary xylem (black), primary phloem (stippled), and the first cambial elements. Note that the entire center of the root has become primary xylem. (C) Later stage, showing the development of secondary xylem and phloem. (D) Still later stage, showing more extensive secondary phloem and xylem with cambium now entirely encircling the xylem.

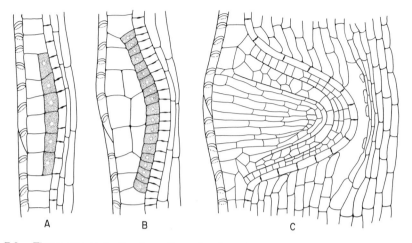

Fig. 7.6. Three stages in the formation of a lateral root. Meristematic cells (stippled) arise in the pericycle and the cells of the lateral root develop from these. Redrawn from Holman and Robbins, *A Textbook of General Botany* (1938), after van Tieghem.

acteristic root cap, region of cell division, and so forth. As this meristem develops, the endodermis and tissues exterior to it are first compressed and later ruptured. The elongating lateral root penetrates through the tissues external to it, partly by mechanical pressure and probably partly by digesting the tissues through which it passes. Eventually the lateral rootlet emerges from the root of which it forms a branch and becomes an externally visible part of the root system. The rupture of older suberized tissues by such a penetration of elongating lateral roots could in part explain the uptake of water through the suberized portions of the root, as discussed previously.

Root Hairs.   These projections of epidermal cells are confined to the root-hair zone, which may be from a few millimeters to many centimeters in length, depending on the species and the conditions under which the root develops. Since the root-hair zone lies back of the region of cell enlargement, there is no forward progression through the soil of the individual root hairs along the axis of the root. In any rapidly growing root tip new hairs are continually developing just back of the zone of elongation. New root hairs are thus constantly developing in contact with different portions of the soil, a fact which is of fundamental significance in the absorption of water and of mineral salts. Many root hairs may remain alive for an entire growing season, and in a few species they may become suberized or lignified and persist for a year or longer.

The distribution and abundance of root hairs on the root systems of plants appears to be highly variable. Some species characteristically bear few or no root hairs, while the number on other root systems is markedly influenced by the environment in which the roots develop, and still other root systems bear abundant root hairs under a wide range of conditions.

A root hair is essentially a tubular outgrowth of the peripheral wall of an epidermal cell, closed at its distal extremity, projecting more or less at right angles from the long axis of the epidermal cell of which it is an integral part (Fig. 7.7). Root hairs develop only from certain of the epidermal cells. They range in length from less than a millimeter to about a centimeter and are usually about 10 μ in diameter. On some roots as many as several hundred root hairs may be borne on a square millimeter of root surface, although the number per unit area is usually less. The presence of root hairs on a given area of a root often increases the exposed surface of that area from threefold to tenfold.

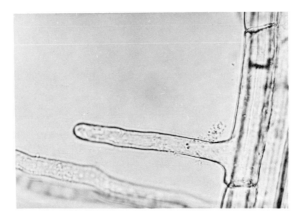

Fig. 7.7. A young root hair. Photomicrograph by Tillman Johnson.

The cell wall of a root hair is composed principally of cellulose and pectic compounds. The outer lamella of the wall seems to be composed entirely of pectic compounds, largely calcium pectate. The tenacity with which root hairs and soil particles adhere to each other is accounted for by this pectic coating. Because of this intimate contact between the root hairs and the soil particles, it is difficult to separate the two by washing or in any other way without injuring or destroying most of the root hairs. The inner wall of a root hair is lined with a thin layer of cytoplasm which is continuous with the cytoplasm of the epidermal cell of which the root hair is a part.

The Pathway of Water Through the Root.   Water enters the roots principally through the walls of the root hairs and the epidermal cells of the root tips. From the epidermal cells the water passes through successive rows of thin-walled cortical cells, and then through the cells of the endodermis (Fig. 7.4). The structure of the walls of the endodermal cells is often unusual. In one type of such cells (Fig. 7.8) a thickened strip is present on the inner surface of the radial and transverse walls. These *Casparian strips,* as they are called, are often suberized, and differ in width and configuration from one

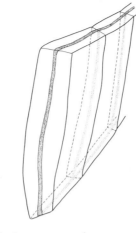

stele ⟵⟶ cortex

Fig. 7.8. Perspective diagram showing location of the Casparian strip in an endodermal cell.

species to another. Their presence impedes the movement of water through the walls in which they occur. In some species the endodermal walls are unusually thick; in such species there are present, opposite the outer end of each area of xylem tissue as seen in cross section, isolated thin-walled endodermal cells called *passage cells*. Such cells may facilitate the movement of water and solutes through the endodermis. The passage of water and solutes through the endodermis is probably also facilitated by the presence of lateral root initials which usually have developed by the time the endodermis is mature.

After passing through the endodermis, water moves into the xylem ducts, in most species after traversing a few intervening layers of pericyclic cells. There is considerable evidence that a substantial amount of the water crossing the roots in a radial direction moves through the cell walls, which are an important component of the so-called "free-space" (Chapter 15) rather than through the cytoplasm and vacuoles. When water is absorbed through older, suberized roots, a process which occurs in many woody and probably some other kinds of plants, the tissues which it traverses are different from those described here. It must move, in turn through suberized cells (cork), secondary phloem, and cambium before reaching the xylem tissues.

Mycorrhizas.   The roots of many species of plants are regularly infected with the mycelia of fungi. Such a root, together with its associated fungal hyphae, is called a mycorrhiza. In ectotrophic mycorrhizas the mycelium is chiefly external to the root, forming a sheath-like mantle around the roots. Some hyphae also penetrate into the cortical zone of the root. In the endotrophic mycorrhizas the hyphae are intracellular, being found principally within the cells of the epidermis and cortex. The fungal infection generally causes a cessation in the formation of root hairs by the host. In some species the infection results in the dichotomous branching of the host plant root tips. Many authorities believe that mycorrhizas are present on the roots of the majority of vascular plants.

Ectotrophic mycorrhizas are found on many forest tree species such as beeches, oaks, hickories, and many conifers. They are particularly abundant on trees growing on soils rich in humus. The fungal associates in forest tree mycorrhizas are generally members of the basidiomycete group of fungi. Frequently host roots infected with ectotrophic fungi are so densely sheathed with fungal hyphae that no part of the host root ever directly touches the soil.

Endotrophic mycorrhizas are found on many species of the orchid, heath, and gentian families, as well as in some tree species like maple and walnut.

Because the fungal associates appear to lack the enzyme systems required to degrade cellulose or lignin, the fungal infection does not cause the rotting of the host tissue. The hyphae are able to penetrate young, growing root cells of the host. Thus the fungus is able to utilize carbohydrates and amino acids absorbed from host cells without killing the host tissue. The mycorrhizal association is considered to be a symbiotic relationship. The hyphae of the fungi greatly increase the surface area of the roots, thus presumably increasing both water and mineral salt absorption.

The increase in the water absorbing potential afforded by the fungal hyphae may be quite important to the existence of certain host species. The ability of various species of pines to live in sandy soils, which have a relatively poor water retention capacity, may be related to the presence of mycorrhizal associations. Pines not infected with fungi are estimated to have three times as much transpirational surface area as root surface area. This ratio would not appear to favor the existence of such species in an area having frequent semistress soil moisture conditions. The presence of mycorrhizas on the roots of a pine tree is estimated to increase its absorption area many-fold, thus producing a more favorable transpiration surface/absorption surface ratio.

In addition to the increase of water absorption by mycorrhizas, there is strong evidence that mineral salt uptake is also greatly enhanced. For example, in experiments with pine trees grown in prairie soils, it was found that trees with mycorrhizal roots absorbed 234 per cent more phosphorus, 86 per cent more nitrogen, and 75 per cent more potassium than pines with uninfected roots.

The Absorption of Water by Roots.   From the standpoint of the absorption of water by plants, a clear distinction should be made between conditions under which such a movement of water occurs readily in a soil and those under which such a movement of water is slow or nonexistent. If capillary movement occurs readily, the translocation of water may take place toward the young roots whenever they are absorbing water. There are two principal conditions under which the capillary translocation of water can occur in soils at appreciable rates: (1) in any zone of soil which is not more than a few meters above a water table, and (2) in the upper layers of any soil after a heavy rain or irrigation, but before the water content of the soil has decreased to its field capacity. As the water in the films surrounding the soil particles with which the root tips are in contact becomes depleted, more water moves toward those particles by capillarity. The actual rate of such a capillary movement of water through the soil may become a factor influencing the rate of absorption. Root systems, however, are not static, but are more or less continually growing through the soil. The rate of the root growth of most species decreases, as a general rule, with the increasing wetness of the soil above the field capacity, because of the corresponding reduction in soil aeration. Hence in soils in which the capillary movement of water occurs—which are necessarily relatively wet—the rate of the elongation of roots is generally less than in otherwise similar but somewhat dryer soils. This continued growth of root tips through the soil brings them into contact with other portions of the soil water, so that even if capillary movement to certain parts of the soil ceases, the capillary movement of water to the roots may be reestablished by the extension of the root tips themselves into the zones of the soil that have not yet been depleted of water which can move by capillarity.

Many plants, much of the time, grow in soils at water contents between the permanent wilting percentage and the field capacity. In this range of soil-water content, capillary movement of water is very slow. Once most of the film water present on the soil particles with which the root tips are in contact has been absorbed, it cannot be replaced in any significant quantity by capillary movement from adjacent regions of soil if the soil water content is below the field capacity.

Since, in soils at a water content below the field capacity, the movement of water toward the roots is very slow, the principal method by which the roots come in contact with additional increments of water is by continually growing through the soil. Mature root systems of many species of plants bear millions of root tips. Each of these numerous root tips may be pictured as progressing through the soil and absorbing most of the water present in the smaller interstices between the soil particles with which it comes in contact. Relatively large quantities of water can be absorbed in this way, at least by some species of plants. For example, the total length of all the roots on a four-month-old rye plant has been calculated to be 623 kilometers. On the average, therefore, the aggregate daily increase in the length of the roots on this plant was about 5 kilometers. In addition it was estimated that nearly 90 km of new root hairs were formed, on the average, each day. Such a rate of root extension would permit the absorption by such a plant of 1.6 liters of water daily from a sandy soil or about 2.9 liters from a heavy clay loam, when both soils are at the field capacity.

Frequently different regions of the root system of a given plant are in zones of soil with different water potentials, some of which may be below the permanent wilting percentage. Many kinds of plants survive and even grow under such conditions.

Mechanism of the Absorption of Water.   An increase in the negativity of the water potential of the mesophyll cells of a leaf as a result of transpiration causes the water in the xylem vessels to pass into a state of tension (negative pressure) which results in a decrease in the water potential in the xylem ducts (Chapter 8). As soon as the water potential in the xylem ducts in the absorbing region of the roots exceeds that in the contiguous cells, a gradient of water potentials is established across the root, increasing in negativity from cell to cell from its epidermal layer to the xylem conduits. The nature of this gradient is discussed more fully in Chapter 8.

Whenever the water potential in the peripheral walls of the young root cells becomes more negative than that of the water in the soil, water will move from the soil into the root. Since the osmotic potential of the soil solution in most soils is negative by only a small fraction of a bar, the water potential of the absorbing cells of a root does not have to be very negative before water can enter them from any soil with a water content equal to or greater than the field capacity.

The absorptive process which has just been described is often called "passive absorption," because the entry of water into the roots is brought about by conditions which originate in the top of the plant and the root cells seemingly play only a subsidiary role. Transpiration is the process usually responsible for an increased negativity of the water potential of the mesophyll cells (Chapter 6), although the growth of aerial organs such as stem tips, young leaves, and young fruits may also engender more negative water potentials in the apical regions of the plants.

The mechanism of absorption just described undoubtedly accounts for the uptake of most of the water which enters the roots of the plants, but it is not the only mechanism of water absorption which is known to operate in plants. In many species an internal pressure known as *root pressure* often develops in the xylem ducts (Chapter 8). The occurrence of sap exudation resulting from root pressure can be demonstrated with some kinds of plants by immersing the root system of a detopped plant in a potometer (Fig. 7.9). After a time a dilute sap will begin to ooze from the cut end of the stem, and the absorption of water will be indicated by the movement of the meniscus on the capillary arm of the potometer. If the volume of water exuded is measured, it will be found to be about equal to the volume absorbed. In other words, water is being absorbed and is moving in an upward direction through the plant as a result of processes which take place in the root cells. This type of absorption in which the mechanism involved is localized in the root system, is often called "active absorption." Guttation (Chapter 6), root pressure, and active absorption are usually considered to be different aspects of the same phenomenon.

There is good evidence for the existence of a relatively simple osmotic

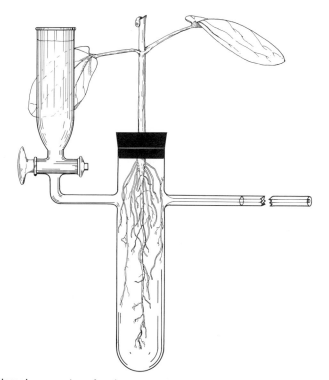

Fig. 7.9. Potometer as set up for demonstrating sap exudation as a result of the active absorption of water by the root system of a young plant.

mechanism of active absorption, but it is not certain that this is the only mechanism responsible for root pressure, xylem sap exudation, and guttation. In a number of species it has been shown that the osmotic potential of the sap in the xylem ducts and thus its water potential is seldom more negative than −2 bars. However, this is more negative than the water potential of most soils at the field capacity or higher water content. The osmotic movement of water from the soil to the xylem ducts could therefore occur through the "multicellular membrane" of the intervening root cells in spite of the fact that such cells have a more negative osmotic potential than either the soil solution or the xylem sap. The mechanism of such a movement of water can be interpreted in terms of the establishment of a consistently more negative gradient of water potentials from cell to cell across the root from the epidermis to the xylem ducts. The possibility that some such movement of water may occur through the cell walls should not be overlooked (*cf.* discussion of "free space" in Chapter 15).

Although there is little doubt of the existence of an osmotic mechanism of active absorption, the results of certain experiments are difficult to reconcile with the view that this process can be accounted for entirely by such a mechanism. It has therefore been postulated that a second mechanism of active absorption exists which is dependent for its operation upon the metabolic activity of the root cells.

The influence of metabolic processes on the active absorption of water may, however, be largely or entirely an indirect one. The absorption of mineral salts by the cells is markedly influenced by the metabolic conditions prevailing within them, especially their rate of aerobic respiration (Chapter 15). Furthermore, there are good reasons for believing that the passage of salts into the xylem ducts is largely controlled by the metabolic conditions in the adjacent living cells. Metabolic conditions, therefore, may influence the steepness of the osmotic gradient between the xylem sap and the soil because of their effect on the absorption and the cell-to-cell movement of mineral salts: and their influence on the active absorption of water may be exerted largely or entirely in this or in other indirect ways.

Although some plants, under certain conditions, exude relatively large quantities of water as a result of active absorption (Chapter 8), in general the volumes of water moving into plants as a result of this process are small compared with the volumes entering plants as a result of passive absorption. The effectiveness of active absorption in reducing the water content of a soil also appears to be considerably less than that of passive absorption. The active absorption mechanism does not absorb water against a water potential more negative than -1 or -2 bars; whereas the passive absorption mechanism results in the absorption of water at a relatively rapid rate until the water potential of the substrate approaches the permanent wilting percentage (approximately -15 bars).

Environmental Factors Influencing the Rate of Absorption of Water. Any factor which influences the water potential of the water in the peripheral walls of the young roots or the water potential in the soil will influence the rate of the absorption of water. Furthermore, the roots of a plant are more or less continually growing through the soil, and when the water content of the soil is less than the field capacity, the absorption of water at any appreciable rate can occur only if such growth of roots continues. Factors which influence the rate of root growth may therefore also have important effects on the amount of water which can be absorbed.

For reasons which should be clear from the preceding discussion, the rate of the absorption of water is greatly influenced by the rate of transpiration. Hence any factor which influences the rate of transpiration will indirectly also affect the rate of the absorption of water. Conversely, as already mentioned in the discussion of transpiration, any factor which influences the rate of the absorption of water will also influence the rate of transpiration. The more important soil factors which influence the rate of the absorption of water will now be discussed.

Available Soil Water.   In general, this term is used to refer to that fraction of the soil moisture in excess of that present at the permanent wilting percentage. As the soil moisture content decreases from the field capacity to the permanent wilting percentage, water becomes decreasingly available to plants, tending to decrease the rate of the absorption of water. In some species, at least, this effect is to some extent offset by the fact that the osmotic potential of the leaves becomes more negative as the soil-water content decreases. In the more extreme situations, water in the leaf cells may pass into a state of tension, as described in the next chapter. Hence as the soil-water potential decreases, more negative water potentials are achievable in the leaves. This permits an increase in the effectiveness of the passive absorption mechanism.

At soil-water contents below the permanent wilting percentage, the water intake is so slow that the turgidity of the leaf cells ordinarily cannot be maintained. Relatively high soil-water contents, in the range above the field capacity, result in a diminution in the rate of absorption by many species, because of the accompanying decrease in soil aeration (as discussed later).

Soil Temperature. The reduction in the absorption rate of water occurs in many kinds of plants at soil temperatures well above freezing, but the exact magnitude of this effect differs according to the species. In general, plants native to warm climates undergo a greater reduction in the rate of water intake when the soil is chilled than plants which are habitants of cooler climates. For example, watermelon and cotton, warm-season crops, absorbed only 20 per cent as much water at 10°C as at 25°C, while collards, a cool-season crop, absorbed 75 per cent as much water at the lower of these two temperatures as at the higher. The relationship between soil temperature and the rate of the absorption of water by white clover plants at several soil-water potentials, as inferred from the rate of transpiration, is shown in Fig. 7.10. Similar relations doubtless hold for other species. For obvious reasons, virtually no water is absorbed by the roots from frozen soils.

The mechanism whereby low soil temperatures cause a retardation in the rate of absorption of water is doubtless a complex one. The principal factors involved are: (1) the retardation in the rate of root elongation, (2) the increased viscosity of water, (3) the decreased rate of the movement of water through the soil, (4) the decreased permeability of the membranes of root cells to water which in turn is related to the increased viscosity of such membranes and, (5) the decreased metabolic activity of the root cells.

Soil temperatures sometimes also become high enough to exert a retarding effect on the rate of the absorption of water. For example, it has been found that the absorption of water by lemons, oranges, and grapefruits decreased when soil temperatures exceeded 30 or 35°C.

Aeration of the Soil. In general, the absorption of water by the roots of most species of plants proceeds more rapidly in soils which are well aerated than in those which are not. In poorly aerated soils oxygen concentrations may approach a zero value, and carbon-dioxide concentrations ranging up to 12 per cent are not uncommon. The saturation or near-saturation of a soil with water is the most common cause of deficient soil aeration. Under such condi-

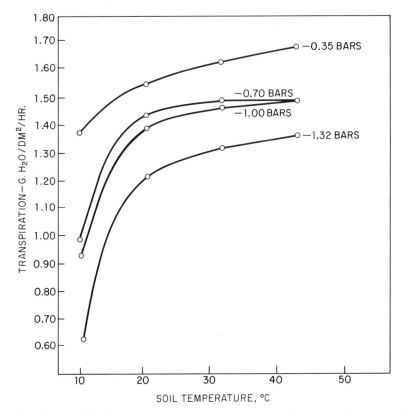

Fig. 7.10. Relationship between the soil temperature and the rate of transpiration in white clover plants at several different soil water potentials. Data of Cox and Boersma, *Plant Physiol.* **42,** 1967:550.

tions it appears probable that the reduced rates of water absorption which are evinced by most plants result more from a severe oxygen deficiency than from an accumulation of carbon dioxide. A drastic reduction in the supply of available oxygen in the soil reduces the rate of the respiration of the root cells, and this in turn influences other metabolic processes as well as the rate of root growth. One of the results of this sequence of disturbed physiological conditions is a lower rate of absorption of water. While the roots of most kinds of plants can survive for at least short periods in soils practically devoid of oxygen, presumably as a result of the occurrence of anaerobic respiration (Chapter 13), maintenance of this process for any considerable period of time leads to the death of the roots of many species. Tolerance of a condition of deficient soil aeration differs considerably from one terrestrial species to another. When deficient aeration is caused by flooding, for example, maize and tobacco show severe injury within a very few days, whereas sunflower and sorghum are much less affected. At the other extreme among crop plants is rice, which normally grows in a flooded soil. Likewise some kinds of trees, such as cypress, tupelo, and mangrove grow normally with their roots in saturated or submerged soils, while others such as many kinds of oaks and pines, and yellow poplar *(Liriodendron)* are unable to survive such conditions very long without suffering injury or death.

In contrast with most other species the roots of hydrophytes naturally grow in water-saturated soils and absorb water regularly from such soils. Some species of hydrophytes have well-developed intercellular spaces which are continuous from the leaves through the stems into the roots. It has been shown in some such species that oxygen moves to the roots through such channels, and such a movement of oxygen doubtless occurs in many other kinds of hydrophytes. In other species which grow with their roots submerged, however, no such prominent system of air passages is present. Even in such species, however, some movement may occur to the roots from aerial organs through intercellular spaces of the usual dimensions. The roots of those species in which little or no oxygen becomes available through downward movement from the tops apparently are able to carry on their metabolic processes at relatively low oxygen concentrations.

Concentration of the Soil Solution. The concentration of the soil solution in most soils in humid regions is so low that it has only a very slight effect on the soil-water potential. In alkali or saline soils, on the other hand, the concentration of dissolved salts in the soil water is often sufficient to result in a highly negative osmotic potential, which in extreme situations may attain −100 bars or even less. The copious applications of fertilizers to greenhouse or agricultural soils, especially if sandy, or irrigation with water containing dissolved salts in considerable concentration, often result in reducing the osmotic potential of the soil solution to a negative value of several bars or more.

The soil-water potential, except in soils below the field capacity, is essentially equal to the osmotic potential of the soil solution. In general the rate of the absorption of water decreases as the osmotic potential of the substrate becomes more negative. The effect of solutes on the movement of water into roots appears to be principally an osmotic one, specific effects of ions playing only a secondary role. Most plants develop normally only when the osmotic potential of the substrate solution is not more negative than a few bars. Halophytes, i.e., plants indigenous to saline or alkali soils or substrates, are the only important exception to this statement.

Absorption of Water by Leaves. The leaves and other aerial parts of plants frequently become wet as a result of rain, dew, or fog. Floods also result in the temporary immersion of aerial plant organs. Many species of plants can absorb at least limited amounts of water through the leaves. The turgidity of the wilted leaves of many species can be restored by immersing them in water. Wilted leaves are often observed to regain their turgidity during the night hours. This may sometimes result partly from the absorption of dew although a rate of water absorption in excess of the rate of transpiration is the main factor accounting for the nocturnal recovery of leaf turgor (Chapter 8).

SUGGESTED FOR COLLATERAL READING

Black, C. A. *Soil-Plant Relationships, 2nd Edition.* John Wiley & Sons, Inc., New York, 1968.

Burges, A., and F. Row, Editors. *Soil Biology.* Academic Press, Inc., New York, 1967.

Esau, Katherine. *Plant Anatomy, 2nd Edition.* John Wiley & Sons, Inc., New York, 1965.

Harley, J. L. *The Biology of Mycorrhiza.* Leonard Hill Books, London, 1969.

Hillel, D. *Soil and Water.* Academic Press, Inc., New York, 1971.

Kohnke, Helmut. *Soil Physics.* McGraw-Hill Book Company, New York, 1968.

Kozlowski, T. T., Editor. *Water Deficits and Plant Growth,* 3 vols. Academic Press, New York, 1968, 1968, 1972.

Kramer, P. J. *Plant and Soil Water Relationships. A Modern Synthesis.* McGraw-Hill Book Company, New York, 1969.

Russell, E. W. *Soil Conditions and Plant Growth, 9th Edition.* Longmans, Green and Company, London, 1961.

8 | INTERNAL WATER RELATIONS

An overwhelmingly large proportion of the water absorbed by the roots of land plants is lost in the process of transpiration. Smaller quantities are utilized in growth and in photosynthesis and, in some species, limited amounts of water may be lost by guttation. Water must therefore move through the intervening tissues and organs from the absorbing regions of the root to the tissues in which it is utlized, or from which it passes out of the plant. The process whereby water moves through the plant is termed the *conduction, transport,* or *translocation* of water.

The most striking illustrations of the upward transport of water occur in trees. A number of kinds of trees are known which grow to heights of at least 100 meters. Among these are the coast redwoods (*Sequoia sempervirens*) and big trees (*Sequoiadendron gigantea*) of California, the Douglas Fir (*Pseudotsuga douglasii*) of western North America, and the blue gums (*Eucalyptus* spp.) of Australia. Since the root systems of trees always penetrate at least a few meters into the ground, the actual distance through which at least part of the water absorbed is elevated is more than the above ground height of the tree.

The mechanism by which this feat is accomplished in tall trees has been the subject of much experimentation and even more speculation. The ensuing discussion of the ascent of sap through plants will be principally in terms of its movement through trees, because much of the experimental work on this problem has been performed on the woody species. Any explanation of this phenomenon which can be shown to be adequate for tall trees should also prove satisfactory for the vascular species of lesser stature.

The Path of Water Through the Plant.   A principal route of entry of water into most plants is through the epidermal cells and root hairs at or near the tips of the roots (Fig. 7.4). In many trees and shrubs it appears that considerable amounts of water may also be absorbed through the older, suberized portions of the root system. Regardless of the intervening tissues through which it passes, the water eventually moves into the vessels or tracheids of the root xylem. Once in the xylem ducts its general direction of movement is upward. The xylem tissue is continuous from just back of the tips of the roots, through the roots, into and through the stems, the petioles of the leaves, and ultimately, usually after much branching, terminates in the mesophyll of the leaf (Fig. 8.1).

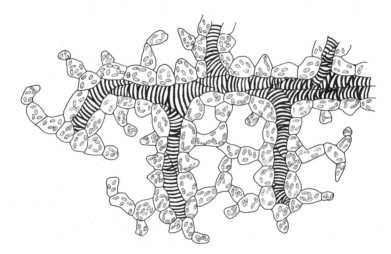

Fig. 8.1. Termination of vessels in the mesophyll of an apple leaf.

Most cells in a leaf are only a few cells distant from a vein or vein ending. The xylem tissue through which the water moves is thus a continuous unit system within the body of the plant. Along most of its course the water moves *en masse* through the vessels or tracheids. From the xylem ducts in the leaves the water passes into the mesophyll cells. In the mesophyll it moves from cell to cell, eventually most of it being lost from the cells by evaporation into the intercellular spaces. The movement of the water through the cells of the root and leaf mesophyll must be regarded as integral parts of the process of the translocation of water.

Although the great bulk of the water which passes through the plant follows the route just described and is lost in the transpirational process, small quantities escape this fate. All along the path of its movement small amounts of water pass into adjacent living cells and are utilized in cell division and enlargement, especially in cells differentiating from the cambium layer. Actively growing stem tips, root tips, and fruits also utilize considerable quantities of water, principally in the enlargement phase of growth, while chlorophyllous cells utilize water in the photosynthetic process. In most species, however, not more than 1 or 2 per cent of the water which enters a plant is utilized in growth and metabolic processes, the remainder being lost from the plant in transpiration.

That the xylem is the principal water-conducting tissue of plants has been recognized at least since the time of the girdling experiments of Malpighi in 1671. Girdling (or "ringing") a stem, so that a narrow band of all the tissues external to the xylem are removed, does not prevent the movement of water to the organs attached to that stem above the ring. On the other hand, cutting through the outer layers of the xylem tissue of a stem results in an almost immediate wilting of the leaves attached to the stem above the ring.

Anatomy of Stems. Since the mechanism of the movement of water can scarcely be understood intelligently without some knowledge of the anatomy of the tissues through which it moves, a brief review of stem structure is desirable before proceeding further with a discussion of this process. Stems vary greatly in their structure, every species possessing some anatomical features which are peculiar to itself. Nevertheless certain general patterns of tissue arrangement have been found to prevail, and the stem structure of most species approximates one or another of these general arrangements.

The stem anatomy of two representative species is shown in cross section in Figs. 8.2 and 8.3, which are self-explanatory. The maize stem represents an herbaceous monocot type of structure; the structure of the tulip poplar is typical of that of woody dicot stems.

The structure of woody stems cannot properly be appreciated merely from a consideration of one-year-old stems. *The primary tissues* of such stems develop from tissues formed at the apical growing tip during the growth of the stem in length. Nearly all perennial stems also grow in diameter as a result of the development of *secondary tissues* from the cambium. As a result of the division, enlargement, and differentiation of cambial cells, additional layers of xylem (*secondary xylem*) are laid down on the inner face of the cambium, and new layers of phloem tissue (*secondary phloem*) on its outer face. The secondary growth of woody stems is initiated during the first season of their development and continues during each growing season thereafter. Hence after a few years the great bulk of any woody stem is composed of secondary tissues. The apical and lateral growth of woody stems and the relation of these processes to the formation of the primary and secondary tissues of the stem are considered in more detail in Chapter 19.

The spring-formed xylem tissue, as viewed in cross section, is usually distinctly different in aspect from that formed later in the season. In many angiosperms the "spring" wood contains more and larger vessels, and the cell walls are generally thinner than in the subsequently formed "summer" wood. In the

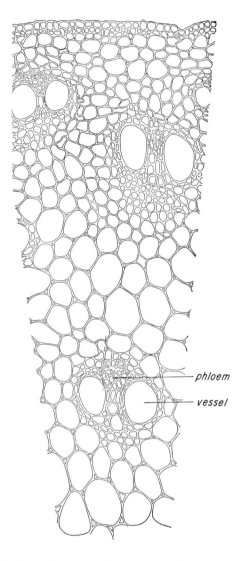

Fig. 8.2. Segment of young maize stem as seen in cross section. Three vascular bundles and parts of two others' are shown. The large-celled tissue between the bundles is sometimes classed as pith, sometimes as parenchyma.

conifers the spring-formed tracheids are thinner-walled and of larger cross-sectional diameter than those formed later in the growing season. The transition from spring to summer wood is often a very gradual one. On the other hand, the more open xylem tissues formed each spring abut directly upon the denser tissues which were produced during the preceding summer, thus giving rise to an abrupt line of demarcation between the zones of xylem formed in any two

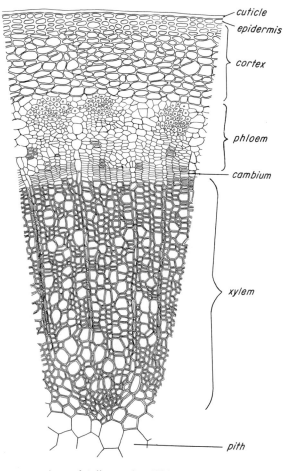

Fig. 8·3. Segment of a young stem of tulip poplar (*Liriodendron tulipifera*) as seen in cross section.

successive seasons. The result of this growth behavior is that a cross section of the trunk or branch of any tree appears as a system of concentric layers, the so-called *annual rings,* each representing an annual increment of growth. In rare cases no annual ring or more than one annual ring may be produced in a season, but usually each ring represents the xylem resulting from the activity of the cambium during one year.

In many woody species, as the xylem tissues increase in age, important changes occur in the color, composition, and structure of the various elements, resulting in the conversion of *sapwood* into *heartwood.* As sapwood ripens into heartwood the walls of any remaining living cells of the xylem become increasingly lignified, the death of these cells soon following. The water content of the tissues is generally reduced, and such compounds as oils, resins, gums, and tannins accumulate in the cells or cell walls. The darker coloration of the heart-wood of most species as compared to the sapwood is caused by such accumulations.

In mature trees the heartwood becomes merely a central supporting column surrounded by a cylinder of sapwood which varies in thickness from a few to many annual layers, depending upon the species and the environmental conditions under which the tree was growing. In some species (apple, elm) the heartwood remains virtually saturated with water, while in others (ash) it becomes relatively dry. The water in the heartwood of such species as apple and elm appears to be largely static and is not directly involved in translocation.

Coincident with the development of secondary xylem, secondary phloem tissues also develop from the cambium. Cork cambiums are also initiated in the bark, which produce cork layers (Chapter 19). Profound modifications therefore occur in the outer tissues as well as in the xylem of woody stems as they grow older.

A somewhat more detailed concept of the structure of the cells and elements of the stems of angiosperms through which the movement of water occurs is presented in Figs. 8.4 and 8.5, which represent longitudinal, tangential and radial sections from the xylem of a tulip tree which may be taken as representative for this group of plants. Figures 8.6 and 8.7 illustrate in a similar way the structure of the xylem tissues of the white pine—a representative gymnosperm. In arborescent monocots, such as palms, the conducting systems are very complicated and will not be described in this book.

The xylem tissues of the wood of angiosperms are composed of vessels, tracheids, fibers of several types, wood parenchyma, and xylem ray cells. There is a tremendous variability in the proportional distribution and arrangement of these tissues according to the species.

The most characteristic elements in the xylem tissue of angiosperms are the vessels. These are, in general, more or less tubular structures which may extend through many meters of the xylem. In some species cross walls, usually perforated, are of frequent occurrence in the vessels, in others such cross walls are infrequent or lacking. The vertical walls of vessels are lignified, the lignified thickening of the *primary xylem* often forming characteristic patterns such as rings or spirals. In trees, vessels may range in diameter from about 20 μ to about 400 μ. In vines they may be as much as 700 μ in diameter. The vessels branch extensively in certain regions of the plant, especially at nodes, within the leaf lamina, and in the parts of root systems where root branching occurs.

Thin areas called pits occur, not only in all parts of the vessel walls which are contiguous with other vessels or cells, but also in the walls of the majority of plant cells. Three main types of pits, *simple pits, bordered pits,* and *half-bordered pits,* are found in plant cell walls (Fig. 8.8). Strictly speaking, the term pit refers only to the opening in the secondary wall of one cell, the term *pit pair* often being used to designate the two complementary pits of adjacent cells.

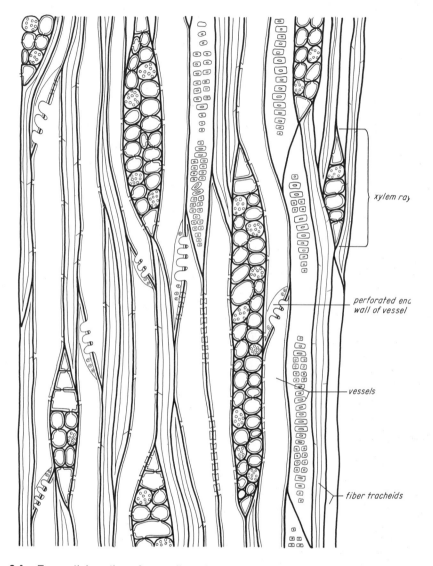

xylem ray

perforated end
wall of vessel

vessels

fiber tracheids

Fig. 8.4. Tangential section of a small portion of the wood of a tulip tree (*Liriodendron tulipifera*). This is a plane section except that the end walls of the vessels have been shown in perspective in order to indicate their structure clearly. Most of the vessels and fiber tracheids show in sectional view, but a few show as the surface view of the tangential wall. Surface and sectional views of bordered pits show in the vessel walls. Simple pits show in the end walls of the ray cells and between ray cells. Half-bordered pits show between vessels and ray cells. After L. G. Livingston.

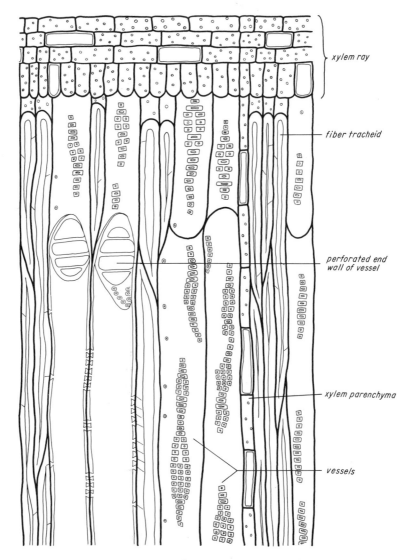

xylem ray

fiber tracheid

perforated end
wall of vessel

xylem parenchyma

vessels

Fig. 8.5. Radial section of a small portion of the wood of a tulip tree (*Liriodendron tulipifera*). Most of the vessels show as a surface view of the walls; two are in sectional view. Surface and sectional views of bordered pits show in vessel walls. Simple pits show in surface view in the walls of the xylem ray and xylem parenchyma cells. After L. G. Livingston.

The stages in the development of a vessel, which is a more complex phenomenon than the formation of the other elements of the wood, are indicated in Fig. 8.9. The original cell resulting from the division of a cambium cell increases rapidly in diameter, simultaneously developing a rather prominent vacuole. Secondary lignified layers develop on the longitudinal walls of the *vessel element,*

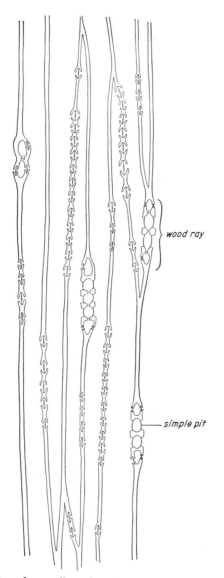

Fig. 8.6.   Tangential section of a small portion of the wood of white pine (*Pinus strobus*). The vertically oriented elongated elements are tracheids. Bordered pits show in sectional view in the tracheid and marginal ray cells. Simple pits show in sectional view in the other ray cells. Half-bordered pits show between marginal ray cells and other ray cells.

following which the disintegration of the protoplasm and the dissolution of the end walls occur. The result of this series of processes is the formation of a typical tubular, nonliving *vessel* by the coalescence of a number of vessel elements, each of which has been differentiated from a single cell originating from

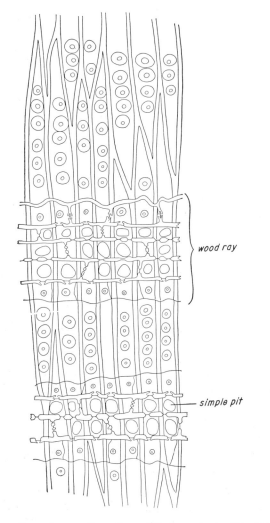

Fig. 8.7. Radial section of a small portion of the wood of white pine (*Pinus strobus*). The vertically oriented elongated elements are tracheids. Bordered pits show in face view in the tracheids; in sectional view in the marginal ray cells. Simple pits show in surface and sectional views in the ray cells. Half-bordered pits show between marginal ray cells and other ray cells. Three rows of marginal ray cells show in outline only.

a division of a cambium cell. In many species, especially woody plants, vessel formation does not occur in as regular a manner as just described. In many such species the ends of the vessel segments overlap and openings develop in the side walls near the ends (Fig. 8.4).

Tracheids are found in the wood of many but not all species of angiosperms. They are typically more or less spindle-shaped cells with thick walls

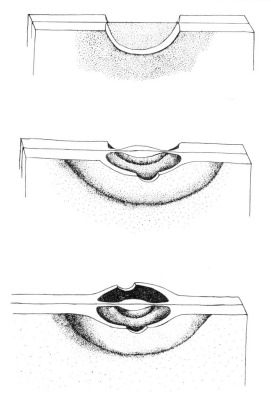

Fig. 8.8. Perspective diagram illustrating types of pits in white pine *(Pinus strobus):* bottom, full-bordered pit; middle, half-bordered pit; top, simple pit.

which are usually angular. Mature tracheids contain no protoplasm and hence are nonliving like the xylem vessels. The largest tracheids are about 5 mm in length, and about 30 μ in diameter. The walls of tracheids are pitted like those of the vessels. Tracheids are water-conducting cells, but in most angiosperms they are relatively of much less importance as channels through which water moves than the vessels.

Tracheids are developed from the cambium by a process essentially similar to that by which vessel elements are formed, except that the size and shape of the resulting cells are different, and that there is no linear coalescence of the individual elements such as occurs in the formation of vessels.

The xylem tissue of practically all angiosperms also contains wood parenchyma cells which, unlike the elements already described, remain alive for some time after their differentiation. Uusually the death of the wood parenchyma cells does not occur until the wood of which they are a part is converted into heartwood. Wood parenchyma cells are generally somewhat elongate, and occur in the xylem as a vertical series of cells placed end to end. The strands of the

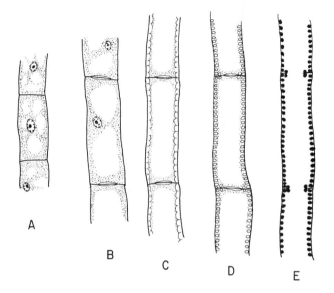

Fig. 8.9. Stages in the development of a vessel in the petiole of celery. (A) Enlarged cambial cell. (B) Young vessel segment showing lenticular thickenings on the end walls. (C) End walls still present in the vessel segment with spiral thickenings forming on the lateral walls. (D) Protoplasm disintegrating in vessel segments, spiral thickenings on thin portions of end walls. (E) Mature vessel in which the end walls have disappeared. After Esau, *Hilgardia* **10**, 1936:488.

wood parenchyma thus constituted often extend for long distances in a vertical direction through the wood. The distribution of the wood parenchyma strands throughout the annual ring of the xylem tissue varies with the species. In some they are more or less scattered through the xylem, in others they occur in the last layer or two of the cells which are produced in the summer wood—in other words, at the termination of the season's growth—and in others they are only in contact with the vessels, or in contact with other wood parenchyma cells which are themselves in contact with the vessels.

All of the xylem elements previously described are oriented with their long axes in a vertical direction. In addition to this vertical system there is also present in the xylem a transverse, radiating system in which the long axis of the cells is at right angles to the long axis of the stem. These transversely oriented tissue units are known as *vascular rays*. In the stems of most species they are continuous from the outer extremity of the phloem through the cambium into the xylem, which they penetrate to a greater or lesser distance. The portion of the vascular ray found in the xylem is termed the *xylem ray* or *wood ray*; that found in the phloem the *phloem ray*. Xylem rays may vary from one to many cells in thickness and likewise in height, certain types usually being characteristic of any one species (Figs. 8.4–8.7). The cells of the xylem rays, like those of the wood parenchyma, usually remain alive until the woody tissue is converted into heartwood. The xylem rays probably serve as routes along which the lateral movement of water occurs from the xylem to the cambium and phloem, and along which the translocation of soluble foods takes place from the phloem to the living cells of the xylem.

The living ray cells are in contact at various points with the strands of living wood parenchyma cells. The vertically oriented wood parenchyma strands and the transversely oriented xylem ray strands thus form a unit system of living cells within the woody cylinder. Hence there is present a continuous intermeshing network of living cells throughout the greater mass of nonliving vessels, tracheids, and fibers in the younger portions of any woody angiosperm stem. There are probably few if any of the conducting elements—vessels and tracheids—which are not in contact at one or more points with this continuous system of living cells.

The wood of gymnosperms is simpler in its structure than that of the angiosperms, this group of woody species also showing, in general, a greater uniformity in stem structure than the latter group. The only cell types universally present in the wood or coniferous trees are the tracheids and wood ray cells. Wood parenchyma cells are also present in the wood of most species of conifers, while in many species tracheid-like fiber cells are also present. The most important distinction between the wood of conifers and that of angiosperms is the total absence of vessels in the former.

The tracheids are the distinctive element in conifer wood, constituting as they do the great bulk of all the woody tissues present in such species. In conifers the tracheids form a densely packed type of woody tissue, composed of interlocking cells. Vertically contiguous tracheids always overlap along their tapering portions (Fig. 8.7). The movement of water and solutes from one tracheid to another is facilitated by means of the pits in the adjacent walls. Because of the numerous cross walls in a xylem tissue composed almost entirely of tracheids, water encounters a greater resistance in moving through such tissues than in traversing woody tissues which contain vessels. Nevertheless it is interesting to note that the tallest trees in the world are conifers in which the upward translocation of water occurs through tracheids.

Theories of the Mechanism of the Translocation of Water Through Plants.
A number of theories of the mechanism by which the ascent of xylem sap is brought about in plants have been suggested, and it is probable that more than one mechanism is involved in the process. The present state of our knowledge justifies a discussion of three possible mechanisms of the upward transport of water through plants.

"Vital Theories." Although the vessels and tracheids through which the longitudinal transit of water occurs are nonliving, they are always in more or less intimate contact with living cells (wood parenchyma and xylem rays). Sug-

gestions have therefore been made from time to time that the upward translocation of water is brought about in some manner by the living cells of the stem, although there is almost no direct evidence in favor of such a view.

Numerous experiments demonstrate quite clearly that the primary mechanism of the rise of sap in trees operates independently of the living cells of the stem. In one experiment a 75-year-old tree about 22 meters high was used. This was sawed off close to the ground and the cut end of the trunk was immersed in a solution of picric acid, which is toxic to living cells. The picric acid solution slowly moved up the stem. Fuchsin, a red dye, added to the liquid in which the basal end of the tree was immersed 3 days after the picric acid, also ascended to the top of the tree through tissues in which the living cells had been killed by the picric acid. Water also continued to ascend through similar stems after they had been completely killed by exposure to a temperature of 90°C.

In all experiments of the type just described, it has been observed that the leaves at the top of a stem, part or all of which has been killed, sooner or later wilt and wither, although this effect is by no means an immediate one, often appearing only after several days. Proponents of "vital theories" have accepted this as evidence that the living cells of a stem are essential for the passage of water through it. This delayed lethal effect on the leaves, however, could result from either or both of two causes. The killing of the stem tissues often causes the formation of substances which plug the vessels or tracheids, thus impeding the upward movement of the water. Furthermore, the death of the cells in the treated regions causes the release into the conducting channels of toxic compounds which, when transported to the leaves, also cause the death of the leaf cells.

Root Pressure. An exudation of xylem sap often occurs from the stub of a freshly cut stem (Fig. 8.10), from the stump of a newly cut tree, or from the incisions or borings made into plants. Most such exudations result from the development of a pressure in the dilute sap of the xylem ducts resulting from the operation of a mechanism in the roots (Chapter 7), hence the term "root pressure." Guttation in intact plants (Chapter 6) also results from root pressure.

The magnitude of the root pressure in any plant is usually measured by means of a manometer. With few exceptions the xylem sap exudation pressures which have been recorded for plants do not exceed 2 bars and most of them are less than this. In trees of some height in which root pressure occurs, such as the birch, a gravity hydrostatic gradient exists in the xylem ducts. In other words, with increasing height in the tree, the magnitude of the root pressure decreases, at the rate of about 0.1 bar per meter of height.

Although it is undoubtedly true that root pressure does, in some species of plants and under certain conditions, account for the movement of some water in an upward direction through plants, there are several reasons why this process cannot be considered to be the principal or even an important mechanism by which water is moved through plants. In the first place, there are many species in which this phenomenon has never been observed. In the second

Fig. 8.10. Experimental arrangement for the demonstration of roof pressure from the cut stump of a potted plant.

place, the magnitude of the pressure developed is seldom sufficient to force water to the tops of any except a relatively low-growing species of plants. Neither is the rate of flow, as it occurs in most species, adequate to compensate for the known rates of transpiration. In the third place, root pressures are usually negligible, in temperate regions at least, during the summer period when transpiration is most rapid. During periods of rapid transpiration, the cut surfaces of most plants not only fail to exude sap, but usually absorb water if it is supplied at the cut surface.

The Cohesion of Water Theory. Molecules of water, although ceaselessly in motion, are also strongly attracted to each other. In masses of liquid water the existence of such intermolecular attractions is not obvious, but when water is confined in long tubes of small diameter, their existence can often be demonstrated. If the water at the top of such a tube is subjected to a "pull," the resulting stress will be transmitted all along the column of water because of the mutual attraction ("cohesion") between the water molecules. Furthermore, because of an attraction between the water molecules and the molecules of the wall of the tube ("adhesion"), putting the water column under stress does not result in pulling the water away from the encompassing wall. The water-conducting system of plants constitutes just such a system, enclosing continuous, often intermeshing, threadlike columns of water which extend from the top to the bottom of the plant. Because of the cohesion between the water molecules and their adhesion to the walls of the xylem ducts, a stress applied at any point in this system will be propagated to all its parts. Such a stress tautens the water columns and creates within them a state of tension.

Whenever evaporation is occurring from the walls of the mesophyll cells into the intercellular spaces, the water potential of the mesophyll cell walls becomes more negative. Such cell wall water potentials are primarily imbibitional in origin (Chapter 5). Water therefore moves into the walls from the adjacent cytoplasm, resulting in turn in a movement of water from the vacuole into the cytoplasmic layer. The resulting greater negativity of the water potential is propagated to all parts of the cell. Within the lamina of the leaf, gradients of water potentials, gradually increasing in negativity from cell to cell in the directions in which the water is moving, are established between the xylem ducts and the cells from which the evaporation is occurring. Water therefore moves from a given vessel or tracheid into the adjacent cells, which results in the development of a tension in the water column occupying that element of the xylem. Concurrently tensions are similarly developing in other xylem ducts within the leaf. The later potential (ψ_w) in the xylem elements will be decreased by the amount of this tension. Water under a tension (which is equivalent to a negative pressure) of 13 bars, for example, has a water potential (ψ_w) of -13 bars relative to pure water at the same temperature and under atmospheric pressure (Chapter 5). The osmotic potential (ψ_s) of the xylem sap is seldom lower than -1 or at most -2 bars; hence this is not an important factor in determining its water potential.

Whenever conditions are such that the entry of water into the lower ends of the xylem ducts from the adjacent root cells is retarded, tensions of considerable magnitude develop in the water columns and the movement of water from the xylem conduits into the mesophyll cells will be relatively slow. If, under such conditions, evaporation continues from the mesophyll cells, the volume of water within them may continue to diminish even after their turgor pressures have decreased to a zero value. Under such conditions the walls of a cell may be pulled inward as a result of the adhesion between them and the shrinking volume of water, and the counterpull exerted by the walls on the water will throw it into a state of tension (Chapter 5). The water potentials of such cells are more negative than their osmotic potentials would indicate to be possible. It seems probable that water is more likely to be sustained under tension in mesophyll cells in leaves of the sclerophyllous type than in leaves of the herbaceous type.

The tension developed in the conducting elements is transmitted along their entire length to their lower termination just back of the root tips and probably, very often at least, across the root tissues as well. The magnitude of the tension which develops in the xylem conduits is increased by conditions which favor a rate of water loss from the leaves considerably in excess of the rate at which water enters the roots from the soil.

The radial movement of water across the cells of the root from the soil into the lower ends of the water-conducting elements must be considered an integral part of its translocation process. At their lower terminations the xylem vessels or tracheids are in contact with the pericycle, or more rarely, directly with the endodermal cells of the root (Fig. 7.4). Water moves from the adjacent root cells into the conducting elements because the tension (water potential) developed in these elements is more negative than the water potential of the root cells. When the tension in the water columns is relatively low, gradients of water potentials will be established across the root cells similar to those which are established in the mesophyll cells when water is moving through them. The water potential will become more and more negative from cell to cell along this gradient in the direction in which the water is moving, *i.e.,* from the periphery of the root toward the xylem. Water will enter the peripheral cells of the root whenever the water potential of such cells is more negative than that of the soil water (Chapter 7). If high tensions are generated in the xylem ducts, they may be so much more negative than the osmotic potentials of the root cells through which water passes that a tension develops in such cells in a manner similar to that already described for the mesophyll cells. Because of their generally less negative osmotic potentials, the development of such tensions is more likely in root cells than in leaf cells. When the entire hydrodynamic system of the plant, including the root and leaf cells, has passed into a state of tension, the movement of the water through the system from the absorbing surfaces of the roots to the evaporating surfaces of the leaves occurs by mass flow.

In most trees practically all the upward movement of water occurs in the vessels or tracheids of a few of the outermost annual rings of the sapwood; and in some species, especially those with ring-porous wood, it is almost entirely confined to the outermost ring. As is well known, in some species of trees (beech, sycamore, redwood, etc.) the heartwood of the trunk may disappear completely by decay. The fact that such hollow trees continue to live and thrive is conclusive evidence that the heartwood plays no essential role in the upward movement of water in such species.

Magnitude of the Cohesion of Water. If water is to move through the xylem ducts as postulated by this theory, its cohesion must be of sufficient magnitude to resist the stresses which are imposed on it. The maximum height to which water moves in plants does not exceed about 120 meters. This height is equivalent to about 12 bars which represents the maximum stress to which the water columns are subjected as a result of their own weight. Water, however, encounters a certain amount of resistance in moving through the conducting tissues. The greater the velocity of the current of water, the greater the resistance to which it will be subjected. At a velocity corresponding to the usual rates of transpiration the resistance has been estimated to be equal to the pressure required to support water to the height of the plant. The minimum adequate value for the cohesion of water, therefore, becomes 24 bars. To this must be added the resistance encountered by water in crossing the tissues of the root and the mesophyll cells of the leaf which is equivalent to only a few bars. The estimated minimum cohesion required to lift water to the top of the tallest tree is therefore about 30 bars. Experimentally determined values for the cohesion of water, although quite variable, range up to 350 bars.

The existence of cohesion in water can be demonstrated by means of an apparatus such as that shown in Fig. 8.11. As evaporation proceeds from the porous clay cup, water moves up the vertical glass tube, followed by mercury from the reservoir. In a successful demonstration the mercury will continue to rise above the level to which it will stand in a barometer. The water is now being pulled up the tube, a phenomenon which is possible only because of the cohesion between water molecules and the adhesion between the water molecules and the glass wall of the tube. The water in the tube is thrown into a state of tension which is transmitted to the mercury column below it because of the strong adhesion between water and mercury. The pull on the water

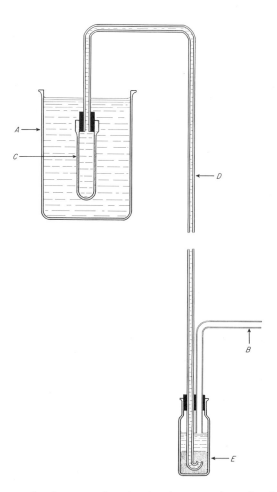

Fig. 8.11. Apparatus for demonstrating the development of tension in a water column as a result of evaporation. (A) Beaker of boiled water. (B) Glass tube. (C) Porous clay cup filled with water. (D) Capillary glass tube filled with water. (E) Mercury layer in bottle. For successful operation it is essential that the water in (C) and (D) be free of air bubbles. This is accomplished by boiling the water in (A) and allowing the hot water to siphon through the apparatus and out at (B). To perform the demonstration, the beaker (A) is removed from around the porous cup (C).

column originates at the evaporating surface of the cup, and results from the attractions between molecules operative in maintaining numerous microscopic water menisci in the pores of the clay cup. A rise of mercury to a height of 226.6 cm has been demonstrated in such an apparatus. This is approximately three times as high a column of mercury as will be supported by atmospheric pressure acting alone. While the maximum tension developed in the water column in these experiments does not exceed 2 bars, this demonstration illustrates a physical system which is analogous to that which is believed to operate in the plant.

Development of Tension in the Water Columns. Convincing evidence that the water in the xylem vessels is often in a state of tension has been obtained by the direct observations of vessels under a microscope. The stems of some species of herbaceous plants, especially the cucurbits, are highly suitable for such observations. Such a stem of an intact, rapidly transpiring plant can be fastened in position across the stage of a microscope and, by careful dissection, the vessels can be examined individually. If one of the vessels under observation is jabbed with the point of a fine needle, an immediate jerking apart of the water column at the point of the rupture will be seen to occur, indicating that the water in the intact vessel was in a state of tension.

Interesting evidence that the water in the xylem ducts of woody stems is, at times, under tension has been obtained by means of an instrument known as the *dendrograph*. This is a self-recording instrument which measures the variations in the diameters of tree trunks. It is constructed so that its sensitivity is very great, and its recordings are not influenced by temperature effects upon the instrument. Dendrographs are used principally to measure the periodic variations in the diameter growth of trees. However, even in trees in which the diameter growth has ceased, slight diurnal periodic variations in their diameters are of regular occurrence.

A record of the periodic variations in the diameter of a tree trunk for a period of several days at a season when little diameter growth was occurring is pictured in Fig. 8.12. The trunk attained its minimum diameter during the afternoon hours, at a period when the water columns are undoubtedly under

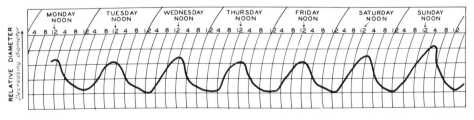

Fig. 8.12. Daily variations in the diameter of the trunk of a Monterey pine (*Pinus radiata*) as measured with a dendrograph. Data of MacDougal, *Carnegie Inst. Wash. Publ.* No. 462, 1936:55.

their maximum tension. While under tension the water columns become taut and decrease in diameter. Because of the enormous adhesion between the water and the walls of the ducts, a slight contraction occurs in their diameter. Such diurnal changes in the diameter of a tree trunk result from the alternate contraction of the vessels or tracheids when the water in them is under tension, followed by their dilation when the tension is slackened.

Although the existence of tensions in the water of the xylem ducts has long been recognized, not until 1964 was a suitable technique devised for measuring them quantitatively. A branch of the plant in which it is desired to measure the tension is inserted into this apparatus as shown in the illustration (Fig. 8.13). A short length of the stem, from which the bark has been peeled back, is allowed to protrude from the pressure chamber through a compression gland. When the branch was cut, water was withdrawn from the cut surface into the twig as a result of the tension existing in the water columns. By the gradual application of nitrogen gas pressure a compression is exerted on the tissues of the leaves and the twig. When water just appears at its cut surface, as observed under a suitable lens, the water columns have been attenuated to the same degree that they were under the tension that existed at the time the twig was cut. The pressure which must be applied to just force the water to the cut end of the twig is equal to the tension (negative pressure) which existed at the

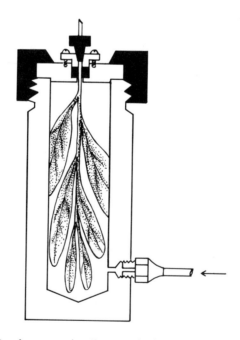

Fig. 8.13. Apparatus for measuring the magnitude of the tension which develops in the xylem ducts. Redrawn from Scholander, *et al., Plant Physiol.* **41**, 1966:530.

time the twig was cut. While this procedure cannot be regarded as a precise method of measuring tensions in the water columns, it undoubtedly yields values which are at least approximately correct.

The hydrodynamic status of the water in the xylem ducts ranges from small positive pressures, when a root pressure prevails, to tensions of about −100 bars which have been measured in some desert plants. The tensions in xylem water usually fall within the narrower range of from −5 to −30 bars. Because of the hydrostatic gravity effect, tensions are greater towards the top of a tree than in its more basal portions.

Vessels and tracheids normally contain water at the time of their differentiation and remain filled with water for varying periods of time thereafter. Ultimately most of the water columns in a plant break for one reason or another, although in any normally functioning plant at least some of the water columns remain intact. Extreme or prolonged drought is one cause of the breaking of the water columns. In colder climates the water in woody stems is subject to alternate freezing and thawing during the winter months. The freezing of water in the xylem conduits forces any dissolved gases out of solution causing cavitation, *i.e.,* the formation of gas bubbles in the water columns, thereby interrupting their continuity. It is not known how the continuity of such broken water columns is reestablished, although in those few species in which it is adequate, root pressure may provide the necessary mechanism.

The Relation of Transpiration to the Movement of Water Through the Plant.   In most discussions of the cohesion of water theory, the relation of transpiration to this process is so greatly emphasized that it has come to be almost indelibly associated with this concept. Transpiration is not, however, the immediate cause of the upward movement of water in plants. Water ascends in the xylem ducts only when water potentials of sufficient negativity have been created in the vacuoles or the walls of the cells in organs in the upper parts of the plant. Since the evaporation of water from the walls of the mesophyll cells is the most frequent cause of the generation of such negative water potentials, the process of transpiration has been generally linked in discussions with the mechanism of the ascent of sap. It is only because of its effect in generating relatively negative water potentials in the mesophyll cells that transpiration sets in motion the entire train of water through the plant. Any other process which results in generating relatively negative water potentials in cells at or near the terminus of any plant axis can also motivate the upward translocation of water. The upward movement of water continues during the night hours after transpiration has virtually ceased. This lag in the cessation of the upward movement results from the residual negative water potentials of the leaf cells at the end of the daylight period. Water will continue to enter these cells until they reattain the maximum turgidity of which they are capable under the conditions prevailing. Similarly the movement of water will occur into any rapidly growing stem tip or fruit, because the binding up of water in certain phases of the growth process increases the negativity of the water potential in the cells of such organs thus motivating the migration of water toward such centers of growth activity. It is possible that the movement of water into the growing regions may be motivated by metabolic mechanisms as well as by physical mechanisms, but the evidence for the existence of such metabolic mechanisms is not very convincing.

Lateral Movement of Water.   A cell-to-cell lateral movement of water in a radial direction undoubtedly occurs along the vascular rays in the stems of most species of plants. In woody stems there is probably also a lateral movement of water around the stem in a tangential direction. Except in trees in which the grain of the wood is twisted, the conducting vessels on one side of the tree generally connect with branches on that side of the tree at their upper extremity, and with roots on the same side of the tree at their lower extremity. If no lateral movement of water occurs in the woody stems, it would be expected that the removal of the roots from one side of a tree would result in a dearth of water, or perhaps even death of the leaves or branches on that side of the tree.

Experiments have been performed on apple, peach, oak, and other woody species in which the roots on one side of the plant were removed in an attempt to ascertain whether or not the lateral movement of the water occurs. Although the water content and growth of the plants treated in this manner diminished, there was no difference in the moisture content of the leaves on the two sides of the tree. Neither did the leaves on the side from which the roots had been removed show any greater tendency to wilt on clear, warm days than those on the other side. These results indicate very strongly that the lateral movement of water occurs in woody stems, and that the water conductive system of plants acts as a unit system.

Downward Movement of Water.   The cohesion theory of the movement of water will account equally well for the conduction of water in either an upward or a downward direction through the plant. Whenever conditions are such that the water potential of the cells of an organ which is basally situated relative to another is more negative than that of the cells of a more nearly apical organ, a reversal in the direction of the movement of the water may occur. It has been shown experimentally that during a heavy rain, especially after a period of high transpiration, water may be absorbed by the leaves of woody plants and move in a downward direction through the trunk. The movement of water out of lemon fruits during the daylight hours, described later in this chapter, is another example of a reversal in the direction of the water movement.

Wilting.   One of the commonest of observations among agriculturalists and gardeners is that the leaves of many species of plants often wilt on hot summer afternoons, only to regain their turgidity during the night even if the plants are not provided with additional water by rainfall or irrigation. In hot, dry regions, or during hot weather in more temperate regions, such a phenomenon may be a regular daily occurrence, even during periods when the soil is well supplied with water. This familiar reaction of plants is called *temporary* or *transient wilting* and clearly results from a temporary excess of the rate of transpiration over that of absorption. As a result the total volume of water in the plant shrinks, although not equally in all the tissues. In general, the diminution of water content is greatest in the leaf cells. The condition commonly called wilting is induced whenever the shrinkage in the volume of water in the leaf cells is sufficient to cause them to lose all or most of their turgor.

Wilting as a visible phenomenon is confined chiefly to species in which the leaf tissues are composed largely of thin-walled, parenchymatous mesophyll cells, and in which the leaves are maintained in their usual firm, expanded condition principally by the turgidity of such cells. In many species of plants the leaves are supported largely by lignified tissues. Examples of species bearing such leaves include many of the evergreens and the numerous sclerophyllous species common in the semiarid regions of many parts of the world. Such leaves wilt just as do parenchymatous ones in the sense that a marked loss of turgor may occur in the leaf cells. The wilting of such leaves is not usually characterized, however, by the drooping, folding, or rolling which are the visible symptoms of wilting in leaves composed principally of parenchymatous tissues.

Even on days upon which visible wilting does not take place, *incipient wilting* is of frequent occurrence. Incipient wilting is the term applied when the loss of turgor by the leaf cells is only partial; it does not result in the visible drooping, rolling, or folding of the leaves. Incipient wilting is of almost universal occurrence in the leaves of terrestrial plants on bright, warm days whenever environmental conditions are not severe enough to induce the more extreme and visibly discernible temporary wilting. Leaves passing into a state of transient wilting always first pass through the stage of incipient wilting. Both incipient and transient wilting are to be distinguished from permanent wilting (see later discussion) which results from a deficiency of water in the soil rather than from a temporary excess of transpiration over absorption.

As a general rule the leaves wilt first, because they are the organs from which the great bulk of water loss occurs, but the decrease in turgor gradually spreads throughout the plant as the internal deficiency of water becomes more severe. The loss of turgor is thus general, although not usually equal, throughout all of the tissues of a plant whenever wilting of any duration occurs. Any living cell in a plant may wilt, if this term is used to designate an approximately complete loss of turgor, which will be the general sense in which it will be employed in this discussion. The longer the condition of wilting persists, the more pronounced such a systemic loss of turgor will be. We may speak, therefore, not only of wilted leaves or other plant parts, but also of wilted plants.

Because of its effect on the dynamics of the internal hydrostatic system, wilting initiates a train of far-reaching effects upon physiological conditions and processes. Some of these effects have received attention in Chapter 6 in the discussion of the factors influencing the periodicity of transpiration. Such effects upon photosynthesis and growth are considered in Chapters 11 and 20, respectively.

Comparative Daily Periodicities of Transpiration and the Absorption of Water.   The phenomenon of wilting, as well as the results of many experiments which show that a daily diminution in the water content of leaves and other plant organs is of frequent occurrence, are both indirect evidence that the transpiration rate frequently exceeds the rate of the absorption of water during the daylight hours. Only a few investigations have been undertaken, however, in which simultaneous measurements have been made of the transpiration and absorption rates of the same plant over periods of 24 hours or longer.

The results of one experiment of this type in which loblolly pine (*Pinus taeda*) was used under approximately "standard day" conditions are shown in Fig. 8.14. As shown in this figure there was a distinct lag in the rate of absorption as compared with the rate of transpiration during the daylight hours—*i.e.,* during the period of relatively high transpiration rates. There was also a fairly well-marked tendency, shown clearly on the second day of the experiment, for the peak absorption rate to occur somewhat later in the day than the peak transpiration rate. Similar results have been obtained with other species, and with some of them the absorption lag was even more pronounced. During the night hours the rate of absorption was continuously higher than the rate of transpiration. In other words, the tissues of the plant were being progressively depleted of water during the daylight hours, while their water supply was being replenished at night. The lag in the rate of water absorption behind the rate of transpiration results largely from the relatively high resistance of the living root cells to the passage of water across them.

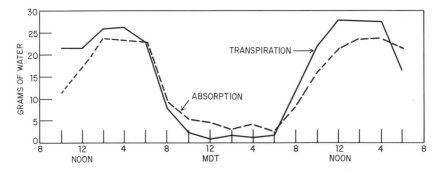

Fig. 8.14.   Comparative daily periodicities of transpiration and the absorption of water in the loblolly pine (*Pinus taeda*). Data of Kramer, *Am. J. Botany,* **24,** 1937:12.

Daily Variations in the Water Potential of Plant Cells.   Many determinations have been made of daily variations in the water content of leaves and other plant organs. Such measurements have been valuable in elucidating the principle that the development of an internal water deficit is a phenomenon of almost daily occurrence in most plants during their growing season. However, such determinations do not give a valid picture of the dynamic aspects of the internal water relations of plants. The same change in water content, for example, which induces a certain shift in the water potentials or turgor pressures of the leaf cells of one species may have a very different effect on the water potentials or turgor pressures in the leaf cells of another species. The influence of fluctuations in water content upon the internal movements of water and upon physiological processes can only be fully interpreted if the status of the water present is expressed in terms of water potentials or analogous dynamic units.

Daily variations in the water potential of the leaves of grapes have been studied under approximately standard day conditions by means of the pressure chamber apparatus described earlier (Fig. 8.15). The leaf water potential in this plant which is not far below zero in the morning hours, becomes increasingly more negative during the day until mid-afternoon, and then decreases in negativity until approximately the early morning values are reattained during the night hours. This pattern of the daily variation in leaf water potential is doubtless a representative one for many kinds of plants. Similar, although probably less marked, diurnal variations undoubtedly also occur in the water potentials of the cells in other plant organs.

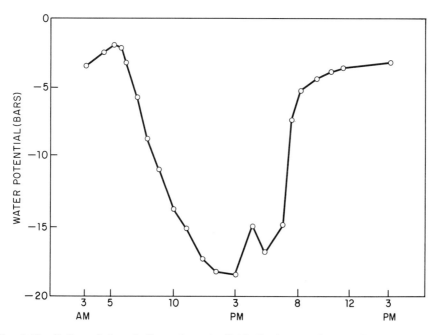

Fig. 8.15. Daily variations in the water potential in the leaves of grape. Data of Klepper, *Plant Physiol.* **43**, 1968:1933.

Permanent Wilting. This term refers only to wilting from which a plant will not recover unless the water content of the soil is increased. It is engendered by the development in the soil water of a negative water potential so great that the rate of movement of water into the plant is inadequate for the maintenance of turgor. As with transient wilting, the visible symptoms of permanent wilting are apparent only in thin-leaved species of plants, but physiologically equivalent conditions may develop in practically all terrestrial species.

In a soil that is slowly drying out, temporary wilting slowly grades over into permanent wilting. Under such conditions, each night recovery of the plant from temporary wilting occurs more slowly and is less complete, until finally even the slightest nocturnal recovery fails to take place and the plant passes into a continuous state of permanent wilting which grows progressively more drastic the longer it persists.

As the available water in the soil becomes depleted, the continuity of the soil water with the water in the plant is interrupted, and the water mass in the plant becomes essentially an isolated unit hydrodynamic system. When this condition prevails, the stress in the hydrodynamic system gradually becomes intensified, since even if the stomates are closed as they usually are in permanently wilted plants, cuticular transpiration continues, thus gradually reducing the total volume of water within the plant. The maintenance of plants in a state of permanent wilting for more than a few days often results in the death of the root hairs as a result of a deficiency of water. This is one reason why the recovery of many plants from permanent wilting takes place very slowly, if at all, even after water again becomes available in the soil.

Reference has previously been made to the fact that in a rapidly transpiring plant tensions may spread to all parts of the hydrodynamic system. A tension generated in the water columns, if of sufficient magnitude, will be propagated into the root cells and the cells of the aerial organs. The continued gradual reduction in the volume of water in a plant, which occurs during permanent wilting, throws the entire residual mass of water into a state of tension. Such a situation will be sustained as long as a state of permanent wilting prevails.

Internal Redistributions of Water in Plants. Whenever a plant is replete with water, differences in the water potential from one organ or tissue to another, sink to a minimal value. But whenever the absorption of water by a plant is occurring at a rate which does not compensate for the rate of transpiration, an internal water deficit develops and marked differences in the water potential may be engendered in some parts of the plant as compared with others. Under such conditions a redistribution of water may occur from one organ of the plant to another. The greater the internal water deficit, the greater the likelihood that such internal movements of water will occur.

Internal movements of water from fruits to leaves or vice versa seem to be of especially common occurrence. The diurnal expansion and contraction of mature lemon fruits illustrate this phenomenon (Fig. 8.16). As shown in this

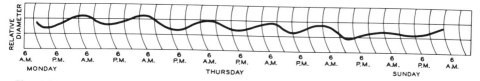

Fig. 8.16. Daily variations in the diameter of lemon fruits. Data of Bartholomew, *Am. J. Botany*, **13**, 1926:107.

figure, the lemon fruit begins to contract in volume each day at about 6:00 A.M. and continues to shrink until about 4:00 P.M. Evidently during this part of the day, which corresponds to the period of high transpiration, water is moving out of the fruits to the other organs of the tree, since direct transpirational water loss from the fruit itself is negligible. Between the hours of 4:00 P.M. and 6:00 A.M. the next morning the volume of the fruit gradually increases, indicating that water is moving back into it during this period. These results illustrate strikingly that water can move in either direction through the xylem, depending upon the relative magnitudes of the water potentials in different parts of the plant.

The behavior of cotton bolls, however, during their enlargement period of about 16 days is quite different from that just described for mature lemon fruits. The diameters of cotton bolls increase consistently, although at a somewhat greater rate during the daylight than during the night hours (Fig. 8.17). The bolls continue to enlarge even during periods when the leaves are severely wilted, indicating that even under such conditions, water continues to move into the growing fruits. After the enlargement of the cotton bolls has ceased, however, reversible changes in diameter occur similar to those which take place in the well-developed lemon fruits. Such daily variations in volume doubtless are of frequent occurrence in many other kinds of succulent fruits in which the enlargement growth has ceased.

Growing stem tips may continue to obtain water even when the older parts of the stem are losing it. The growth in the length of the stem above the first node in tomato plants, for example, has been found to occur at approximately the same rate both day and night. The stem below the first node, however, showed a measurable shrinkage in length during the daytime, and an equal elongation at night, undoubtedly corresponding to reversible changes in the turgidity of the stem cells. Obviously the meristematic stem tip cells continued to obtain water during the daylight hours while the rest of the stem was losing water, and some of the water utilized in their growth probably came from the older stem cells.

It seems probable that, as a general rule, actively meristematic regions such as growing stem and root tips, and enlarging fruits, under conditions of internal water deficiency, acquire more negative water potentials than other tissues. There are experimental indications that the water potentials of such

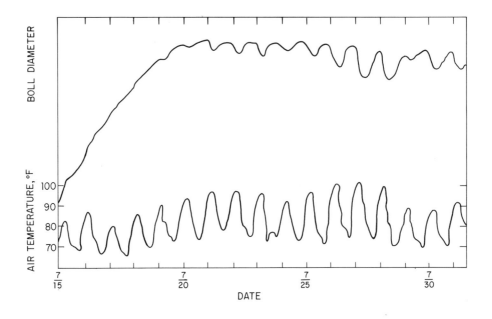

Fig. 8.17. Daily variations in the diameter of a cotton boll. During the first five days the boll was still growing. After attaining full size a marked shrinkage occurred in the boll during the warm part of each day. Data of Anderson and Kerr, *Plant Physiol.*, **18**, 1943:266.

tissues are largely of imbibitional rather than osmotic origin (Chapter 5). Even under conditions which induce the temporary wilting of the leaves, water often continues to move into meristematic regions in quantities sufficient to permit a continuation of growth. Under conditions of severe internal water deficiency, however, approaching or corresponding to a state of permanent wilting, the growth of all meristems is greatly retarded or entirely checked (Chapter 20).

Drought Resistance. Some species of plants are better able than others to survive and develop in habitats in which a dearth of water is frequent or usual. This capacity of surviving periods of drought with little or no injury is termed *drought resistance*. All perennial species of plants native to semiarid regions are more or less drought-resistant. The same is true of those species indigenous to local habitats which for one reason or another are unduly dry, even in humid climates. Drought-resistant species or varieties of plants are important in the agricultural economy of certain regions, such as the dry-farming areas of western North America. Certain varieties of crop plants are much more productive in dry regions than other varieties of the same species. Examples of this are the durum and emmer varieties of wheats.

Most species which grow in semiarid regions, such as the deserts of the southwestern United States and adjacent Mexico or in locally dry habitats, can be conveniently classified into three groups: (1) ephemerals, (2) succulents, and (3) drought-enduring species.

The ephemerals are a prominent feature of the vegetation of all semiarid regions which are characterized by definite rainy seasons. With the advent of rains, the seeds of such species germinate, and the entire life cycle of the plant is completed within a few weeks. The new crop of seeds survives the intervening dry period until the next rainy season. Such plants have been termed "drought-escaping." They are no more drought resistant than many mesic annual plants.

Succulents constitute a considerable proportion of the vegetation of most semiarid regions (Fig. 8.18) and are frequently found in locally dry habitats such as sand dunes and beaches in regions of humid climate. The most conspicuous succulents of the American semidesert regions mostly belong to the cactus family (Cactaceae). The other more important families of plants which include a number of succulent species are the Euphorbiaceae, Liliaceae, Crassulaceae, Aizoaceae, and Amaryllidaceae. The succulents are a distinctive group of plants not only in structure but in metabolism (Chapter 13) and water

Fig. 8.18. Succulents of the semideserts of southern Arizona. Left, sahuaro (*Cereus giganteus*); right, century plant (*Agave parryi*) in fruit. Photographs by B. S. Meyer.

economy as well. Species of the succulent habit of growth are able to survive dry periods because of the relatively large quantities of water which accumulate in the inner tissues of the fleshy stems or (in some species) in the fleshy leaves. A relatively thick cuticle and the fact that in many succulents the stomates are generally open only at night are important factors in permitting the conservation of water by such species. Many cacti can live for months on this stored water, even if entirely uprooted from the soil.

A few species of plants, of which alfalfa and mesquite are examples, can live in dry climates because they develop very deep root systems, often reaching to a water table.

None of the categories of plants described so far can be regarded as truly drought-resistant in the sense that their cells can endure a severe reduction in water content for a sustained period of time without serious injury. This is true only of species which are commonly designated as "drought-enduring." One of the more extreme examples of such a species among the higher plants is the creosote bush (*Larrea divaricata*), which is the dominant plant through large areas of the semiarid regions of the southwestern United States and northern Mexico. This species carries the same set of leaves through both the wet and dry seasons. During drought periods the water content of the leaves of the creosote bush is sometimes less than 50 per cent of their dry weight. The water contents of the leaves of most woody mesic species, on the other hand, generally range between 100 and 300 per cent of their dry weight.

Some species of plants, including especially many mosses, lichens, and algae, can be reduced to a virtually air-dry condition during drought periods, yet remain viable and resume their life processes very quickly when they are again provided with a supply of moisture. The seeds of many species are drought resistant in the sense that they may be reduced to a nearly air-dry condition without losing their viability.

Some structural and behavioral features of xerophytes doubtless contribute in some degree to their ability to survive in dry habitats. Many such species bear relatively small leaves; in others the leaves abscise with the advent of the dry season. Anatomically the leaves of many plants characteristic of xeric habitats are characterized by a thick cuticle and hypodermal sclerenchyma tissue.

All of the features mentioned here would tend to reduce transpirational water loss. However, the low rates of transpiration evinced by most xerophytes much of the time results principally from the low water content of the soil in which they are rooted. When the soil-water supply becomes adequate, the transpiration rates of most xerophytes are comparable with those of many mesophytes.

During a prolonged period of soil-water deficiency the store of water in a plant is gradually depleted. Although during such periods the stomates of xerophytes are closed most or all of the time, cuticular transpiration continues, resulting in a steady intensification in the internal water deficiency. One result of this gradual depletion in the water content of a plant is a progressive increase in the severity of the stress in the internal hydrodynamic system resulting in a steadily increasing negativity of the water potential in all parts of the plant. The persistence of such a condition sooner or later results in the death of plants or at least of parts of plants which possess little or no drought resistance. Many drought-enduring species, on the other hand, can tolerate this condition for months at a time without suffering irrecoverable injury.

One of the basic factors in drought resistance of plants is the capacity of the cells to endure a high degree of desiccation without suffering irreparable injury. The death of plant cells as a result of drying does not appear to result primarily from desiccation of the protoplasm as such, but from various accompanying mechanical disruptions.

As a cell dries out, the vacuole usually contracts more than the cell wall, thus leading to the distortion and tearing of the protoplasm. Such a destruction of the structural integrity of cells appears to be a main cause of death in cells which lack the property of drought resistance. A hydrolytic breakdown of proteins and other metabolites often accompanies the extreme desiccation of plant cells. Such metabolic dislocations may be injurious in themselves or may damage the submicroscopic structure of the protoplasm in non-drought-resistant species.

SUGGESTED FOR COLLATERAL READING

Esau, Katherine. *Plant Anatomy, 2nd Edition.* John Wiley & Sons, Inc., New York, 1965.

Kozlowski, T. T., Editor. *Water Deficits and Plant Growth,* 3 vols. Academic Press, New York, 1968, 1968, 1972.

Kramer, P. J. *Plant and Soil Water Relationships, A Modern Synthesis.* McGraw-Hill Book Company, New York, 1969.

Ledbetter, M. C., and K. R. Porter. *Introduction to the Fine Structure of Plant Cells.* Springer-Verlag, New York, 1970.

Maylan, B. A., and B. G. Butterfield. *Three-dimensional Structure of Wood.* Syracuse University Press, Syracuse, 1972.

Troughton, J., and Lesley A. Donaldson. *Probing Plant Structure.* McGraw-Hill Book Company, New York, 1972.

Zimmerman, M. H. *The Formation of Wood in Forest Trees.* Academic Press, Inc., New York, 1964.

9 | KINETICS AND ENERGETICS OF METABOLISM

Enzymes. A characteristic of living cells is their capacity for accomplishing at relatively rapid rates and within a physiological range of temperatures (for most plants from about 10 to about 40°C), chemical transformations which would occur only very slowly or not at all in nonliving systems. Every actively metabolizing cell is a site of hundreds of such reactions, the directions of which are controlled along complex but integrated pathways. Reaction follows reaction in regular order, each setting the stage for the next. The inherent regulation of metabolic systems is so delicate that physiological processes of great complexity move smoothly and swiftly to completion.

This orderly progress of each of the many chemical reactions in living cells is achieved through the agency of certain specific compounds called *enzymes*. All enzymes are basically proteins (Fig. 9.1) although most of them, as discussed later, have nonproteinaceous constituents bound to them or more or less closely associated with them. A simple example of an enzymatic reaction is the digestion of the disaccharide sucrose to D-glucose and D-fructose under the agency of the enzyme *sucrase:*

$$C_{12}H_{22}O_{11} + H_2O \xrightarrow{\text{sucrase}} C_6H_{12}O_6 + C_6H_{12}O_6$$
$$\text{sucrose} \qquad\qquad\qquad \text{D-glucose} \quad \text{D-fructose}$$

The compound on which an enzyme acts is called the *substrate*. Some enzymes,

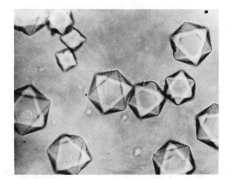

Fig. 9.1. Crystals of the enzyme urease from the jack bean (*Canavalia ensiformis*). Photograph from J. B. Sumner.

in effect, act simultaneously on two substrates. Good examples are the enzymes involved in the transamination reactions discussed in Chapter 16.

At least a thousand different enzymes have been found to occur in living organisms. Some cells undoubtedly contain hundreds of different enzymes. They are not, however, indiscriminately distributed in cells. Some are definitely known to be associated with certain organelles. For example, the many enzymes involved in aerobic respiration (Chapter 13) are largely, if not entirely, restricted to the mitochondria. Some kinds of enzymes are found in more than one kind of organelle. Those enzymes involved in electron transport are present, for example, in both the mitochondria and in the chloroplasts (Chapters 10 and 13). On the other hand some kinds of enzymes are located principally, if not entirely, in the ground cytoplasm of the cells in which they occur. This appears to be true, for example, of the enzymes involved in glycolysis (Chapter 13).

Basically all enzymes are catalysts and, in fact, are often referred to as organic catalysts. An enzyme cannot initiate a reaction which is inconsistent with the laws which govern the energy exchanges between reacting molecules, but can only accelerate the rate of a reaction. In one important way, however, enzymes can also control the direction of metabolism. If a given kind of enzyme is totally lacking in a cell, no metabolism can take place in the direction of the reaction catalyzed by that enzyme. Most cells of higher plants, for example, lack the enzyme cellulase, hence cannot digest cellulose, as can many bacteria and fungi.

Each enzyme is specific in the sense that it can only act upon a certain substrate or group of substrates. The enzyme urease is an example of one which exhibits absolute specificity by acting only on urea as a substrate. Apparently each kind of enzyme can affect only one particular type of chemical bond, or if two substrates are acted on simultaneously, the enzyme must be specific for a pair of bonds, one present in each substrate. When a number of compounds possess a bond in common, it is possible that they can be acted on by the same enzyme, For example, the enzyme emulsin can hydrolyze any β-glucoside (Chapter 14) since the chemical linkage between the sugar and nonsugar groups in all such glucosides is the same.

Enzyme specificity is also illustrated by the fact that different end products are formed from the same substrate under the influence of different enzymes. The trisaccharide raffinose (Chapter 14) is constructed from moieties of three hexose sugars: fructose, glucose, and galactose. In the presence of sucrase, raffinose is hydrolyzed into fructose and melibiose, while in the presence of emulsin it is hydrolyzed into sucrose and galactose. Sucrase acts on the bond between the fructose and glucose moieties of raffinose; emulsin on the bond between the glucose and galactose moieties.

Enzymes are metabolic products of the living organisms in which they occur, and their biosynthesis as specific kinds of proteins is considered in Chapter 16.

Classification of Enzymes.   Many enzymes have been named for the substrate on which they act, *ase* being commonly, but not invariably, the terminal ending of the name. Examples are sucrase, cellulase, and maltase. Some enzymes have been so named as to indicate the kind of reaction in which they participate, as well as the substrate. Examples are succinic acid dehydrogenase and cytochrome oxidase. The names of some enzymes have not been derived as described so far. One of these is emulsin, previously mentioned. Two others are papain and bromelin, both protein-hydrolyzing enzymes which are found in plants.

A comprehensive classification and systematic nomenclature of enzymes has been recommended by the Commission on Enzymes of the International Union of Biochemistry in 1961. This Commission has suggested that enzymes be classified into six main divisions, as follows:

(1) *Oxidoreductases.*   These are enzymes which catalyze oxidation-reduction reactions. Oxidases and dehydrogenases are examples (Chapter 13).

(2) *Transferases.*   These are enzymes which catalyze the transfer of chemical groups. Transaminases (Chapter 16) are good examples.

(3) *Hydrolases.*   These are enzymes which catalyze the hydrolysis of various compounds by water. Sucrase, cellulase, and the lipases (fat-hydrolyzing enzymes) are examples.

(4) *Lyases.*   These are enzymes that reversibly catalyze the removal of chemical groups from substrates by non-hydrolytic means. Decarboxylases (enzymes that split CO_2 out of molecules) are examples (Chapter 13).

(5) *Isomerases.*   These are enzymes which catalyze the conversion of a compound into one of its isomers. Triose phosphate isomerase, an enzyme involved in glycolysis (Chapter 13), is an example.

(6) *Ligases.*   These are enzymes which catalyze the linking together of two molecules in a reaction which also involves the breaking of a phosphate bond in ATP (see later in this chapter) or a similar compound. Examples are some of the thiokinases (Chapter 14).

The commission further recommended the adoption of a specific "trivial" or common name and also a systematic name for each enzyme. The trivial names for most enzymes are those which have already become established by usage. The systematic name, insofar as feasible, indicates the action of the enzyme. A lipase, for example, under this system would be given the systematic name of "glycerol ester hydrolase." The trivial names of enzymes are used in this book, partly because they are more likely to be familiar and the systematic nomenclature is not yet in wide usage.

Chemical Nature of Enzyme Systems.   As previously mentioned, all enzymes are basically proteins which are compounds constructed of amino acid residues arranged in exact sequences (Chapter 16). The sequential arrangements of amino acids are different from one kind of protein to another, and have been worked out for some proteins, including certain enzymes. Some enzymes consist of protein molecules as such. Examples are urease and papain. In such enzymes it must be assumed that the arrangement and the orientation of the amino acid residues endow the enzyme with its catalytic properties. Many enzymes, however, require that nonprotein constituents be more or less tightly bound to or closely associated with the protein component in order for the enzyme to exert its catalytic effect. The general term *cofactor* is often used to designate such constituents of enzymes or enzyme systems.

Among such cofactors are a number of metals, specifically iron, copper, zinc, magnesium, manganese, molybdenum, and potassium. Some of these are the structural components of the enzyme such as the copper in tyrosinase and ascorbic acid oxidase, and the iron in catalase and peroxidase. When such nonprotein groups are integral parts of the enzyme molecule, they are commonly referred to as *prosthetic* groups. Many enzymes require the presence of certain cations such as Mg^{++}, Mn^{++}, Ca^{++}, or K^+ if they are to exert their catalytic effects. Such ions, often referred to as *metal activators,* may be only loosely bound to the enzyme protein in some systems; they are more or less firmly bound in others. The most important role of the micrometabolic elements in plants (Chapter 15) is that of acting as prosthetic groups of enzymes, or as metal activators in enzyme systems.

Many enzymes cannot exert their catalytic effects unless there are also present certain complex, nonproteinaceous organic molecules. Some of these are so tightly bound to the enzyme proper as to constitute prosthetic groups. Others are more loosely associated with the enzyme and are called coenzymes. Whether serving in the capacity of a prosthetic group or as a coenzyme, these complex organic compounds all have some moiety in their structure which has been derived from the compounds ordinarily classified as vitamins (Chapter 18). This is a major reason why vitamins are essential in the nutrition of living organisms, although it should not be inferred that all vitamins function as the constituents of enzyme systems. The more important of these coenzymes will now be described in greater detail.

Pyridine nucleotides. Of the utmost importance in cellular energetics are the two coenzymes known as the pyridine nucleotides. A molecule of nicotinamide-adenine-dinucleotide (NAD+) is constructed from one molecule of nicotinamide, one molecule of adenine, two molecules of D-ribose, and two molecules of phosphoric acid. Nicotinamide-adenine-dinucleotide phosphate carries an additional phosphate group as shown in the formula given by the succeeding. Nicotinamide is one of the compounds classified under the heading of the vitamin *niacin,* the other being nicotinic acid.

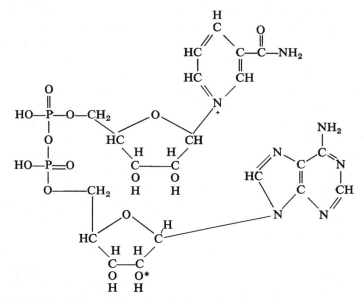

Nicotinamide-adenine-dinucleotide (NAD+). In nicotineamide-adenide-dinucleotide-phosphate (NADP+) the —OH group marked with an asterisk is replaced by a —H_2PO_4 group.

In their reduced forms these coenzmes are designated as NADH + H+ and NADPH + H+ respectively. The reason for such a notation should be clear from the following structural formulas, in which all except the nicotinamide moiety of the molecule is designated by R:

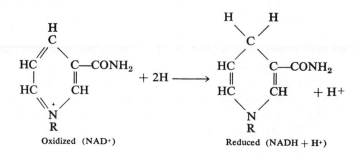

Oxidized (NAD+)   Reduced (NADH + H+)

In its nonreduced (oxidized) form NAD$^+$ is deficient in one electron, usually indicated by the plus sign on the nitrogen atom. The reduction of this compound involves the removal of the equivalent of two hydrogen atoms from the substrate and their transfer to NAD$^+$. This is equivalent to two protons (2H$^+$) and two electrons. One electron goes to satisfy the electron deficiency in the nitrogen atom of the nicotinamide moiety; the other electron plus a proton, together corresponding to a hydrogen atom, is added at the number 4 (top) carbon atom. The remaining proton (H$^+$) is released into the solution. In its reduced form therefore this compound is written as NADH+H$^+$. The reduction of NADP$^+$ occurs in exactly the same way.

The reduction forms of these two compounds are sometimes written as NADH$_2$ and NADPH$_2$, respectively, or as NADH and NADPH, respectively. For a number of years NAD$^+$ was called diphosphopyridine nucleotide (DPN$^+$) while NADP$^+$ was referred to as triphosphopyridine nucleotide (TPN$^+$), and these names are still frequently encountered in discussions of these coenzymes.

These two pyridine nucleotides are the coenzymes of dehydrogenase and reductase enzymes which catalyze many basically important oxidation-reduction reactions in living organisms. In their reduced forms they are "high-energy" compounds (see later). Numerous examples of the reactions in which these two coenzymes participate are discussed later in the book, especially in Chapters 10 and 13.

Flavin nucleotides. Two well-known coenzymes of this type are flavin mononucleotide (FMN) and flavin adenine dinucleotide (FAD), the molecules of both of which are constructed in part from riboflavin (vitamin B$_2$). These are examples of coenzymes which are firmly bound to the protein component of the enzyme.

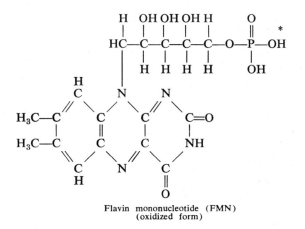

Flavin mononucleotide (FMN)
(oxidized form)

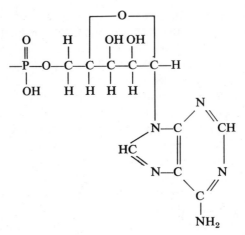

In flavin adenine dinucleotide (FAD) the above group is substituted for the H atom
marked with an asterisk in FMN

In general, flavin nucleotides, like the pyridine nucleotides, are principally involved in oxidation-reduction reactions. Examples of reactions in which they participate will be found later in this chapter and in Chapters 10 and 13. In their reduced forms these compounds are designated as $FMNH_2$ and $FADH_2$.

Thiamin. This well-known vitamin (B_1) serves as a coenzyme in the form of the compound thiamin pyrophosphate:

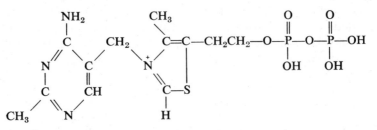

The presence of a sulfur atom in the thiamin molecule is noteworthy. Thiamin pyrophosphate serves as the coenzyme in reactions catalyzed by certain decarboxylases, oxidases, and ketolases. Examples of the action of such enzymes will be found in Chapter 13.

Pyridoxal. This compound and the closely related ones, pyridoxine and pyridoxamine, are collectively referred to as vitamin B_6:

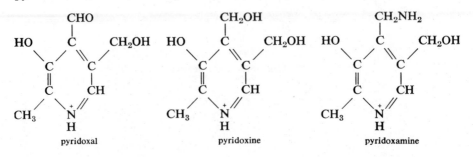

pyridoxal   pyridoxine   pyridoxamine

The best known coenzyme derivative from among these compounds is pyridoxal phosphate. This coenzyme participates in transaminations, decarboxylations, and other important reactions involved in amino acid metabolism (Chapter 16):

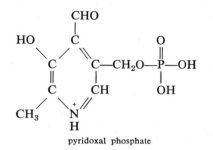

pyridoxal phosphate

Pantothenic acid. The importance of this B vitamin lies in the fact that it is one of the compounds out of which the complex sulfur-containing molecule known as Coenzyme A is constructed. The chemistry and energy relations of this compound are discussed later in this chapter, and its roles in certain crucial metabolic reactions are considered in Chapters 13 and 14.

Enzyme Induction.   It has long been known that the synthesis of certain enzymes in microorganisms often occurs only when the substrate for that enzyme is present and does not occur in the absence of the substrate. Such enzymes are said to be inducible. Only a few such enzymes have been identified in the higher plants, although it is possible that more occur than have been recognized. One of the best known examples of such an enzyme in higher plants is *nitrate reductase,* which catalyzes the conversion of nitrate to nitrite (Chapter 16). This enzyme can be extracted, at least from the plants of some species, if they are grown with nitrate in the substrate medium, but not if an ammonium compound is the substrate source of nitrogen.

Mechanism of Enzyme Action.   It is axiomatic that, in order for molecules or ions to interact, they must come into contact. As a result of their kinetic energy the molecules or ions in any system frequently collide with each other. However, not every such collision results in a chemical reaction between the two particles involved, even if the tendency for such a reaction to occur is present. If the impact is a relatively gentle one, the particles may merely bounce apart. At any given moment some of the molecules or ions in a system are moving very slowly, others are moving very rapidly, and others exhibit various intermediate velocities. If the impact between two particles is sufficiently violent, their electron shells interpenetrate to such an extent that a chemical reaction may be engendered. Atoms and electrons become re-arranged, and there is a reshuffling of the chemical bonds resulting in the formation of different kinds of atoms, ions, or molecules.

The excess energy required for interaction above that required to simply bring the molecules or ions into contact is called *activation energy.* The magnitude of this quantity varies greatly from one kind of reaction to another. When the activation energy for a given reaction is small, it takes place readily; if large, the reaction takes place slowly or not at all. The mechanism of enzymatic action, like that of other catalysts, appears to be one of lowering the activation energy of a reaction. When the activation energy is reduced, a larger proportion of the particles in a system possess sufficient energy to react (Fig. 9.2).

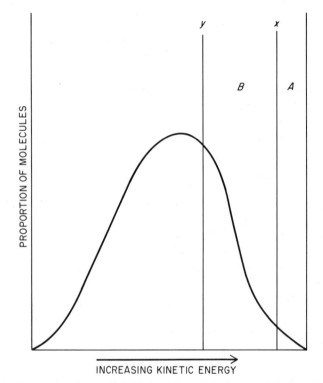

Fig. 9.2. Diagram to illustrate the effect of an enzyme on activation energy. (A) Fraction of the molecules which reacts in the absence of the enzyme. (B) Additional fraction which reacts in the presence of the enzyme, because the activation energy of the molecules has been reduced from x to y.

Factors Affecting Enzymatic Reactions.   The effects of various factors upon the reactions catalyzed by the enzymes present in higher plants have been studied almost entirely *in vitro*. The usual procedure has been to prepare a more or less purified extract of the enzyme to be studied, then to bring this extract into contact with the substrate under controlled conditions, and finally to measure the rate of the ensuing reaction. Enzymatic reactions as they occur in living cells are influenced by the structural characteristics of the organelles with which they are associated and furthermore *in vivo* factors such, for example, as *pH* are often different from those under which enzyme extracts are tested. Hence it is necessary to have some reservations about the application of results obtained *in vitro* to enzymatic behavior *in vivo*. It seems probable that the effects of various factors on an extracted enzyme may be quantitatively unlike the effects they would have on the same enzyme *in vivo,* although qualitatively the effects are probably similar.

Temperature. There are two distinctly separate and often opposing effects of temperature upon enzymatic reactions. In general the rate of such a reaction increases with a rise in temperature, but temperature also exerts an inactivating or destructive effect on enzymes, beginning in most plant tissues at about 40°C. This latter effect is an irreversible thermal denaturation phenomenon which is characteristic of proteins in general and which occurs readily in a moist or liquid medium. Enzymes in dry seeds or spores can often endure temperatures of 100°C or even higher for a considerable period without suffering any deleterious effects. The optimum temperature for an enzymatic reaction is that at which any further accelerating effect of the temperature increase is just offset by the destructive action of the temperature which has been attained upon the enzyme system (Fig. 9.3).

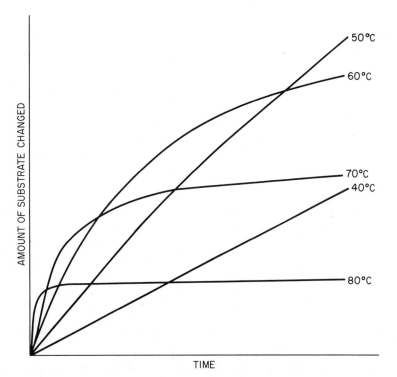

Fig. 9.3. Generalized curves illustrating the influence of temperature upon the rate of an enzymatic reaction over a period of time. Redrawn from Dixon and Webb, Academic Press, Inc., New York, *Enzymes,* 1958:151.

Hydrogen Ion Concentration. The activity of an enzyme is greatly influenced by the hydrogen ion concentration of the medium in which the reaction occurs. This is one of the most important known effects of hydrogen ion concentration in the realm of biological phenomena. In general, the activity of an enzyme appears to be at its maximum at a certain optimum hydrogen ion concentration and to decrease, usually rapidly, on each side of this value (Fig. 9.4). The optimum and limiting *p*H values for activity differ from one enzyme to another. Optimum values may range from as low as *p*H 1.5 for some enzymes to as high as *p*H 10 for others, although for most enzymes the optimal values are in the range of *p*H 6 to *p*H 8. The optimum value for a given enzyme may also vary, however, with the source of the enzyme, the substrate, the nature of the other substances in the medium, and temperature. Since enzymes are proteins, the effects of the *p*H are doubtless largely on the ionic character of the constituent amino and carboxylic acid groups (Chapter 16). In addition, very high or very low *p*H values may result in partial to complete inactivation of the enzyme protein.

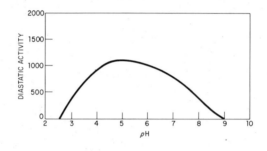

Fig. 9.4. Relation between *p*H and the activity of diastase from malt. Data of Sherman *et al., J. Am. Chem. Soc.,* 41, 1919:234.

Concentration of the Enzyme. As long as other factors such as the substrate concentration, the *p*H, and the temperature do not become limiting, the rate of an enzymatic reaction usually increases linearly with the enzyme concentration (Fig. 9.5). A few apparent exceptions to this general rule are known, most of which result from the phenomenon of feedback inhibition, further mentioned later.

Concentration of the Substrate. At a given enzyme concentration an increase in the concentration of the substrate results at first in a rapid rise of the reaction rate until a point is reached at which further increases in this concentration result in no further increase in the velocity of the reaction (Fig. 9.6). The rapid rise portion of this curve represents the zone of concentrations in which the number of substrate molecules is not adequate to satisfy all the reactive sites on the enzyme. When this concentration is reached the curve flattens off to a plateau, and there is essentially no further increase

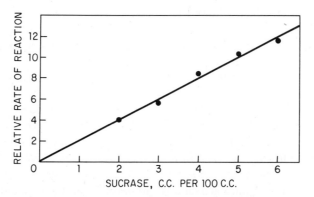

Fig. 9.5. Relation between the concentration of sucrase and its hydrolytic activity. Data of Nelson and Hitchcock, *J. Am. Chem. Soc.*, **43**, 1921:2640.

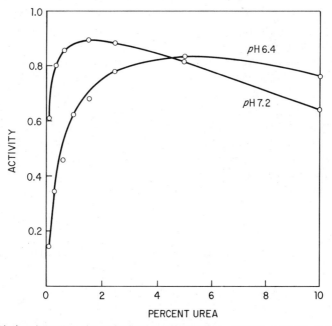

Fig. 9.6. Relation between the substrate concentration and the activity of the enzyme urease. Data of Howell and Summer, *J. Biol. Chem.*, **104**, 1934:620.

in activity with an increase of substrate concentration. In other words, above a certain substrate concentration it is the enzyme concentration which becomes the limiting factor in the reaction.

Concentration of the End Products. Enzymatic reactions, like all other chemical reactions, are subject to the principle of mass action. Hence, when an enzymatic reaction is allowed to occur in a test tube, as the end products of a reaction accumulate, the apparent rate of the reaction decreases. In living cells, however, most enzymatic reactions are followed by others, resulting in further conversions of the products of the first reaction. Hence the products of the first reaction do not accumulate and cannot retard their own synthesis by a mass action effect.

In some sequences of enzymatic reactions the products of the reaction inhibit their own syntheses by combining with one of the enzymes which operates earlier in the metabolic pathway leading to the synthesis of that product. This phenomenon is referred to as *feedback inhibition*.

Cases of *feedback activation* are also known to exist in plant metabolism. Feedback inhibitions and activations probably operate as important control mechanisms in cells.

Hydration. The effect of increased hydration upon the enzymatic activity of plant tissues is most easily demonstrated during the germination of seeds. The enzymatic activity, measured in terms of any specific enzyme known to be present, is low and often nonexistent in dry but viable seeds. As imbibition of water proceeds during germination, the activity of the enzymes increases more or less progressively, an effect which can be ascribed, at least indirectly, to the increase in the hydration of the tissues of the seed.

Inhibitors. These are compounds which retard or inhibit the activity of an enzyme. Two main types of enzyme inhibitors can be distinguished: *competitive* and *noncompetitive*. A competitive inhibitor is one which is structurally so similar to the substrate molecule that it preempts some of the active sites on the enzyme, thus temporarily blocking the further progress of the reaction at these sites. For example, in the presence of FAD, succinic acid is converted into fumaric acid under the influence of the enzyme succinic dehydrogenase:

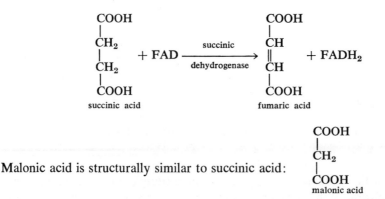

Malonic acid is structurally similar to succinic acid:

If malonic acid is introduced into the system it temporarily blocks at any one moment in time some of the reactive sites on the enzyme molecules which would otherwise be effective in transforming succinic acid to fumaric acid and the rate of the reaction is slowed. Competitive inhibitory effects, of which this is an example, can be increasingly nullified by increasing the concentration of the substrate in the system. The frequency of enzyme unions with the normal substrate relative to unions with the competitive inhibitor is a chance phenomenon which is determined by the relative concentrations of the substrate and competitive inhibitor.

In noncompetitive inhibition the inhibitor is bound to sites on the enzyme in such a way that it cannot be displaced. Since enzyme molecules which have complexed with noncompetitive inhibitor molecules remain inactive, the addition of more substrate does not increase the rate of the reaction. Under these conditions the depleted population of enzyme molecules is the limiting factor and not the substrate concentration. Noncompetitive enzyme inhibitors include iodacetates, fluorides, cyanides, azides, and carbon monoxide.

Free Energy. This intrinsic property of the component of a system has already been referred to in developing the concept of the water potential in Chapter 4. The free energy (F) of any component of a system may be expressed by the following equation:

$$F = E + PV - TS, \text{ in which}$$

$E =$ the internal energy within the molecules which includes translational, rotational, vibrational, electronic, and nuclear energies.
$PV =$ the pressure-volume product (one liter-atmosphere equals 24.2 calories).
$T =$ the absolute temperature.
$S =$ the entropy, a quantity which expresses the degree of randomness of the molecules, or of "incapacity to do work."

When dealing with water potential (ψ_w) the E component drops out of the equation, because this quantity (ψ_w) does not involve energy intrinsic in the molecules. When considering metabolic processes, however, it is the E component which is of significance. The quantity of energy resident in each molecule differs from one kind to another. Because the values for E and S are unknown, it is not possible to calculate the values of F in absolute values. However it is possible to calculate the *change* in free energy which occurs when one compound is converted into or reacts with another. For example, the complete oxidation of glucose to carbon dioxide and water has a ΔF (change in free energy) value of about -673,000 cal per mole of glucose. A negative ΔF value means that the reaction takes place with a decrease in free energy.

Reactions which occur spontaneously do so with a decrease in free energy and are called *exergonic* reactions. Reactions which occur with an increase in free energy are termed *endergonic* reactions. Such reactions take place only if energy is supplied to the system in such a way as to "drive" the reaction. Examples are: (1) photochemical reactions, of which photosynthesis is a prime example, energy being supplied by photons, and (2) endergonic metabolic reactions which are "coupled" with exergonic metabolic reactions as described later.

High Energy Compounds. In many metabolic reactions, energy made available in an exergonic reaction is used to "drive" an endergonic reaction. This can only occur, however, if there is a common reactant:

$$A \xrightarrow[\text{exergonic reaction}]{} B$$

$$B \xrightarrow[\text{endergonic reaction}]{} C$$

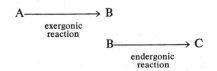

Characteristically the compounds coupling exergonic reactions (B in the above diagram) are "high-energy" compounds. When such compounds undergo hydrolytic or oxidative reactions, they exhibit a large decrease in free energy. When participating in a coupled sequence of reactions at least a part of the intrinsic energy of such compounds is transferred to the compound or compounds which are the end products of the endergonic reaction.

The best-known of these high energy compounds is the nucleotide adenosine-5′-triphosphate (ATP) which is of widespread and probably universal occurrence in living organisms. A molecule of adenosine is constructed from one molecule of the purine base, adenine, and one molecule of D-ribose. The addition of a phosphate group results in the compound adenosine-5′-monophosphate (AMP). Adenosine-5′-diphosphate (ADP) is formed by the addition of a second phosphate group to the one in AMP, adenosine-5′-triphosphate (ATP) by the addition of a third phosphate group to the two in ADP. The chemistry of these compounds is discussed further in Chapter 16. In the following representation of the molecular structure of these three compounds, R represents the adenosine moiety:

ATP ADP AMP

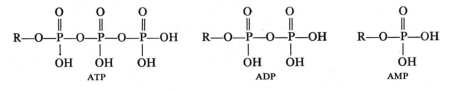

The hydrolysis of ATP to ADP occurs as follows:

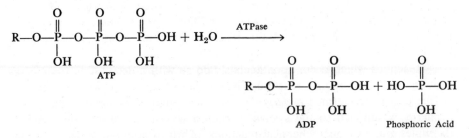

The change of free energy involved in this reaction is approximately −8000 cal/mole at *p*H 7.0. The further hydrolysis of adenosine diphosphate (ADP) to adenosine monophosphate (AMP) and phosphoric acid also results in a large change in free energy (about -6500 cal/mole at pH 7). On the other hand the hydrolysis of the AMP to adenosine and phosphoric acid, which involves the breaking of an ester instead of a pyrophosphate bond, results in a free energy change of only about -2200 cal/mole at pH 7. Examples of the roles of ATP in metabolic reactions will be found in later chapters.

While ATP and ADP appear to be the most abundant and most important of the nucleotide type of high-energy compounds, other such compounds also play a role in living organisms. Several of these, which participate significantly in plant metabolism, are guanosine triphosphate (Chapters 13, 14, and 16), uridine triphosphate (Chapters 14 and 16), and inosine triphosphate (Chapter 10).

Thioesters formed between Coenzyme A and certain organic acids are another group of high-energy compounds which participate in metabolic processes. Coenzyme A, which is constructed in part from the B vitamin, pantothenic acid, contains a highly reactive sulfhydryl group:

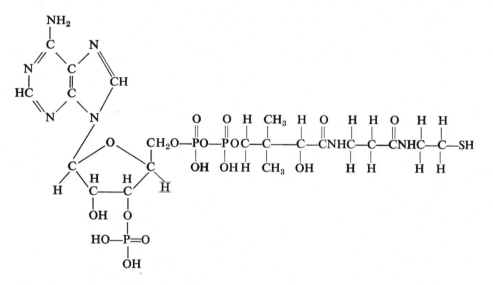

For convenience this compound is usually written CoA-SH. The best known thioester in this group is acetyl-S-CoA which, upon hydrolysis, shows a free energy change of about -8000 cal/mole at pH 7:

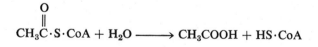

Acetyl-S-CoA plays important roles in the Krebs cycle phase of respiration (Chapter 13) and in the oxidation of fats (Chapter 14). Two other important compounds of this type are succinyl-S-CoA (Chapter 13) and malonyl-S-CoA (Chapter 14).

In their reduced forms, the pyridine nucleotides (NADH + H+ and NADPH + H+) are high-energy compounds from the standpoint of plant metabolism. They play extremely important roles in the energy transfers within the cell. Calculations show that the oxidation of NADH + H+ to NAD+ would result in a free energy change of about −52,000 cal/mole. NAD+ participates in many coupled reactions of which the following is an example (see also Chapter 13):

$$\underset{\text{3-phosphate}}{\text{glyceraldehyde}} + \text{H}_3\text{PO}_4 + \text{NAD}^+ \xrightarrow{\underset{\text{dehydrogenase}}{\text{triose phosphate}}} \underset{\text{glyceric acid}}{\text{1,3 diphospho-}} + \text{NADH} + \text{H}^+$$

$$\text{NADH} + \text{H}^+ + \text{acetaldehyde} \xrightarrow{\underset{\text{dehydrogenase}}{\text{alcohol}}} \text{ethanol} + \text{NAD}^+$$

In effect some of the energy intrinsic in the NADH + H+ which is obtained by the oxidation of the substrate of the first reaction (glyceraldehyde-3-phosphate) is transferred by reduction of the acetaldehyde and becomes incorporated in the end product (ethanol) of the second reaction.

In a similar manner NADPH + H+ serves as a high energy compound in many types of reactions. This is also true of FMNH₂ and FADH₂. The oxidized and reduced forms of the flavin moiety of these compounds are as follows:

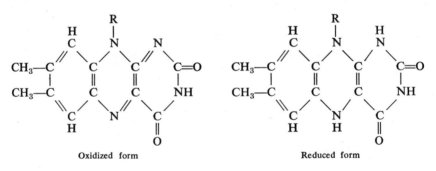

Oxidized form   Reduced form

The cytochromes are another group of compounds which, in their reduced state, operate in effect as high-energy compounds. A number of these compounds, similar in chemical structure, are known to occur in both plant and animal cells. They are large, complex molecules, consisting of an iron porphyrin prosthetic group, similar to the magnesium porphyrin group in the chlorophylls (Chapter 10), conjugated with a protein moiety. The cytochromes occur in both the chloroplasts and mitochondria and play important roles in photosynthesis (Chapter 10) and respiration (Chapter 13).

Oxidation-Reduction Reactions.   Many of the reactions which are component steps in metabolic processes are of the oxidation-reduction type. An understanding of the nature and implications of such reactions is a prerequisite to an understanding of the significance of many of the individual biochemical steps which make up metabolic pathways.

If a given molecule gains an electron in a reaction in which it participates, this molecule is said to become *reduced*. As chemical equations are usually written it is not always obvious whether gains (or losses) of the electrons occur. In a large proportion of the reductions involved in metabolic pathways, a proton is gained along with an electron by the molecule being reduced. This is equivalent to saying that a hydrogen atom (1 proton + 1 electron) is gained by the molecule. Along with the gain of an electron or a hydrogen atom, a molecule being reduced is also gaining in chemical energy which is one of the components of the quantity E in the free energy equation previously discussed.

Since energy cannot be created or destroyed, it follows that some other molecules in the reaction must be losing chemical energy and electrons at the same time the molecules gaining in chemical energy are being reduced. The molecules to which this loss of electrons and loss of energy occur are said to become *oxidized*. Loss of a hydrogen atom or atoms is characteristic of many oxidation reactions since a proton and an electron often move simultaneously from one molecule to another.

Some oxidation reactions are characterized by an increase in the number of oxygen to carbon bonds. The relative degrees of oxidation of a series of carbon compounds can be presented as follows:

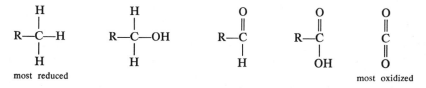

most reduced   most oxidized

In this series, from left to right, the number of bonds between carbon and oxygen are 0, 1, 2, 3, and 4 respectively. The greater the number of such bonds, the more oxidized the molecule and the less chemical energy the molecule contains.

Many oxidation-reduction reactions are of the coupled type, previously described. A number of such reactions are component steps in the Krebs cycle (Chapter 13). One of them, the conversion of malic acid to oxalacetic acid will be used to illustrate the nature of this kind of reaction:

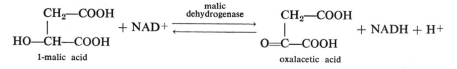

Malic acid loses two hydrogen atoms in this reaction, indicating that it is being oxidized. The fact that oxalacetic acid contains one more carbon to

oxygen bond than malic acid is further evidence that oxidation has occurred. Since malic acid is being oxidized it follows that it loses some chemical energy in the reaction. One molecule of the coenzyme NAD^+ is reduced by gaining two hydrogen atoms (protons plus electrons). The resultant $NADH + H^+$ has gained energy at the expense of the malic acid.

Some compounds, of which the previously mentioned cytochromes are examples, are oxidized or reduced by the loss or gain of free electrons, respectively. The iron component of the cytochromes can alternately gain and lose electrons, which results in alternate shifts between the Fe^{++} (ferrous) and the Fe^{+++} (ferric) form. The equation for the reduction of iron from the ferric to the ferrous form is as follows:

$$Fe^{+++} + e^- \leftrightarrows Fe^{++}$$

The significance of these reversible shifts in the oxidation state of the iron portion of the cytochromes is considered in Chapters 10 and 13.

SUGGESTED FOR COLLATERAL READING

Bernhard, S. *The Structure and Function of Enzymes.* W. A. Benjamin, Inc., New York, 1968.

Conn, E. E., and P. K. Stumpf. *Outlines of Biochemistry, 2nd Edition.* John Wiley & Sons, Inc., New York, 1966.

Davies, D. D. *Intermediary Metabolism.* Cambridge University Press, New York, 1961.

Dixon, M., and E. C. Webb. *Enzymes.* Academic Press, Inc., New York. 1964.

Giese, A. C. *Cell Physiology, 3rd Edition.* W. B. Saunders Company, Philadelphia, 1968.

Neilands, J. B., and P. K. Stumpf. *Outlines of Enzyme Chemistry.* John Wiley & Sons, Inc., New York, 1958.

———. *Enzyme Nomenclature. Recommendations of the International Union of Biochemistry on the Nomenclature and Classification of Enzymes.* Elsevier Publishing Company, New York, 1965.

Street, H. E. *Plant Metabolism.* Pergamon Press, Elmsford, New York, 1963.

10 | PHOTOSYNTHESIS

The biological world, with negligible exceptions, runs at the expense of the material and energy capital accumulated as a result of photosynthesis. From the products of photosynthesis and from a few simple inorganic compounds obtained from the environment, there are built up in living organisms all of the multifarious and complex kinds of molecules which constitute the cellular structure of plants and animals or are otherwise essential to their existence.

Not only do most of the materials which enter into the cellular structure of plant and animal bodies derive from photosynthetic capital, but the energy expended in the operation of living organisms also derives from this source. Plants and animals are constituted so that only energy released as a result of the oxidation of foods can be used in metabolic processes, growth, locomotion, or in other ways. The chemical energy of foods, releasable upon oxidation, all represents the converted energy of sunlight which was originally entrapped within the molecules of organic compounds during photosynthesis.

A goodly share of what economists term wealth originates directly or indirectly as a consequence of photosynthesis. This is true not only of all plant and animal products but also of our heritage of coal, oil, and gas from past geological ages. These latter products all derive from the remains of living organisms and hence also represent photosynthetic capital. The energy released from them upon combustion represents the sunlight of past geological ages which was captured and converted into chemical energy by the photosynthesis of the plants which flourished during geological epochs long antedating the advent of man.

Man's stake in photosynthesis is thus even greater than that of any other living organism. Not only is he, like the plants and all other animals, dependent for his very existence upon this process, but he is also indebted to it for many of the goods and most of the energy which contribute to the maintenance of his standard of living above a mere subsistence level.

Radiant Energy.   An elementary knowledge of the physical properties of light and other kinds of radiant energy is essential for a proper understanding of photosynthesis, the synthesis and properties of chlorophyll, and many other photobiological processes. Radiant energy, as judged from some of its properties, appears to be propagated across space as undulatory waves. Ordinary sunlight or "white light" from any artificial source seems homogenous to the human eye, but after it has passed through a prism, it appears as a spectrum of colors as was first demonstrated by Newton in 1667. The order of the more prominent colors in a spectrum of sunlight is red, orange, yellow, green, blue-green, blue, and violet. Each of these colors corresponds to a different range of *wavelengths* of light (Fig. 10.1). A wavelength is the distance between two successive crests of a wave. The wavelengths which induce the sensation of light range from about 390 nm in the violet to about 760 nm in the far-red.

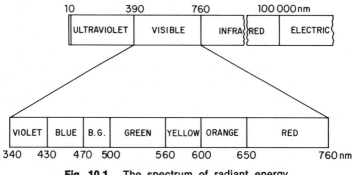

Fig. 10.1. The spectrum of radiant energy.

Visible light, however, constitutes only a very small part of the spectrum of radiant energy. Beyond the visible red lies the long zone of infrared radiations which range up to a wavelength of about 100,000 nm. Electric waves are still longer and range up to a kilometer or more in length. The waves used for radio and television transmission are in this portion of the electromagnetic spectrum.

Just below the region of visible light in the radiant energy scale lies the ultraviolet zone, which ranges down to wavelengths as short as 10 nm. Even shorter are the x-rays (0.01–10 nm) much used for their therapeutic effects in medicine and in x-ray photography. Below them on the scale lie the gamma rays (0.0001–0.01 nm) which are emitted by radium, also used in medical therapy. Shortest of all are the cosmic rays, which are less than 0.0001 nm in wavelength.

The wavelengths which reach the earth's surface from the sun range from about 300 nm in the ultraviolet to about 2600 nm in the infrared. About half of the solar energy reaching the earth's surface is in the invisible infrared region. In their natural habitats plants are also subjected to bombardment by the extremely long electric waves, and the extremely short cosmic waves, but there is no experimental evidence that either of these kinds of radiant energy has any effect upon plants.

In the preceding discussion we have referred to radiant energy as a wave phenomenon with an unjustifiable air of finality. Some radiant energy phenomena, such as the behavior of light in optical systems, can be satisfactorily explained only in terms of the postulate that light travels as waves. Other effects, however, appear completely unintelligible in terms of this hypothesis. The most important of these are photochemical reactions such as the effect of light upon sensitized photographic paper and its role in the process of photosynthesis. At the present time such phenomena can only be explained satisfactorily by the assumption that light is particulate in nature. According to this concept a beam of light is pictured as a stream of tiny particles. Each of these particles is called a *photon*. When such photons impinge against a suitable substance, their energy may be transferred to the electrons which they strike, thus inducing photochemical reactions.

The energy manifestation of a photon is called a *quantum*. The energy value of a quantum *(q)* is related to its wavelength according to Planck's law, which is expressed by the equation:

$$q = h\nu = \frac{hc}{\lambda}$$

In this equation h represents Planck's constant (6.624×10^{-27} erg sec), ν is the frequency of radiant energy in waves per second, c is the velocity of light (3×10^{10} cm per sec in air or vacuum), and λ is the wavelength in centimeters. Since h and c are constants it is evident that the energy value of quanta varies inversely with their wavelengths. A quantum of ultraviolet radiation with a wavelength of 100 nm, for example, has four times the energy value of a quantum of violet light with a wavelength of 400 nm and eight times that of a quantum of infrared radiation with a wavelength of 800 nm.

A commonly used unit in discussions of the quantum aspects of radiant energy is the *einstein* which is defined as one mole (6.02×10^{23}) of photons (or quanta). According to the principle of photochemical equivalence, only one molecule or atom can be activated, in absorption phenomena, by one quantum, regardless of the energy value of the quantum. Hence one einstein can activate one mole of a compound. An einstein is a quantity unit only and not an energy unit, since the energy value of quanta, and hence of the corresponding einsteins, varies with their wavelengths.

Radiant energy varies in several different ways, the most important of which are: (1) *irradiance* ("intensity"), (2) *quality,* and (3) *duration.*

Irradiance is the radiant energy receipt per unit area per unit of time. It is commonly expressed as ergs per cm² per sec, microwatts per cm², or as g-cal per cm² per minute. One g-cal per cm² per min is called a *langley* and is equivalent to about 4.2×10^7 ergs per cm² per minute or about 7×10^5 ergs per cm² per second. On a clear summer's day in the mid-temperate zones the impinging solar radiant energy has an irradiance of between 1.2 and 1.5 langleys at noon. A representative irradiance of 1.4 langleys is equivalent to about 5.9×10^7 ergs per cm² per minute. The irradiance of light and its "brightness" are not the same, the latter being an index of the illuminating capacity of light as perceived by the human eye. Neither are they expressed in the same units. Nevertheless there is usually a fairly close correlation between these two quantities. Illumination values corresponding to an irradiance of 1.2 to 1.5 langleys are about 8,000 to 10,000 foot-candles. On a cloudy winter's day in the same latitudes, irradiance and illumination values are commonly only 10 per cent or even less of those at corresponding hours on a clear summer's day (Fig. 10.2).

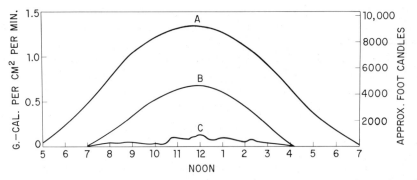

Fig. 10.2. Daily variations in light intensity at horizontal incidence. (A) On a clear day in early July. (B) On a clear day in early January. (C) On an early January day with heavy clouds. Curves based on measurements made at Columbus, Ohio (40° N. latitude).

In interpreting photobiological phenomena in plants, irradiance should preferably be expressed in energy units, although foot-candles have often been used because such illumination values can be measured more conveniently than energy values.

Quality refers to the wavelengths composition of light. The quality of light from a tungsten filament bulb, for example, is very different from that of sunlight. The latter is proportionately richer in blue light and poorer in infrared than the former. Similarly the quality of light from a fluorescent tube, is different from that from a tungsten bulb, the latter being proportionately richer in the red region of the spectrum than the former.

Duration as applied to the light-relations of plants, refers to the number of hours in a 24-hour day during which a plant is exposed to illumination.

The Overall Process of Photosynthesis.   The basic facts regarding photo-synthesis at the organismal level were well recognized by the early years of the twentieth century. It was known that carbon dioxide and water were used in the process, that the necessary energy came from light, and that the process occurred only in chlorophyll-containing cells. It was further recognized that oxygen was released in the process and that the first readily detectable product was a carbohydrate.

A *summary* equation representing the process of photosynthesis can be written as follows:

$$6CO_2 + 6H_2O \xrightarrow[\substack{\text{chlorophyll-containing} \\ \text{cells}}]{\substack{\text{radiant energy} \\ \text{673 kg-cal}}} C_6H_{12}O_6 + 6O_2$$

This equation does not indicate the mechanism of the process but is merely a statement of the net physico-chemical exchanges which occur in the shorthand of chemical symbols. As usually written, a hexose sugar is commonly designated as the first carbohydrate formed in photosynthesis, but in actuality it is neither the first nor the only product of the process. The value 673 kg-cal is, strictly speaking, valid only for glucose; the energy requirement of the reaction is approximately the same, however, per CH_2O group, for the synthesis of any hexose, or for the synthesis of any disaccharide or polysaccharide derived from hexoses.

Leaf Anatomy in Relation to Photosynthesis.   In the vascular plants photo-synthesis occurs chiefly in the leaves, which in the majority of species are thin, expanded organs possessing a large external surface in proportion to their volume. This type of structure permits the display of a large number of chloro-plast-containing cells to light in proportion to the volume of the leaf. The labyrinth of intercellular air spaces in the interior of the leaf is so extensive that practically every green cell is in contact with the internal atmosphere of the leaf. This is true of leaves of the needle type as well as leaves with expanded blades. As a result of this loose cellular structure, the internal leaf surface (surface of the leaf cells in contact with the intercellular spaces) is much greater than the surface of the epidermal cells exposed to the outside atmos-phere. In a lilac leaf, for example, the internal surface is about thirteen times as great as its external surface. Most of the carbon dioxide absorbed directly by the mesophyll cells passes into them from the intercellular spaces after first having diffused into the leaf through the stomates, although some may be ab-sorbed directly through the epidermal cells (Chapter 11). The presence of intercellular spaces in a leaf therefore provides a much more extensive carbon dioxide absorbing surface than if all of this gas were absorbed directly through the external leaf surfaces. Since the walls of all of the cells within a leaf are normally more or less saturated with water, the vapor pressure of the internal air spaces is usually higher than that of the outside atmosphere. This makes it possible for the leaf cells to absorb atmospheric carbon dioxide without being exposed to the usually relatively dry external atmosphere. Whenever the stomates are open, the internal atmosphere of the intercellular spaces is continuous with that of the outside atmosphere, and carbon dioxide can diffuse with little impediment from the outside air into the intercellular spaces. After passing into solution in the water saturating the mesophyll cell walls, part of the carbon dioxide reacts with water, forming carbonic acid (H_2CO_3). Some of the carbon dioxide diffuses to the chloroplasts in this form and some as dissolved carbon dioxide.

The Chloroplast Pigments.   Green is the predominant color of the plant kingdom. With few exceptions, leaves are green, as are also many other plant organs such as herbaceous and young woody stems, young fruits, and the sepals of flowers. The green coloring matter of plants is often given the blanket term of chlorophyll, although actually a number of different kinds of chlorophyll occur in plants.

Less evident is the fact that the leaves and many other organs of plants also contain yellow pigments. These are seldom apparent except in leaves in which the chlorophyll fails to develop or in which it is destroyed as a result of senescence or other physiological changes. Maize plants that have been grown from grains in the dark, for example, do not synthesize chlorophyll, but are usually yellow in color because of the presence of yellow pigments. In the autumn, the disintegration of the chlorophyll in the leaves of many woody species unmasks the yellow pigments also present in the leaves. The yellow chloroplast pigments or leaves belong to the group of compounds called the *carotenoids*.

In some kinds of algae, reddish, or bluish pigments called the *phycobilins* also occur in the chloroplasts or, in species in which organized chloroplasts are not present, in structural material corresponding to the chloroplasts.

The Chlorophylls.   The chlorophylls constitute a family of closely related chemical compounds. The structural formulas for two of the chlorophylls are shown in Fig. 10.3. Chlorophyll is a pollywog-shaped molecule with a complex porphyrin ring ("head") to which is attached a long phytyl ($C_{20}H_{39}$) chain ("tail"). The porphyrin ring is composed principally of four pyrrole nuclei linked together bearing side chains. The metal constituent of the molecule, magnesium, is incorporated in this ring. The hydrocarbon phytyl grouping upon hydrolysis gives rise to the alcohol phytol ($C_{20}H_{39}OH$), which is an isoprenoid derivative (Chapter 14).

Chlorophyll *a* is most nearly of universal occurrence, being present, as far as is known, in all photosynthetic organisms except the photosynthetic bacteria. Chlorophyll *b* is found in all higher plants and in the green algae, but is not present in algae of most other phyla. Chlorophyll *c* is found in the brown algae and diatoms, which do not contain chlorophyll *b*. Similarly the red algae con-

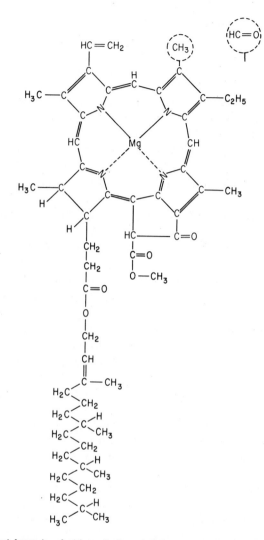

Fig. 10.3. Structural formula of chlorophyll *a*. Another arrangement of the double bonds in this formula is possible. The formula for chlorophyll *b* is the same except that an HC=O group occurs in place of the CH₃ group enclosed in the dotted circle.

tain chlorophyll *d* but no chlorophyll *b*. In the purple photosynthetic bacteria another kind of chlorophyll called *bacteriochlorophyll* is present, whereas the green photosynthetic bacteria contain still another similar pigment called *Chlorobium* chlorophyll. All of these chlorophylls are very similar in chemical composition and all of them are magnesium-containing compounds. *Protochlorophyll*, discussed later in this chapter, also belongs in this group of compounds.

The chlorophylls are chemically very closely related to several other kinds of physiologically fundamental compounds. The most important of these are the heme of blood hemoglobin, the cytochromes in both plant and animal cells, and the enzymes catalase and peroxidase. All of these related compounds contain iron rather than magnesium as the metallic constitutent of the porphyrin ring.

All of the chlorophylls exhibit *fluorescence*. This is the property manifested by certain substances which, when illuminated, radiate light of other wavelengths, usually longer than those which are absorbed. Chlorophyll *a* in ethanol solution exhibits a deep blood-red fluorescence, best seen by viewing the solution in reflected light. Similar solutions of chlorophyll *b* exhibit a brownish-red fluorescence. Chlorophylls in living cells also exhibit fluorescence.

Chlorophyll Synthesis.   Chlorophylls are products of the synthetic activities of plants. A number of conditions are known to be necessary for or at least to influence greatly the synthesis of chlorophyll in plants. The absence of any one of these factors will inhibit or drastically check chlorophyll synthesis, resulting in the condition known as *chlorosis*. This term is most frequently applied when the failure of chlorophylls to develop is the result of a deficiency of one of the essential mineral elements. Different types of chlorosis may develop in the leaves of any species depending upon the factor limiting the chlorophyll formation (Chapter 15). The following discussion of the factors influencing chlorophyll synthesis refers primarily to the formation of chlorophylls *a* and *b* in the higher plants.

Genetic factors.   That certain genetic factors are necessary for the development of chlorophylls is shown by the behavior of some varieties of maize in which a certain proportion of the seedlings produced cannot synthesize chlorophylls, even if all the environmental conditions are favorable for their formation. As soon as the food in the grain is exhausted such "albino" seedlings die. This trait is inherited in such strains of maize as a Mendelian recessive and hence is apparent only in plants homozygous for this factor. More or less similar genetic effects on chlorophyll synthesis have been demonstrated in a number of other species.

Light.   Light is usually necessary for the development of chlorophyll in the angiosperms. In the algae, mosses, ferns, and conifers, however, chlorophyll synthesis can occur in the dark as well as in the light, although the quantity produced is often less in the absence of light than in its presence. In a few angiosperms such as the seedlings of the water lotus *(Nelumbo)* and in the cotyledons of citrus embryos, chlorophylls can also develop in the absence of light.

Relatively low intensities of light are generally effective in inducing chlorophyll synthesis. All wavelengths of the visible spectrum will, if their energy value is adequate, cause chlorophyll development in etiolated seedings (chlorophyll-free seedlings which have been grown in the dark) except those wavelengths longer than 680 nm.

Like other complex compounds synthesized in plants, chlorophyll represents the terminal product of a long chain of reactions. The dark-grown seedlings of higher plants are usually yellow in color, but they actually contain traces of a green pigment, called *protochlorophyll*. Chemically this compound is closely related to the other chlorophylls, differing from chlorophyll *a* only in having two less hydrogen atoms in the molecule. There is good evidence that protochlorophyll is the immediate precursor of chlorophyll *a*. The reduction of protochlorophyll to chlorophyll *a* appears to be the last step in the synthesis of this latter compound. This reaction occurs only in the light, and the protochlorophyll is the light-absorbing agent for its own transformation to chlorophyll *a*.

Intense light brings about the disintegration of the chlorophylls in leaves, although at a less rapid rate than in chlorophyll solutions (*cf.* the discussion of photo-oxidation in Chapter 11). In leaves exposed to intense light, therefore, the synthesis and decomposition of chlorophylls are probably going on simultaneously. In accord with this concept it has been found in a number of species of plants that the chlorophyll content per unit leaf weight or per unit leaf area increased with decreasing light intensity until a relatively low intensity was reached. A further decrease in light intensity below this value caused a decrease in the chlorophyll content.

Oxygen.   In the absence of oxygen, etiolated seedlings fail to develop chlorophylls even when illuminated under conditions otherwise favorable for chlorophyll formation. This fact implies that aerobic respiration is required for the occurrence of at least some of the reactions whereby chlorophyll is synthesized.

Carbohydrates.   Etiolated leaves which have been depleted of soluble carbohydrates fail to turn green even when all of the other conditions to which they are exposed favor chlorophyll synthesis. When such leaves are floated on a sugar solution, however, sugar is absorbed and the chlorophyll formation occurs rapidly. A supply of carbohydrate foods is therefore essential for the formation of chlorophyll.

Nitrogen.   Since nitrogen is a part of chlorophyll molecules, it is not surprising to find that a deficiency of this element in the plant retards chlorophyll formation. The failure of chlorophylls to develop is one of the commonly recognized symptoms of nitrogen deficiency in plants (Chapter 15).

Magnesium.   Like nitrogen, this element is also a part of chlorophyll molecules. The deficiency of magnesium in plants results in the development of a characteristic mottled chlorosis of the older leaves (Chapter 15).

Iron.   In the absence of iron in an available form, green plants are unable to synthesize chlorophylls and the leaves soon become blanched or yellow in color (Chapter 15). While not a constituent of chlorophyll molecules, iron is essential for their synthesis.

Other Mineral Elements.   In the absence of manganese, copper, or zinc, more or less characteristic chloroses develop in plants (Chapter 15). These substances apparently play at least indirect roles in the synthesis of chlorophyll.

Temperature.   In general, chlorophyll synthesis can occur over a wide range of temperatures. In etiolated wheat plants, the synthesis of chlorophylls occurs at any temperature within the range 3–48°C, but it is most rapid between 26 and 30°C. Somewhat similar results have been obtained on the chlorophyll synthesis in ,potato tubers except that this synthesis appears to be the most rapid in the approximate range 11–19°C. The range of temperature over which chlorophyll synthesis occurs, as well as the temperatures of the most rapid synthesis, undoubtedly vary considerably from one species to another.

Water.   The desiccation of leaf tissues not only inhibits the synthesis of chlorophylls but seems to accelerate the disintegration of the chlorophylls already present. A familiar example of this effect is the browning of grass during droughts.

The Carotenoids.   This is a group of orange, red, yellow, and brownish pigments. The carotenoids of leaves are restricted to the chloroplasts, and this also appears to be true of the carotenoids in the cells of most algae. In the photosynthetic bacteria they are present in the chromatophores. In certain other plant organs, such as the petals of some flowers, the carotenoids occur in chromoplasts.

At least sixty different carotenoids have been found in plants. Chemically they are isoprenoid derivatives (Chapter 14). Best known of the pigments in this group is the orange-yellow carotene, first isolated from carrot roots. This is a hydrocarbon with the formula $C_{40}H_{56}$ and exists in three isomeric forms. β-carotene is the most abundant of these in plants and appears to be invariably present in the chloroplasts (Fig. 10.4). This compound is the precursor of vitamin A which is formed in the animal body by a hydrolytic splitting of the molecules of β-carotene, as follows:

$$C_{40}H_{56} + 2H_2O \longrightarrow 2C_{20}H_{29}OH$$

Lycopene, a red pigment found in the fruits of tomatoes, red peppers, roses, and other species, is isomeric with carotene (Fig. 10.4).

The xanthophylls are mostly yellow or brownish pigments, many, but not all, of which have the molecular formula $C_{40}H_{50}O_2$. Chemically the xanthophylls are closely related to carotene, and transformations from one kind of carotenoid to another occur readily in plants. In general the xanthophylls are more abundant in plants than the carotenes.

Lutein (Fig. 10.4) is by far the most abundant xanthophyll in leaves; other leaf xanthopylls include zeaxanthin, violaxanthin, cryptoxanthin, and neoxanthin. Numerous other xanthophylls occur in the various phyla of algae. Among the most abundant of these pigments are the fucoxanthins which impart to brown algae their distinctive color. Several brownish pigments of this type are also present in the diatoms, including fucoxanthins, diatoxanthin, and diadinoxanthin.

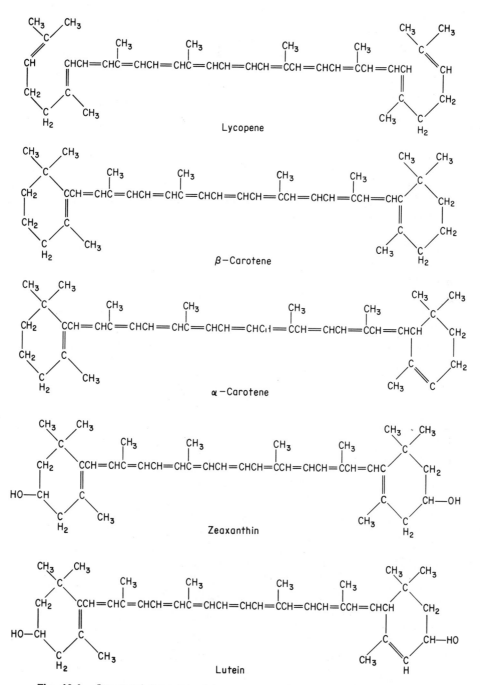

Fig. 10.4. Structural formulas of lycopene, carotenes, zeaxanthin, and lutein.

The Phycobilins.   These constitute a group of reddish or bluish pigments found in many kinds of algae. They are present in the chloroplasts in those species which contain them, and in the cellular material corresponding to the chloroplasts in those species which do not contain organized chloroplasts. These pigments are tightly bound to protein moieties. The phycoerthyrin of most red and some blue-green algae and the phycocyanin of most blue-green and some red algae are the best-known examples of such pigment-protein complexes. Phytochrome (Chapter 20), an enzymatically active pigment in higher plants, also appears to be a phycobilin. Although basically different from the chlorophylls in other respects the phycobilins resemble them in chemical structure inasmuch as the molecule contains four pyrrole rings. These rings occur in linear arrangement in the phycobilins, however, in contrast to their cyclic arrangement in the chlorophylls.

Structure of the Chloroplasts.   The chloroplasts are the seat of photosynthesis, except in the blue-green algae in which the units of protoplasm which function in this process are not organized into plastids and in the photosynthetic bacteria in which chromatophores fulfill this role. In the algae chloroplasts exhibit a wide range of shapes and sizes, but in the higher plants they are similar in outward appearance from one species to the next. The mature chloroplast of a higher plant is typically a flattened ovate spheroid with the longer axis

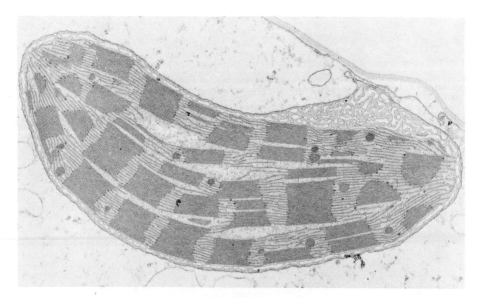

Fig. 10.5.   Section through a chloroplast from a corn (maize) leaf showing internal structure as discussed in the text. × 41,200. Electron micrograph by L. K. Shumway. From Jensen and Park, *Cell Ultrastructure,* Copyright 1967, Wadsworth Publishing Company, Inc.

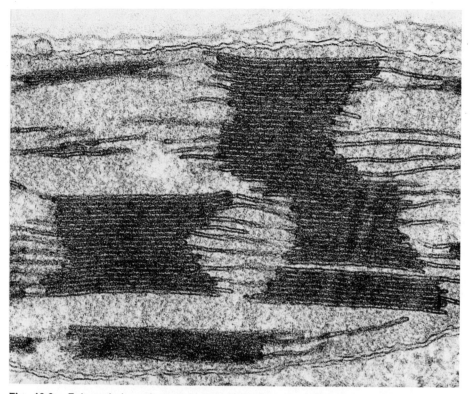

Fig. 10.6. Enlarged view of grana in the chloroplast of a leaf of wild coffee (*Psychotria bacteriophyla*). × 94,000. Electron micrograph from H. T. Horner, Jr.

ranging between 2 μ and 4 μ in length. The number of chloroplasts present in a cell of a higher plant may range from a few to over a hundred, and the number present may be quite different at different stages in the development of a given cell. In some algae and mosses at least some of the chloroplasts originate by the division of preexisting chloroplasts. Such a fission of chloroplasts is rare or nonexistent in higher plants, in which they seem to arise from simpler bodies, called *proplastids,* which develop in the cytoplasm.

Electron microscope studies have revealed the elaborate internal structure of a chloroplast, including the denser *grana* (Fig. 10.5). The chloroplast is surrounded by an outer double membrane which exhibits the property of differential permeability. Within the chloroplast there is a series of parallel dark layers called *lamellae,* which extends for its full length. The denser areas which occur in the lamellae and which, as seen in cross section, are stacked one above the other, are the *grana lamellae,* also called the small thylakoids. The term granum refers to the entire stack. A more highly magnified view of grana is shown in Fig. 10.6. The less dense lamellae which interconnect the grana lamellae are called the *intergrana* or *stroma lamellae* or large thylakoids. The space between the lamellae is occupied by a light-colored, largely proteinaceous material called the *stroma,* which may also interpenetrate the lamellae themselves. The chlorophylls and other pigments which may be present are located in or on the entire lamellar structure. Ribosomes are present in the stroma, but no mitochondria.

Scrutiny of the denser portions of a lamella under very high magnifications of the electron microscope reveals its even finer structure. The inner surface of the membrane of a granum lamella is composed of tightly packed units called *quantosomes* (Fig. 10.7), which may be arranged in either a very regular or a random fashion. The approximate size of each quantosome in the lamellae of spinach chloroplasts is 18 by 16 nm and they are about 10 nm thick.

The chemical composition of a chloroplast, as might be expected, is exceedingly complex. In addition to the pigments, previously discussed, the com-

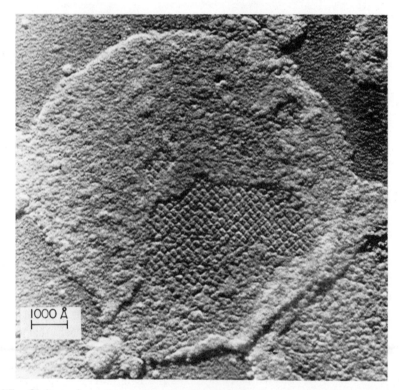

1000 Å

Fig. 10.7. Surface view of a granum lamella. The small quantosomes are arrayed in an orderly fashion in the lower portion of the picture. Electron micrograph by R. B. Park. From Jensen and Park, *Cell Ultrastructure,* Copyright 1967, Wadsworth Publishing Company, Inc.

pounds present include quinones, sterols, lipids, phospholipids, glycerides, proteins, amino acids, DNA, carbohydrates, and numerous enzymes.

Neither the fine structure of the chloroplast as such nor its chemical composition as such shed much light on its dynamic operation. A knowledge of its sub-plastid architecture in terms of the composition of its component parts should give better clues to the manner in which a chloroplast functions. Such knowledge does not yet exist in any definitive sense, but several hypotheses have been advanced regarding the probable structure of chloroplast lamellae at the molecular level. Most theoretical models of chloroplast structure postulate the presence of alternate layers of aqueous proteins and lipids. An example is the model proposed in Fig. 10.8. It is considered that the chlorophyll molecules are oriented with the hydrophilic porphyrin head of each molecule extending into the aqueous protein layer and its lipophilic phytyl tail extending into the lipoid layer. Carotenoid molecules are considered to be present only in the lipoid layer.

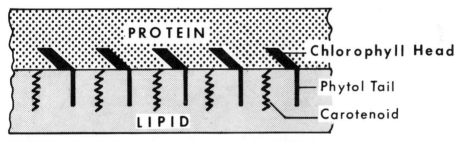

Fig. 10.8. Postulated arrangement (diagrammatic) of chlorophyll and carotenoid molecules in the protein and lipid layers of a chloroplast lamella.

Roles of the Chloroplasts. For some years it was believed that only that part of photosynthesis represented by the Hill reaction, to be discussed shortly, could be carried on by the chloroplasts and that other constituents of the living cell system besides the chloroplasts were essential for the occurrence of the complete photosynthetic process. In 1954, however, Arnon and co-workers showed that chloroplasts which had been completely isolated from plant cells could, if suspended in a suitable medium, carry out the entire process of photosynthesis. The chloroplast is so structured and equipped with enzyme systems that the entire multiphase process of photosynthesis and cognate processes can occur in it whether it is functioning within a cell or in a suitable medium outside of the cell.

The primary roles of the chloroplasts in photosynthesis are as light absorbers and energy transformers. Dark reactions subsequent to the photochemical phase of photosynthesis also occur in the chloroplasts. From experiments in which chloroplasts have been fractionated into constituent parts it has been concluded that the photochemical transactions of photosynthesis occur only in the pigmented lamellar structure, while the dark reactions of photosynthesis and other processes are largely or entirely confined to the stroma.

Only absorbed light can be utilized in photosynthesis. A self-evident property of any pigment is that it absorbs wavelengths in some parts of the visible spectrum to a greater degree than in others. Many studies have been made of the absorption spectra of the chloroplast pigments after their extraction from leaves or other photosynthetic tissues by suitable organic solvents. From such data it is possible to estimate fairly accurately the proportions of incident light absorbed by each pigment which thereby become a potential source of energy in photosynthesis. The prominent absorption bands of chlorophylls *a* and *b* are in the blue-violet and short-red regions of the spectrum (Fig. 10.9). β-carotene absorbs only wavelengths in the blue-violet portion of the spectrum (Fig. 10.10). The phycobilins on the other hand, absorb visible light principally in the green-orange part of the spectrum (Fig. 10.11).

The chlorophylls are the primary light absorbers of the wavelengths used in photosynthesis, but the carotenes (and phycobilins in blue-green and red algae) play accessory roles by absorbing the light which is used in this process. Light absorbed by these accessory pigments must be transferred to chlorophyll before it can become utilized as an energy source in photosynthesis. However no kind of cell is known which is able to accomplish photosynthesis if it does

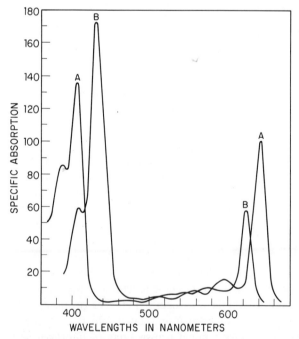

Fig. 10.9. Absorption spectra of chlorophylls *a* and *b* in ether solution. Data of Zscheile and Comar, *Botan. Gaz.* **102**, 1941:468.

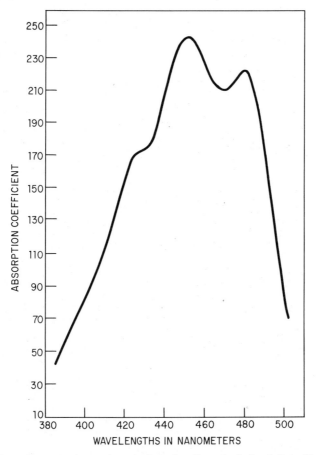

Fig. 10.10. Absorption spectrum of β-carotene in ether-alcohol solution. Data of Miller, *et al., Plant Physiol.,* **10,** 1935:377.

not contain one or more of the chlorophylls. The chlorophylls must therefore function in other roles in the process of photosynthesis besides that of light absorbers.

The "Light" and "Dark" Reactions of Photosynthesis. Over a temperature range of about 10–25°C, if the light intensity and carbon dioxide concentration are not limiting, the Q_{10} of photosynthesis is approximately two. Strictly chemical reactions characteristically have a Q_{10} of from two to three. This fact indicates that at least one of the reactions involved in photosynthesis is of a purely chemical type. Since this fact was first pointed out by Blackman in 1905, this reaction has often been called the *Blackman reaction.* It has also frequently been referred to as the *dark reaction,* since it does not require light, and therefore may take place in either the light or the dark. Modern interpretations of

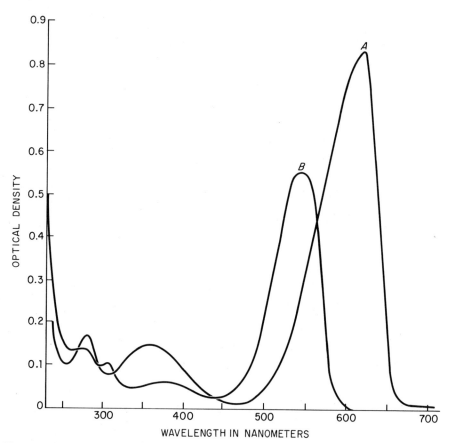

Fig. 10.11. Absorption spectra of two representative phycobilins. (A) Phycocyanin. (B) Phycoerythrin. Data of O hEocha, *Physiology and Biochemistry of the Algae*, Academic Press, Inc., New York, 1962.

the mechanism of photosynthesis indicate that a number of dark reactions are involved in the process.

A chemical reaction which proceeds only at the expense of absorbed light is called a *photochemical reaction*. That photosynthesis involves such a reaction or reactions can be inferred from the fact that it occurs only in the light. The Q_{10} of a photochemical reaction is approximately one. Under low light intensities, even with a relatively high carbon dioxide concentration and other conditions favorable for photosynthesis, the temperature coefficient of the process is about one, indicating that under such conditions the rate of photosynthesis is limited by a photochemical phase. It is now considered that there are at least two photochemical reactions involved in the process.

That photosynthesis involves both photochemical and chemical ("dark") reactions is also shown by the results of investigations in which plants are exposed to intermittent light. As shown by Emerson and Arnold in 1932 when cultures of the green alga *Chlorella* were exposed to intermittent illumination at the rate of 50 flashes per second, the periods of illumination being much shorter (0.0034 sec) than the intervening dark periods (0.0166 sec), the photosynthetic yield *per unit of light* was increased about 400 per cent as compared with the rate in continuous light.

Assuming that a photochemical reaction or reactions come first, the results just described can be explained as follows. When illumination is continuous the products of a light reaction are formed faster than they can be utilized in the relatively slower dark reactions. When the light is intermittent, however, all or most of the products of a photochemical reaction or sequence of such reactions are removed by a dark reaction during the intervening dark period, and the photosynthetic output per unit of light is considerably greater.

These results indicate that the rate of photosynthesis is limited by the stage of the process which occurs at the slowest rate. With low light intensities and an adequate carbon dioxide supply, a photochemical phase is limiting and temperature will have little effect on the rate of the process. With high light intensities and an adequate carbon dioxide supply but relatively low temperatures, the rate of photosynthesis is limited by a dark phase and will increase considerably with a rise in temperature.

The Photosynthetic Unit.   One approach to an understanding of the possible mechanisms of action of the chloroplasts in photosynthesis is to attempt to ascertain the size of the so-called "photosynthetic unit," which is the number of molecules of chlorophyll which must be present for the reduction of one CO_2 molecule to occur when the dark reactions of photosynthesis, which are much slower than the light reactions, are operating at maximum efficiency. Evidence from various kinds of experiments indicates that this number is surprisingly large. A recent estimate is that there are almost 500 chlorophyll molecules in each such photosynthetic unit.

The significance of the photosynthetic unit seems to be that while many chlorophyll molecules act as light absorbers, only a few of them participate in the photochemical phase of photosynthesis. In other words energy is transferred from many chlorophyll molecules which in a sense act as accessory pigments, to the very few which act as the reactive centers. The quantosome, previously described, may possibly represent the morphological expression of this physiological unit. Since there is substantial evidence for the existence of two photochemical reactions in photosynthesis, it is generally considered that two different types of photosynthetic units are present in the chloroplasts.

Such a pooling of the light energy absorbed by many chlorophyll molecules into a few reactive centers would be especially effective in increasing the efficiency of photosynthesis when plants are exposed to only low intensity light, which is often the situation under natural conditions. Under high light intensities, however, the absorption of light by the many chlorophyll molecules take place much more rapidly than the photochemically active molecules can utilize it, and photosynthesis under such conditions is a relatively inefficient process.

The Hill Reaction.   The English biochemist R. Hill showed in 1937 that the illumination of suspensions of chloroplasts in water in the presence of a suitable hydrogen acceptor results in the release of oxygen. The reaction can be represented as follows, A standing for a hydrogen acceptor:

$$2H_2O + 2A \xrightarrow[\text{chloroplasts}]{\text{light}} 2AH_2 + O_2$$

Some of the componds which can act as hydrogen acceptors in this reaction are 2,6 dichlorophenol, benzoquinone, and certain indophenols.

Subsequently, the very important discovery was made that $NADP^+$ and NAD^+ can act as hydrogen acceptors (oxidants) in a Hill type reaction:

$$2H_2O + 2NADP^+ \xrightarrow[\text{chloroplasts}]{\text{light}} 2NADPH + 2H^+ + O_2$$

The basic features of the Hill reaction are the photolysis of water and the generation of reduced pyridine nucleotides which have gained in free energy (reducing power) at the expense of radiant energy. This reaction or one closely comparable to it is a constituent step in the overall process of photosynthesis.

Source of the Oxygen Released in Photosynthesis.   For many years it was more or less tacitly assumed that the oxygen released in photosynthesis comes from the carbon dioxide utilized, an assumption based upon the equivalence between the quantity of oxygen liberated and the quantity of oxygen in the carbon dioxide used. By using water and carbon dioxide labelled with the heavy isotope of oxygen ($^{18}O_2$) as the raw materials used in photosynthesis by *Chlorella,* Ruben and co-workers showed in 1941, however, that the oxygen set free in the process comes from the water molecules. In other words water molecules must be split in photosynthesis, a finding which is consistent with the implications of the Hill reaction. An obvious corollary of this finding is that more water molecules must participate in the overall photosynthetic reaction than is indicated in the conventional summary equation given earlier, because more moles of oxygen are released than can be generated from only six moles of water.

Bacterial Photosynthesis.   Although the discussion in this book deals primarily with photosynthesis in the higher plants, investigations of the process as it occurs in the photosynthetic bacteria by van Niel and others during the mid-part of the twentieth century have contributed substantially towards an understanding of its mechanisms. Photosynthesis in the anaerobic green and purple bacteria occurs by an overall reaction similar to that in the higher plants except that hydrogen is obtained from sources other than water. For example, green sulfur bacteria can use hydrogen sulfide as a source of hydrogen in photosynthesis, the overall reaction being:

$$6CO_2 + 12H_2S \longrightarrow C_6H_{12}O_6 + 6H_2O + 12S$$

No oxygen is evolved in this or in other types of bacterial photosynthesis, further evidence that the oxygen released in photosynthesis comes from water.

Photophosphorylation.   When illuminated chloroplasts in aqueous suspension are supplied with ADP, inorganic phosphate, and $NADP^+$ in the absence of CO_2; ATP and $NADPH + H^+$ are synthesized and O_2 is evolved. The net chemical effect, although not the mechanism (see Fig. 10.13) of this process can be represented by the following equation:

$$2NADP^+ + 2ADP + 2H_3PO_4 \xrightarrow[\text{chloroplasts}]{\text{light}} 2NADPH + 2H^+ + 2ATP + O_2$$

This type of photophosphorylation represents a sumultaneous phosphorylation of ADP along with a Hill type reaction involving $NADP^+$. If the $NADP^+$ as well as the CO_2 is omitted from the system, only ATP is synthesized and no O_2 is evolved:

$$ADP + H_3PO_4 \xrightarrow[\text{chloroplasts}]{\text{light}} ATP + H_2O$$

The first of these reactions is termed *noncyclic photophosphorylation;* the second, *cyclic photophosphorylation.* The reasons for these terms will become evident from the later discussion. Both are anaerobic processes.

Photosynthetic Enhancement.   In 1943 Emerson and Lewis discovered that the light of wavelengths longer than about 680 nm in the far-red region of the spectrum is extremely inefficient in photosynthesis in the green alga *Chlorella,* although some such light is absorbed by chlorophyll *a.* Later it was found that the low efficiency of far-red light could be increased by superimposing light of shorter wavelengths on the far-red light. More photosynthesis was accomplished per unit of light energy used when the far-red and shorter wavelength were absorbed simultaneously than the sum of the amounts of photosynthesis accomplished when each waveband of light was absorbed separately. This phenomenon is referred to as *photosynthetic enhancement* and is taken as evidence that two independent but interacting light reactions participate in the photochemical phase of photosynthesis. The phenomenon of photosynthetic enhancement has also been demonstrated in higher plants (Fig. 10.12).

194   CHAPTER TEN

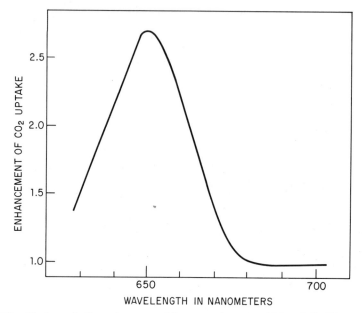

Fig. 10.12. Photosynthetic enhancement in a species of goldenrod (*Solidago virgaurea*) under conditions of weak light. Data of Björkman, *Brittonia* **18**, 1966:224.

Conversion of Radiant Energy into Chemical Energy. The unique feature of photosynthesis is the conversion of radiant energy into chemical energy during the photochemical phase of the process. The mechanism of this energy conversion is far from being completely understood, but enough is known about it that plausible hypotheses regarding this process can be formulated. Any acceptable hypothesis must incorporate the important facts and concepts already discussed including the structure and role of the chloroplasts, the absorption spectra of the plant pigments, the photosynthetic unit, the Hill reaction, water as a source of the oxygen evolved in the process, photosynthetic enhancement, and photophosphorylation.

A working hypothesis of the photochemical phase of photosynthesis which seems to be emerging is that of two light reactions, which will be referred to as System I and System II, tied together by a chain of electron carriers. This postulation, which almost certainly will undergo modification as further knowledge accumulates, is depicted in a somewhat simplified form in Fig. 10.13. The relative oxidation-reduction potential of the various constituents relative to hydrogen (H_2) is shown on the vertical scale to the left. A negative value on this scale indicates reducing potential; a positive value, oxidizing potential.

In addition to the phenomenon of photosynthetic enhancement, other evidence from differential inhibition indicates the existence of two photoreactions in the process. System II can be inhibited with dichlorodimethylurea, a compound which does not interfere with the operation of System I.

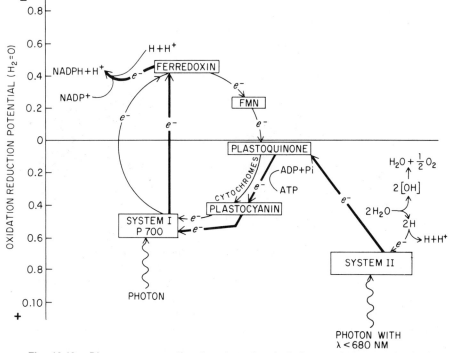

Fig. 10.13. Diagram representing the photochemical phases of photosynthesis. Bold type arrows indicate pathway when Systems I and II are both operative. Lighter arrows show pathway when System I is operating alone.

The operation of each system is mediated by a different pigment complex. System I is characterized by the presence of a form of chlorophyll *a* designated as P700 from its wavelength absorption maximum. This form of chlorophyll *a* also appears to be the photochemically active agent in the photosynthetic units operative in this system. The absorption spectra of chlorophyll *a* and other pigments vary somewhat depending upon their physical state. Hence different components ("forms") of the same pigment within the chloroplast may differ slightly in their absorption spectra.

Other pigments in System I, which include other forms of chlorophyll *a*, some forms of chlorophyll *b*, and carotenoids, act in the accessory capacity of transferring the light energy which they absorb to the photochemically active centers of P700. All photosynthetically active wavelengths of light are absorbed and used by this system. Cyclic photophosphorylation is associated with System I.

System II is characterized by the presence of a form of chlorophyll *a* having an absorption maximum of about 672 nm, perhaps other forms of chlorophyll *a*, some forms of chlorophyll *b*, some forms of chlorophyll *c* (in the brown algae), and carotenoids. Most of these pigments probably function in an accessory light-capturing role rather than as the photochemically active centers of the photosynthetic unit. Phycobilins also play an accessory pigment role in this system in the red and blue-green algae. They may be especially important in this capacity in such plants because they absorb principally in the mid-portion of the visible spectrum. The absorption of these wavelengths by chlorophyll is relatively low, and furthermore they are the wavelengths which penetrate deepest in water.

Only light of wavelengths shorter than about 680 nm is absorbed and utilized by System II. When only light of wavelengths longer than about 680 nm impinges on the chloroplasts only System I is operative. When light of shorter wavelengths also impinges on the chloroplasts, the enhancement of photosynthesis, as previously described, occurs. Noncyclic photophosphorylation is associated with System II.

Whether there is any morphological separation of the two photosynthetic units (Systems I and II) or whether they are intermeshed in the lamellar structure of the chloroplast is not known.

The postulated operation of System I is as follows. The absorption of a quantum of light energy by an electron of the chlorophyll molecule results in transforming the electron into an "excited" state; *i.e.,* it now carries more than its usual quota of energy. Because of its excess energy, such an excited electron leaves the chlorophyll molecule and transfers successively to ferredoxin, FMN, plastoquinone, cytochromes, plastocyanin, and perhaps other as yet unidentified compounds. In the course of its shuttling from one acceptor molecule to another, the electron loses its excess energy which is, however, conserved, at least in part, by contributing to the synthesis of ATP from ADP and inorganic phosphate, probably as a result of coupled reactions as it passes along the cytochrome chain. When the electron returns to the chlorophyll molecule after completing its cyclic path it resumes its "ground" or unexcited state. No oxygen is released in the operation of System I and no $NADP^+$ is reduced.

The postulated operation of System II indicates that this system sequentially precedes and eventually connects with System I. The absorption of light quanta by Pigment System II results in the photolysis of water in a Hill type reaction and in the generation of energy-surcharged ("excited") electrons. The photolysis of water results in its dismutation into hydrogens (H), protons (H^+), hydroxyl radicals (OH), and electrons (e^-). Electrons from the water molecules replace the electrons lost from the chlorophyll molecules or possibly are themselves the electrons which become excited. The combination of the OH radicals occurs by a reaction which in summary effect is as follows:

$$2(OH) \longrightarrow H_2O + 1/2\ O_2$$

The actual mechanism of this reaction is unknown, but it is worthy of note that this is the stage of photosynthesis which results in the release of oxygen.

The excited electrons first transfer to plastoquinone. From this compound they pass along the cytochrome chain during which some of their excess energy is utilized in coupled reactions in which ATP is formed from ADP and inorganic phosphate. This is the process of noncyclic photophosphorylation. The electron then joins a chlorophyll molecule of Pigment System I which has just lost an electron to ferredoxin as a result of its excitation by light energy. When System I follows System II, the electron is not transferred to FMN but goes on to reduce $NADP^+$ to $NADPH + H^+$. The electron is free to follow this path, because, unlike in System I, it or its equivalent came from a source outside of a chlorophyll molecule so all such molecules retain their full quota of electrons.

Regardless of how accurately Fig. 10.13 may depict the operation of the photochemical phase of photosynthesis, there is general agreement that its basic result is the synthesis of $NADPH + H^+$ and ATP, representing the creation of what is sometimes called "assimilatory power." The molecules of these two kinds of compounds and perhaps of other similar but as yet unidentified ones serve as the vehicles whereby the chemical energy derived from radiant energy is transferred to other molecules.

The Carbon Pathway in Photosynthesis.   It might seem logical to adopt the viewpoint that the term photosynthesis should be restricted to the photolytic phases which result in the conversion of radiant energy into the chemical energy of ATP and $NADPH + H^+$ since this is the distinctive and unique phase of the process. However, the reduction of carbon dioxide and the related syntheses of the numerous important carbon-containing metabolites in immediately subsequent reactions have also long been considered an integral part of the overall process of photosynthesis. These syntheses proceed by a complex of enzyme-catalyzed "dark" reactions which go on in the chloroplasts. Although these reactions do not require light, they do not proceed appreciably in its absence, because they are dependent on the energized products of the photochemical phase of photosynthesis.

The Calvin-Benson Cycle.   Starting in the late 1940's investigations were undertaken by Calvin, Benson, and others on the "carbon pathway" of photosynthesis, *i.e.,* on the series of compounds into which the CO_2 utilized in the process is sequentially converted. These experiments were performed by supplying carbon dioxide or a bicarbonate made with radioactive ^{14}C to the experimental plants, usually the unicellular algae *Chlorella* or *Scenedesmus,* for measured periods of time, then killing the cells with boiling methanol and ascertaining in what kinds of compounds the radioactivity was present (Fig. 10.14). The identification of the products formed was accomplished by chromatographic and autoradiographic technique.

Radioactivity was found to appear in a great variety of compounds even within a very short time. Among the compounds which became radioactive within one minute of the time that $^{14}CO_2$ was supplied to the cells were sugars, sugar phosphates, organic acids, and amino acids. When photosynthesizing algae were exposed to the radioactive CO_2 for only 5 sec or less, however, it was found that the first compound to become radioactive was phosphoglyceric acid,

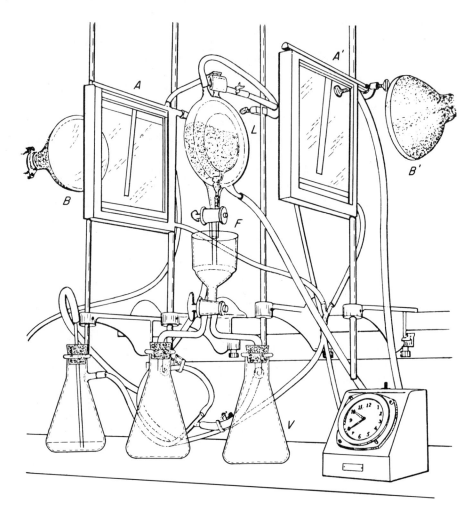

Fig. 10.14. An apparatus for determination of the first products of photosynthesis. A, A', heat shields filled with water; B, B', photoflood lamps; L, the "lollipop" containing the cell suspension; F, a sintered-glass filter for removing the cells from suspension after they have been illuminated for various times; V, vessels for receiving the filtrate of the cell suspension. Courtesy of M. Calvin, From Phillips, E. A., *Basic Ideas in Biology*, p. 308. Copyright, 1971, Edwin A. Phillips, The Macmillan Company.

a three-carbon acid. Furthermore, it was found that almost all the radioactivity in the phosphoglyceric acid molecules, after a short exposure to $^{14}CO_2$, was located in the carboxyl group of these molecules. Hence it seemed logical to assume that such carboxyl groups had been formed by the addition of CO_2 to a 2-carbon "acceptor" molecule. Search for such a molecule proved fruitless, however, and further investigation revealed that the CO_2 acceptor is actually a 5-carbon atom sugar phosphate, ribulose 1, 5-diphosphate.

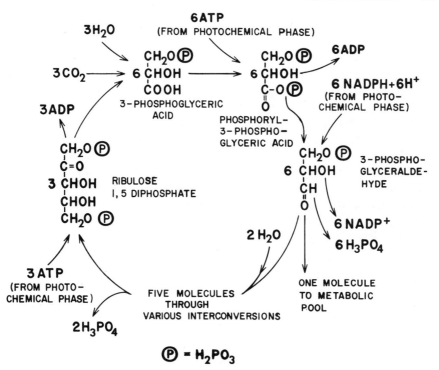

$$\circledP = H_2PO_3$$

Fig. 10.15. Carbon reduction cycle.

By tracing the pathway of radioactive carbon through the metabolic vortex, Calvin and other workers in the field were able to identify the sequence of reactions which results in the reduction of carbon, a sequence which in part is cyclic (Fig. 10–15). Each of three molecules of ribulose 1, 5-diphosphate picks up ("fixes") one molecule of carbon dioxide which, with the addition of three molecules of water, results in the formation of six molecules of 3-phosphoglyceric acid. As previously mentioned this compound is the first readily detectable product in this phase of photosynthesis. These six molecules are converted in a reaction involving 6 ATP to six molecules of phosphoryl-phosphoglyceric acid, with the accompanying conversion of each ATP to ADP. The molecules of phosphoryl-3-phosphoglyceric acid are reduced by NADPH + H+ to six molecules of 3-phosphoglyceraldehyde, a phosphorylated triose sugar. One molecule of NADP+ and one of H_3PO_4 are formed in this reaction per molecule of 3-phosphoglyceraldehyde reduced. The ATP and the NADPH + H+ used in these last two reactions are generated during the photolytic phase of photosynthesis. It is principally in these two reactions that the photolytic and carbon-reducing phases of photosynthesis are linked and that the energy originally derived from light is incorporated into an important metabolite.

Of each six molecules of 3-phosphoglyceraldehyde formed, one enters the metabolic pool and represents a net gain of both substance and energy to the metabolic system. Five of each six molecules of 3-phosphoglyceraldehyde go through a series of interconversions involving various phosphorylated sugars which lead to the synthesis of three molecules of ribulose-5-phosphate. Water is utilized in certain of these reactions, and phosphoric acid is released in others. The regeneration of the CO_2-acceptor, ribulose 1, 5-disphosphate is accomplished by further phosphorylation of the ribulose-5-phosphate by a reaction with ATP. This latter compound is considered to come from the photolytic phase of photosynthesis.

In brief the carbon-reduction Calvin-Benson cycle constitutes a complex of reactions whereby (1) carbon-dioxide fixation occurs, (2) the carbon-dioxide fixation product is reduced through the agency of NADPH + H$^+$ and ATP which derived the necessary energy from light, (3) the metabolic pool gains substance and energy to the extent of one molecule of 3-phosphoglyceraldehyde or equivalent for each three molecules of CO_2 reduced, and (4) the CO_2-acceptor molecules, ribulose 1, 5-diphosphate, are reconstituted in undiminished amounts. A summary equation for this phase of photosynthesis can be written as follows:

$$3CO_2 + 9ATP + 5H_2O + 6NADPH + 6H^+ \longrightarrow$$
$$1 \text{ glyceraldehyde-3-phosphate} + 9 \text{ ADP} + 6NADP^+ + 8H_3PO_4$$

As previously mentioned, other compounds besides those participating directly in the carbon reduction cycle rapidly become radioactive when $^{14}CO_2$ is supplied to photosynthesizing algal cells. Among these are carbohydrates such as sucrose, certain organic acids, various lipoidal substances, and a number of different amino acids. The synthesis of amino acids usually involves the reduction of nitrates, a process which, in photosynthesizing cells, derives the necessary energy from photophosphorylation (Chapter 16). Such compounds are synthesized from the intermediates of the carbon reduction cycle and can be regarded as direct photosynthetic products inasmuch as their synthesis is driven by the ATP and NADPH + H$^+$ generated in the photolytic phase of photosynthesis. Most of these compounds are also made in non-photosynthetic tissues by other synthetic processes, although even in such tissues all of the energy and most of the material substance used in such processes derive ultimately from photosynthesis.

The basic facts regarding the operation of the carbon reduction cycle as discussed so far were discovered as a result of experimentation with unicellular green algae. Substantial evidence has been obtained, however, that this cycle, or at least one very similar to it, also operates in a number of species of higher plants.

The Hatch-Slack Pathway. There is also good evidence for the existence of at least one other metabolic pathway of photosynthesis which has been named for two of its major investigators. When sugar cane leaves are allowed to photosynthesize in $^{14}CO_2$ for only a few seconds, the great bulk (80 per cent or more) of the radioactive carbon is found to be present in malic and aspartic acids, and very little in phosphoglyceric acid. Subsequently this route of carbon dioxide fixation was also found to occur in maize, sorghum, and some kinds of tropical grasses as well as in certain dicots such as some species of *Chenopodium, Amaranthus,* and *Atriplex.* It probably occurs in many other species of plants.

It is probable that carbon-dioxide fixation occurs in these plants by its reaction with phosphoenol pyruvic acid (Chapter 13) forming oxalacetic acid, which also can be shown to become radioactive very promptly when $^{14}CO_2$ is supplied:

$$
\begin{array}{c}
\text{COOH} \\
| \\
\text{C—O—H}_2\text{PO}_3 + \text{CO}_2 + \text{IDP} \\
\| \\
\text{CH}_2
\end{array}
\xrightarrow[\text{carboxylase}]{\substack{\text{phosphoenol}\\ \text{pyruvic}}}
\begin{array}{c}
\text{COOH} \\
| \\
\text{C=O} + \text{ITP} \\
| \\
\text{CH}_2 \\
| \\
\text{COOH}
\end{array}
$$

Phosphoenol pyruvic acid   Oxalacetic acid

The conversion of oxalacetic acid to malic acid could occur by the reversal of one of the well-known reactions of the Krebs cycle (Chapter 13):

$$
\begin{array}{c}
\text{COOH} \\
| \\
\text{C=O} \\
| \\
\text{CH}_2 \\
| \\
\text{COOH}
\end{array}
+ \text{NADH} + \text{H}^+
\xrightarrow[\text{dehydrogenase}]{\text{malic}}
\begin{array}{c}
\text{COOH} \\
| \\
\text{CHOH} \\
| \\
\text{CH}_2 \\
| \\
\text{COOH}
\end{array}
+ \text{NAD}^+
$$

Oxalacetic acid   Malic acid

Oxalacetic acid can also be converted into the amino acid, aspartic acid, by a reductive amination (Chapter 16). It is probable, however, that oxalacetic acid lies in the direct synthetic pathway leading to photosynthesis, the malic and aspartic acids acting only as metabolic reservoirs by way of reversible reactions with oxalacetic acid.

An interesting feature of the Hatch-Slack pathway is the spatial separation of the two phases of the overall process of photosynthesis. In all species that fix carbon by this route the chloroplast-containing cells of the leaf are radially arranged around the vascular bundles. A single layer of large parenchymatous cells sheaths the bundle, and these cells contain distinctively large chloroplasts. A layer of smaller mesophyll cells surrounds the bundle sheath, the cells of this layer containing smaller chloroplasts than those in the bundle sheath cells.

The stages in the process already described, which in essence result only in carbon-dioxide fixation, appear to occur in the mesophyll cells. The further steps, which involve the energy-storing reactions of photosynthesis, appear to take place in the bundle sheath cells.

The oxalacetic acid is decarboxylated, releasing CO_2 and pyruvic acid. The latter can serve again as a carbon dioxide acceptor, thus operating in a *cyclic* manner. The carbon dioxide is picked up by molecules of a second acceptor, most likely the same ribulose 1, 5-diphosphate which plays a key role in the Calvin-Benson cycle. The further steps of the photosynthetic process appear to be the same as those in the Calvin-Benson cycle, or at least very similar to them. The main feature of the Hatch-Slack pathway is that the CO_2 utilized in photosynthesis is delivered to the energy-storing reactions by a more circuitous route than that followed in the Calvin-Benson pathway.

The photosynthetic rates at atmospheric carbon-dioxide concentrations and nonlimiting light intensities for a number of species in which the Hatch-Slack pathway of carbon-dioxide fixation is followed have been found to be considerably higher than in many species in which only the Calvin-Benson cycle is operative. The explanation may lie wholly or largely in the low rates of photorespiration of the Hatch-Slack pathway plants. This relationship is discussed more fully in Chapter 12. The Hatch-Slack pathway plants also have higher temperature optimums for photosynthesis than those in which only the Calvin-Benson cycle prevails.

SUGGESTED FOR COLLATERAL READING

Chichester, C. O., Editor. *Advances in the Chemistry of Plant Pigments.* Academic Press, New York, 1972.
Clayton, R. *Molecular Physics in Photosynthesis.* Blaisdell Publishing Company, New York, 1965.
Devlin, R. M., and A. V. Barker. *Photosynthesis.* D. Van Nostrand Company, New York, 1971.
Gibbs, M., Editor. *Structure and Function of the Chloroplasts.* Springer-Verlag, New York, 1971.
Goodwin, T. W. *Chemistry and Biochemistry of Plant Pigments.* Academic Press, Inc., New York, 1965.
Goodwin, T. W., Editor. *Biochemistry of the Chloroplasts,* 2 vols. Academic Press, Inc., New York, 1966, 1967.
Hatch, M. D., C. B. Osmond, and R. O. Slatyer. *Photosynthesis and Photorespiration.* John Wiley Interscience, New York, 1971.
Ledbetter, M. C., and K. R. Porter. *Introduction to the Fine Structure of Plant Cells,* Springer-Verlag, New York, 1970.
Rabinowitch, E., and Govindjee. *Photosynthesis.* John Wiley & Sons, Inc., New York, 1969.
San Pietro, A., F. A. Green, and T. J. Army, Editors. *Harvesting the Sun.* Academic Press, Inc., New York, 1967.
Shibata, K., A. Takamiya, A. T. Jagendorf, and R. C. Fuller, Editors. *Comparative Biochemistry and Biophysics of Photosynthesis.* University Park Press, Baltimore, 1970.
Vernon, L. P., and G. R. Seely, Editors. *The Chlorophylls.* Academic Press, Inc., New York, 1966.
Zelitch, I. *Photosynthesis, Photorespiration, and Plant Productivity.* Academic Press, Inc., New York, 1971.

11 | FACTORS AFFECTING PHOTOSYNTHESIS

The Principle of Limiting Factors. Earlier investigators of the effects of various conditions upon the rate of photosynthesis attempted to distinguish among minimum, optimum, and maximum values for each factor in relation to photosynthesis. In evaluating the effect of temperature upon photosynthesis, for example, it was generally considered that there was a minimum temperature below which no photosynthesis occurred, an optimum at which the process takes place most rapidly, and a maximum above which photosynthesis ceases. Advocates of this point of view, however, soon found themselves confronted with the anomalous situation of a fluctuating "optimum." The "optimum" carbon-dioxide concentration was found to be greater at high light intensities than at low ones, the "optimum" temperature was found to vary with the light intensity, and the "optimum" light intensity was different for plants well supplied with water than for those which were inadequately supplied.

The first important step in the clarification of this problem of the influence of various factors upon photosynthesis was taken in 1905 when F. F. Blackman enunciated the "principle of limiting factors." This principle is essentially an elaboration of Liebig's law of the minimum (Chapter 20) and was stated by its author as follows: "When a process is conditioned as to its rapidity by a number of separate factors, the rate of the process is limited by the pace of the 'slowest' factor."

The explanation of this principle can best be presented in terms of the illustration given by Blackman (Fig. 11.1). Assume the intensity of light to be just great enough to permit a leaf to utilize 5 mg of carbon dioxide per hour in

203

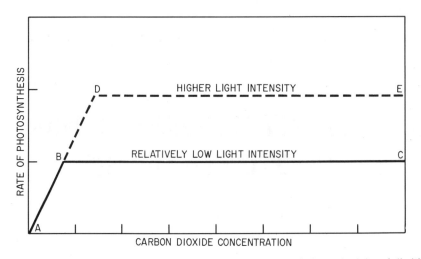

Fig. 11.1. Diagram to illustrate Blackman's interpretation of the principle of limiting factors.

photosynthesis. If only 1 mg of carbon dioxide can enter the leaf in an hour, the rate of photosynthesis is limited by the carbon dioxide factor. As the carbon dioxide supply is increased, the rate of photosynthesis is also increased until 5 mg of carbon dioxide enter the leaf per hour. Any further increase in the supply of carbon dioxide will have no influence upon the rate of photosynthesis, unless a sufficient concentration is present to bring about retarding effects, because insufficient light energy is available to permit its utilization. Light has now become the limiting factor, and a further increase in the rate of photosynthesis can be brought about only by an increase in the intensity of light. These results are indicated graphically as *ABC* in Fig. 11.1. This theory assumes a progressive increase in the rate of the process with a quantitative increase in the limiting factor (in this example, carbon dioxide) until the point is reached at which some other factor becomes limiting (in this example, light intensity). At this point the increase stops abruptly (point *B* in Fig. 11.1), and the rate of photosynthesis becomes constant (*BC* of Fig. 11.1). According to this concept, when the magnitude of photosynthesis is limited by one of a set of factors, only a shift in that factor toward a condition more favorable for the process will result in an increase in the rate of photosynthesis.

If the light intensity is sufficient to permit the leaf to utilize 10 mg of carbon dioxide per hour, then the rate of photosynthesis will rise with an increase in the carbon dioxide concentration to a value about twice as great as that at which the maximum rate of photosynthesis was attained at the lower light intensity. The results under these conditions can be indicated graphically by *ADE* in Fig. 11.1.

Light and carbon dioxide are not only the factors which can be limiting in the photosynthetic process. Theoretically, as examples given later in the chapter will show, any of the factors which influence this process can, under certain conditions, become limiting.

Most subsequent workers have been unable to accept the principle of limiting factors in quite the simple form in which it was first proposed by Blackman. Most investigators have found that when the rate of increase of photosynthesis is plotted along the ordinate with the quantitative variations in some one factor as the abscissa, the resulting curve is not found to show an abrupt transition to the horizontal (points *B* and *D*, Fig. 11.1), as postulated by Blackman's formulation of this principle, but shows instead a gradual transition to a position approximately parallel to the abscissa (Fig. 11.4). Within this transition region it is evident that an increase in either of the two factors involved will result in an increase in the rate of photosynthesis.

The explanation for this gradual transition in the direction of the curve, rather than the abrupt change postulated by the original Blackman theory, is that it results at least in part from the fact that the seat of photosynthesis is in the chloroplasts, of which there are millions in even a small leaf. Obviously it is impossible that each and every chloroplast will be subjected to exactly the same conditions at exactly the same time. All of the chloroplasts are not equally exposed to light; neither are all of them equally well supplied with carbon dioxide. As a factor approaches a limiting value, it may check the rate of photosynthesis in some chloroplasts sooner than in others. It is quite possible therefore for light to be the limiting factor for some chloroplasts, while carbon dioxide is simultaneously the limiting factor for other chloroplasts. Similar comments apply to most of the other factors influencing photosynthesis. Hence the rate of photosynthesis, as measured in terms of entire organs, will exhibit only a gradual change when factors affecting the process are modified, and there exist well-defined regions in curves such as those shown in Fig. 11.4, in which two or even more factors may be considered to act simultaneously as limiting factors.

Another qualification of the theory of limiting factors not envisaged in the original Blackman formulation should also be mentioned. When present in a high enough intensity or concentration, as discussed later in this chapter, most of the factors influencing the rate of photosynthesis exert a depressing or inhibitory effect on the process.

The modifications which have been imposed upon the original concept of limiting factors do not invalidate this principle as a good approximation to the facts, nor do they destroy its value as a point of view from which to interpret the influence of various factors upon the rate of photosynthesis. The significant fact is that the rate of the process, except in relatively narrow transitional regions, is usually determined in the main by the least favorable factor, which may for convenience be spoken of either as the limiting factor or as the factor in relative minimum.

The principle of limiting factors is applicable to all physiological processes and will receive further evaluation in relation to growth phenomena in Chapter 20.

Apparent and True Photosynthesis.   The measurements of photosynthesis are complicated by the fact that certain other processes involving the same materials are proceeding in the cells at the same time. The process of respiration (Chapters 12 and 13) is continuously in progress in all cells, resulting in an oxidation of part of the carbohydrates synthesized in photosynthesis. This introduces an error which is inherent in the usual methods of measuring photosynthesis. Measurements of the quantity of photosynthate formed in a given time are always less than the true value of the amount which has been consumed in respiration. In many measurements of photosynthesis the simultaneous occurrence of respiration is disregarded, and the results obtained are designated as the *apparent* or *net* photosynthetic rate—in other words, as the rate of photosynthesis minus the rate of respiration. Since in rapidly photosynthesizing tissues the rate of photosynthesis is often ten to twenty times as great at the rate of respiration, the apparent photosynthetic rate is often not greatly less than the true rate. For many purposes measurements of the rate of apparent photosynthesis have as great or greater significance as determinations of the "true" rate. Presumed values for the "true" rate of photosynthesis are sometimes calculated by correcting apparent rates for the amount of respiration considered to occur during the period of measurement. The values used for such corrections are usually obtained by measuring the respiration rate of the same plant or organ when it is subjected to complete darkness. This is not always a valid procedure, however, because in some kinds of plants a light-enhanced respiration called *photorespiration* (Chapter 12) occurs, which complicates the gas exchanges of the leaf.

The rate of carbon-dioxide consumption by plants, which is an index of the apparent photosynthesis, under favorable field conditions, is usually between 10 and 30 mg CO_2 per dm² of leaf area per hour. Another index of apparent photosynthesis is the gain in the dry weight of leaves, which under conditions favorable for photosynthesis, is usually between 0.5 and 2.0 g per square meter per hour.

The Role of Carbon Dioxide.   All of the carbon dioxide used by green plants reaches the chloroplasts as dissolved carbon dioxide, carbonic acid, or one of the salts of the latter. In land plants the atmosphere is the only important source of carbon dioxide. Carbon dioxide released in respiration may be utilized in photosynthesis without leaving the plant, but under conditions favorable for photosynthesis this does not constitute a very large fraction of the total used. Submersed water plants use carbonates and bicarbonates as well as dissolved carbon dioxide and carbonic acid as substrates of photosynthesis.

The atmosphere is composed chiefly of two gases, nitrogen (about 78 per cent) and oxygen (about 21 per cent), but also contains, in addition to a variable but never large amount of water vapor, small quantities of other gases. One of its lesser constituents, carbon dioxide, which constitutes on the average only about 0.03 per cent by volume (300 ppm) of the atmosphere, plays a role of the greatest significance in the biological world. As a result of the photosynthetic activity of green plants, the carbon dioxide from the air becomes chemically bound for periods of indefinite length in the organic molecules which are the basis of all life. In view of its important biological role, the proportion of carbon dioxide in the atmosphere seems precariously small. For reasons to be discussed shortly, however, its concentration in the atmosphere is maintained at a near constant value.

Sources of the Atmospheric Carbon Dioxide.   While green plants are continually removing carbon dioxide from the atmosphere, other processes are continually replenishing the atmospheric reservoir with this gas. Carbon dioxide is continually being returned to the atmosphere as a product of the respiration of plants and animals.

Carbon dioxide is released into the atmosphere as a result of the respiration of both green and non-green plants. For the purposes of this part of the discussion, bacteria are classified as non-green plants. The relatively great importance of this latter group of organisms as generators of this gas is not always appreciated. The organic residues of plants and animals are decomposed as a result of the activities of bacteria and fungi. During such decay processes the carbon of these residues is mostly released in the form of carbon dioxide as a result of the metabolic activities of these organisms and escapes into the air. The evolution of carbon dioxide gas from soils is often very considerable and is frequently referred to as "soil respiration"; this represents largely the respiration of microorganisms. Such respiration probably results in a greater return of carbon dioxide to the atmosphere than the respiration of all animals.

Carbon dioxide is also released into the atmosphere from volcanoes, mineral springs, and the combustion of coal, oil, gasoline, wood, and other fuel materials, but the total annual increment from these sources is small relative to the amount present in the atmosphere.

Oceans are much more important reservoirs of carbon dioxide than the atmosphere. The oceans occupy nearly three-fourths of the earth's surface and are estimated to contain about eighty times as much carbon in forms available to plants as the atmosphere. The carbon dioxide in ocean waters is involved in a complex series of chemical and biological cycles. Marine plants consume carbon dioxide in photosynthesis and release it in respiration. Marine animals feed either upon marine plants or other animals, but ultimately, as for land animals, all of their food comes from the process of photosynthesis. A part of the carbon in the food consumed by such organisms is released into the water as carbon dioxide in the process of respiration. Aquatic microorganisms accomplish the decay of dead plants and animals, releasing most of the carbon in the organic remains in the form of carbon dioxide. Complex equilibria between the dissolved carbon dioxide, carbonates, and bicarbonates also exist. Large quantities of carbonates are tied up by certain marine animals in the formation of shells. Other marine animals and some marine plants precipitate large quantities of carbon dioxide in a chemically-combined form as the calcium carbonate of calcareous rocks of which coral reefs are the most familiar example. The conversion of bicarbonates into carbonates results in the release of carbonic acid and thus increases the available carbon dioxide content of the water. Eventually such rocks (limestones, etc.) may be raised above sea level, and the carbon dioxide tied up in the form of carbonates again is released to the atmosphere or dissolved in running water during the dissolution of the rock.

Similar although not quite such complex cycles of carbon dioxide exist in the bodies of fresh water.

There is also a constant exchange of carbon dioxide between the oceans and the atmosphere. In fact, on theoretical grounds, there is good reason to suppose that the carbon dioxide concentration of the atmosphere is more or less effectively maintained in dynamic equilibrium with that of the oceans. Carbon dioxide probably escapes from the oceans whenever its atmospheric concentration falls below the usual value, and dissolves in the oceans whenever a contrary shift in atmospheric carbon dioxide concentration occurs. The maintenance of such a large-scale dynamic equilibrium between the oceans and the atmosphere is probably the principal factor accounting for the constancy of the carbon-dioxide concentration of the atmosphere.

The cycle of carbon in nature is shown diagrammatically in Fig. 11.2.

The Entrance of Carbon Dioxide. It has been definitely shown in some species that the bulk of the carbon dioxide which enters a leaf passes in through the stomates, and this has generally been assumed to be true, probably correctly, for most species. However, substantial amounts of this gas do enter the epidermal cells of some kinds of plants directly through the cuticle. In some species, of which avocado and begonia are examples, this seems to be the chief route of entry even when the stomates are fully open. All entry of dissolved carbon dioxide, bicarbonates, or carbonates into the leaves of submersed vascular hydrophytes occurs directly through the epidermis.

The rate of entrance of carbon dioxide through the stomates is very considerable in proportion to the aggregate area of the stomatal pores. Under conditions favorable for photosynthesis, carbon dioxide has been found to diffuse into a catalpa leaf from the atmosphere at the rate of 0.07 ml per cm^2 of leaf surface per hour. Since the stomates in this leaf occupy only 0.9 per cent of the surface, the diffusion of carbon dioxide gas through them took place at a rate of 7.77 ml per cm^2 of stomatal aperture per hour, the assumption being made that all carbon dioxide entered through the stomates. Under the same conditions a normal solution of sodium hydroxide absorbs from the atmosphere, even in rapidly moving air, only 0.177 ml of carbon dioxide per cm^2 per hour. In other words, carbon dioxide gas diffused through the stomates at a rate approximately fifty times as fast as it diffused into an efficient absorbing surface.

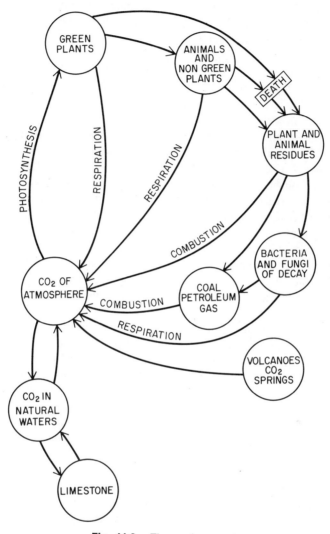

Fig. 11.2.   The carbon cycle.

*Effects of Variations in the Atmospheric Concentration of Carbon Dioxide
upon the Rate of Photosynthesis.*   Although the average carbon dioxide content of the atmosphere is 0.03 per cent, the actual range in plant habitats is
from about half to several times this value. Lower than average values are not
infrequent in relatively quiet air in zones in which high rates of photosynthesis
prevail. In a field of maize, for example, the carbon dioxide content of the air
surrounding the plants may become measurably less than the average atmospheric value during daylight hours whenever high rates of photosynthesis
occur. Similarly, in the atmosphere on the level with the crowns of trees in a

dense forest the carbon dioxide content is sometimes considerably less than the average atmospheric value during hours when rapid photosynthesis is in progress. In a tightly closed greenhouse the carbon dioxide content of the air may decrease measurably during the course of a bright day.

A considerable part of the carbon dioxide utilized by plants in many habitats may be released locally as a result of "soil respiration," *i.e.,* in the respiration of soil microorganisms. Such a release of carbon dioxide is especially pronounced in well-fertilized soils, soils rich in organic matter, and many forest soils. "Soil respiration," when marked, may result in a local enrichment of the carbon dioxide concentration in the air stratum close to the surface of the ground. Such a rise in carbon dioxide content is greatest during the night hours when the offsetting effect of photosynthesis in low-growing plants is absent. On a well-fertilized field the formation af carbon dioxide as a result of "soil respiration" during a 24-hour period may equal or exceed the consumption in photosynthesis during the daylight hours.

The carbon dioxide content of the air is also influenced by fogs and mists. The quantity present per unit volume of air is measurably higher on foggy mornings than on clear mornings, and rates of apparent photosynthesis are correspondingly increased in a foggy as contrasted with a clear atmosphere if no other conditions are limiting.

In general, at least over relatively short periods, with an increase in the concentration of atmospheric carbon dioxide there is an increase in the rate of photosynthesis until some other factor, most commonly light, becomes limiting. The results of an experiment on the relation of carbon-dioxide concentration to the rate of photosynthesis are shown in Fig. 11.3, in which, it should be noted the highest light intensity used was only about 10 per cent of noonday summer sunlight.

If no other factor is limiting, the rate of photosynthesis rises in most species, at least for a while, with an increase in the carbon-dioxide concentration up to 15–20 times the usual atmospheric concentration and sometimes more. Relatively high concentrations of carbon dioxide in general retard photosynthesis. The concentration at which such depressant effects are initiated varies with the kind of plant, the stage of development of the photosynthetic tissue, the length of exposure to carbon dioxide, and other prevailing environmental conditions. In the leaves of *Hydrangea otaksa,* for example, the rate of photosynthesis, although considerably retarded, is not entirely inhibited in short-time experiments at concentrations as great as 20 per cent carbon dioxide.

Under natural conditions in moist regions during seasons of favorable temperature, the carbon dioxide concentration of the atmosphere is most frequently the limiting factor in photosynthesis for all photosynthetic tissues which are well exposed to light. There seems to be little doubt that an increase in the atmospheric concentration of this gas to at least several times its average value of 0.03 per cent has a continuously favorable effect on photosynthesis, as long as no other factors are limiting. There is evidence, however, that only slightly

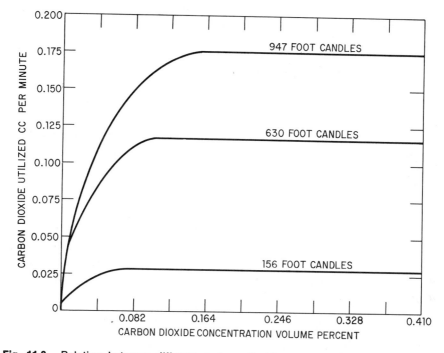

Fig. 11.3. Relation between different carbon dioxide concentrations **and the rate of** photosynthesis in wheat at three different light intensities. Data of Hoover, *et. al., Smithsonian Inst. Misc. Collections,* **87,** No. 16, 1933:16.

higher concentrations of carbon dioxide than those which favor enhanced photosynthesis over periods of a few hours or days may have a detrimental effect on the process over longer periods. Tomato plants, for example, exposed to ten times the atmospheric concentration of carbon dioxide during daylight hours show detrimental effects within less than two weeks.

The Role of Light.   The energy stored by green plants in the molecules of carbohydrates during the photosynthesis can be supplied only by light. Any source of radiant energy which includes wavelengths within the range of the visible spectrum will induce photosynthesis, provided its intensity is sufficiently great. Although a few of the longer wavelengths within the range of the effective in photosynthesis, and some of the shorter wavelengths of infrared can be used by photosynthetic bacteria, in general this process can occur only in the visible part of the spectrum. Under natural conditions sunlight, either direct or reflected from the sky or other objects, is the only source of radiant energy including wavelengths which can be used in photosynthesis. Photosynthesis will occur under artificial sources of illumination if of sufficient intensity. Electric lights of various kinds are often used in experimental work on photosynthesis and to some extent in greenhouses as supplementary sources of illumination.

Light, like all forms of radiant energy, varies in intensity (irradiance), quality, and duration, and the influence of this factor upon photosynthesis will be discussed under these three headings.

Optical Properties of Leaves. Of the visible light which falls on leaves, as with radiant energy in general, a part is reflected, a part is transmitted through the leaf, and a part is absorbed by the leaf. The proportion of the visible light which is absorbed varies considerably according to the kind of leaf and the intensity of the light but is frequently in the neighborhood of 80 per cent. A clear distinction should be drawn between the proportion of the visible light absorbed by a leaf and the proportion of the total radiant energy absorbed. Because of the small fraction of the incident infrared radiation absorbed, the total radiant energy absorbed is in the neighborhood of 50 per cent for many kinds of leaves (Chapter 6). Within the visible spectrum absorption is relatively high in the blue-violet and orange-red regions and relatively low in the yellow-green region which, of course, accounts for the green color of most leaves. The absorption spectrum of most leaves can be accounted for largely in terms of the absorption spectra of the chlorophylls and the carotenoids, described in the preceding chapter.

Effects of Variations in Irradiance ("Light Intensity") upon the Rate of Photosynthesis. In general, with an increase in irradiance there is an increase in the rate of photosynthesis until some other factor, most commonly the carbon-dioxide concentration, becomes limiting (Fig. 11.4). At relatively low

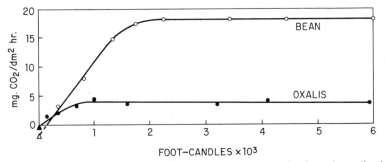

Fig. 11.4. Rates of apparent photosynthesis in bean (sun plant) and oxalis (shade plant) in relation to light intensity. Data of Böhning and Burnside, *Am. J. Botany*, **43**, 1956:599, 600.

light intensities, as long as no other factor is limiting, the rate of photosynthesis is approximately proportional to the irradiance. At the usual atmospheric carbon-dioxide concentration, and with no other factors limiting, maximum photosynthetic rates are attained when leaves are exposed to light intensities substantially less than maximum sunlight intensity, which at noon on a clear summer day is usually equivalent to 8,000–10,000 foot-candles.

The minimum irradiance under which photosynthesis reaches its maximum rate, often called the saturation light intensity, at atmospheric carbon-dioxide concentration and otherwise nonlimiting conditions, differs considerably between the "sun" and "shade" species of plants. The peak photosynthetic rate for bean, a representative sun plant, was found to be attained at a little more than 2,000 foot-candles (Fig. 11.4). The peak rates as determined by the same investigators for other sun plants—soybean, cotton, tobacco, sunflower, castor bean, and tomato—were in the range of from one-fourth to one-third maximum noonday summer sunlight. The great majority of crop plants are sun plants.

In *Oxalis tubra,* a typical shade plant, the maximum photosynthetic rate was found to be attained, at atmospheric carbon-dioxide concentration and otherwise nonlimiting conditions, at approximately 1000 foot-candles (Fig. 11.4). Other shade species found by these investigators to show approximately the same behavior include African violet, *Philodendron,* and several species of ferns. Coffee (some varieties) and cacao are two of the few crop plants which are shade species.

Results such as those discussed here will be obtained only when a single leaf or a small plant, in which there is little or no shading of one part by another, is used as the experimental material. When the effect of light on photosynthesis is considered in terms of an entire tree, a different relation holds. The rate of photosynthesis for an entire apple tree, for example, increases progressively with an increase in light intensity up to or nearly that of full sunlight. This is undoubtedly because many of the interior leaves on a large tree are heavily shaded. Although, in general, the maximum rates of photosynthesis are attained in the leaves of most (probably all) species at light intensities considerably below that of full sunlight, investigations show that many of the interior leaves of an apple tree receive as little as 1 per cent or less of the sunlight received by peripheral leaves. Even in full sunlight, therefore, many of the leaves on an apple tree do not photosynthesize at their maximum capacity. The lower the light intensity, the greater the proportion of the leaves of which this will be true. Hence, the greater the intensity of the incident light, the greater the average rate of photosynthesis is per unit of leaf area. The total photosynthesis per tree, therefore, increases progressively with increased illumination up to or at least close to the maximum possible sunlight intensity.

Internal shading effects on the rate of photosynthesis have been demonstrated experimentally for such diverse plants as the loblolly pine (*Pinus taeda*) and the hornwort (*Ceratophyllum demersum*), a submersed aquatic, and doubtless occur in many other kinds of plants with densely arranged foilage. Similar effects exist in compact stands of plants such as fields of some crop species. The mutual shading of the plants in plots of alfalfa is sufficiently marked that the photosynthetic rate of such plots considered as a unit is highest at peak light intensities, or at intensities close thereto (Fig. 11.9), as in an apple tree.

The intensity of the sunlight incident on the earth's surface varies from hour to hour and from season to season, as well as with meteorological conditions. Clouds, fogs, dust, and atmospheric humidity, all influence the intensity of the radiation which reaches the surface of the earth. The exposure and pitch of a slope are also factors influencing the intensity of the light impinging upon a given location, and are particularly of importance in hilly or mountainous country. Other conditions being equal, the intensity of sunlight also increases with an increase in altitude. In aquatic habitats, the light intensity decreases with depth below the surface of the water. Most variations in the intensity of natural light are accompanied by variations in light quality of greater or lesser magnitude. Usually, however, under out-of-doors conditions, variations in the intensity component of light are of greater physiological influence than the accompanying variations in the quality component.

Ecologically one of the most important factors causing different species of plants to be exposed to differences in light intensity is the effect of taller plants in shading those of lesser stature. Some species thrive and photosynthesize efficiently only in fully exposed locations; others can complete their normal life cycle in intensely shaded habitats.

Even under a tree with a rather open crown the light intensity is only one-tenth to one-twentieth that of full sunlight. Hence light is usually the limiting factor for photosynthesis in most species of plants when growing in the shade of trees. Species which normally grow in deep shade are often exceptions to this statement. On cloudy days, light is also usually the limiting factor in photosynthesis. Because of the prevalence of cloudy weather in many regions during the winter, plants under glass often photosynthesize at very low rates during these months. The short length of the daylight period also contributes to the low daily rates of photosynthesis during the winter season.

Extremely high light intensities exert an inhibitory effect on photosynthesis, a phenomenon called *solarization*. Solarization effects appear to result principally, and probably entirely, from the phenomenon of *photo-oxidation,* in which leaves consume oxygen in the light, and use it in the oxidation of certain cell constituents, carbon dioxide being released in the process. Photo-oxidation is an abnormal process and should not be confused with respiration, especially with the kind of respiration called photorespiration (Chapter 12). The continuation of this process for any appreciable period of time results in the oxidation of cellular components, including chlorophylls, and a generally destructive effect on the cells in which it occurs. It is generally considered that the presence of the carotenoids in a cell exerts a protective effect against the destruction of the chlorophylls by photo-oxidation.

Light may exert indirect as well as direct effects upon photosynthesis. Low light intensities favor stomatal closure and hence may sometimes check photosynthesis by restricting the entrance of carbon dioxide as well as by acting as a direct limiting factor. Similarly, high light intensities often cause increased rates of transpiration. Indirectly this causes a reduced water content and more a negative water potential in the leaf cells, which may in turn cause a diminished rate of photosynthesis.

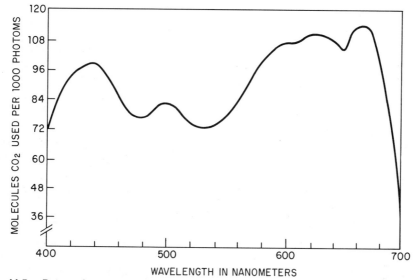

Fig. 11.5.   Rates of photosynthesis in bean under different wavelengths of light. Data of Balegh and Biddulph, *Plant Physiol.,* **46,** 1970:4.

Effects of Different Light Qualities upon Photosynthesis.   The results of an investigation of the effect of different wavelengths of radiation upon the rate of photosynthesis of bean plants are presented in Fig. 11.5. The results indicate the occurrence of maximum photosynthesis in the orange-red region of the spectrum, a secondary maximum in the blue-violet region of the spectrum, and a lower rate in the mid-portion of the spectrum. In general, photosynthetic maxima correspond with the maximum absorption bands of the chlorophylls and the carotenoids. Similar results have been obtained with other terrestrial species, and with the green alga *Ulva lactuca.* In a red alga (*Porphyra* sp.), however, maximum photosynthesis was attained in the mid-portion of the visible spectrum, evidence that phycobilin pigments are playing an important accessory pigment role in such organisms.

There are a number of conditions under which plants growing in their natural habitats are exposed more or less continuously to light of a different quality from that of full sunlight at the earth's surface. On cloudy days, for example, the intensity of light is not only less than on clear days, but it is proportionately richer in blue and green wavelengths.

Light which has been filtered through the crown of a tree is usually proportionately richer in green rays than direct sunlight because of the greater proportionate absorption by the leaves in the red and blue portions of the spectrum. This effect upon the light quality is most marked in hardwood forests in which the tree crowns form an almost continuous canopy. The herbs, shrubs, and smaller trees growing in such forests are subjected to light which is not only of much lower intensity than full sunlight but is also different in quality from the light impinging upon the forest canopy.

In habitats of submerged aquatics both the intensity and quality of the light are usually very different from the intensity and quality of the sunlight at the earth's surface. Pure water absorbs radiations in the red-orange portion of the spectrum much more effectively than in the blue-green region. While the absorption coefficients of natural waters for the various wavelengths of light vary somewhat, depending upon the substances dissolved or dispersed in the water, in general the shorter wavelengths penetrate to greater depths than the longer wavelengths. Hence with increasing depth in either fresh or ocean water not only is the intensity of the light reduced, but its quality is greatly modified. Aquatic plants growing at a depth of 20 meters, for example, will be exposed to light proportionately much richer in blue-green rays, although of lower intensity, than those at a depth of 1 meter.

Alpine plants are also exposed to light of a different composition than species at lower altitudes. The atmosphere absorbs the shorter wavelengths of the sun's radiation more effectively than the longer ones. Because of the shorter column of atmosphere through which it passes, sunlight at high altitudes is therefore not only more intense than at lower elevations, but is also relatively richer in the shorter wavelengths of visible radiation and the ultraviolet.

Effects of Duration of the Light Period upon Photosynthesis. In general a plant will accomplish more photosynthesis in the course of a day if exposed to illumination of favorable intensity for ten or twelve hours than if suitable light conditions prevail for only four or five hours. In arctic regions photosynthesis may occur continuously throughout the 24-hour day of the summer months, although the rate may vary cyclically throughout the day. A considerable capacity for sustained photosynthesis under continuous light is indicated by the results of certain experiments. Leaves on young apple trees, exposed to a continuous illumination of about 3,200 foot-candles at 25°C and the usual atmospheric concentration of carbon dioxide, were shown to photosynthesize at an undiminished rate for periods of at least 18 days.

Temperature Effects on Photosynthesis.   The measurement of the effect of temperature upon photosynthesis in terrestrial plants is complicated by the fact that the leaf temperatures of such plants are seldom the same as atmospheric temperatures. Whenever leaves are exposed to direct illumination, which is almost invariably the situation when photosynthesis is occurring at a rapid rate, their temperatures exceed those of the surrounding atmosphere. It is difficult, therefore, if not impossible, to maintain the temperature of the leaves of land plants at a desired value while they are exposed to light of any considerable intensity. Evaluation of the effect of the temperature of leaves upon photosynthesis is possible only if the actual leaf temperature is measured. For this reason many of the more critical studies of the effect of temperature upon photosynthesis have been made with submerged water plants, in which a close thermal equilibrium is maintained between the plant body and the surrounding water.

Temperature Limits of Photosynthesis.   Photosynthesis can take place over a wide range of temperatures. The photosynthetic rate may exceed the rate of respiration in several species of conifers at temperatures as low as $-6°C$. Tropical plants cannot carry on photosynthesis at temperatures as low as those at which many temperate-zone plants do. In most tropical species, photosynthesis apparently will not occur at temperatures below $5°C$. At the other end of the temperature range for photosynthesis stand the species of algae indigenous to hot springs, which can survive $75°C$ and probably carry on photosynthesis at temperatures close to this value. Many semi-desert and tropical species can withstand air temperatures of $55°C$ and probably photosynthesize at temperatures not far below this. In most plants of temperate regions the range of temperatures within which photosynthesis occurs at a relatively rapid rate is about 10 to $35°C$.

The Effect of Temperature on the Rate of Photosynthesis.   If neither carbon dioxide nor light nor any other factor is limiting, the initial rate of photosynthesis increases with a rise in temperature up to a point which varies somewhat from one kind of a plant to another. A further increase in temperature results in a rapid decline in the initial rate of photosynthesis. Moreover, the rate of photosynthesis at all higher temperatures within physiological limits decreases with time (Fig. 11.6). The higher the temperature within these limits the sooner this decline in rate sets in and the more rapidly it occurs. This diminution in the rate of photosynthesis with time is probably a result of enzyme inactivation (Chapter 9).

It should be emphasized again that the relation between photosynthesis and temperature described here holds only when no other environmental factor is limiting. Such experiments on the overall temperature characteristics of the process must therefore be conducted with the plants exposed to an atmosphere in which the carbon dioxide concentration is considerably higher than the usual atmospheric concentration of 0.03 per cent.

In nature the maximum rate of photosynthesis which might be achieved under the prevailing temperature is often not realized because of the limiting effect of some other factor. Temperature, within the ordinary physiological range for plants, has little effect on the rate of the photosynthesis of plants growing in deep shade or of the plants exposed to the low light intensities of cloudy days. Under such conditions, light is the limiting factor.

Similarly, on warm days, well-watered plants exposed to bright light often do not photosynthesize at the maximum rate possible under the prevailing temperature, because carbon dioxide at atmospheric concentration is the limiting factor. It has been found, for example, that temperature has very little influence on the rate of the photosynthesis of alfalfa under field conditions within a range of approximately 16 to $29°C$. In other words the temperature coefficient (Q_{10}) of the process under such conditions is approximately 1.

The Role of Water in Photosynthesis.   Less than 1 per cent of the water absorbed by a land plant is used in photosynthesis. Thus it seems virtually cer-

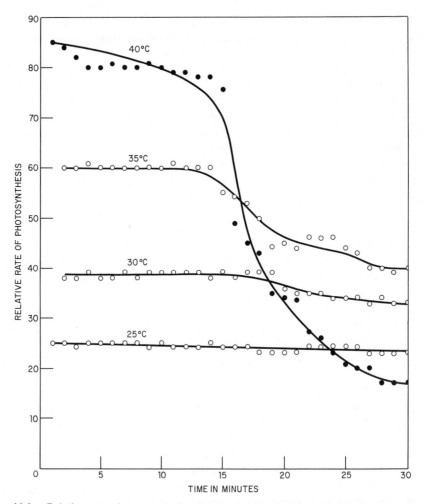

Fig. 11.6. Relative rate of apparent photosynthesis in waterweed (*Anacharis canadensis*) at different temperatures over a period of 30 minutes under nonlimiting conditions of light intensity and carbon dioxide concentration.

tain that the indirect effects of the water factor upon photosynthesis are more pronounced than its direct effects. In other words, the deficiency of water as a raw material is rarely if ever a limiting factor in photosynthesis. Nevertheless a reduction in the water content of leaves results in a decrease in the rate of photosynthesis, as is illusltrated by the results of certain studies on the effects of withholding water upon the rate of apparent photosynthesis in apple trees. When the soil was allowed to dry out gradually, starting from the field capacity, reduction in the rate of apparent photosynthesis became evident within a few days at an air temperature of 26.7° C and was pronounced before visible wilting took place. When the soil-water content had fallen to the permanent wilting percentage, and the leaves were distinctly wilted, apparent photosynthesis was 87 per cent less than the initial rate. Upon watering the soil the leaves regained turgidity within a few hours, but the original rates of apparent photosynthesis were not regained until 2–7 days had passed. An important implication of this last observation is that the retarding effects of drought on photosynthesis may linger for some time after water has again become available in the soil. Similar results of the effect of soil moisture on the rate of photosynthesis in sugar cane have also been obtained (Fig. 11.7).

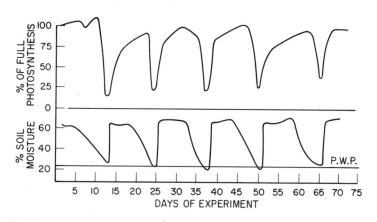

Fig. 11.7. Variations in the rate of the apparent photosynthesis in sugar cane and percentage soil moisture during five drying cycles. Each time the soil dried to approximately the permanent wilting percentage (P.W.P.) its moisture content was raised to approximately the saturation value by irrigation. Data of Ashton, *Plant Physiol.*, **31,** 1956:270.

The influence of a reduction in the water content of leaves upon the rate of photosynthesis probably results from either or both of two principal causes: (1) a reduction in the diffusive capacity of the stomates, and (2) a decrease in the hydration of the chloroplasts and other parts of the protoplasm which in some manner diminishes the effectiveness of the photosynthetic mechanism. This latter effect is discussed further later in this chapter.

Effect of Oxygen Concentration on Photosynthesis.   The photosynthetic organs of terrestrial plants are seldom exposed to oxygen concentrations which deviate appreciably from the usual atmospheric concentration of about 21 per cent. Hence studies of the effect of oxygen concentration on the rate of photosynthesis are primarily of theoretical interest. In many plants, an increase in the oxygen concentration results in a decrease in the rate of photosynthesis, and the normal atmospheric concentration is sufficiently high to induce a slower photosynthetic rate than prevails at lower oxygen concentrations (Fig. 11.8). This effect is often called the *Warburg effect* after the German biochemist who first

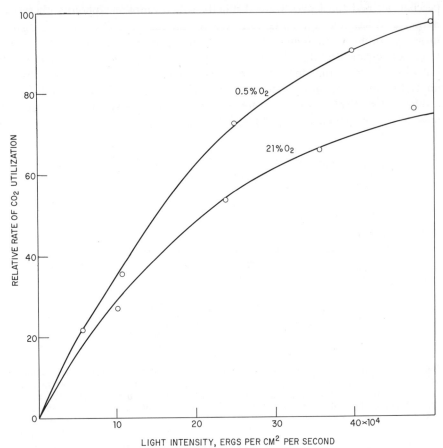

Fig. 11.8. Relative rates of photosynthesis in wheat at low and high oxygen concentrations under different light intensities. Data of McAlister and Myers, *Smithsonian Inst. Misc. Collections,* **99**, No. 6, 1940.

discovered it in the alga *Chlorella* in 1920. The operation of the Warburg effect has also been demonstrated in a number of land plants, of which wheat and soybeans can be cited as examples. Other terrestrial species, such as maize, sugar cane, and sorghum do not show this effect.

There appears to be a close relation between the Warburg effect on photosynthesis and the process of photorespiration (Chapter 12). In species showing this effect, relatively high oxygen concentrations result in sidetracking a larger proportion of the photosynthetic intermediates into the synthesis of glycolic acid than do lower concentrations of oxygen. Glycolic acid is an important intermediate compound in photorespiration. As discussed more fully in Chapter 12, species with a high photorespiration rate are less efficient photosynthetically than those with a low photorespiration rate.

The Effects of Internal Factors on the Rate of Photosynthesis.   In addition to the environmental factors which influence photosynthesis, the rate of the process is also influenced by certain factors within the plant. In general, because of the experimental difficulties encountered in their investigation, the effects of such internal factors on photosynthesis are less well understood than those of external factors. A brief discussion of several of these factors is warranted, however, by the present state of our knowledge.

Chlorophyll Content.   The results of a number of experiments seem to point to the conclusion that there is no proportional relationship between chlorophyll content and photosynthesis in the leaves of vascular plants. In other words, it appears that the chlorophyll content of the leaves is seldom the limiting factor in photosynthesis in such species even when all external conditions are favorable for the process.

Hydration of the Protoplasm.   That the hydration of protoplasmic constituents is an important internal factor influencing photosynthesis has already been mentioned. This is especially true of the degree of hydration of the chloroplasts. As the hydration of the chloroplasts diminishes, their water potential becomes more negative. A water potential of -8 bars in sunflower, and -12 bars in pea, is sufficient to inhibit oxygen evolution by isolated chloroplasts and, by inference, the process of photosynthesis. Such water potential values are of common occurrence in leaves under natural conditions.

Leaf Anatomy.   The rate of photosynthesis in any leaf is partly conditioned by the anatomy of that leaf. The size and distribution of the intercellular spaces, the relative proportions and distribution of the palisade and spongy layers, the size, position, and structure of the stomates, the thickness of the cuticular and epidermal layers, the amount and position of chlorenchyma, the proportion and distribution of non-green mesophyll tissues, and the size, distribution, and efficiency of the vascular system all influence the rate of photosynthesis. The effects of the structure of leaves upon the rate of photosynthesis result principally from influences upon the rate of the entrance of carbon dioxide, upon the intensity of the light penetrating to chlorenchyma cells, upon the maintenance of the turgidity of the leaf cells, and upon the rate of the translocation of the soluble carbohydrates out of the photosynthesizing cells.

Microstructure of Cell Organelles.   The microstructure of cell constituents, especially the chloroplasts, may influence the rate or even the course of photosynthesis. For example, in maize chloroplasts, the membranes bounding the lamellae appear to be continuous with the outer membrane of the chloroplast, a structural arrangement not found in the chloroplasts of many other species. It has been postulated that this type of chloroplast structure may be characteristic of plants in which the Hatch-Slack pathway prevails.

Daily Variations in the Rate of Photosynthesis.   Representative curves of the daily variation in rates of apparent photosynthesis for small plots of alfalfa are presented in Fig. 11.9. These data were obtained from well-watered plots of alfalfa, about 2 meters square, enclosed in transparent celluloid cabinets with air circulating through each cabinet at rates ranging up to several thousand liters per minute. The conditions to which the plants were exposed approximated those of a "standard day" (Chapter 6). The net consumption of carbon dioxide by the plants was ascertained by measuring the difference in the concentration of this gas in the inflowing and outflowing streams of air. In general, as with the apple tree previously described, the rate of photosynthesis under these conditions showed a close correlation with the light intensity.

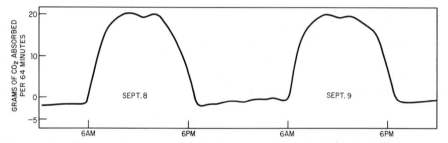

Fig. 11.9. Daily variations in the rate of the apparent photosynthesis of alfalfa. Data of Thomas and Hill, *Plant Physiol.,* **12,** 1937:292.

Innumerable other types of daily periodicities of photosynthesis are possible. The pattern of any such periodicity depends in part on the kind of plant and on the unit of plant material for which photosynthesis is measured; whether a single leaf, a small plant, a large plant such as a tree, or a plot of vegetation. It will also obviously vary with the daily cycle of environmental conditions, which not only differs in a general way from one climatic center to another but also shows seasonal and daily variations within any climatic region.

Magnitude and Efficiency of Photosynthesis.   Because of its irreplaceable role in the biological economy of the earth, enormous interest attaches to the world magnitude of photosynthesis and to the efficiency of the process in converting solar energy into chemical energy. All of the figures given in this section are necessarily estimates but they are probably close enough to the true order of magnitude that meaningful conclusions can be drawn from them.

A generally accepted estimate is that 2 to 3×10^{10} tons of carbon are fixed in organic form by the photosynthesis of all the terrestrial plants on the earth's surface per year. Estimates for aquatic plants, mostly marine and mostly phytoplankton, are more uncertain but the amount of carbon fixed by them is probably at least equal to that fixed by terrestrial plants, and may be considerably more. It is estimated that about one per cent of the photosynthate produced by land plants each year is utilized by man as food.

Of the total amount of solar radiation falling on the earth's surface, estimated at about 5×10^{20} kg-cal per year, only about 0.1 to 0.3 per cent is converted into the energy of organic compounds in photosynthesis. One reason for the low conversion percentage is that about half of the solar radiation is in the infrared region of the spectrum, none of which can be used in photosynthesis. In terms of the total radiant energy flux, the entire biological world appears to be a marginal operation! The absolute amounts of photosynthate produced, however, as estimates given in the preceding paragraph show, is enormous.

The efficiency of photosynthesis is not intrinsically as low as indicated here. Efficiencies as high as 20 per cent have been obtained in *Chlorella* cultures under laboratory conditions and efficiencies of 7 to 10 per cent in large scale algal cultures. Furthermore, efficiencies as high as about 15 per cent have been measured for the leaves of some terrestrial plants under laboratory conditions. Reasons for the low efficiency of global photosynthesis are fairly obvious, especially for terrestrial plants. Large areas of the earth's surface, principally snow and ice fields and deserts, are practically destitute of plant life. Other large areas, principally semidesert regions, are only sparsely occupied by vegetation. A large proportion of the oceans is also, in effect, desert or at least semidesert, because prevailing conditions of one kind or another are unfavorable to photosynthesis. Furthermore, except in tropical or semitropical regions, photosynthesis goes on in vegetation-covered areas for only a part of each year. This is true both of crop plants and natural vegetation. A further loss of photosynthetic efficiency in crop plants stems from the fact that they do not fully occupy an area during the earlier stages of their growth cycle. Some of the more efficient crop plants, including maize, sugar beet, sugar cane, and wheat convert 2 ± 0.5 per cent of the photosynthesizable solar energy received during a growing season on the field which they occupy into chemical energy. Calculations indicate that the photosynthetic efficiency of at least some species of forest trees is of the same order of magnitude.

SUGGESTED FOR COLLATERAL READING

Devlin, R. M., and A. V. Barker. *Photosynthesis.* D. Van Nostrand Company, New York, 1971.

Hatch, M. D., C. B. Osmond, and R. O. Slatyer. *Photosynthesis and Photorespiration.* John Wiley Interscience, New York, 1971.

Heath, O. U. S. *The Physiological Aspects of Photosynthesis.* Stanford University Press, Stanford, California, 1969.

Rabinowitch, E., and Govindjee. *Photosynthesis.* John Wiley & Sons, Inc., New York, 1969.

San Pietro, A., F. A. Greer, and T. J. Army, Editors. *Harvesting the Sun.* Academic Press, Inc., New York, 1967.

Zelitch, I. *Photosynthesis, Photorespiration, and Plant Productivity.* Academic Press, Inc., New York, 1971.

12 | RESPIRATION

When seeds germinate in a dark room, the total weight of the developing seedlings increases for a number of days, but their dry weight consistently decreases. By chemical analysis it can be shown that the loss of the dry weight of seedlings growing in the absence of light results entirely from the disappearance of a portion of the foods which had accumulated in the seed. The gain in total weight of such seedlings is almost entirely a consequence of the absorption of water which occurs during the early stages of germination in quantities far surpassing any loss of dry weight resulting from the disappearance of the foods. The quantity of mineral salts absorbed by young seedlings in the course of a week or two is usually too small to have any appreciable effect upon either their dry or total weight.

If seedlings developing in the dark are enclosed in a chamber which is constructed so that a slow, continuous stream of air can be passed through it and if frequent analyses are made of the air, it can be demonstrated that the air emerging from the chamber contains a lower percentage of oxygen and a higher percentage of carbon dioxide than the air which entered. Furthermore, if such seedlings are enclosed in a calorimeter, it can also be shown that heat, which is one kind of energy, is continuously escaping from them.

All of these phenomena, the disappearance of food resulting in a decrease in the dry weight, the absorption of oxygen, the evolution of carbon dioxide, and the liberation of energy, are different external manifestations of the process of respiration which occurs not only in germinating seedlings but in all living cells.

The gaseous exchanges accompanying respiration were discovered and extensively studied before any special significance was attached to them. This situation was true of plants as well as animals, in both of which oxygen is usually consumed and carbon dioxide is usually released during respiration. The term respiration has therefore long been used to refer to these externally apparent gaseous exchanges and is still sometimes employed in this sense.

However, as will be shown later, gaseous exchanges of the usual type are not invariable accompaniments of respiration. Carbon dioxide is not always released nor is oxygen always used in respiration. For these reasons plant physiologists use the term "respiration" primarily to refer to the oxidation of foods in living cells, with the resulting release of energy. A part of the energy is transferred to compounds other than those which are oxidized and some is used in the activation of certain cell processes.

Aerobic Respiration.   Respiration of the type described in the preceding paragraphs is, strictly speaking, called *aerobic respiration,* since it proceeds at the expense of atmospheric oxygen. A type of respiration known as *anaerobic respiration* is also of common occurrence in plant cells (Chapter 13). Anaerobic respiration does not require atmospheric oxygen but may occur in its presence. The basic difference between the two kinds of respiration is that atmospheric oxygen participates as a reactant in some of the stages of aerobic respiration but not at any stage of the process when the respiration is strictly anaerobic. When the term "respiration" is used without qualification it usually refers to aerobic respiration.

On the assumption that hexose sugar is the substrate, the summary chemical equation for aerobic respiration is:

$$C_6H_{12}O_6 + 6O_2 \longrightarrow 6CO_2 + 6H_2O + 673 \text{ kg-cal}$$

Obviously such a complicated reaction as the oxidation of a hexose does not take place in a single step as indicated in this convenient summary equation. The involved series of reactions through which this stepwise oxidation proceeds in the metabolic system of the plant is described in the following chapter. The summary equation merely tells us that for the oxidation of one mole of a hexose, six moles of oxygen are required; that six moles each of carbon dioxide and water result from this oxidation and that 673 kg-cal of energy are made available as a result of the reaction. Since equimolar weights of gases occupy the same volume (Avogadro's hypothesis), the volume of oxygen utilized when a hexose is oxidized is equal to the volume of carbon dioxide released. The water formed as a result of this respiration becomes a part of the general mass of water present in the respiring cells. Such water is often termed *metabolic water.*

Hexose sugars are most commonly the sustrates which are oxidized in the cells of the higher green plants. Insoluble carbohydrates, such as starch, can be used in respiration only after they have been converted into soluble carbohydrates. The metabolic pathways followed when carbohydrates are the substrates of respiration are discussed in Chapter 13. When plant cells contain both carbohydrates and fats, the former apparently are consumed first in respiration, before any inroads are made on the fats. When fats serve as the respiratory substrate they, being insoluble, must first be converted into soluble compounds before any oxidation of the breakdown products can proceed (Chapter 14). The utilization of proteins in the respiration of plant cells does not appear to occur commonly except in tissues which have been depleted of carbohydrates and fats. In starved leaves, for example, proteins are converted into amino acids which are then oxidized, a process which is commonly accompanied by the synthesis of asparagine, glutamine, and perhaps other amides. Under extreme conditions of starvation, the subsequent oxidation of such amides may result in the release of ammonia in plant tissues (Chapter 16).

As shown in several following chapters the process of respiration results in the generation of numerous highly reactive compounds in the cell from which other compounds, critical in the metabolism of the plant, are constructed by the operation of further synthetic pathways. This generation of reactive compounds is one of the important roles of the process. The other important role of respiration is the conservation of energy in forms which can be utilized by the cell. Energy becomes manifest in the living cells as well as in the inorganic systems as chemical energy, heat energy, radiant energy, surface energy, mechanical energy, potential energy, etc. All of the energy released from molecules during respiration can ultimately be traced back to radiant energy which was previously entrapped in the process of photosynthesis.

The release of heat energy during respiration can be readily demonstrated by enclosing the rapidly respiring plant materials such as germinating seeds, opening buds, or young floral parts in a calorimeter. One hundred grams of germinating wheat grains, for example, may release heat with sufficient rapidity to acquire temporarily a temperature as much as 20°C higher than that of the same weight of dead seeds in a comparable calorimeter. Although the energy released in respiration is generally expressed in terms of heat units—calories or kilogram calories—a substantial portion of it is not evolved as heat. That portion of the energy which is released as heat is all lost to the cells in which the respiration occurs. It is pure waste from the standpoint of the plant and is roughly analogous to the frictional loss of energy in a machine. In warm-blooded animals, in contrast to plants, the heat released in the respiratory process is important in maintaining the bodily temperature.

The most important way in which the energy made available in respiration is used, especially in younger cells, is in various synthetic processes. The synthesis of both glycerol and various fatty acids (Chapter 14) are specific examples of such reactions. The energy required to drive such synthetic endergonic reactions is derived from ATP or other high energy compounds (Chapter 9). Adenosine triphosphate is the principal molecular vehicle whereby energy derived from respiratory substrate molecules is transferred in energy-requiring synthetic reactions. A major role of respiration is the synthesis of ATP molecules. As much as 50 per cent and sometimes even more of the releasable energy in the respiratory substrate molecules may be conserved, at least temporarily, in ATP molecules (Chapter 13).

Other energy-requiring processes in plant cells, apart from endergonic synthetic reactions, are the maintenance of protoplasmic structure, the migration of the chromosomes and other cell constituents during cell division, the streaming of the cytoplasm, the accumulation of solutes by cells (Chapter 15), the translocation of solutes (Chapter 17), the growth of stems in opposition to gravity, and the growth of root tips against the frictional resistance of the soil. Although most, if not all, of the energy used in such processes comes from respiration, the amounts used in any of these processes is probably relatively small compared with the quantities utilized in endergonic synthetic processes.

However, not all plant processes rely upon respiratory energy for their motive power. Transpiration is the most important exception. The energy required for the vaporization of water in this process comes either directly from the radiant energy of sunlight or from the heat energy of the surrounding atmosphere.

Comparative Rates of Respiration.   Rates of respiration as expressed either in terms of oxygen consumption or carbon dioxide liberation vary greatly, depending upon the plant organ or tissue and the environmental conditions. Since the seat of respiration lies in the protoplasm, a correlation often exists between the proportion of protoplasm present in a tissue and the intensity of the respiration process in that tissue. As a general rule, respiration rates are found to be the greatest in meristematic tissues, such as growing root or stem tips or the embryos of germinating seeds. It is precisely in such tissues that the proportion of protoplasm is greatest in relation to the total dry weight of the tissue. In mature tissues, such as photosynthetically active leaves, a larger proportion of the dry weight of the tissue mass is composed of inert cell-wall materials, hence the respiration rates of such tissues, expressed in the usual terms, are almost invariably less than those of meristems under comparable conditions. Senescent tissues, such as yellowing leaves or ripe fruits in which the proportion of protoplasm to dry weight is still smaller, generally have lower rates of respiration than the same tissues had when in a metabolically active condition. The lowest rates of respiration are found in dormant seeds and spores, a marked increase in the rate of respiration being one of the striking physiological aspects of germination. The relatively slow rate of respiration in such structures is not, however, primarily the result of a low proportionate amount of protoplasm, but of other factors, among which the deficient hydration of the tissues is one of the most important.

The representative rates of respiration for a number of plant organs are listed in Table 12.1. Even for the plant parts tabulated these rates are to be regarded as only approximations, since the rate for any one plant organ or tissue is subject to marked fluctuations resulting from the influence of various internal and external factors.

TABLE 12.1   RESPIRATION RATES OF PLANT TISSUES IN TERMS OF
VOLUME OF OXYGEN ABSORBED OR VOLUME OF CARBON
DIOXIDE RELEASED IN 24 HOURS PER GRAM
OF DRY WEIGHT[a]

Plant	Organ	Tempera-ture °C	Respiration rate
			ml O_2 absorbed
Wheat *(Triticum sativum)*	young roots	15–18	67.9
Red clover *(Trifolium pratense)*	leaves	20–21	27.2
Rice *(Oryza sativa)*	young roots	14–17	44.4
Mint *(Mentha aquatica)*	roots	18–19	37.2
			ml CO_2 liberated
Lilac *(Syringa vulgaris)*	leaf buds	15	35.0
Linden *(Tilia europa)*	leaf buds	—	66.0
Lettuce *(Latuca sativa)*	germinating seeds	16	82.5
Poppy *(Papaver somniferum)*	germinating seeds	16	122.0
Mold *(Aspergillus niger)*	mycelium	—	1800.0

[a]Data of Kostychev, *Plant Respiration,* C. J. Lyon; Editor. P. Blakiston's Son and Co., Philadelphia, 1927, p. 5.

The Respiratory Ratio.   The ratio of the volume of CO_2 released to the volume of O_2 absorbed in the respiratory process is termed the respiratory ratio or respiratory quotient. When the complete oxidation of a hexose sugar occurs, as already pointed out:

$$\frac{6 \text{ moles CO}_2}{6 \text{ moles O}_2} = 1$$

The respiratory ratio for any plant or plant part can be determined by making parallel measurements of the rates of carbon dioxide release and oxygen consumption.

The respiratory ratio of germinating seeds in which the accumulated foods are principally in the form of carbohydrates is invariably found to be approximately 1, as long as oxygen has free access to such seeds. This is true, for example, of the germinating grains of practically all of the cereals (wheat, maize, oats, etc.). Similarly the respiratory ratios for the leaves of many species of plants and most flowers have been found to be in the neighborhood of 1.

The proportion of the volume of carbon dioxide evolved to the volume of oxygen absorbed may vary greatly from a unit value, however, when the primary respiratory substrate is other than a hexose or when the oxidation is incomplete. The respiratory ratio for seeds in which the foods present are principally oils is invariably less than 1, because the proportion of oxygen to carbon is less in fats than in carbohydrates. The respiratory ratio for the complete oxidation of tripalmitin ($C_{51}H_{98}O_6$), a representative fat (Chapter 14) for example, is approximately 0.7. For the same reason the respiratory ratio when the primary respiration substrate is a protein is less than 1 (usually 0.8–0.9). The opposite situation prevails in the leaves of some plants, especially succulents, in which certain dicarboxylic acids such as malic, tartaric, and oxalic, accumulate and serve as the primary substrates of respiration. Such compounds are relatively rich in oxygen as compared with carbohydrates, and their respiratory ratio is therefore greater than 1.

As discussed more fully in the following chapter, the release of carbon dioxide without any corresponding utilization of oxygen is characteristic of anaerobic respiration. Strictly speaking such a process does not have a respiratory ratio. Very often, however, especially when the oxygen supply is deficient, both aerobic and anaerobic respiration occur simultaneously in a plant tissue. Under such conditions the volume of carbon dioxide evolved may be very large in proportion to the volume of oxygen absorbed, and the respiratory ratio is much greater than 1.

Factors Affecting the Rate of Aerobic Respiration.   A number of factors, some internal, others external, are known to influence the rate of respiration of plant cells. Under some conditions the metabolic pathway of the respiration process, as well as its rate, may be influenced by the prevailing factors. The failure of oxygen to gain access to a tissue in adequate quantities, for example, may force some of the respiration along anaerobic pathways which would have the effect of increasing the respiratory ratio.

Protoplasmic Conditions.   As pointed out previously, young meristematic tissues, which are relatively rich in protoplasm, usually have higher rates of respiration per gram of dry weight than older tissues in which the proportion of the cell wall material is greater, and which often also contain nonliving cells. The gross amount of protoplasm present is only a crude index, however, of the respiratory capacity of a tissue. The internal structural relationships of the various organelles and the occurrence and distribution of enzyme systems differ from one kind of cell to another and undoubtedly have important effects on respiration rates. Because they are the centers of aerobic respiration, special attention focuses on the mitochondria in this connection. The number of mitochondria per cell is an important factor in controlling the rate of respiration in that cell.

Temperature.   As is true of most other biological processes, temperature effects upon the rate of respiration are rather complex. In general, within certain limits an increase in temperature results in an increase in the respiration rate (Fig. 12.1). In the temperature range between 0 and 45°C, an increase in temperature resulted in an increase in the initial rate of respiration of the pea seedlings. At temperatures above approximately 30°C, the rate of respiration showed a decrease with time, which became more marked the higher the tem-

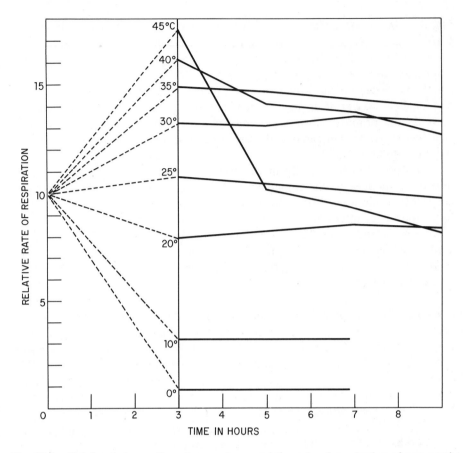

Fig. 12.1. Relation between time, temperature, and the rate of respiration of pea seedlings. Dotted lines represent the period during which the temperature of seedlings was changed from 25°C to the indicated temperatures. Data of Fernandes, *Rec. trav. Botan. néerland* **20**, 1923:107-111.

perature. For the pea seedlings used in this experiment the optimum temperature would appear to be about 30°C, as this is approximately the maximum temperature at which there is a steadily maintained rate of respiration. The optimum temperature for respiration, considered in this sense, is not the same for all plant tissues. For some it is clearly higher than the value obtained in this experiment with pea seedlings and for others it is lower. The decreasing rate of respiration with time, which becomes progressively more pronounced with an increase in temperature, is considered to result from the inactivation of enzymes.

As the temperature is decreased below 0°C, the rate of respiration gradually diminishes until it becomes imperceptible. Measurable rates of respiration have been recorded, however, in some plant tissues at temperatures as low as −20°C.

The Q_{10} of respiration for plant tissues within the temperature range 10 to 30°C is usually between 2.0 and 2.5. At temperatures below 10°C, higher Q_{10} values have been found for a number of tissues. Above 30°C, the determination of the Q_{10} values for plant respiration with any degree of certainty is difficult because the rate commonly is in part a function of the time interval over which it is measured.

The temperature to which a plant organ is exposed sometimes has important *indirect* effects on the rate of respiration. When the temperature of a potato tuber is lowered from a few degrees above to about 0°C, the respiration rate *increases*. This is probably the result of the effect of low temperatures in causing a shift in the starch-sugar equilibrium toward the sugar side. An increase in the quantity of respiratory substrate in plant cells results in an increase in the rate of respiration whenever it is the limiting factor, a situation which apparently pertains to potato tubers under these conditions. Similar indirect effects of temperature upon the rate of respiration are probably of frequent occurrence in other plant tissues.

Food. As a general rule, an increase in the soluble food content (usually carbohydrates) of plant cells results in an increase in the respiration rate up to a certain point at which some other factor becomes limiting. One effect of the concentration of foods in cells upon the rate of respiration has just been described in the preceding section. The effect of this factor on respiration rates can also be demonstrated in etiolated leaves. For example, it has been reported that 100 g of carbohydrate-deficient etiolated bean leaves released an average of 89.6 mg of carbon dioxide per hour at room temperature. After floating the same leaves upon a sucrose solution for two days in the dark during which a considerable absorption of sugar occurred, the average rate of carbon-dioxide release increased to 147.8 mg per hour.

Oxygen Concentration of the Atmosphere. The effect of the oxygen concentration of the atmosphere on the rate of respiration varies with the kind of tissue, concentration of oxygen, period of exposure, and other prevailing environmental conditions. The apparent magnitude of the effect, and sometimes also its direction (increase or decrease), may also differ, depending upon whether the carbon-dioxide output or the oxygen intake is used as the index of respiration. Variations in the oxygen content of the aerial atmosphere are too slight to have any appreciable effect on respiration rates, but this is not true of the soil atmosphere (Chapter 7). In general unless the oxygen concentration deviates by at least 5 per cent from the usual atmospheric concentration, however, the effects on the rates of respiration are small or negligible.

The rate of the carbon-dioxide output of potato tubers has been found to be essentially the same over a range of oxygen concentrations from 6.2 to 98.6 per cent. With artichoke (*Helianthus tuberosus*) tubers, the rate of the carbon dioxide release was essentially the same in oxygen concentrations above atmospheric value, as in air; but in oxygen concentrations below atmospheric value the rate showed a progressive lowering with a decrease in concentration. With carrot roots, irregular results were obtained at oxygen concentrations below atmospheric; but at oxygen concentrations above atmospheric, the higher the concentration, the greater the rate of carbon dioxide output. The rate of respiration, measured either as carbon dioxide output or as oxygen intake, was progressively reduced as oxygen concentrations were lowered below atmospheric in the following plant materials: asparagus stalks, bean fruits, spinach tops, shelled peas, and carrot roots.

One of the factors accounting for such differences in the relation between oxygen concentration and the rate of respiration is the usual atmospheric rate of respiration. The respiration rate of tissues in which the usual atmospheric rate is relatively low—for example, potato tubers—appears to be less affected by a lowering of oxygen concentration than that of tissues in which the atmospheric rate is higher.

Another factor accounting for differences in the effect of oxygen concentration on respiration is the greater capacity of some tissues than others for fermentation (anaerobic respiration). An example of this effect is the contrasting behavior of wheat and rice seedlings (Fig. 12.2). At low oxygen concentrations, the carbon-dioxide output by rice seedlings was much greater than that by wheat seedlings largely because of the greater capacity of the former for fermentation than the latter.

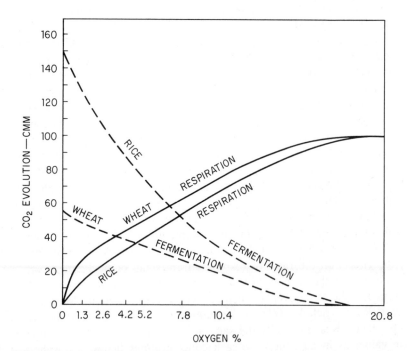

Fig. 12.2. Relation between oxygen concentration and the rates of fermentation and respiration by wheat and rice seedlings. Data of Taylor, *Am. J. Botany* **29**, 1942:726.

Carbon Dioxide Concentration of the Atmosphere.   The effect of the carbon-dioxide concentration on respiration, like that of oxygen, differs with its concentration, the kind of tissue, the period of exposure, and other prevailing environmental conditions. As with oxygen, the usual atmospheric fluctuations in carbon-dioxide concentration of the aerial atmosphere are not sufficient to have appreciable effects on the respiration rates.

In a study of white mustard seedlings it was shown that the rate of respiration decreased with increase in the carbon-dioxide concentration (Table 12.2). This effect was shown whether respiration was measured in terms of carbon-dioxide release or oxygen absorption. The decreasing effect on the rate of carbon-dioxide release was more marked than on the rate of absorption of oxygen. Hence the higher the carbon dioxide concentration of the atmosphere, the lower the respiratory ratio.

TABLE 12.2   THE EFFECT OF CARBON DIOXIDE CONCENTRATION UPON THE RATE OF RESPIRATION OF GERMINATING WHITE MUSTARD SEEDS[a]

	Percentage of CO_2 initially present					
	0	10	20	30	40	80
CO_2 evolved (ml)	58	48	38	33	26	17
O_2 absorbed (ml)	71	57	49	45	38	32
Respiratory ratio	0.82	0.84	0.77	0.73	0.69	0.53

[a] Initial concentration of O_2 in each experiment is 20 per cent. The duration of the experiments is 14 hours. Data of Kidd, *Proc. Roy. Soc.* (London), B 89, 1915: 148.

The rate of respiration of some plant tissues, on the contrary, is increased when they are exposed to relatively high concentrations of carbon dioxide. For example, the exposure of potato tubers to concentrations of carbon dioxide in excess of about 20 per cent for periods of greater than 20–24 hours resulted in a marked increase in the respiration rate as measured in terms of oxygen consumption. In 60 per cent carbon dioxide the rate of respiration sometimes exceeded that of the controls by more than 200 per cent. The increased rate of this respiration is undoubtedly to be explained, at least in part, by the increase in the concentration of sugar which occurs in the cells under such conditions. Shorter periods of exposure of the potato tuber to high concentrations of carbon dioxide resulted in a decreasing rather than an increasing effect upon the respiration. Similar effects of high concentrations of carbon dioxide were found for onion and tulip bulbs and for beet roots. With asparagus shoots and shelled lima beans, on the other hand, high concentrations of carbon dioxide resulted in a reduction in the respiration rate as in germinating mustard seeds.

The most marked effects of variations in the oxygen and carbon dioxide concentrations of the atmosphere upon respiration are doubtless those upon roots, underground stems, and seeds. The carbon dioxide concentration of the soil atmosphere may sometimes be as much as 10 per cent and occasionally even higher, while the oxygen content may, in some soils, at some times, approach a zero value (Chapter 7). The usual effect of either or both of these conditions is a retardation in the rate of respiration. As a result of such diminished respiration rates, such fundamental processes as the absorption of water (Chapter 7) and of mineral salts (Chapter 15) by roots, the growth of roots (Chapter 20), and the germination of seeds (Chapter 24) may be retarded or even entirely inhibited.

Hydration of the Tissues. This is basically a protoplasmic factor, but its clearly demonstrable effects on the rate of respiration justify its separate consideration. Its effect can most clearly be demonstrated in germinating seeds (Fig. 12.3). The marked rise in the rate of respiration as the water content of wheat grains increases from 16 to 17 per cent, as contrasted with the small effects of increases in water content in the range below 16 per cent, is probably a result of practically all of the water in the grains being in the adsorbed condition at water contents below 16 per cent. Minor variations in the water content of well-hydrated plant tissues do not have any great influence on the rate of respiration. However when leaves or other plant organs approach a wilt-

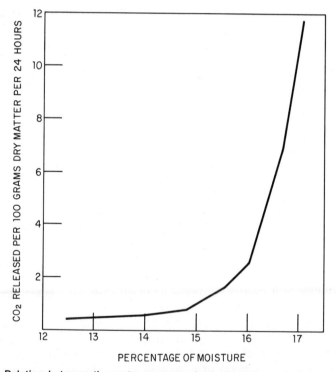

Fig. 12.3. Relation between the water content of **wheat** grains and the rate of respiration. Data of Bailey and Gurjar, *J. Agr. Research,* **12,** 1918:692.

ing condition they often exhibit an increase in the rate of respiration. As leaves become wilted, accumulated starch is often converted into sugars and the resulting increase in the soluble carbohydrate content of the cells may account for this rise in the respiration rate.

Light.   The analysis of the effects of light on respiration is complicated by the occurrence in many plants of a light-mediated process called *photorespiration,* which is discussed later in this chapter. Light effects on "dark respiration," using this term to contrast it with photorespiration, are indirect. In chlorophyllous organs light may affect the rate of respiration because of its influence upon the supply of the respiratory substrate resulting from photosynthesis. The direct illumination of plant organs commonly results in an increase in their temperature which may in turn result in an increase in respiration rate.

Injury.   The wounding of plant tissues almost invariably results in a temporarily increased rate of respiration. If a potato tuber is cut in half, for example, the loss of carbon dioxide from the two halves will be considerably greater than from the intact tuber. Similar results have been observed for many other plant tissues. The increased respiratory activity of wounded or otherwise injured plant tissues gradually rises to a maximum which is generally attained within a day or two, after which a diminution in rate sets in until approximately the rate which prevailed in the uninjured tissues is reestablished.

This increased respiration of potato tubers following wounding may be correlated with an increase in the sugar content of the tuber. An increase, which amounted in some experiments to from 53 to 68 per cent of the sugar originally present, occurred gradually, the maximum not being attained until several hours after the injury. The increase in sugar content is greater in the cells close to the cut surface than in those which are more remote from it. This increase in the quantity of the respiratory substrate is apparently an important factor in accounting for the increased loss of carbon dioxide by potato tubers following wounding, and probably of many other tissues as well.

Mechanical Effects.   A purely mechanical "stimulation" of respiration has been demonstrated in the leaves of a number of species. A gentle rubbing or bending of the leaf blade was sufficient to induce a marked rise in the respiration rate (often over 100 per cent) which persisted for some hours. No such effect was found for leaves in an atmosphere of nitrogen, indicating that the influence is on the aerobic phases of respiration. The mechanism of this effect is unknown, but it is obvious that it should be taken into account in any experiments on respiration which require handling of the plant material.

Effects of Certain Chemical Compounds on Respiration.   Although the rate of respiration, like that of other metabolic processes, may be influenced by many different kinds of compounds, particular interest attaches to certain substances which act more or less as enzymatic inhibitors when present in very low concentrations at one stage or another of the respiratory mechanism. Among these are cyanides, azides, carbon monoxide, fluorides, malonates, iodoacetate, and 2, 4-dinitrophenol (Chapters 9 and 13).

The Light Compensation Point.   In leaves or other chlorophyllous tissues the rate of photosynthesis usually exceeds the rate of respiration during the daylight hours. In maize, for example, the rate of photosynthesis during the daylight hours is, on the average, about eight times the rate of respiration. (See Fig. 11.10 for similar data on alfalfa.) The carbon dioxide released in respiration is re-utilized by the cells in photosynthesis, but since the latter process is occurring more rapidly than the former, additional carbon dioxide is continuously diffusing into the plant from the outside environment. Similarly, photosynthesis produces more oxygen than is used in respiration, the surplus diffusing out of the plant. Hence during the daylight hours, as long as conditions favorable for photosynthesis prevail, there is a net movement of carbon dioxide into the green parts of plants, and a net loss of oxygen from them. Under such conditions the occurrence of the gaseous exchanges accompanying respiration in green leaves is completely masked.

At night or in the dark the reverse conditions obtains, oxygen moving into the green parts of the plant and carbon dioxide passing out of them. Similar gaseous exchanges are characteristic of the nongreen organs of a plant, whether in the light or in the dark. The magnitude of the gaseous exchanges occurring between a green plant organ and its environment in the absence of light are usually less than those which generally take place—but in the opposite direction —in its presence.

Since at low intensities light is usually the limiting factor in photosynthesis it is evident there should be a certain light intensity at which the rate of photosynthesis and the rate of respiration in a leaf or other chlorophyllous organs are exactly equal. At this light intensity, called the *light compensation point,* the volume of carbon dioxide being released in respiration is exactly equal to the volume being consumed in photosynthesis, while the opposite is true of oxygen. In other words, at the light compensation point apparent photosynthesis is zero. The light intensity corresponding to this compensation point varies considerably with different species of plants. The light compensation point for any one species is also influenced by various environmental factors, especially temperature, and is markedly affected by the conditions to which the leaves or other photosynthetic organs have been exposed during their development. The light compensation point for a number of aquatic species is of the order of 1–2 per cent (100–200 foot candles) of the intensity of full midday summer sunlight. The leaves of some shade species of land plants have light compensation points of approximately 50 to 100 foot candles. In sun species the compensation point is usually slightly higher, being of the order of 100–200 foot candles.

No plant can survive indefinitely in nature at the light intensity of the compensation point. Under such conditions there is no photosynthesis which compensates for night respiration. Also, the light compensation point is usually measured only for the leaves or aerial organs of the plant, no allowance being made for the respiration of the roots or other underground organs. Hence the actual minimum light intensity at which any species could survive in nature would necessarily be somewhat greater than its compensation point as usually measured.

The Carbon-Dioxide Compensation Point.   Under a nonlimiting light intensity a carbon-dioxide concentration exists under which photosynthesis just compensates for respiration, and the value of apparent photosynthesis is zero. The atmospheric carbon-dioxide concentration under which this condition prevails is called the *carbon-dioxide compensation point*. The value of this quantity varies considerably with environmental conditions, especially temperature, and also from one kind of plant to another. Some species characteristically have very low carbon-dioxide compensation points (close to 0 ppm); examples are maize, sugar cane, and sorghum. Many others have much higher carbon-dioxide compensation points (50 ppm or even higher); examples are tomato, soybean, and wheat.

Photorespiration.   This term refers to a distinctive type of respiration exhibited by many kinds of plants when they are exposed to light. It takes place only in chlorophyllous plant tissues and occurs in part through the photosynthetic mechanism in the chloroplasts, in part in a special kind of organelle known as the peroxisomes, and in part in the mitochondria. The peroxisomes are small spherical organelles about 0.2 to 1.5 μ in diameter which seem to be primarily associated with this process.

There is a close correlation between the carbon-dioxide compensation point and the incidence of photorespiration. Species having very low carbon-dioxide compensation points, such as maize, exhibit little or no photorespiration. Species with relatively high carbon-dioxide compensation points, such as tomato, have relatively high rates of photorespiration. The magnitude of photorespiration can be ascertained by measuring the carbon-dioxide output of a leaf or leafy shoot when exposed to light in a carbon-dioxide-free atmosphere. Such measurements have shown the photorespiration rate in tobacco is about four times as great as its mitochondrial respiration rate. By contrast, in maize, there is only a slight increase in respiration in the light as compared with the dark, although the mitochondrial respiration rate of this species is closely comparable with that of tobacco.

Glycolic acid is a key metabolite in photorespiration. This compound is a side product in photosynthesis, being formed in the chloroplasts, presumably from some of the sugar intermediates. The first step in the oxidation of glycolic acid occurs in the peroxisomes as follows:

$$
\begin{array}{c}
\text{CH}_2\text{OH} \\
| \qquad\qquad + \text{O}_2 \\
\text{COOH} \\
\text{Glycolic acid}
\end{array}
\xrightarrow{\text{glycolic acid oxidase}}
\begin{array}{c}
\text{CHO} \\
| \qquad\qquad + \text{H}_2\text{O}_2 \\
\text{COOH} \\
\text{Glyoxylic acid}
\end{array}
$$

The H_2O_2 formed is dismutated into water and oxygen by the enzyme catalase. The further steps of glycolic acid metabolism are not entirely clear, but there is evidence that the glyoxylic acid is next converted into glycine in the peroxisomes, and that two molecules of glycine in turn are converted into serine, this latter step apparently taking place in the mitochondria. The carbon dioxide release appears to occur during this step in the reaction chain. The serine apparently is recycled back into the pool of photosynthetic intermediates.

A close interrelationship exists between the processes of photosynthesis and photorespiration. Plants with a high rate of photorespiration are, in general, photosynthetically less efficient than plants with a low rate. Photorespiration involves the side-tracking of the intermediates of photosynthesis into glycolic acid, the oxidation of which does not involve any phosphate or phosphate esters. Hence, unlike dark respiration, photorespiration does not result in the generation of any high energy compounds such as ATP. Plants exhibiting a pronounced Warburg effect (Chapter 10) usually have relatively high photorespiration rates, because a relatively high oxygen concentration favors the synthesis of glycolic acid, which, in turn, favors photorespiration. Their higher rate of photorespiration lowers the photosynthetic productivity of such plants.

Lack of measurable photorespiration is a characteristic of plants in which the Hatch-Slack pathway of photosynthesis is followed. Some evidence suggests that the inability to detect photorespiration in such plants is not the result of the absence of such a process, but rather because the carbon dioxide generated is recycled into photosynthesis and thus is retained in the metabolic system. This may largely account for the fact that plants with the Hatch-Slack pathway are more efficient photosynthetically than those which follow the Calvin-Benson pathway. For example, when beans, a Calvin-Benson pathway plant, were subjected to an atmosphere containing 2 to 5 per cent oxygen they grew about twice as fast as in oxygen at the usual atmospheric concentration of 21 per cent. When, after a while, such plants were placed back in air, their growth rate slowed. The growth rate of maize, a Hatch-Slack pathway plant, on the other hand, was not affected by subjecting the plants to a 2–5 per cent atmosphere of oxygen. The explanation of these results with bean seems to be that at the lower than atmospheric oxygen concentration less glycolic acid is synthesized. This leads to a lower rate of photorespiration and hence a higher rate of net photosynthesis. The increased availability of photosynthate favors an increased growth rate. In Hatch-Slack pathway plants, for anatomical and metabolic reasons that are not wholly understood, photorespiration is so eliminated or sufficiently reduced that it does not result in a diminution of net photosynthesis.

Photorespiration should not be confused with photo-oxidation, discussed in Chapter 11. The former is a normal process, occurring in many species of plants under moderate irradiances. The latter is an abnormal and destructive process, occurring only under light intensities which are excessive, for a given kind of plant, relative to those to which the plant is usually subjected.

SUGGESTED FOR COLLATERAL READING

Giese, A. C. *Cell Physiology, 3rd Edition.* W. B. Saunders Company, Philadelphia. 1968.

Stiles, W., and W. Leach. *Respiration in Plants.* John Wiley & Sons, Inc., New York, 1960.

Zelitch, I. *Photosynthesis, Photorespiration, and Plant Productivity.* Academic Press, Inc., New York, 1971.

13 | METABOLIC PATHWAYS OF RESPIRATION

The summary chemical equation commonly used to represent the process of respiration:

$$C_6H_{12}O_6 + 6O_2 \longrightarrow 6CO_2 + 6H_2O + 673 \text{ kg-cal}$$

represents equally well the combustion of sugar. It might be taken to imply that the release of energy in respiration is achieved in a manner analogous to that by which heat energy is liberated in combustion. Actually the two processes are not closely similar in mechanism. Sugar and atmospheric oxygen are much too stable to unite directly within any temperature range in which living organisms could survive.

Unlike the combustion type of oxidation which would result from a direct combination of oxygen with sugar, the respiration of a hexose to carbon dioxide and water involves long sequences and cycles of reactions in which many kinds of organic molecules participate. Some of these chains of reactions are known, at least for some tissues; others seem very probable, but are less well supported by experimental evidence. These reaction chains are not the same for all tissues, and for a given tissue may differ depending upon various conditions and especially on whether or not atmospheric oxygen is available. Regardless of the exact pathway which the reaction sequence follows, however, oxidation occurs very gradually, step by step, with only small units of energy being released or transferred at a time.

As will be shown in due course the component reactions of respiration are of a number of different kinds. The most significant from the standpoint of energy release of transfer are the oxidation-reduction reactions, many of which are achieved by transfers of electrons or hydrogen atoms from molecule to molecule (Chapter 9).

Many enzymes and coenzymes play a role in the stepwise oxidation of carbohydrates or of other molecules which serve as respiratory substrates. A substantial number of these enzymes and coenzymes have been isolated, and their roles in the respiratory mechanism have been ascertained. Most of our understanding of the respiratory process has resulted from the study of these enzymatic reactions. The action of a specific enzyme may be checked or blocked by certain inhibitors (Chapter 9). By the use of a suitable inhibitor it is possible to block the action of one enzyme without checking the activity of others. The respiratory sequence of reactions can therefore be stopped at a given point and the products which accumulate studied. By systematically interrupting the chain of reactions at various points in this way, considerable information can be obtained regarding the nature of different steps in the overall process.

Anaerobic Respiration.   The external aspects of aerobic respiration have been considered in the preceeding chapter; a similar consideration will now be given to anaerobic respiration before the mechanism of the overall process is discussed.

Alcoholic Fermentation.   The term "fermentation" is applied loosely to a variety of oxidation processes which are features of the metabolism of various species of bacteria and fungi. Some of the better-known fermentations are alcoholic fermentation, acetic acid fermentation, lactic acid fermentation, butyric acid fermentation, oxalic acid fermentation, and citric acid fermentation. Some fermentation reactions are widely used in industry for the commercial production of certain chemical compounds. The chemical aspects of many of the fermentation processes occurring in bacteria and fungi have been investigated much more thoroughly than the similar processes which take place in the tissues of higher plants.

Most fermentation reactions are anaerobic, but some are aerobic. In all of them incompletely oxidized compounds accumulate as end products, and this is the most characteristic feature of fermentation processes. The term fermentation is also used to apply to the closely analogous processes which occur in the tissues of higher plants under certain conditions as discussed later.

Of the many known kinds of fermentations the most thoroughly investigated is the process of alcoholic fermentation. Although this process has been familiar to the human race for time out of mind, it was not until the classical researches of Pasteur, begun in 1857, that it was recognized that alcoholic fermentation resulted from the metabolic activities of yeast plants and that it was an anaerobic process. Yeasts, it should be recalled, are single-celled fungi belonging to the Ascomycetes. Yeasts can multiply by budding and under certain conditions produce ascospores (Fig. 13.1).

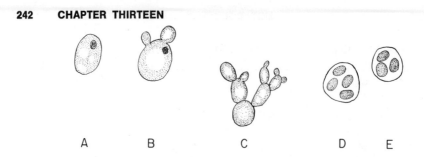

A   B   C   D   E

Fig. 13.1. Yeast plants. (A) Vegetative cell. (B) Budding of a yeast cell. (C) "Colony" of cells resulting from budding. (D) and (E) Asci containing ascospores.

A further step in the understanding of alcoholic fermentation was furnished by Buchner's demonstration in 1897 that an active agent or enzyme ("zymase") could be extracted from yeast cells and that this enzyme could catalzye the process in the total absence of yeast cells. Zymase is now recognized to be a complex of enzymes rather than a single enzyme.

Alcoholic fermentation may occur in almost any moist sugar-containing medium or sugar solution, such as a fruit juice, which is inoculated with yeast or left exposed to the air. Since various species of wild yeast are blown about through the atmosphere, inoculation of such media will occur without human intervention.

The following summary equation represents the net chemical changes occurring in alcoholic fermentation:

$$C_6H_{12}O_6 \xrightarrow{\text{"zymase"}} 2C_2H_5OH + 2CO_2 + 21 \text{ kg-cal}$$

As this equation shows, the fermentation of 1 mole of a hexose sugar results in the production of 2 moles of ethanol and 2 moles of carbon dioxide, energy to the amount of approximately 21 kg-cal being released in the process. The carbon dioxide evolved escapes as a gas, accounting for the effervescence of a fermenting liquid. Certain by-products such as glycerol, succinic acid, amyl alcohol, and other compounds are also usually produced in small quantities as a result of the subsidiary reactions. The chemical mechanism of this process is considered later in the discussion.

Yeasts can ferment glucose, fructose, galactose, and mannose directly. Since yeast cells also contain the enzymes sucrase and maltase, the disaccharides sucrose and maltose can also be fermented after being hydrolyzed to hexose sugars. On the other hand, most kinds of yeasts cannot ferment starch because they do not synthesize amylases. This is the reason that germinated barley (malt) is used rather than the ungerminated grains in the brewing industry, since the sugar content of the grains increases greatly during germination.

Alcoholic fermentation is an anaerobic process, occurring without any utilization of atmospheric oxygen. Oxidation is accomplished by intramolecular atomic shifts which take place in such a manner that the sum total of the energy remaining in the resulting compounds is less than that present in the original substrate. The flammable nature of alcohol attests to the fact that alcoholic fermentation results in only an incomplete oxidation of hexose molecules. Hence the quantity of energy released in anaerobic respiration is much less than in aerobic respiration. In spite of its relative inefficiency the process of fermentation is the method by which yeast plants obtain all the necessary energy under anaerobic conditions.

It might be supposed that the efficient aeration of a sugar solution containing yeast plants would result in the complete oxidation of the sugars by a process of aerobic respiration. On the contrary, ethanol and carbon dioxide are the principal end products whether the reaction occurs in the presence or absence of oxygen. Some aerobic respiration (as much as one-third of the total, according to some investigators) does occur when the yeast cells have access to oxygen. The predominance of fermentation even in the presence of oxygen is generally ascribed to the possession by yeast cells of a relatively ineffective respiratory enzyme mechanism, as compared to a highly active fermentation system. In the presence of oxygen, however, the multiplication of yeast cells occurs at a much more rapid rate than in its absence. This is presumably a consequence of the markedly greater energy release resulting from the occurrence of some aerobic respiration.

When sugar in solution is fermented by yeast, one of the end products, ethanol, accumulates in the solution, while carbon dioxide escapes as a gas. However, there is a definite limit to the accumulation of ethanol. When its proportion in the liquid reaches a certain value, which may range from 9 to 18 per cent depending upon the species or strain of yeast, the cells are poisoned by the alcohol and the fermentation process stops.

Anaerobic Respiration in the Tissues of Higher Plants.   Under anaerobic conditions, and sometimes even in the presence of oxygen, processes identical with or similar to alcoholic fermentation occur in many of the tissues of the higher plant. The several terms fermentation, intramolecular respiration, and anaerobic respiration are in common use to designate such processes. The occurrence of such a process in any higher plant tissue results in the evolution of carbon dioxide and often, although by no means invariably, in the accumulation of ethanol within the cells. In some higher plant tissues ethanol is not a product of anaerobic respiration, and in others the quantity of ethanol formed does not correspond quantitatively to the amount of hexose broken down. Various organic acids, such as oxalic acid, tartaric acid, malic acid, citric acid, and lactic acid, are also common end products of anaerobic respiration in the tissues of higher plants.

At least a few tissues of higher plants possess such an effective anaerobic respiration mechanism that this process predominates over aerobic respiration even in the presence of oxygen in considerable concentrations. In germinating rice grains, for example, in oxygen concentrations as high as about 8 per cent, the rate of anaerobic is equal to the rate of aerobic respiration; and even in germinating wheat germs—a more markedly aerobic tissue—a considerable amount of anaerobic respiration occurs at this oxygen concentration. Anaerobic respiration also takes place in the presence of oxygen in at least some kinds of tissues in which the usual course of aerobic respiration has been interrupted by cyanide or some other specific enzyme inhibitor.

In the majority of the tissues of the higher plants, however, anaerobic respiration is induced only when the access of atmospheric oxygen to the tissues is largely or entirely cut off. This process can be initiated in almost any tissue of a higher plant by subjecting it to an atmosphere which is devoid of oxygen or to an atmosphere in which the oxygen concentration is below a certain relatively low critical value. Different tissues of higher plants differ greatly in their toleration of lack of oxygen and the resultant anaerobic respiration which takes place in the cells. Some plants or plant organs can survive under such conditions for long periods; others succumb within a day or two. Maize seedlings, for example, do not remain alive much more than a day in an atmosphere devoid of oxygen. Apple and pear fruits, on the other hand, can survive storage in an atmosphere of pure oxygen or pure nitrogen for months without injury. These fruits continue to evolve carbon dioxide under such conditions, thus indicating the occurrence of a type of respiration for which atmospheric oxygen is not necessary.

Many of the known examples of anaerobic respiration in higher plants result from structural features of plant organs which prevent ready access of oxygen to the interior tissues. For example, the seed coats of a number of species are only slightly permeable to oxygen. During the earlier stages of the germination of such seeds, before the coats are ruptured, anaerobic preponderates over aerobic respiration. The best known example of this is in pea seeds, which in the early stages of germination evolve a volume of carbon dioxide three or four times as great as the volume of oxygen absorbed. Similarly, anaerobic respiration also occurs in maize grains, oat grains (especially if the glumes are left intact), and sunflower fruits during the early stages of germination. A similar condition probably obtains in many other seeds and dry fruits.

Anaerobic respiration also occurs naturally in some fleshy fruits. The "skin" of some fruits, of which the grape is the most familiar example, is relatively impermeable to oxygen; hence this process undoubtedly accurs in such organs.

The adverse effect of a flooding of the soil upon many species of plants appears to result from the substitution of anaerobic for aerobic respiration, since such a flooded soil is practically devoid of oxygen. The submergence of a field of maize, for example, soon results in serious injury to the plants, even if only the roots are immersed. If the flooded condition of the soil continues very long, the death of the plants commonly results. The plants often exhibit many of the symptoms of desiccation, suggesting a likelihood that the physiological processes of the roots have been altered in such a fashion that the absorption of water is no longer occurring at an adequate rate (Chapter 7).

At least two probable reasons can be advanced for the injurious effects of the substitution of anaerobic for aerobic respiration in normally aerobic tissues.

One of these is the much lower energy output of the former process as compared with the latter. Anaerobic respiration makes available only a small fraction as much energy per molecule of hexose oxidized as aerobic respiration. In metabolically active tissues, especially, the curtailed rate of energy transfer is probably inadequate for the normal maintenance of cell processes and deleterious effects are soon engendered within the cells. Another probable cause of the impairment of cells as a result of fermentation is the accumulation of substances which exert toxic effects on the protoplasm. During anaerobic respiration ethanol and other more or less toxic compounds accumulate in the cells in which this process is occurring and may be translocated to other parts of the plant which are still under aerobic conditions. In tissues in which anaerobiosis commonly occurs, considerable concentrations of such compounds are tolerated without harm. In normally aerobic tissues, however, the tolerance for such substances is much less and their accumulation within the cells soon leads to injurious effects.

In contrast with most terrestrial plants are the many aquatic species in which the rhizomes and roots, and in some species other organs as well, are continuously submerged. In some such species considerable quantities of oxygen diffuse into the submerged organs through aerenchyma tissue from the aerial organs (Chapter 7). At times such an internal movement of oxygen gas is adequate to maintain a completely aerobic respiration in the submerged organs, but at many times, at least in some species, it is inadequate, and some anaerobic respiration occurs. This has been shown to be of regular or frequent occurrence in the roots and rhizomes of the spatterdock (*Nymphaea advena*) and several other aquatic species. In such organs, respiration is at times preponderantly aerobic and at times preponderantly fermentative, although there is seldom a time at which some anaerobic respiration is not occurring at least in those tissues which are most remote from the supply of oxygen.

When many tissues are transferred from anaerobic to aerobic conditions, fermentation is largely or entirely suppressed and the rate of consumption of the respiratory substrate is reduced. This phenomenon is called the *Pasteur effect*. Although the rate of utilization of the substrate is diminished by the shift to aerobic conditions, the energy made available in the cells is usually increased because of the greater efficiency of aerobic as compared with anaerobic respiration as an energy source in metabolism. The existence of the Pasteur effect has long been known in yeasts and in animal tissue. Such a mechanism has also been reported to operate in various tissues of the higher plants, for example, in barley leaves, apple fruits, carrot and parsnip roots, and potato tubers.

Glycolysis. For the purposes of discussion it is convenient to distinguish two major phases of the main route of respiration: (1) the oxidation of carbohydrate to pyruvic acid, a process called *glycolysis,* and (2) the subsequent oxidation of pyruvic acid. The first of these phases appears to take place in many kinds of tissues, both plant and animal, in either the presence or absence of atmospheric oxygen. The second phase may follow a number of diverse pathways, however, as described later. Another important pathway of respiration, the pentose phosphate pathway, is also described later in this chapter.

Glycolysis is considered to start with phosphorylated hexose molecules. The origin of some kinds of phosphorylated monosaccharide molecules as an indirect result of photophosphorylation has already been discussed in Chapter 10. The additions of phosphate are most commonly accomplished by the transfers of phosphate groups from ATP but sometimes directly from phosphoric acid. Throughout most of the reactions of glycolysis the molecules of carbohydrates or derivatives therefrom remain in the phosphorylated state. The generally recognized stages of glycolysis, also called the Embden-Meyerhof-Parras (EMP) pathway, starting with glucose and ending with pyruvic acid, are shown in Fig. 13.2.

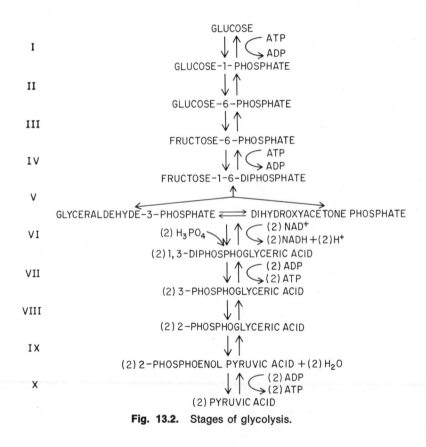

Fig. 13.2.   Stages of glycolysis.

Reaction I results in the phosphorylation of D-glucose in the 1 position as follows:

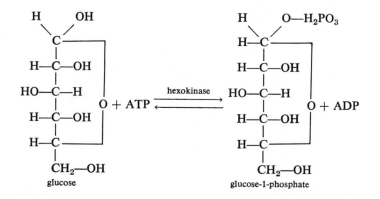

glucose

glucose-1-phosphate

In reaction II the phosphate group of glucose-1-phosphate is shifted to the 6 position. Actually much of the glucose-1-phosphate in plant cells originates as a result of the phosphorolysis of starch (Chapter 14) rather than directly from glucose.

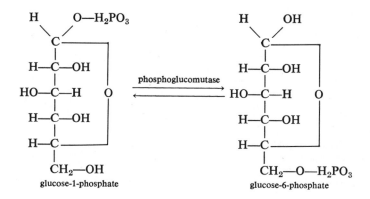

glucose-1-phosphate

glucose-6-phosphate

Reaction III results in the conversion of glucose-6-phosphate to fructose-6-phosphate:

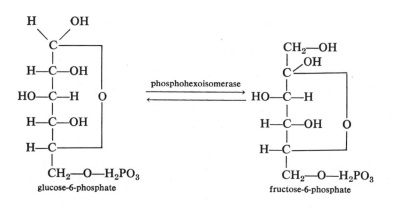

glucose-6-phosphate

fructose-6-phosphate

By interaction with ATP, fructose-6-phosphate is then converted into fructose-1,6-diphosphate (reaction IV):

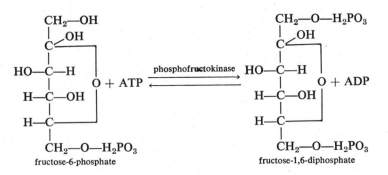

fructose-6-phosphate   fructose-1,6-diphosphate

In reaction V fructose-1, 6-diphosphate is split into molecularly equivalent amounts of two trioses, glyceraldehyde-3-phosphate and dihydroxyacetone-phosphate:

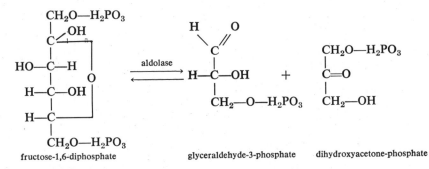

fructose-1,6-diphosphate   glyceraldehyde-3-phosphate   dihydroxyacetone-phosphate

The further reactions of glycolysis involve only the glyceraldehyde-3-phosphate. However, an equilibrium exists between this compound and the dihydroxyacetone-phosphate:

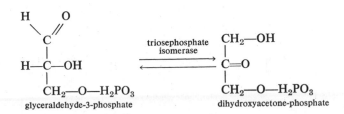

glyceraldehyde-3-phosphate   dihydroxyacetone-phosphate

Hence as the glyceraldehyde-3-phosphate is utilized, more of it can be formed from dihydroxyacetone-phosphate by this reaction. There is thus a net production of two glyceraldehyde-3-phosphate molecules for each fructose-1,6-diphosphate molecule entering glycolysis. There will therefore be two such molecules traversing the subsequent steps in glycolysis for each initial glucose molecule.

The next reaction (VI) is one of the most complicated in the glycolytic sequence, so much so in fact that some authorities consider it to take place in two steps:

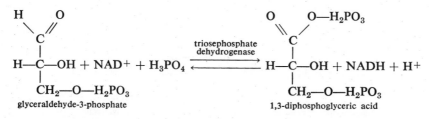

glyceraldehyde-3-phosphate

1,3-diphosphoglyceric acid

The glyceraldehyde-3-phosphate is both oxidized (by the loss of 2 H's to NAD$^+$) and phosphorylated by a reaction with phosphoric acid. The added phosphate is in the form of a high energy acyl phosphate group.

In reaction VIII the 1,3-diphosphoglyceric acid reacts with ADP forming 3-phosphoglyceric acid and ATP.

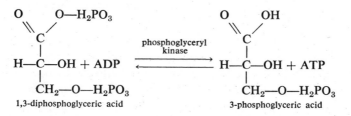

1,3-diphosphoglyceric acid

3-phosphoglyceric acid

In effect much of the high energy of the acyl phosphate group is transferred to ATP in this reaction.

The next reaction (VIII) is simply a transfer of the phosphate group from the 3 to the 2 position of the phosphoglyceric acid:

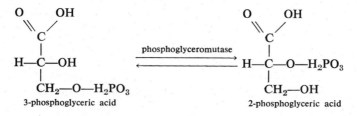

3-phosphoglyceric acid

2-phosphoglyceric acid

In reaction IX, 2-phosphoglyceric acid is dehydrated, forming phosphoenol pyruvic acid which contains a high energy enolic phosphate group:

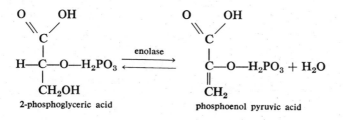

2-phosphoglyceric acid

phosphoenol pyruvic acid

Pyruvic acid is formed by the transfer of the phosphate group of the phosphoenol pyruvic acid to ADP (reaction X). In effect the high energy of the enolic phosphate group is incorporated in ATP.

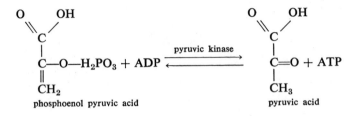

A summary equation for glycolysis, based on Fig. 13.2, can be written as follows:

$$C_6H_{12}O_6 + 2\ H_3PO_4 + 2\ NAD^+ + 2\ ADP \longrightarrow$$
$$2\ CH_3 \cdot CO \cdot COOH + 2\ ATP + 2\ NADH + 2\ H^+ + 2\ H_2O$$

A net of two moles of ATP (four generated minus two used in converting the original hexose to hexose diphosphate) and two moles of NADH + H$^+$ are generated per mole of hexose broken down. Furthermore, by operation through the process of oxidative phosphorylation, as discussed later, each mole of the NADH + H$^+$ gives rise to three moles of ATP, a total of six in all. A net gain of eight moles of ATP, which can be used as energy sources in other metabolic reactions, is therefore generated from each mole of hexose oxidized to pyruvic acid.

The pyruvic acid, which is also a product of glycolysis, is a highly reactive compound. It occupies a pivotal position in the sequence of respiratory reactions, because from it further reactions diverge in a number of directions. The chemical pathway along which this compound undergoes further modifications depends in part upon whether prevailing conditions are aerobic or anaerobic. Some of the better known compounds into which pyruvic acid can be converted, in one kind of organism or another, are acetaldehyde, lactic acid, oxalacetic acid, the amino acid alanine (Chapter 16), and acetyl-S-CoA (see later).

Anaerobic Oxidation of Pyruvic Acid.   In the absence of oxygen, and under certain other conditions, the anaerobic oxidation of pyruvic acid usually occurs, but the course of the reaction differs in different tissues and organisms. In general, the products of the anaerobic respiration of pyruvic acid are incompletely oxidized compounds such as alcohols and organic acids.

In yeasts and many other fungi, and in at least some higher plants under some conditions, anaerobic respiration results in the formation of ethanol.

The first step in this reaction is the decarboxylation of pyruvic acid as follows:

$$CH_3 \cdot CO \cdot COOH \xrightarrow{\text{pyruvic carboxylase}} CH_3 \cdot CHO + CO_2$$

pyruvic acid   acetaldehyde

A second step is the reduction of acetaldehyde to ethanol, which in yeast occurs as follows:

$$CH_3 \cdot CHO + NADH + H^+ \xrightarrow[\text{dehydrogenase}]{\text{alcoholic}} CH_3 \cdot CH_2 \cdot OH + NAD^+$$

acetaldehyde   ethanol

In the muscle tissue of animals, and in some bacteria, anaerobic respiration results in the conversion of pyruvic acid to lactic acid:

$$CH_3 \cdot CO \cdot COOH + NADH + H^+ \xrightarrow[\text{dehydrogenase}]{\text{lactic acid}} CH_3 \cdot CHOH \cdot COOH + NAD^+$$

pyruvic acid   lactic acid

The utilization of NADH + H$^+$ in these last two reactions is an important aspect of anaerobic respiration. There is a one to one relationship between the production of NADH + H$^+$ (step VI of glycolysis) and the utilization of NADH + H$^+$ in the generation of ethanol or lactic acid. Either of the latter two processes regenerates NAD$^+$ which can then be reused in step VI of glycolysis. In aerobic respiration, discussed later, the NADH + H$^+$ molecules ultimately pass their two hydrogens to oxygen in the process of oxidative phosphorylation. In the production of ethanol the NADH + H$^+$ molecules pass their hydrogen acceptor, both alcoholic and lactic acid fermentations are anaerobic. passed to pyruvic acid. Since there is no requirement for oxygen to act as a hydrogen acceptor both alcoholic and lactic acid fermentations are anaerobic.

Relatively little energy becomes available to cells as a result of the further conversions of pyruvic acid to other compounds under anaerobic conditions.

Aerobic Oxidation of Pyruvic Acid. Under aerobic conditions the oxidation of pyruvic acid usually occurs along entirely different pathways than under anaerobic conditions and the end products of its oxidation are usually carbon dioxide and water. This kind of oxidation is basically accomplished by a cyclic series of steps commonly called the Krebs cycle, after Hans Krebs, the biochemist who first postulated it, in 1937. It is often called the tricarboxylic acid cycle or the citric acid cycle. Aerobic respiration is accomplished in mitochondria by this cycle operating integrally with the electron transport-oxidative phosphorylation processes which will be discussed later in the chapter. This pathway of aerobic respiration is of widespread occurrence in both plants and animals.

For the moment we will direct our attention only to the specific compounds involved in this cyclic series of reactions and defer discussion of its intrinsic energy relations until later. As shown in Fig. 13.3 a first reaction (I) is an oxidative decarboxylation of pyruvic acid involving NAD$^+$ and CoA $\cdot$ SH (Chapter

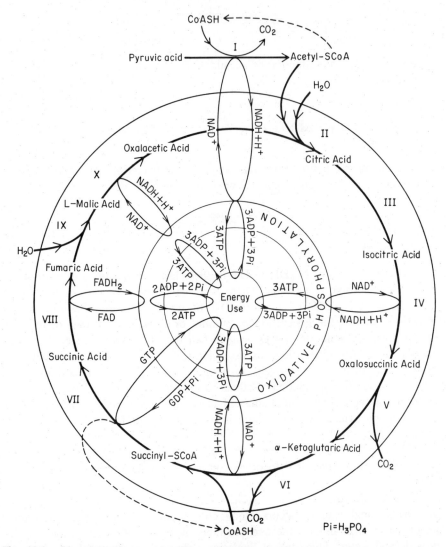

Fig. 13.3. Diagrammatic representation of stages in the Krebs cycle and oxidative phosphorylation. Modified from "Plant Metabolism" by J. E. Varner, McGraw-Hill Encyclopedia of Science and Technology **10**, 1971:402–405. Copyright 1971, McGraw-Hill Book Company. Used with permission of McGraw-Hill Book Company.

9). This reaction requires several other cofactors and proceeds in several steps, but can be summarized as follows:

$$CH_3 \cdot CO \cdot COOH + CoA \cdot SH + NAD^+ \xrightarrow{\text{pyruvic oxidase}}$$

pyruvic acid

$$CH_3 \cdot CO \cdot S \cdot CoA + NADH + H^+ + CO_2$$

acetyl-S-CoA

Some of the acetyl-S-CoA which enters the Krebs cycle also comes from the oxidation of fatty acids (Chapter 14).

Acetyl-S-CoA then reacts with *oxalacetic acid,* the end product of a previous turn of the Krebs cycle and with water, forming *citric acid* (reaction II), CoA·SH being reconstituted in the process:

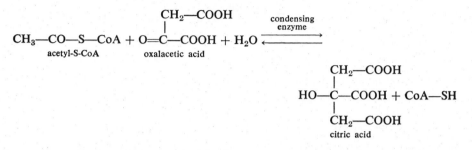

Citric acid is then isomerized to *iso-citric acid* under the influence of the enzyme *aconitase* (reaction III):

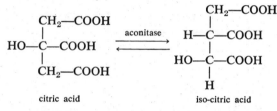

Some authorities consider that citric acid is first converted into aconitic acid which in turn is converted into iso-citric acid.

The iso-citric acid is then converted to *oxalosuccinic acid* by an oxidative reaction involving NAD+ (reaction IV):

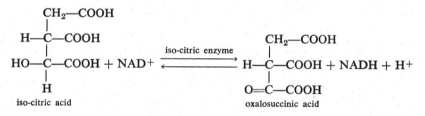

In the following decarboxylation reaction (V) oxalosuccinic acid is converted into *α-ketoglutaric acid:*

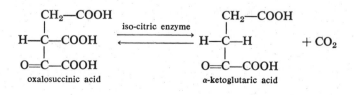

These last two reactions are so closely linked that it is considered that they are both activated by an enzyme complex designated as the *iso-citric enzyme*.

The next step in the cycle (reaction VI) is the oxidative decarboxylation of α-ketoglutaric acid to *succinyl-S-CoA*.

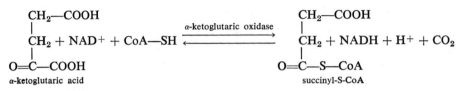

This reaction is similar to the one whereby pyruvic acid is converted to acetyl-S-CoA. It is catalyzed by *α-ketoglutaric acid oxidase;* several cofactors are also required.

Under the influence of the enzyme *succinic thiokinase* reaction VII occurs, resulting in the formation of *succinic acid,* the conversion of guanosine diphosphate (Table 16.2) to guanosine triphosphate, and reconstitution of CoA-SH:

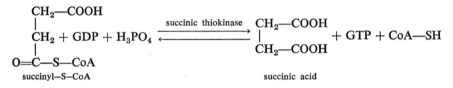

Succinic acid is then converted into *fumaric acid* (reaction VIII) in the oxidative reaction in which FAD (Chapter 9) serves as the hydrogen acceptor, the reaction being catalyzed by *succinic dehydrogenase*:

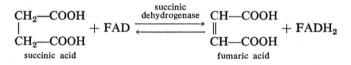

Fumaric acid, with a gain of water, is converted into *l-malic acid,* a reaction (IX) catalyzed by the enzyme *fumarase*:

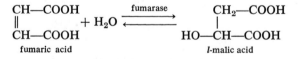

The final step in the cycle (reaction X) is the oxidative conversion of malic acid into oxalacetic acid, a reaction in which NAD^+ participates, and which is catalyzed by the enzyme *malic dehydrogenase*:

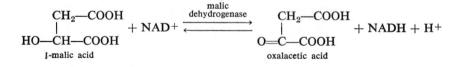

The oxalacetic acid may then react with acetyl-S-CoA starting a second cycle of this series of reactions. Reactions I and VI of the cycle are not readily reversible, but all of the others are.

The net result of one turn of the cycle is the complete oxidation of one molecule of pyruvic acid. Carbon dioxide is eliminated in this cycle in the formation of acetyl-S-CoA, α-ketoglutaric acid, and succinyl-S-CoA, respectively (reactions I, V, VI). In five reactions (I, IV, VI, VIII, X) a pyridine nucleotide or equivalent compound is reduced by acquiring 2 hydrogens per molecule. As discussed more fully later these five pairs of hydrogens are eventually oxidized to water which requires $2\frac{1}{2}$ O_2. Five H_2O would therefore result from these reactions, but H_2O is used in reactions II, VII (here coming from the H_3PO_4), and IX. The net release, therefore, is of two molecules of H_2O. The summary equation for this phase of respiration may therefore be written:

$$CH_3 \cdot CO \cdot COOH + 2\tfrac{1}{2} O_2 \longrightarrow 3\ CO_2 + 2\ H_2O$$

A similar summary equation for glycolysis would be:

$$C_3H_6O_3 + \tfrac{1}{2} O_2 \longrightarrow CH_3 \cdot CO \cdot COOH + H_2O$$

When the NADH + H^+ generated in glycolysis is not utilized in the production of ethanol or lacetic acid in fermentation processes, oxygen is required (as it is with Krebs cycle generated NADH + H^+) as the ultimate acceptor of the pair of hydrogens from this high energy compound. This accounts for the oxygen in the summary equation for glycolysis.

Adding these two equations gives a summary equation for the complete oxidation of a triose sugar:

$$C_3H_6O_3 + 3\ O_2 \longrightarrow 3\ CO_2 + 3\ H_2O$$

It should be remembered that each hexose molecule entering glycolysis will yield two triose molecules, both of which may proceed through glycolysis and the Krebs cycle.

Electron Transport and Oxidative Phosphorylation.   The Krebs cycle has two main roles. One of these is the production of metabolic pools of glycolytic and Krebs cycle intermediates which can serve as source molecules in the synthesis of other important organic compounds. For example, as discussed in Chapter 16, several important amino acids are synthesized from intermediate compounds formed in the Krebs cycle and glycolysis. Among these are alanine from pyruvic acid, glutamic acid from α-ketoglutaric acid, and aspartic acid from oxalacetic acid. The reverse situation also prevails; at least some of the intermediates of the Krebs cycle can be fed into it from other reactions which are integral in other phases of cellular metabolism.

The other important role of the Krebs cycle and closely associated processes is the formation of energy-rich molecules of ATP or equivalent compounds which serve to power the energy-requiring processes in cells, especially endergonic metabolic processes. Such energy transfers are the mainspring of life. In reaction VII an energy-rich triphosphate (GTP) is synthesized. Of even greater significance is the fact that in five of the reactions of the Krebs cycle NAD^+ or FAD is reduced, one of the substrate intermediates being correspondingly oxidized. The reduced pyridine and flavin nucleotides are eventually oxidized to their nonreduced state. This oxidation proceeds as a series of coupled reactions, oxygen being involved only in the terminal step. All details of this chain of reactions are not understood but a probable, although somewhat simplified version, is shown in Fig. 13.4. NAD^+, previously reduced to $NADH + H^+$ by transfer of 2 H from a Krebs cycle intermediate, is oxidized to NAD^+ by transfer of 2 H to FAD, which in turn is reduced to $FADH_2$. Each H is the equivalent of one proton (H^+) plus one electron (e^-). $NADPH + H^+$ can also participate in such a reaction with FAD.

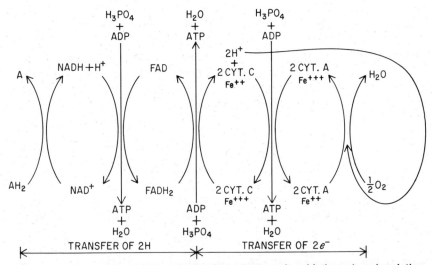

Fig. 13.4. Diagram illustrating electron transport and oxidative phosphorylation.

The next steps involve compounds known as the cytochromes, complex iron compounds which are capable of being alternately oxidized and reduced by a loss or gain of electrons, respectively. In the oxidized form of cytochromes (Chapter 9) the iron is in the ferric state; it is reduced to the ferrous state by incorporation of one electron into the valence shell of the iron atom. The $FADH_2$ loses two protons (2 H^+), and the two corresponding electrons reduce the iron in two cytochrome molecules from the Fe^{+++} to the Fe^{++} state:

$$FADH_2 \longrightarrow FAD + 2\,H^+ + 2\,e^-$$

$$2\,e^- + 2 \text{ cytochrome } Fe^{+++} \longrightarrow 2 \text{ cytochrome } Fe^{++}$$

A number of cytochromes (as many as seven in some systems) participate in this *electron transport chain,* the reduced form of one transferring electrons to the oxidized form of the next one in the series, the former becoming oxidized in the process. For simplicity only two cytochromes are represented in Fig. 13.4, the last one in the series being the one known as *cytochrome oxidase.* The final steps in the transport of electrons involve the formation of water and the reduction of oxygen, the 2 H^+ coming from the preceding reaction:

$$2 \text{ cytochrome } Fe^{++} \longrightarrow 2 \text{ cytochrome } Fe^{+++} + 2\,e^-$$

$$2\,H^+ + 2\,e^- + \tfrac{1}{2}\,O_2 \longrightarrow H_2O$$

The energy level of the hydrogen or electron acceptors in the electron transport system decreases from left to right as written in Fig. 13.4. If the fall in energy value between the adjacent hydrogen carriers is great enough (at least 8,000–12,000 cal per mole), part of this energy drop is conserved in the production of ATP from ADP and inorganic phosphate.

The available evidence indicates that three moles of ATP are synthesized for each mole of NADH + H^+ entering the electron transport system. This process whereby ATP is synthesized at the expense of energy coming indirectly from the Krebs cycle and glycolytic intermediates and more directly from NADH + H^+ and similar reduced nucleotides is called *oxidative phosphyorylation.* It should be contrasted with the process of photophosphorylation which occurs as an integral part of photosynthesis and should be also distinguished from substrate level phosphorylation discussed in the next paragraph.

Efficiency of the Respiration Process. We can now estimate how effective the overall process of respiration is in conserving energy which can be used in endergonic processes and other cellular activities. Starting with the glycolytic series of reactions (Fig. 13.2) we note in reaction I and in reaction IV each, that one mole of ATP is converted to one mole of ADP for each mole of glucose involved. Also that in reactions VII and X one mole of ATP is generated for each mole of phosphoglyceric acid or pyruvic acid formed, respectively. Such a generation of ATP molecules is termed *substrate level phosphorylation,* since the phosphate added to ADP comes from substrate molecules and not from inorganic phosphate as is true in oxidative and photophosphorylation. Since two moles of phosphoglyceric acid or pyruvic acid are derived from each mole of glucose, the total number of moles of ATP generated per mole of glucose is four. There is therefore a net gain of two moles of ATP as a result of substrate level phosphorylation per mole of glucose participating in the reaction. Also in reaction VI of the glycolytic sequence one mole of NADH + H^+

is formed per mole of diphosphoglyceric acid reacting, which is equivalent to two moles of NADH + H+ formed per mole of glucose. Through the electron transport and oxidative phosphorylation sequence of reactions this will ultimately result in the generation of six moles of ATP. A total of eight moles of ATP per mole of glucose oxidized is therefore generated during glycolysis.

In the aerobic phase of respiration (Krebs cycle) oxidative phosphorylation results in the generation of three moles of ATP in four of the NAD+ reduction reactions (I, IV, VI, X) and two from the reduction of FAD (reaction VIII). Furthermore in reaction VII one mole of GTP is synthesized as a result of the substrate level phosphorylation. This totals fourteen moles of ATP plus one mole of GTP per mole of pyruvic acid oxidized, which in turn is equivalent to thirty moles of ATP plus GTP per mole of glucose oxidized. Adding to this the eight moles of ATP generated during glycolysis, we arrive at an estimated total of thirty-eight moles of ATP or equivalent generated as a result of the complete oxidation of one mole of glucose.

The energy released by the complete oxidation of a mole of glucose is about 673 kg-cal. Each mole of ATP represents, as a minimum estimate, about 8 kg-cal of releasable energy. Hence a minimum of 304 kg-cal of the energy available in a mole of glucose is conserved in the form of energy-rich phosphates as a result of the glycolytic, Krebs cycle, and oxidative phosphorylation series of reactions. This figure, which is a minimal one, represents a conservation efficiency of 45 per cent. The following equation summarizes the material and energy changes in the overall process of respiration as it occurs in the living cell.

$$C_6H_{12}O_6 + 6\ O_2 + 38\ ADP + 38\ H_3PO_4 \longrightarrow$$

$$6\ CO_2 + 38\ ATP + 44\ H_2O + 369\ \text{kg-cal (approx.)}$$

The 369-kg-cal quantity which appears in this equation represents that part of the energy derived from glucose which is liberated as heat (673 − 304), and is hence unavailable in metabolic or other cellular processes.

The Pentose Phosphate Pathway.   The *pentose phosphate pathway* or *hexose monophosphate shunt* is a series of reactions leading from glucose-6-phosphate, which constitutes an alternate pathway in respiration. This overall pathway can be divided into two major phases. The first phase, which involves only the first three reactions, constitutes the entire oxidative portion of the pathway. The phase which follows involves a multitude of nonoxidative rearrangement steps.

The oxidative phase (Fig. 13.5) begins with a conversion of glucose-6-phosphate to glucolactone-6-phosphate. In this reaction carbon 1 of the glucose-6-phosphate is oxidized while NADP+ is reduced to NADPH+ + H+. A nonoxidative intermediate step involving the addition of a molecule of water and the breaking of the pyranose ring then follows, yielding gluconic acid-6-phos-

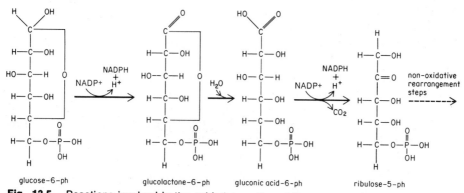

Fig. 13.5.   Reactions involved in the oxidative portion of the pentose phosphate pathway.

phate. This reaction is followed by an oxidative step involving the decarboxylation of carbon 1 of the gluconic acid-6-phosphate which is coupled to the reduction of another NADP+ to NADPH + H and yields ribulose-5-phosphate as the
product. This step terminates the oxidative portion of the pentose phosphate
pathway.

In oxidation, by way of the glycolysis-Krebs cycle system, glucose-6-phos-
phate is degraded ultimately to 6 CO_2, while in the pentose phosphate system
it is never degraded oxidatively further than a five carbon compound and 1 CO_2.
In the latter system it would require the oxidation of six different glucose-6-
phosphate molecules to yield 6 CO_2 molecules. The oxidation of these 6 glucose-
6-phosphate molecules would also yield 6 ribulose-5-phosphate molecules which
can proceed through the nonoxidative rearrangement steps, discussed later,
which ultimately convert the 6 ribulose-5-phosphate molecules to 5 glucose-6-
phosphate molecules. Thus for every 6 glucose-6-phosphate molecules entering
the pentose phosphate pathway there is a yield of 6 CO_2 molecules and 5 glu-
cose-6-phosphate molecules. In a "net reaction" sense, this might be thought of
as the oxidation of 1 glucose-6-phosphate to 6 CO_2.

In terms of energy conservation, there is a production of an NADPH +
H+ at each of two steps per glucose-6-phosphate oxidized via the pentose phosphate system. Thus there would be a production of 12 NADPH + H+ for every
6 CO_2 molecules yielded or glucose equivalent oxidized in this system. If these
12 NADPH + H+ were to proceed through the cytochrome oxidative phosphorylation pathway, there would be an ultimate production of approximately
36 ATP's for each glucose-6-phosphate equivalent oxidized via the pentose
phosphate pathway. This compares very favorably with the 38 ATP's produced
per glucose-6-phosphate via the glycolysis-Krebs cycle-cytochrome system pathway. In actuality it has been found that the cytochrome system is rather inefficient in accepting NADPH + H+ so that it is likely that the ultimate product
of respiration via the pentose phosphate pathway is NADPH + H+ itself, rather

than ATP. This NADPH + H$^+$ is thought to be utilized in synthetic reactions such as those of fatty acid synthesis (Chapter 14) which have a specific requirement for NADPH + H$^+$.

The nonoxidative rearrangement steps which make up the second phase of the pentose phosphate system are very similar to the rearrangement steps in the Calvin-Benson cycle of photosynthesis (Chapter 10). Many reactions are catalyzed by either a transaldolase or transketolase enzyme which essentially removes a 2 or 3 carbon unit from the carbon 1 end of one sugar molecule and transfers this unit to the carbon 1 end of the second sugar molecule. A generalized outline of the rearrangement portion of the pathway is shown in Figure 13.6.

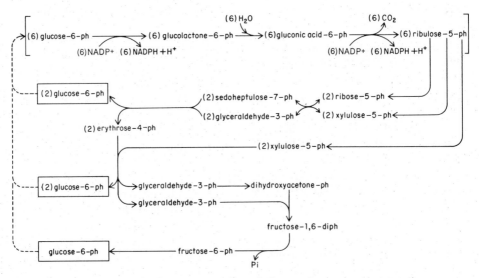

Fig. 13.6. Diagrammatic representation of the pentose phosphate pathway.

An important intermediate compound in the rearrangement sequence is ribose-5-phosphate, which is used in the synthesis of nucleic acids (Chapter 16). If the ribulose-5-phosphate resulting from the oxidative portion of the pathway is funneled exclusively to ribose-5-phosphate, which in turn is utilized in nucleic acid synthesis, the rearrangement steps shown in Figure 13.6 would not be expected to occur and thus there would be no regeneration of glucose-6-phosphate. However, even if the rearrangement steps are "short circuited" at ribose-5-phosphate, there is still a yield of 2 NADPH + H$^+$ for each glucose-6-phosphate entering the pentose phosphate pathway.

Another point at which the rearrangement steps can be "short circuited" occurs following the production of glyceraldehyde-3-phosphate. It is probable that at least a portion of the glyceraldehyde-3-phosphate synthesized in the pentose phosphate system enters the glycolytic pathway of respiration rather than undergoing conversion into hexoses.

Respiration in Relation to the Mitochondria.   Metabolic processes cannot be viewed in their entirety as pure chemistry but only in relation to the cellular structures in which they occur. The mitochondria (Fig. 3.2, Fig. 13.7) have often been designated as the "powerhouse of the cell" because they are the site of many of the respiratory processes. All of the enzyme systems operative in the Krebs cycle, electron transport, and oxidative phosphorylation, are known to be present in the mitochondria. They are also the site of much of the amino acid (Chapter 16) synthesis in the cell, and of the oxidation of fatty acids (Chapter 14). Some of these enzyme systems are also present in other organelles. For example, an electron transport system with its component cytochromes is present in the chloroplasts. Synthesis of at least some amino acids also occurs in the chloroplasts. Some of the processes occurring in a mitochondrion appear to be restricted to certain of its parts and others to other parts. There is evidence that the enzymes of the Krebs cycle are present in the region between the membranes while the electron transport system is located within the membranes.

Fig. 13.7. Artist's conception of a mitochondrion.

The enzyme systems involved in glycolysis and in the pentose phosphate pathway are found principally, although not exclusively, in the undifferentiated cytoplasm of the cell.

Dark Fixation of Carbon Dioxide and Its Relation to Organic Acid Metabolism.   As is clear from the previous discussion, respiration represents principally the metabolism of organic acids. Many organic acids, including some of those which participate in the Krebs cycle, are present in plant cells only as transitory metabolites. Others accumulate in plants in substantial quantities. Some of these arise in side reactions from respiratory intermediates; among such reactions are the dark carboxylations to be described shortly.

The principal organic acids known to accumulate in plants are citric as in citrus fruits; iso-citric, as in blackberry fruits and in the leaves of succulents; malic, as in apple, cherry, and plum fruits and in the leaves of succulents; tartaric, as in grape fruits; aconitic, as in the stems of sugar cane; succinic, as in stem tissue of lucerne (*Medicago sativa*); fumaric, as in sunflower stems; malonic, as in bean (*Phaseolus coccinea*) stems; and oxalic, as in the leaves of spinach and the petioles of rhubarb. Organic acids usually occur in plant cells in solution in the cell sap as the acid itself or as salts or esters thereof. The flavors of many fruits result from the presence of organic acids and their esters. However, one organic acid, oxalic, often occurs in abundance as crystals of its calcium salt in the cytoplasm.

The organic acids integrally involved in the respiratory process are also metabolically related to the synthesis of fatty acids (Chapter 14) and amino acids (Chapter 16). The former are one of the kinds of compounds from which fats and oils are made; the latter are the compounds from which proteins are synthesized. Since the organic acids which serve as respiratory intermediates are derived from carbohydrates, they accupy a metabolic crossroads between carbohydrates on the one hand and fats and proteins on the other.

Dark, non-photosynthetic, fixations of carbon dioxide have been found to occur in a wide variety of plant tissues and are probably of universal occurrence in plants. Such reactions also occur in many animal tissues and in bacteria. These are "dark" reactions only in the sense that they do not require light. Such dark fixations are closely related to certain phases of organic acid metabolism.

Malic acid is the organic acid generated most plentifully as a result of the dark fixation of carbon dioxide. It is synthesized principally according to the following reactions:

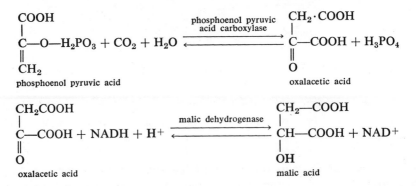

Other organic acids are synthesized in plants by more or less analogous carboxylation reactions, but only one other, iso-citric, occurs in sufficient abundance to warrant specific mention. It appears to be synthesized as follows:

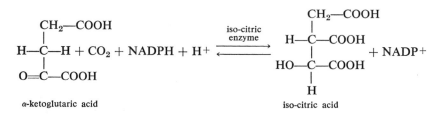

α-ketoglutaric acid   iso-citric acid

Citric acid is readily synthesized from iso-citric acid as shown earlier in this chapter.

While by no means confined to such plants, the fixation of carbon dioxide in the dark by non-photosyntheic processes is an especially prominent feature of the metabolism of succulents (Chapter 8). In *Sedum, Bryophyllum, Opuntia,* and other succulent species the organic acid content of the leaves or phyllodes shows a regular diurnal cycle of increase during the night hours and a decrease during the daylight hours. The principal acid involved is malic acid which is synthesized during the night hours by reactions such as those described previously.

In many succulents the stomates are open at night, facilitating the entry of the carbon dioxide which participates in carboxylation reactions. During the daylight hours the stomates of such succulents are closed. Transpiration is thus held to a low rate which is presumably advantageous to the plant during periods of water deficiency. Many succulents, in fact, are natives of relatively arid regions such as semideserts. Photosynthesis in such succulents uses carbon dioxide released within the cells by the decarboxylation of malic acid as follows:

$$
\begin{array}{ccc}
CH_2-COOH & & COOH \\
| & & | \\
CH-COOH + NADP+ & \xrightarrow{\text{malic enzyme}} & C{=}O + CO_2 + NADPH + H+ \\
| & & | \\
OH & & CH_3 \\
\text{malic acid} & & \text{pyruvic acid}
\end{array}
$$

The decarboxylation of malic acid accounts for the decrease in the acid content of the leaves or other photosynthesizing organs which occurs in the light.

SUGGESTED FOR COLLATERAL READING

Beevers, H. *Respiratory Metabolism in Plants.* Harper and Row, Publishers, New York, 1961.

Conn, E. E., and P. K. Stumpf. *Outlines of Biochemistry. 2nd. Edition.* John Wiley & Sons, Inc., New York, 1966.

Davies, D. D. *Intermediary Metabolism.* Cambridge University Press, 1961.

Geismann, T. A., and D. H. G. Grout. *Organic Chemistry of Secondary Plant Metabolism.* Freeman, Cooper and Company, San Francisco, 1969.

Hatch, M. D., C. B. Osmond, and R. O. Slatyer. *Photosynthesis and Photorespiraration.* John Wiley Interscience, New York, 1971.

Pridham, J. B., and T. Swain, Editors. *Biosynthetic Pathways in Higher Plants.* Academic Press, Inc., New York, 1965.

14 | METABOLISM OF CARBOHYDRATES, LIPIDS, AND RELATED COMPOUNDS

The carbohydrates constitute a large group of chemical compounds many of which are indispensable to the existence of living organisms. They play basic roles in both the metabolism of plants and in their structural organization. Three main groups of carbohydrates are recognized: the monosaccharides, the oligosaccharides, and the polysaccharides. All carbohydrates are polyhydroxy aldehydes, ketones, or derivatives thereof.

The monosaccharides or simple sugars (Fig. 14.1) consist of a chain of two or more carbon atoms to which are attached hydrogen atoms and hydroxyl groups and one aldehydic or one ketonic group. The five carbon atom sugars (pentoses) and six carbon sugars (hexoses) are the most important in plants.

The molecules of oligosaccharides are constructed by the linking together of two to several molecules of simple sugars. The principal kinds are di, tri, and tetrasaccharides. The residues derived from simple sugars are linked together in a specific order in any given oligosaccharide. Sucrose and maltose, both disaccharides, are two of the best known carbohydrates in this group.

The molecules of polysaccharides are constructed by the linkage together of large numbers of monosaccharide units or of units which are slightly modified monosaccharide molecules. The molecules of a polysaccharide are typically long and chain-like and may vary in length, that is, in the number of constituent units. Starch, cellulose, and pectins are the best known polysaccharides occurring in plants.

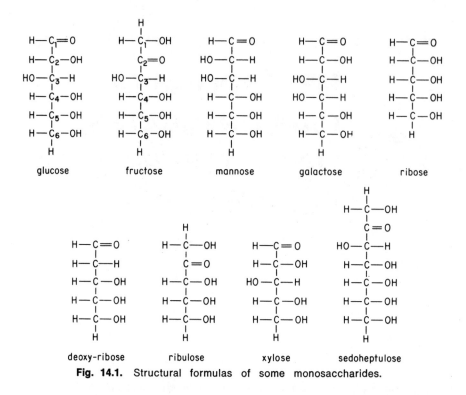

Fig. 14.1. Structural formulas of some monosaccharides.

Isomerism in Monosaccharides.

An important fact about the simple sugars is that many isomers exist with the same molecular formula. Isomers are of two types. In order to discuss this topic it is necessary to refer to the different carbon atoms by number, as shown for glucose and fructose in Fig. 14.1.

Structural isomers result from differences in the basic molecular configuration of the molecules. Glucose and fructose are structural isomers (Fig. 14.1). Although both have the same molecular formula ($C_6H_{12}O_6$) the first is an aldehydic sugar and the second is a ketonic sugar.

Stereoisomers result from the presence of asymmetric carbon atoms in the sugar molecule. An asymmetric carbon atom is one in which each of the four bonds is filled by a different atom or group of atoms. Glucose and galactose are examples of stereoisomers (Fig. 14.1). The only difference between the structures of these two molecules is the relative position of the −OH and −H groups attached to the number 4 carbon atom. Glucose and galactose have identical molecular formulas, both being hexoses, but because of the positional difference of −H and −OH just mentioned, they have different physical and chemical properties. One of the four different atoms or groups attached to carbon 4 in this example is −OH, another is −H, a third is carbons 1, 2, and 3 plus their associated side groups, and a fourth is carbons 5 and 6 plus their associated side groups. Examination of their structure shows that both glucose and galactose have sites of asymmetry at carbons 2, 3, 4, and 5. Because of these multiple sites of asymmetry, it is evident that glucose and galactose represent only two of the many stereoisomers of the six carbon sugars that can exist with identical molecular formulas. Similar families of stereoisomers exist with sugars of other carbon chain lengths.

Molecules with sites of asymmetry will usually bend plane polarized light in an instrument called a polarimeter. A (+) or small case letter *d* preceding the name of a sugar indicates that it is *dextrorotatory* or will bend plane polarized light to the right, while a (−) or *l* indicates that the sugar is *levorotatory* and bends such light to the left. The older names dextrose, for glucose, and levulose, for fructose, were based on this property of rotation.

The lower case *d* mentioned before should not be confused with the upper case D which preceeds the names of many sugars. The letter D indicates a chemical family of sugars and is arbitrarily applied by the chemist to any sugar which has the −OH group written to the right of the asymmetric carbon farthest from the keto or aldo group. In D glucose this would be the −OH group to the right of carbon 5. A molecule of L glucose would be written as the mirror image of D glucose and would therefore have the −OH group of carbon 5 on the left. Although an L family of sugars could exist, in actuality nearly all naturally occurring plant sugars belong to the D family.

The structural formula for glucose given in Fig. 14.1 is adequate for many purposes, but there is strong evidence that in solution the glucose straight chain molecule usually exists in equilibrium with what is called a pyranose ring structure. The formation of the ring structure represents a shift in bonds, but does not change the empirical formula. Two isomers of glucose, α, with the −OH group of carbon 1 written on the right, and β, with this −OH group written on the left, now exist because in the pyranose form carbon 1 becomes a new site of asymmetry:

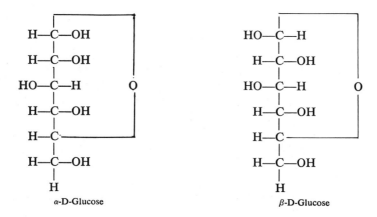

α-D-Glucose   β-D-Glucose

The −OH group of carbon 1 is frequently involved in linkages to other sugar units. Polymers of α-glucose have considerably different chemical and physical properties than polymers of β-glucose.

Frequently the pyranose form of sugars is represented by the Haworth cyclic formulation as exemplified by α-D-glucose, shown by the following. In the α form the −OH of carbon 1 is written in the downward position while in the β form this −OH is written in the upward position.

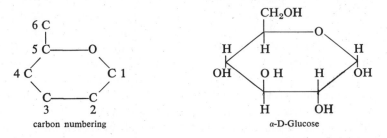

carbon numbering   α-D-Glucose

A ring structure similar to the pyranose configuration, but involving only four carbons and one oxygen is formed by several sugars. This five membered ring, or *furanose* configuration, is exemplified by the hexose fructose and by the pentose ribose shown here.

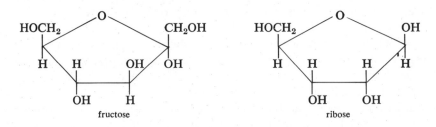

fructose   ribose

Reactive States of Sugars.   Simple sugars appear to exist in living organisms in at least three levels of reactivity: 1) the unmodified simple sugar itself, 2) the simple sugar which has been phosphorylated, and 3) the simple sugar which has become part of a sugar-nucleotide complex (Chapter 16).

The unmodified simple sugar is the least reactive metabolically of these forms. Few, if any, of the sugar interconversions in plants involve only unmodified simple sugars. The more reactive sugar-phosphate form of simple sugars appears to be involved in most reactions at the level of interconversion of one simple sugar to another, while the still more reactive sugar-nucleotide form of simple sugars appears to be generally involved in reactions in which oligosaccharides are formed or polysaccharides are lengthened.

Monosaccharide Interconversions.   Simple sugar interconversions are an integral part of several aspects of plant metabolism. Such interconversions, mostly of sugars in the phosphorylated state, are involved in photosynthesis (Chapter 10), in the early steps in glycolysis (Chapter 13), in the pentose-phosphate pathway (Chapter 13), and in the fabrication of structural or food reserve carbohydrates.

Although a large number of different simple sugars have been found in plants, the majority of them are present in only minute quantities. Notable exceptions are glucose and fructose. The minute quantities of most simple sugars found in plants probably reflects the fact that most of these sugars rapidly become incorporated into large polysaccharide molecules. However, during processes such as seed germination, somewhat larger quantities of monosaccharides may exist because of the rapid degradation of the large polysaccharide molecules. This increase in the monosaccharide level is only temporary, since many of these sugars will enter the respiration pathways or will be reincorporated in the formation of more complex carbohydrates as new cells arise.

The reactions by which one monosaccharide is converted to a second type of monosaccharide nearly always involve sugars in their phosphorylated state. One general scheme of such interconversions can be outlined as follows:

$$\text{Sugar A} + \text{ATP} \xrightarrow{\text{Enzyme X}} \text{Sugar A-phosphate} + \text{ADP}$$

$$\text{Sugar A-phosphate} \xrightarrow{\text{Enzyme Y}} \text{Sugar B-phosphate}$$

The steps in the conversion of glucose to fructose 6-phosphate can be taken as a specific example of such reactions:

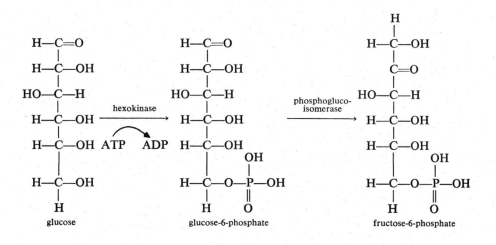

glucose   glucose-6-phosphate   fructose-6-phosphate

Plant Alcohols. Sugars can undergo several kinds of interconversions other than the sugar to sugar type. One such interconversion worthy of note is that from a sugar to a plant alcohol. Two plant alcohols which have been shown to be of some importance in translocation (Chapter 17) are sorbitol and mannitol which are derived from glucose and mannose, respectively, in similar reduction reactions, one of which is shown here:

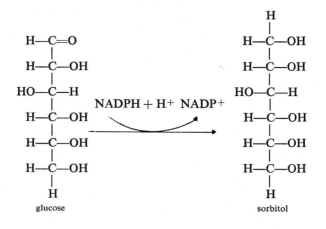

Oligosaccharide Biosynthesis. The synthesis of disaccharides and other oligosaccharides appears to have a prerequisite that one of the sugar units in the reaction be in a sugar-nucleotide-complex state at the time of oligosaccharide synthesis. For example, neither glucose nor glucose-phosphate is reactive enough to combine directly with fructose in the synthesis of the disaccharide, sucrose. Glucose in the form of uridine-diphospho-glucose (UDPG) will, however, react with fructose in the formation of sucrose:

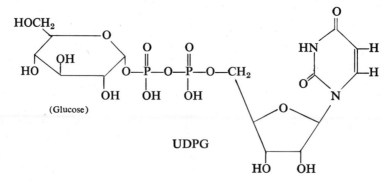

+

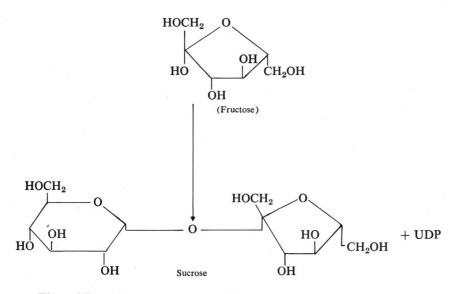

The uridine (Table 16.2) of UDPG contains units of uracil and the 5 carbon sugar, ribose, both of which remain unchanged in the reaction.

A reaction which yields sucrose-phosphate is also known to occur in plants. The fructose must be in the phosphorylated condition to react with UDPG.

$$\text{UDPG} + \text{Fructose-6-phosphate} \rightarrow \text{Sucrose-phosphate} + \text{UDP}$$

The phosphate group can then be hydrolyzed from sucrose-phosphate by a phosphatase enzyme.

A disaccharide of lesser importance in plants is *melibiose*. This sugar, composed of a glucose and a galactose unit, is a product of hydrolysis when raffinose is acted upon by invertase.

The trisaccharide *raffinose* is found in many plants in relatively low concentrations. The structure of raffinose is essentially that of sucrose with the addition of D-galactose to the carbon 6 position of the glucose moiety of the sucrose molecule. The synthesis of raffinose, which involves galactose in a nucleotide complexed form, is outlined as follows:

$$\text{UDP-galactose} + \text{Sucrose} \longrightarrow \text{Raffinose} + \text{UDP}$$

Another fairly common oligosaccharide in plants is the tetrasaccharide *stachyose*. The structure of this molecule is that of raffinose with an additional galactose unit attached to carbon 6 of the galactose moiety of raffinose. While it would seem that UDP-galactose should react with raffinose to yield stachyose, this does not seem to be the case. Instead, a somewhat indirect pathway appears to be involved which includes a molecule called *myo*-inositol as an activator.

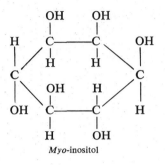

Myo-inositol

The *myo*-inositol reacts with D-galactose to yield a *myo*-inositol-galactose complex called galactinol. The synthesis of stachyose proceeds as follows:

$$\text{Galactinol} + \text{Raffinose} \longrightarrow \text{Stachyose} + \textit{Myo}\text{-inositol}$$

Metabolism of Food Reserve Polysaccharides.   *Starch.*   One type of starch molecule, *amylose,* is composed of many D-glucose units in a straight chain configuration linked by means of α-1,4 glucosidic bonds (Fig. 14.2). The notation 1,4 indicates which carbons of the sugar units are involved in the linkage. Molecular weights of amylose molecules are in the range of 10,000 to 100,000.

The second type of starch molecule, *amylopectin,* has a main chain which, like that of amylose, is composed of α-1,4 glucosidic linkages. However, amylopectin differs from amylose in having numerous side branch chains which attach to the main chain by 1,6 bonds (Fig. 14.2). The amylopectin molecules are usually somewhat larger than amylose molecules, having molecular weights in the range from 50,000 to 1,000,000. Although the proportion of amylose to amylopectin varies in different plants, in general 70 per cent or more of the starch content is usually in the form of amylopectin.

Starch Hydrolysis.   There are two hydrolytic enzymes involved in the degradation of starch molecules. One of these, β-amylase, can hydrolyze both amylose and amylopectin. The mode of action of this enzyme involves the sequential removal of maltose units from the end of a given chain. This enzyme can only degrade chains from the nonreducing end (*i.e.,* the end in which carbon 1 of the terminal glucose is involved in the linkage to the next glucose unit). The disaccharide, maltose, which results from this hydrolysis is composed of two α-D-glucose units in α-1,4 linkage. β-amylase cannot break a 1,6 linkage. This means that amylose can be completely degraded to maltose units, but amylopectin can only be degraded to the points of 1,6 branching. This undegradable core of amylopectin which remains after β-amylase hydrolysis is referred to as β-limit dextrin.

A second hydrolytic enzyme, α-amylase, can also degrade both amylose and amylopectin. This enzyme requires Ca^{++} for activation. The mode of

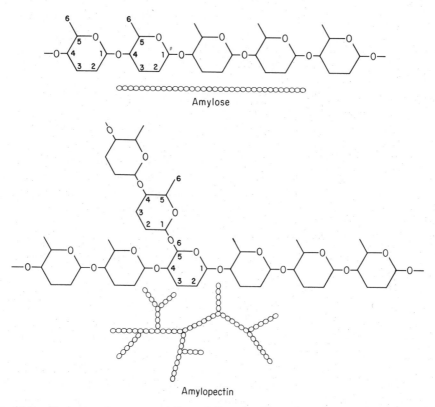

Amylose

Amylopectin

Fig. 14.2. Diagrammatic representation of the structure of amylose and amylopectin.

action of α-amylase reflects, to a large degree, the spatial configuration of the starch molecule it is degrading. The chains of glucose residues making up the amylose or amylopectin molecules appear to be in helical configurations, with six glucose units per turn of the helix. Apparently α-amylase attaches and breaks bonds at the same relative points of two consecutive turns of the helix, thus removing a six glucose unit fragment from somewhere in the middle of the long starch polymer. The point at which the enzyme attaches along the length of the helix appears to be a random one. Thus the 1,6 branch points of amylopectin can be by-passed, although they cannot be broken by this enzyme. The oligosaccharide fragments which result from amylose and amylopectin cleavage by α-amylase are called *dextrins*. A second phase of α-amylase action then further degrades the dextrins to maltose. When α-amylase acts on amylopectin, another of the final degradation products will be fragments containing iso-maltose, which is a disaccharide composed of two α-D-glucose units with a α-1,6 linkage.

The distribution of the amylase enzymes is not uniform throughout the plant. The highest concentrations of these enzymes are found in the organs which accumulate foods. The concentrations of these enzymes appear to increase during periods of rapid metabolic flux in such organs. This occurs, for example, in seeds during germination. There is little evidence however, to indicate that polysaccharide degradation in cells of organs which do not accumulate foods is under the control of either amylase enzyme. A second enzyme system active in these cells will be discussed later.

Starch Biosynthesis. There appear to be at least three distinct pathways leading to the formation of starch in plants.

One such reaction is catalyzed by the enzyme phosphorylase. This reaction involves the sequential lengthening of a small primer glucose chain by adding one glucose unit at a time until a complete starch molecule is built. Although the primer might be a maltose unit *in vivo,* the reaction proceeds much more readily *in vitro* if the primer is a polymer four or five glucose units long at the start of the reactions. The glucose unit which is added must be in the glucose-1-phosphate form. The reaction can be represented in general terms as follows:

$$\text{Glucose-1-phosphate} + \text{Glucose chain } (n \text{ units}) \underset{\longleftarrow}{\overset{\text{phosphorylase}}{\longrightarrow}}$$

$$\text{Glucose chain } (n + 1 \text{ units}) + \text{Phosphate}$$

As this reaction is repeated, the original primer glucose chain reaches dextrin size and eventually becomes an amylose molecule. The phosphorylase enzyme will only catalyze 1,4 additions and therefore cannot be responsible for the complete amylopectin synthesis. The 1,6 branches can be initiated only if a second enzyme called the Q-enzyme is present in addition to phosphorylase. The relative concentrations of the two enzymes appear to be a factor in determining the relative amount of branching which occurs.

It is of great significance that the phosphorylase reaction is a reversible process. At higher *p*H values the breakdown (phosphorolysis) of starch is favored. The phosphorylase reaction may play a more significant role in degrading starch than in synthesizing it. Phosphorylase appears to be the main starch degrading enzyme in cells of organs in which food accumulation is not a major process. As a digestive enzyme it is similar to β-amylase in that it cannot break or by-pass the α-1,6 linkages of amylopectin and therefore results in the formation of a phosphorylase-limit dextrin in addition to glucose-1-phosphate. The glucose-1-phosphate formed by phosphorolysis can be converted to glucose-6-phosphate which could eventually enter the glycolytic pathway.

The other two pathways of starch synthesis both involve sugar nucleotide complexes. As in the phosphorylase reaction, the polymer is increased in length one unit at a time.

One of these reaction pathways involves UDP-glucose (UDPG):

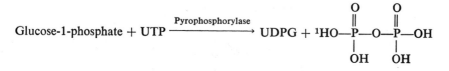

$$\text{Glucose-1-phosphate} + \text{UTP} \xrightarrow{\text{Pyrophosphorylase}} \text{UDPG} + {}^{1}\text{HO}-\overset{\overset{\displaystyle O}{\|}}{\underset{\underset{\displaystyle OH}{|}}{P}}-O-\overset{\overset{\displaystyle O}{\|}}{\underset{\underset{\displaystyle OH}{|}}{P}}-OH$$

$$\text{UDPG} + \text{Glucose chain } (n \text{ units}) \xrightarrow{\text{Transglucosylase}} \text{UDP} + \text{Glucose chain } (n+1 \text{ units})$$

The third starch synthesis pathway is similar to the second but involves ADP-glucose (ADPG) rather than UDPG:

$$\text{Glucose-1-phosphate} + \text{ATP} \xrightarrow{\text{Pyrophosphorylase}} \text{ADPG} + \text{Pyrophosphate}$$

$$\text{ADPG} + \text{Glucose chain } (n \text{ units}) \xrightarrow{\text{Transglucosylase}} \text{ADP} + \text{Glucose chain } (n+1 \text{ units})$$

Both of these reaction sequences proceed strongly in the direction of synthesis and are esentially nonreversible. These reactions appear to result in the formation of only the non-branched amylose type of starch. It has not yet been ascertained whether there is any interaction between the enzymes of the previous two reaction sequences and the Q-enzyme.

The site of starch synthesis is within the chloroplasts and colorless plastids referred to as amyloplasts. During and immediately following periods of rapid photosynthesis there is frequently a build-up of starch. Under certain conditions, one might think of starch as the terminal product of the photosynthetic reaction. A correlation between starch synthesis and photosynthesis seems to be indicated by the recent finding that the pyrophosphorylase enzyme involved with the starch-ADPG sequence (see earlier) is enhanced in activity nearly 50-fold in spinach chloroplasts by the presence of the photosynthetic intermediate, phosphoglyceric acid.

Inulin Biosynthesis. Polymers of fructose are one of the food reserves in some plants. One such *fructosan,* inulin, constitutes a large portion of the acmumulated foods in the underground portions of such plants as the Jerusalem artichoke, dahlia, and chicory, while inulin plus another slightly different fructosan, grass levan, are constituents of the aboveground portions of many grains and other grasses.

The biosynthesis of inulin appears to involve the sequential addition of fructose units as a polymer grows in length to a total of approximately 32 fructose units. The unsual aspect of the synthesis of this polymer is that it appears that sucrose molecules are the donor of each fructose moiety added to the polymer. This trans-fructosidase reaction results in formation of a fructose polymer with β-1,2 linkages. If we use G for glucose and F for fructose, this

---

[1]This double phosphate unit is referred to as pyrophosphate and is sometimes abbreviated PP*i*. The *i* in the abbreviation stands for inorganic.

biosynthesis can be diagrammed as follows:

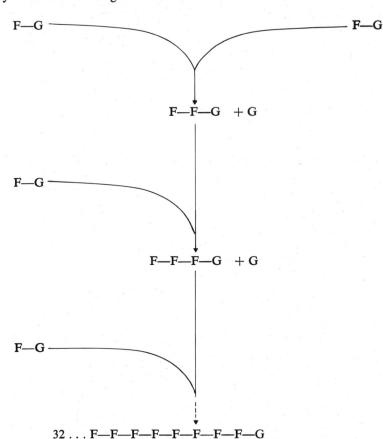

32 ... F—F—F—F—F—F—F—F—G
Inulin

Thus it can be seen that the completed inulin molecule contains a terminal glucose unit. At present, no nucleotide-complex precursors are known for this reaction. It seems likely that when inulin biosynthesis is studied more completely, a nucleotide-complex may be found to be involved.

Metabolism of Structural Polysaccharides. There are several families of polysaccharides which form an integral part of the structural material of plant cell walls.

Cellulose. Cellulose is the most abundant organic compound in the world. This compound is composed of unbranched chains of D-glucose units in β-1,4 linkage, with molecular weights between 50,000 and 400,000. The structure of cellulose is like that of amylose except that the repeating unit of the cellulose molecule is β-D-glucose rather than the α-D-glucose of amylose. A portion of a cellulose molecule is shown here in diagrammatic form.

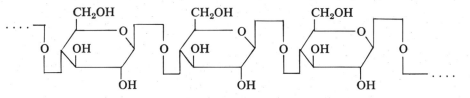

Partial chemical hydrolysis of cellulose yields the disaccharide *cellobiose*. Further hydrolysis yields two units of glucose from each cellobiose molecule. The enzyme cellulase which is present in some bacteria and fungi, but not abundant in higher plants, can also degrade cellulose to cellobiose.

The synthesis of cellulose involves the reacting glucose units in the form of a glucose-nucleotide complex. The nucleotide involved in cellulose synthesis appears to be guanosine-diphosphate (Table 16.2) rather than the UDP or ADP associated with starch synthesis. The synthetic reactions can be summarized as follows:

$$\text{GTP} + \text{Glucose-1-phosphate} \xrightarrow{\text{Pyrophosphorylase}} \text{GDP-glucose} + \text{Pyrophosphate}$$

$$\text{GDP-glucose} + \text{Glucose chain } (n \text{ units}) \xrightarrow{\text{Transferase}}$$
$$\text{Glucose chain } (n + 1 \text{ units}) + \text{GDP}$$

There is considerable uncertainty as to the location of these reactions in the cell. Although the cell wall containing the cellulose is built up externally to the protoplasm, the enzymes required in the cellulose synthesis must have their origin in the protoplasm. Whether there is a migration of enzymes or cellulose precursors or both through the plasmalemma prior to the cellulose synthesis has not been ascertained. One point that is clear is that new cell wall material is deposited at the cell wall-plasmalemma interface, so as a result it is laid down toward the interior of existing cell wall material. Two types of assimilation of cell wall materials are known. The addition of layer upon layer of new cell wall material is referred to as growth by *apposition*. When new cell wall materials impregnate and fill in spaces between existing ones, this type of growth is referred to as *intussusception* (Chapter 19). The additions of compounds such as pectin and lignin (see later) are usually by the latter method.

Callose.   Callose is a compound which is localized in the phloem cells of plants. It is also formed in most other cells following injury and in pollen tubes following the penetration of stigma and style. Callose is a polymer of glucose in β-1,3 linkage which is derived from UDP-glucose.

Pentosans.   Long chain polymers of the pentose sugars arabinose and xylose, known as arabans and xylans, respectively, are also constitutents of some plant cell walls. Arabans are furthermore constituents of certain gums and mucilages. Both arabans and xylans appear to be synthesized from their respective UDP-sugar complexes.

Pectic Acid.   Pectic compounds constitute a large portion of the middle lamella and, in addition, represent a sizable fraction (5–10 per cent) of other cell wall layers. The basic unit of the pectic compounds is the carbon 6 acid form of galactose, known as galacturonic acid. When galacturonic acid molecules are linked in an α-1,4 linkage, the polygalacturonic unit is called pectic acid. A portion of a pectic acid molecule is shown in diagrammatic form here.

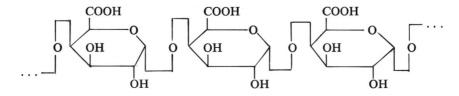

The synthesis of pectic acid involves galacturonic acid in the nucleotide complex UDP-galacturonic acid. As with the other polysaccharide syntheses discussed earlier, the polymer length is increased one unit at a time with the associated release of one UDP unit from the UDP-galacturonic acid complex.

The intermediate steps in the synthesis of pectic acid show the integration of some of the metabolic pathways leading to pectic acid and other polysaccharides:

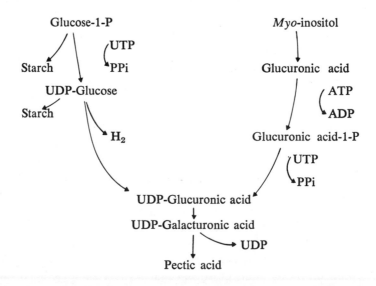

As shown in this diagram, major intermediate compounds in these pathways involve a form of the carbon 6 acid of glucose designated as glucuronic acid. This molecule is a stereoisomer of galacturonic acid and differs from the latter only in the relative positions of the −OH and −H groups at carbon 4. The switching of the relative positions of these two groups can be accomplished enzymatically.

The metabolic pathways indicated show that at least two major routes exist which lead to UDP-glucuronic acid and ultimately to pectic acid. *Myo*-inositol, mentioned earlier, can serve as a source molecule. By a simple enzymatic rupture of the *myo*-inositol ring the molecule can be directly converted to glucuronic acid, which in turn can be activated to the UDP complex form. The second major pathway leading to UDP-glucuronic acid requires UDP-glucose as the source molecule.

Pectin. Pectic acid is a soluble compound which does not form a colloidal gel. It is the precursor of the more widely known compound, pectin. Pectin is a constituent of the middle lamella and the other cell wall layers of plants. It is found in relatively high concentrations in some kinds of fruits. Since pectin does form a colloidal gel, it is commercially extracted from apple and citrus fruits and sold as a gelling agent for jellies. In chemical composition pectin differs from pectic acid in that some of the carboxyl groups are converted to methyl esters. The donor of the methyl group in a transmethylation type reaction is thought to be the amino acid methionine (Chapter 16) as it exists in the S-adenosyl-L-methionine complexed form. Although the details of this reaction are not known, it might be visualized as follows:

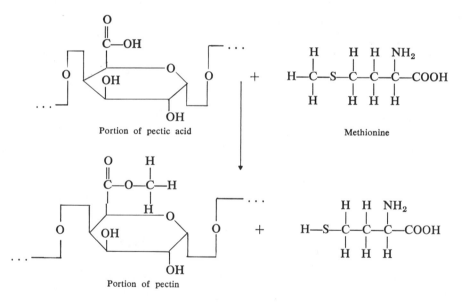

Portion of pectic acid

Methionine

Portion of pectin

Not all carboxyl groups of pectin are methylated. Usually between 50 and 90 per cent of such sites contain the methyl ester. Some of the carboxyl groups of pectin may also be converted to calcium salts. The resulting compound containing such salts is referred to as calcium pectate.

Protopectin. Another member of the pectin family, protopectin, also appears to be derived from pectic acid. Protopectin is an insoluble compound which is found almost exclusively in the cell wall. It is thought to differ from pectin in having greater polymer length, in having fewer carboxyl groups methylated, and also in having a larger proportion of the carboxyl groups converted to the calcium salts. It is known that the presence of calcium salts in the polymer renders it more rigid. Thus the ability of a cell wall to stretch during its growth is likely to be affected by the relative proportion of calcium incorporated into the protopectin.

While the pectins have been discussed as if they are polymers made up exclusively of galacturonic units, some evidence indicates that in actuality they contain a small but consistent fraction of other sugar residues which include galactose, arabinose, and xylose. Whether these other sugar residues exist as part of the main polymer or as side branches is not definitely known.

Studies of the dictyosomes of plant cells have led to the speculation that the synthesis or deposition of pectic compounds or both may be associated with these organelles. The vesicles of the dictyosomes contain high concentrations of pectin precursors such as glucuronic acid. These vesicles migrate physically through the cytoplasm and eventually reach the plasmalemma-cell wall interface.

Hemicelluloses. Hemicellulose is a term applied to a group of poorly defined, heterogeneous cell wall polymers which are made up of many different sugar residues and uronic acids. The hemicelluloses are not chemically related to cellulose but are physically associated with the cellulose molecules of the cell wall. These compounds are usually defined in terms of the relative acidity which is required to hydrolyze them from the cell wall material rather than on the basis of their chemical composition. Structurally the hemicelluloses are known to include residues of glucose, galactose, mannose, xylose, arabinose, glucuronic acid, galacturonic acid, and mannuronic acid. While the proportions of each constituent sugar residue or uronic acid can be determined for a given hemicellulose, little has been elucidated regarding the sequential order of these various units or their types of linkage. Clear-cut distinctions do not exist between hemicelluloses and such other kinds of compounds as the pentosans, some hexosans (mannan, galactan), and the pectic compounds.

Lignin. Lignin is a constituent of the woody tissue of plants, making up from 22 to 34 per cent of such tissue. The rigidity and lack of compressibility of wood can be largely attributed to this compound. Lignin is one of the most abundant organic compounds on earth, ranking second only to cellulose in this regard. Although lignin is not chemically related to the carbohydrates, it will be discussed here because of its relationship to the cell wall structure. Lignin is found in the middle lamella and in the primary and secondary cell walls of woody tissue.

Lignin is known to contain a large number of aromatic benzene-like rings. The exact reaction sequences involved in lignin biosynthesis are not known, but certain aspects of the picture are becoming clear. The original source molecules involved in lignin synthesis appear to be the amino acids phenylalanine and tyrosine (Chapter 16).

These amino acids are converted to one of the main lignin synthesis intermediates, cinnamic acid:

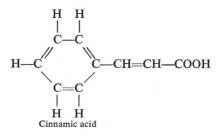

Cinnamic acid

Cinnamic acid undergoes several stepwise modifications and is subsequently converted to molecules of the coniferyl alcohol type, which are thought to be the intermediate precursors of lignin. Lignin appears to be made up of coniferyl alcohol and related compounds in rather involved linkage patterns, with an overall molecular weight of about 8,000.

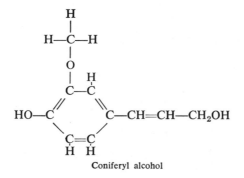

Coniferyl alcohol

The Glycosides. These are compounds formed by a reaction between a sugar and one or more compounds which are nonsugars. When glucose is the sugar component of a glycoside, as it very commonly is, the compound is called a glucoside. All glycosides could exist in two forms, α or β, but all of those known to occur in plants are of the β type. Although of widespread occurrence in plants, the glycosides are never present in large quantities. They may be found in almost any part of the plant. In a pure state they are mostly levorotatory, crystalline, colorless, bitter, and soluble in either water or alcohol. All β-glycosides can be hydrolyzed by the enzyme emulsin or by dilute inorganic acids. A large number of different glycosides have been isolated from plant tissues. Their role in the metabolism of plants, if any, is obscure although it is possible that they may serve in a minor way as reserve foods. Several representative glucosides will be discussed briefly in order to indicate more clearly the general nature of these compounds.

Salicin is found in the bark and leaves of the willow tree. Upon hydrolysis with emulsin it yields glucose and the alcohol saligenol.

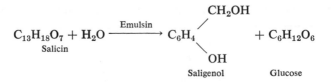

$$C_{13}H_{18}O_7 + H_2O \xrightarrow{\text{Emulsin}} C_6H_4 \begin{matrix} CH_2OH \\ \\ OH \end{matrix} + C_6H_{12}O_6$$

Salicin   Saligenol   Glucose

Amygdalin occurs in the seeds of almonds, apples, peaches, and plums. Upon hydrolysis with emulsin it produces glucose, hydrocyanic acid, and benzaldehyde:

$$C_{20}H_{27}NO_{11} + 2H_2O \xrightarrow{\text{Emulsin}} 2C_6H_{12}O_6 + HCN + C_6H_5CHO$$

Amygdalin   Glucose   Hydrocyanic   Benzaldehyde
                                          acid

Similar cyanogenetic glucosides occur in other plant materials such as leaves of cherries and peaches and in Sudan grass and other sorghums. Under certain conditions domestic animals may be poisoned from eating plant materials containing such glucosides. Such poisoning results from the release of hydrocyanic acid upon the hydrolysis of the glucoside.

Sinigrin is called the mustard oil glucoside. It is found in the black mustard *(Brassica nigra)* and is hydrolyzed as follows:

$$C_{10}H_{16}O_9NS_2K + H_2O \xrightarrow{\text{Emulsin}} C_3H_5CNS + C_6H_{12}O_6 + KHSO_4$$

Sinigrin   Allyl isothio-   Glucose   Potassium
                            cyanate   hydrogen
                            ("Mustard oil")   sulfate

The Anthocyanins. Most of the red, blue, and purple pigments of plants belong to the group known as the anthocyanins. These compounds are glycosides which have been formed by a reaction between a sugar and one of a group of complex, cyclic compounds known as the anthocyanidins. The known sugar components of the anthocyanins are glucose, rhamnose, and gentiobiose.

A number of chemically different anthocyanins have been isolated from seed plants in which they are of widespread occurrence. They are also present in some species of ferns and mosses, but no anthocyanins are known to occur in the algae or fungi. The anthocyanins are water-soluble and are usually dissolved in the cell sap, the cytoplasmic membranes being impermeable to them. Less commonly these pigments are found in plant cells in the form of crystals or amorphous solid bodies. Red pigmentation caused by anthocyanins is of frequent occurrence in flowers, fruits, bud scales, developing leaves, and less commonly, in stems, mature leaves (red cabbage, copper beech, red coleus, etc.) and other plant parts. The reds and purplish reds of autumn foliage also result from the presence of anthocyanins. The blue and purple pigmentation caused by anthocyanins is largely restricted to flowers and fruits.

Anthocyanin pigmentation, especially of flowers, is a very complex phenomenon. Factors affecting the color of plant tissues resulting from anthocyanins include the concentration of the pigment present, the proportions of the different pigments when two or more are present in the cells, the modifying effects on color of the presence of other substances such as tannins and anthoxanthins (see later), the physical state of the anthocyanins (whether in solution or adsorbed), and the pH of the cell sap. Practically all anthocyanins are red in acid solution, and many of them change in color through violet to blue as the pH of the medium increases. Because of the modifying effect of the other factors listed, however, this relation often does not show up in a clear-cut fashion.

Although anthocyanin pigments have been studied extensively in the laboratory from the chemical standpoint, very little is known of the mode of their formation in plants. The genetic capacity for anthocyanin synthesis differs considerably from one kind of plant to another. The synthesis of anthocyanins will not occur in a plant, however, even if the necessary genes are present, unless environmental conditions are also favorable. The formation of anthocyanins seems to be commonly associated with the accumulation of sugars in plant tissues. Any environmental factor such as high light intensity, low temperature, drought, or low nitrogen supply, which favors an increase in the sugar content of a given plant tissue, often favors the synthesis of anthocyanin in that tissue. On the other hand, environmental factors which check the formation or accumulation of sugars often have a similar checking effect on anthocyanin synthesis.

Light influences anthocyanin synthesis directly in some plant tissues. Autumnal red coloration, for example, usually develops its full intensity only in leaves which are directly exposed to light. In some species anthocyanins are synthesized in the absence of light. This is true, for example, in the etiolated seedlings of red cabbage and in the roots of a number of species.

The Anthoxanthins. The exposure of the petals of almost any white flower to ammonia vapor will cause them to turn yellow. This is because of the presence in such tissues of what may be regarded as a colorless form of one of the anthoxanthins. These compounds are chemically quite similar to the anthocyanins and usually occur in plants in the form of glycosides. Most of the anthoxanthins are colorless or nearly so as they occur in the plant but upon extraction and treatment in various ways their typical yellow or orange color develops. Like the anthocyanins they are water soluble, and are usually found in the cell sap. In some plant tissues, the anthoxanthins present are yellow in color. The yellow pigment in the inner bark of the black oak (*Quercus velutina*) is an anthoxanthin called quercitrin. Similar pigments occur in the wood of osage orange, sumac, and other species, and in some fruits such as orange. Some flowers, as for example yellow snapdragons, owe their yellow color to anthoxanthins. The color of most yellow flowers results, however, from carotenoid pigments.

Lipids. The term *fats* and the term *lipids* are sometimes used erroneously as synonyms. Lipids are defined as compounds of plant or animal origin which are soluble in nonpolar solvents such as benzene, ether, and carbon disulfide.

Lipids are therefore a heterogeneous array of compounds which share a solubility characteristic but which are not necessarily related chemically. This array includes neutral fats (more commonly simply called fats or oils), phospholipids, terpenes, steroids, waxes, and a number of other compounds. Lipids are also considered to include the hydrolytic degradation products of any of these compounds.

Fats and oils differ in that fats are solids while oils are liquids at room temperature. The term fats will be used here to refer collectively to both groups. The fats make up one of the food reserves of some plant organs, especially seeds or fruits of such plants as maize, soybean, coconut, peanut, safflower, and castor bean. These food reserves can exist in small amorphous aggregates within the cells as oil droplets. In contrast to animals, the majority of food reserves in most plants is of the carbohydrate type rather than of the fat type. In addition to the role of fats as food reserves, these compounds, and especially the chemically related phospholipids, play an important role in the structure of probably every type of cell membrane, including those of the many kinds of organelles. In contrast to the nonoriented arrangement of fats in oil droplets, the molecules of the lipids of the membrane systems are thought to be oriented in their position relative to one another and relative to the protein molecules contained in these membranes.

Fat Synthesis. *The Synthesis of Glycerol.* The three carbon alcohol, glycerol, is one of the components from which fat molecules are made. Glycerol is synthesized from a compound involved in glycolysis, dihydroxyacetone-phosphate (Chapter 13):

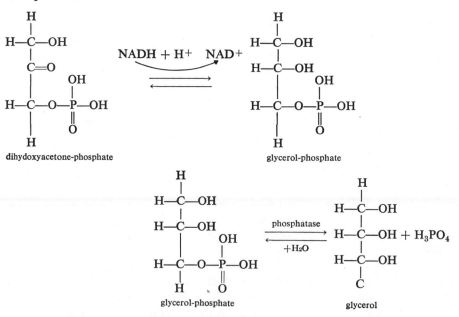

Fatty Acid Synthesis. Fatty acids are another component from which fats are made. There are two principal groups of fatty acids, the saturated and the unsaturated. If a given carbon atom in the carbon chain of a fatty acid has bonds to two adjacent carbon atoms and to two hydrogen atoms it is considered to be saturated with hydrogen or simply "saturated." The general formula for fatty acids of the saturated series is $C_nH_{2n+1}COOH$. If a given carbon in the carbon chain has a double bond to an adjacent carbon it can then have a bond to only one hydrogen atom, and is thus "unsaturated" in terms of hydrogen. Fatty acids having one or more double bonds in the carbon chain belong to the unsaturated group. With very few exceptions the fatty acids found in living organisms contain an even number of carbon atoms.

Some of the fatty acids of the saturated series found most abundantly in plants as glycerides are lauric $(CH_3(CH_2)_{10}COOH)$ found in laurel, coconut, and palm oils; palmitic acid $(CH_3(CH_2)_{14}COOH)$ found in bayberry wax, palm oil, and many other plant and animal fats; and stearic acid $(CH_3(CH_2)_{16}COOH)$, a constituent of many plant and animal fats.

The great bulk of the plant fatty acids are of the unsaturated type. The best known of the unsaturated fatty acids are oleic acid $(CH_3-(CH_2)_7CH=CH(CH_2)_7COOH)$ and linoleic acid $(CH_3(CH_2)_4CH=CH(CH_2)\ CH=CH(CH_2)_7COOH)$ of widespread occurrence in glycerides of plants, and linolenic acid $(C_{17}H_{29}COOH)$ found in linseed oil. A number of other less common unsaturated fatty acids have been isolated from the tissues of plants and animals.

None of the fat-forming fatty acids, saturated or unsaturated, is appreciably soluble in water. The lower members of the saturated series are liquids at ordinary temperatures, while those containing ten or more carbon atoms are solids. Most of the unsaturated fatty acids found in plants are liquids at ordinary temperatures.

The reactions involved in the synthesis of fatty acids are shown in Fig. 14.3.

It had been thought for many years that because fatty acids generally contain only an even number of carbons, that it was probable that their synthesis involved sequential additions of two carbon units to a growing hydrocarbon chain. It is now recognized, however, that this synthesis is an "add three carbons/subtract one carbon" type of reaction which yields a net gain of two carbons. A molecule of major importance in fatty acid synthesis is an activated form of acetic acid, acetyl-S-Co-A, which is also important in the Krebs cycle (Chapter 13).

The activated dicarboxylic acid, malonyl-S-Co-A, is the donor in the "add three carbons" portion of the reaction. Carbon dioxide is the molecule involved in the "subtract one carbon" portion of the reaction. Although originally it is acetyl-S-Co-A which reacts with malonyl-S-Co-A, on subsequent revolutions of the cycle it is the growing fatty acid-S-Co-A molecule which reacts with malonyl-S-Co-A. In the sequence of reactions, one should note the energy consuming steps which require $NADPH + H^+$. The $NADPH + H^+$ used may be generated in the pentose phosphate pathway (Chapter 13).

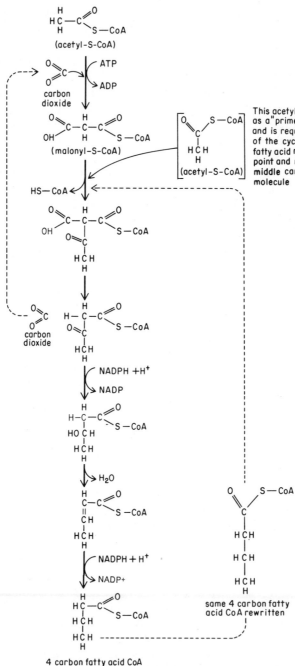

Fig. 14.3. Diagrammatic representation of the pathway of fatty acid synthesis.

Linkage of Fatty Acids and Glycerol. Fat is synthesized from glycerol and fatty acid-S-Co-A molecules. There is some evidence which suggests that glycerol phosphate rather than glycerol may be the molecule reacting with fatty acid-S-Co-A under some conditions. A typical fatty acid-S-Co-A molecule involved in the fat synthesis reaction is shown here:

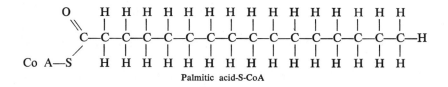

Palmitic acid-S-CoA

Each of three fatty acids forms an ester linkage to an —OH group of glycerol as shown in Fig. 14.4. All three of the fatty acid molecules which become linked to a given glycerol molecule are not necessarily of the same kind. The ester linkage of the fat molecules can be hydrolyzed by the enzyme lipase. Thus food reserve fat can eventually be converted back to glycerol and fatty acids.

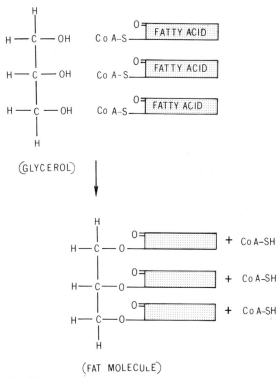

Fig. 14.4.   Diagrammatic representation of fat synthesis.

β-Oxidation of Fatty Acids.   Fatty acids are highly reduced compounds which represent a considerable reservoir of energy which can be released through oxidation reactions. The oxidation of fatty acids occurs in small organelles called mitochondria. The process involves two basic steps; the sequential removal of acetyl-S-Co-A molecules from the carboxyl end of the molecule, and the regeneration of the new Co-A activated carboxyl group at the end of the remaining fatty acid chain (Fig. 14.5). The third carbon from the carboxyl end of the fatty acid molecule is referred to as the β-carbon and because this is the carbon which becomes oxidized to the new carboxyl group, the term β-oxidation is used.

Fatty Acids and Glycerol in Relation to Respiration.   Fatty acids which proceed through the β-oxidation pathway lead to the production of large quantities of acetyl-S-CoA. This acetyl-S-CoA can proceed through the Krebs cycle along the usual aerobic respiration pathway. A single long chain fatty acid molecule can lead to the production of considerably more acetyl-S-CoA molecules than the two acetyl-S-CoA molecules which can be derived from a single hexose molecule. Thus the fatty acids which are constituents of plant fats and oils represent a very large source of respirable energy.

The glycerol derived from the hydrolysis of fats is also a molecule from which energy can be derived. Glycerol can enter the glycolytic pathway via conversion to dihydroxyacetone phosphate (see earlier). The energy which can be derived from the glycerol portion of a long chain fat molecule, however, is obviously much less than can be derived from the fatty acid portion of such a molecule.

Fat to Sugar Conversion.   One might assume that the usual disappearance of reserve oils in seeds during germination could be accounted for by the oxidation of these compounds, as described before, during stages of rapid cell growth. However, there is usually a sizable increase in the sucrose content of the seeds as the oils disappear. This fact is taken as an indication that the oils are not entirely respired, but instead act as the source molecules from which the sucrose is ultimately synthesized. A metabolic pathway starting from acetyl-S-Co-A, resulting from the breakdown of fatty acids, could and probably does account for this sucrose build up. This pathway is essentially the reverse of glycolysis. An important intermediate in this chain of reactions is glyoxylic acid (Chapter 12), and a part of this pathway constitutes the *glyoxalate* cycle. Such fat to sugar transformations occur in the glyoxysomes.

Phospholipids.   Molecules which are similar to fat molecules but which differ in having a complex phosphorylated moiety substituted for one of the usual three fatty acids, are known as phospholipids. As was mentioned earlier, phospholipids are thought to be an integral part of cell membranes.

Sterols.   These are complex cyclic (*i.e.,* containing ring groupings) alcohols of high molecular weight. Cholesterol ($C_{27}H_{45}OH$) is the best known of these

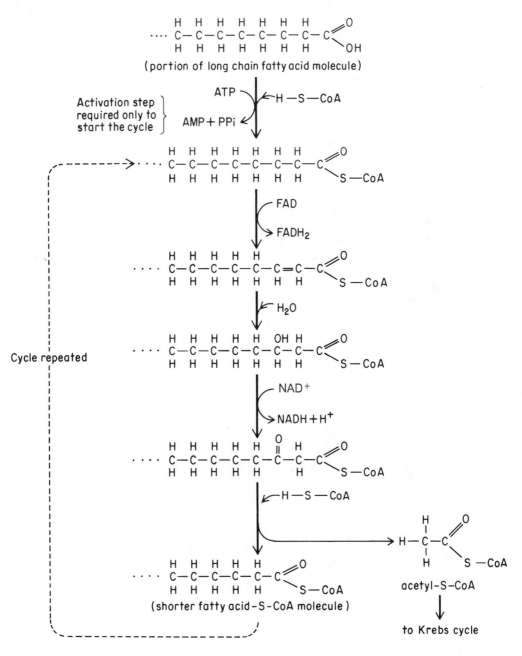

(portion of long chain fatty acid molecule)

Activation step required only to start the cycle

ATP

H—S—CoA

AMP + PPi

FAD

FADH₂

H₂O

Cycle repeated

NAD⁺

NADH + H⁺

H—S—CoA

acetyl-S-CoA

to Krebs cycle

(shorter fatty acid-S-CoA molecule)

Production per β oxidation:
1 FADH₂
1 NADH + H⁺
1 acetyl-S-CoA which can then make one cycle of the Krebs cycle

Fig. 14.5. Diagrammatic representation of the pathway of the β-oxidation of a fatty acid.

compounds and is apparently present in all animal cells, being especially abundant in the brain and nervous tissue. Cholesterol is not known to be widely distributed in plants, but occurs at least in some. A number of sterols, however, have been isolated from plant tissues. One of the most interesting of these is ergosterol ($C_{28}H_{43}OH$). This compound was first discovered in ergot but is now known to be widely distributed in plants and animals. It is abundant in yeast, which serves as its commercial source. Special interest is attached to this compound, since it is a precursor of the antirachitic vitamin D. Upon irradiation, ergosterol is converted through a series of intermediate compounds into this vitamin.

Waxes.   These compounds are usually fatty acid esters of saturated monohydroxyl alcohols such as cetyl alcohol ($C_{16}H_{33}OH$), ceryl alcohol ($C_{26}H_{53}OH$), and micryl alcohol ($C_{31}H_{63}OH$). Waxes are of widespread occurrence in both plants and animals. Examples are beeswax, carnauba wax, and the wax of the bayberry from which candles are made. Physiologically, waxes are important as constituents of the coatings which cover the outer surface of epidermal cell walls. The presence of these waxy coatings greatly reduces the loss of water from exposed plant tissues (Chapter 6). Waxes are rarely found within living plant cells.

Cutin and Suberin.   The chemistry of both of these substances is imperfectly known, although it has long been recognized that their chemical affinities are with the lipids.

Cutin apparently is built up principally of long chain hydroxy-fatty acids, *i.e.,* ones which contain one or more hydroxyl groups in the molecule. Cutin occurs principally on the outer surfaces of epidermal cell walls and, like waxes, greatly reduces water loss from such cells.

Suberin appears to be a mixture of substances consisting principally of condensation products and other modifications of phellonic ($CH_3(CH_2)_{19}\cdot CHOH\cdot COOH$), phloionic ($C_{18}H_{34}O_6$), and other similar acids. The principal chemical distinction between cutin and suberin is that the constituent fatty acids are different in the two materials and that glycerol is one of the hydrolytic products of suberin, but not of cutin. Suberin is found principally in the walls of cork cells.

The Isoprenoids.   The relatively simple hydrocarbon isoprene

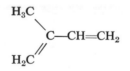

is the unit from which many plant constituents are built, some of which are metabolically very important. Isoprene is synthesized through a series of reactions involving the reduction of acetic acid molecules. The steps of this synthetic pathway are involved and will not be taken up in this book. Isoprenoid molecules are constructed from multiples of C_5H_8 isoprene molecules, having molecular formulas of $C_{10}H_{16}$ (monoterpenes), $C_{15}H_{24}$ (sesquiterpenes), $C_{20}H_{32}$ (diterpenes), $C_{30}H_{48}$ (triterpenes), $C_{40}H_{64}$ (tetraterpenes), or $(C_5H_8)_n$ (poly-terpenes). The isoprene units in isoprenoids may be linked together as straight chains or in the form of rings, some of which may be quite complex. Further modifications in structure may occur as a result of the oxidation, reduction, or other modifications of certain groups.

The essential oils constitute a large group of usually aromatic compounds which are found in many different kinds of plants. Essential oils usually accumulate, in those plants in which they are present, in isolated cells or groups of cells, or in glandular hairs on leaves or stems. Examples are oils of peppermint, lemon, rose, pennyroyal, bergamot, lavender, and sassafras. Such oils are not single compounds but mixtures, composed largely of terpenes or compounds derived from them. Some essential oils are present in the resin ducts of conifers. Turpentine, for example, which is largely composed of pinene ($C_{10}H_{16}$), is obtained by the distillation of the exudate obtained from long leaf pine (*Pinus palustris*).

Rubber is another well-known product of plant metabolism which is also an isoprene derivative. A rubber "molecule" is a polyterpene, consisting of a long straight chain (500–5,000) of isoprene units linked together end to end. Although by no means a universal metabolic product in plants, rubber is known to be present in more than 2,000 species. Practically all of the natural rubber of commerce comes from the tree species *Hevea brasiliensis*.

Rubber occurs as microscopic particles ($0.01-50\ \mu$ in diameter) suspended in the cell sap or latex. Latex is a usually milky-looking liquid confined in those plants in which it occurs to the latex ducts. The latex tubes commonly ramify to all parts of those plants in which they are present. In the trunks of *Hevea* and other woody species they are confined to the bark. The percentage of rubber in latex varies greatly from species to species and from plant to plant within a species. *Hevea* latex is 20–60 per cent rubber by weight. There is no evidence that rubber plays any essential metabolic role in plants. As far as is known, it is merely a metabolic by-product.

Some of the isoprenoids play critical roles in the metabolism of plants. The carotenoids (Chapter 10) are synthesized from a tetraterpene ($C_{40}H_{64}$). Phytol (Chapter 10), the alcoholic component of most chlorophylls, is derived metabolically from a diterpene. Another diterpene, kaurene, appears to be the precursor of the well-known plant hormone, gibberellic acid (Chapter 18).

SUGGESTED FOR COLLATERAL READING

Conn, E. E., and P. K. Stumpf. *Outlines of Biochemistry, 2nd Edition*. John Wiley & Sons, Inc., New York, 1966.
Davies, D. D. *Intermediary Metabolism*. Cambridge University Press, 1961.

Geismann, T. A., and D. H. G. Crout. *Organic Chemistry of Secondary Plant Metabolism.* Freeman, Cooper and Company, San Francisco, 1969.

Harborne, J. B. *Comparative Biochemistry of the Flavonoids.* Academic Press, Inc., New York, 1967.

Hitchcock, C., and B. W. Nichols. *Plant Lipid Biochemistry.* Academic Press, Inc., New York, 1971.

Pridham, J. B., Editor. *Terpenoids in Plants.* Academic Press, Inc., New York, 1967.

Pridham, J. B., and T. Swain, Editors. *Biosynthetic Pathways in Higher Plants.* Academic Press, Inc., New York, 1965.

Robinson, T. *Organic Constitutents of Higher Plants.* Burgess Publishing Company, Minneapolis, 1967.

Schubert, W. J. *Lignin Biochemistry.* Academic Press, Inc., New York, 1965.

15 | ABSORPTION AND UTILIZATION OF MINERAL SALTS

The dry-matter content of any plant tissue can be determined with a fair degree of accuracy by drying a sample of the tissue at a temperature of 100°C. The residue remaining after the evaporation of the water represents the nonaqueous constituents of the tissue. The percentage of dry matter content of plant tissues varies greatly, ranging from 90 per cent or even more in dormant structures such as seeds to 5 per cent or sometimes less in very succulent tissues. That the dry matter fraction of any plant tissue is composed principally of organic compounds can be demonstrated by subjecting it to combustion. This is accomplished by transferring a sample of the dry matter to a crucible and heating it over a flame or in a muffle furnace at a temperature of about 600°C. The small grayish residue resulting from this treatment is called the *ash*. Almost all of the dry matter is oxidized at this temperature and the decomposition products pass off in the form of gases. Practically all of the dry matter that disappears during combustion represents organic compounds which are decomposed as a result of the subjection to high temperatures.

The ash corresponds roughly to the mineral salts that have been absorbed from the soil, but does not include any nitrogen since this element passes off in the combustion process along with carbon, hydrogen, and oxygen. The mineral elements do not occur in the ash in the pure state, but mostly as oxides. The actual values obtained for the ash content of a plant tissue depend upon the ignition temperatures used. A portion of some of the mineral elements present is often lost by sublimation or vaporization. This is especially likely to happen to chlorine and sulfur, but potassium, calcium, phosphorus, and perhaps other elements are sometimes lost in this way. Hence the ash content of a tissue furnishes only a rather crude measure of the mineral element content of that tissue.

The total ash content of plant tissues and organs varies from a fraction of 1 per cent to 15 per cent or even more of the dry weight of the plant material. Fleshy fruits and woody tissues are usually low in ash content, often containing less than 1 per cent, while the ash content of leaves is usually relatively high, often exceeding 10 per cent. Tobacco leaves, for example, contain on the average about 12 per cent of ash on a dry weight basis. The ash content of other plant organs usually lies somewhere between these two extremes.

Elements Found in Plants.   It is probable that there is not a single one of the chemical elements that is not found at least in traces in some species of plants, under certain conditions. Actually about sixty of the known elements have been identified by chemical analysis as occurring in plants. Of these, only fourteen are found regularly in plants in appreciable quantities (Table 15.1) and not all of these appear to be essential. The question of which are the essential plant elements is considered later in this chapter.

TABLE 15.1   ELEMENTAL ANALYSIS OF THE STEM, LEAVES, COB, AND GRAIN OF A MATURE MAIZE PLANT ("PRIDE OF SALINE"), BASED ON THE AVERAGE VALUES FOR FIVE PLANTS[a]

Element	Weight, g	Percentage of total dry weight
Carbon	364.19	43.569
Oxygen	371.42	44.431
Hydrogen	52.17	6.244
Nitrogen	12.19	1.459
Sulfur	1.416	0.167
Phosphorus	1.697	0.203
Calcium	1.893	0.227
Potassium	7.679	0.921
Magnesium	1.525	0.179
Iron	0.714	0.083
Manganese	0.269	0.035
Silicon	9.756	1.172
Aluminum	0.894	0.107
Chlorine	1.216	0.143
Undetermined	7.8	0.933

[a]Data of Latshaw and Miller, *J. Agr. Research*, **27**, 1924: 854.

The composition of plant ash varies both with the species and with the environmental conditions under which the plant has developed. Comparative figures on the percentages of five of the more important mineral elements in several different species of plants growing in the same soil are given in Table 15.2. The data in this table show that even if plants develop under soil and climatic conditions as nearly identical as possible in a greenhouse, different species of plants contain very different proportions of the various elements obtained from the soil.

TABLE 15.2   PERCENTAGE OF CALCIUM, POTASSIUM, MAGNESIUM, NITROGEN, AND PHOSPHORUS IN THE TOPS OF SEVERAL SPECIES OF PLANTS GROWN IN A GREENHOUSE IN AN ALBERTA "BLACK BELT" LOAM SOIL[a]

	Percentage of dry weight				
Species	Ca	K	Mg	N	P
Sunflower	1.68	3.47	0.730	1.47	0.080
Bean	1.46	1.19	0.570	1.48	0.053
Wheat	0.46	4.16	0.225	2.26	0.058
Barley	0.68	4.04	0.292	1.94	0.125

[a]Data of Newton, Soil Sci., 26, 1928: 86.

The composition and other properties of the soil in which a plant is rooted will also have an effect on the proportion of each of the various elements absorbed by that plant. Innumerable examples of this fact can be cited from the practice of fertilizing. The addition to the soil of a compound which can be absorbed by plants usually results in an increased absorption of that substance, although the increase in the amount of the element within the plant tissues is usually not proportionate to the increase in the amount of that element in the soil. Plants often absorb from the soil mineral salts far in excess of their actual metabolic requirements. Potassium, phosphate, sulfate, and other ions often accumulate in plant cells in excess of the quantities actually utilized by the cells.

The three most abundant elements in plants, carbon, oxygen, and hydrogen (Table 15.1) constitute well over 90 per cent of their dry weight. These elements become fixed in plant cells basically through the incorporation of CO_2 and H_2O in the photosynthetic process. Thus the uptake and incorporation of mineral elements by the roots is responsible for only a small fraction of the total plant dry weight. With only minor exceptions, the mineral elements which enter into the composition of terrestrial plants are absorbed by the roots from the soil.

Soil-Associated Cation Exchange.   Most clay particles or micelles of the soil have negatively charged surfaces. Cations in the soil solution therefore have a tendency to be adsorbed in a shell-like layer on the surfaces of these micelles. The cations usually found in soil solution are not attracted with equal tenacity to these surfaces. The aluminum ion, Al^{+++}, is the most strongly adsorbed ion. The ions commonly found in the soil solution, represented in order of decreasing affinities for negatively charged surfaces, are $Al^{+++} > H^+ > Ca^{++} > Mg^{++} > K^+ > NH_4^+ > Na^+$. This series should not be misinterpreted to mean that if H^+ and NH_4^+ ions both are present in the soil solution, H^+ ions will fill all of the adsorption sites and NH_4^+ ions will remain in the soil solution. In actuality, the cited series indicates that if H^+ ions and NH_4^+ ions are present in the soil solution in *equal concentrations,* more adsorption sites will be filled with H^+ ions than with NH_4^+ ions. Furthermore, if a large number of NH_4^+ ions are added to a soil solution formerly containing equal concentrations of H^+ and NH_4^+ ions, some of the NH_4^+ ions displace adsorbed H^+ ions on clay micelle surfaces. A point could be reached such that *more* NH_4^+ ions would be adsorbed to the clay micelles than H^+ ions. The situation in which one species of cation trades position with or displaces a second species of cation at an adsorption site on any negatively charged surface is referred to as *cation exchange.*

In the example described here, as a consequence of the freeing of H^+ ions as excess NH_4^+ ions are added, the soil solution becomes more acidic. The displacement of H^+ ions does not always result in a lowering of the soil solution pH, however. For example, the displacement of H^+ ions by the addition of calcium to the soil in the form of CaO (a practice in agriculture referred to as liming the soil) can have an opposite effect on the pH. As a Ca^{++} ion replaces two H^+ ions on a clay micelle, two H^+ ions unite with the O^{--} ion forming an undissociated water molecule, thus removing hydrogen ions from the solution and raising the soil pH. Hence one result of adding various cations to the soil can be a change in the soil solution pH.

Largely as a result of the negative charge on clay micelles, anions such as NO_3^- are not generally retained in the surface layers of the soil. As nitrate-containing fertilizers are applied, for example, the NO_3^- ions enter the soil solution. Since these ions bear the same charge as the majority of the soil particles, there is little adsorption of them and they therefore sooner or later generally leach down to the water table, which puts them out of reach of many plant roots. In agriculture one is therefore required to apply nitrate-containing fertilizers fairly frequently.

The principal anions other than NO_3^- found in soils are SO_4^{--}, HCO_3^{--}, $H_2PO_4^-$, and OH^-. Like nitrate ions, most of these anions leach out of soils rather readily. Phosphate is an important exception to this statement. Even in soils which have received heavy applications of phosphates, it is usual to find only small quantities of phosphates in the drainage waters. Differences in the retention of various anions in soils are not predictable in terms of ionic charge alone. Obviously there are other factors, many of which are not yet completely understood, which affect the leachability of anions.

Passive Ion Uptake. The uptake (absorption) of ions by diffusion, or other purely physical processes not requiring the metabolic energy of the living plant cells, is referred to as *passive uptake.* Other ions are absorbed at the expense of metabolic energy in a process known as *active uptake,* which will be discussed later in the chapter.

Some passive uptake of ions occurs by simple diffusion, but there are at least three other mechanisms involved in ion absorption which are passive and thus do not require metabolic energy.

Plant Cell-Associated Cation Exchange. Clay particles, as discussed earlier, represent only one of the negatively charged surfaces which are important in cation exchange. The root surfaces of plants generally bear a negative charge. Thus, as a root grows between clay particles of the soil it is possible for cations on the clay micelles to be exchanged for cations already adsorbed on the root cell surfaces. This would account for new cations getting onto root cell surfaces, but not necessarily into the cytoplasm or the vacuole of the cells. Many of the large protein molecules of the cytoplasm and cytoplasmic membranes also bear a negative charge. It is conceivable that after cations enter a cell, they may be adsorbed onto cytoplasmic surfaces by cation exchange. This entrance of cations into cells would be a passive process; that is it would not require metabolic energy. However, a source of energy would be required to facilitate the entrance (inward movement) of a cation if the cation did not enter with an anion partner. We should therefore visualize the cation entering a cell prior to the internal protein-surface cation exchange as being associated with the concomitant entrance of an anion by diffusion.

Donnan Equilibrium. We have indicated previously that charged protein molecules exist within the cytoplasm of a plant cell. Since most protein ions are too large to diffuse out of a cell, we could think of them as fixed ions. In addition to these fixed ions, there is reason to believe that other types of ions that enter a cell as a result of active uptake (see later discussion) also become fixed or "captured" within the cell. Yet there are other types of ions which appear to be able to enter and exit from a given cell in a relatively free manner. Although these latter non-fixed ions might enter a cell basically by diffusion, they will in part be influenced by the presence of the aforementioned fixed ions held within the cell.

The effects of fixed or nondiffusible ions on the entrance of freely diffusible anions and cations might be visualized as follows. Assume that a large number of X^- ions exist within a cell and that because of their size they were unable to diffuse outward through the cell membrane. Assume also that an equal number of diffusible A^+ ions also exist within the cell. Let us now place the cell into a situation such that it is bathed by an external solution containing a high concentration of diffusible A^+ ions and an equal concentration of diffusible B^- ions. Ion pairs (A^+ and B^-) will begin diffusing into the cell. The A^+ and B^- ions will continue to enter the cell in equal numbers until a point is reached such that the product of the concentrations of *diffusible* anions and cations inside the cell will equal the product of the concentrations of *diffusible* anions and cations outside the cell. This point is referred to as the Donnan equilibrium (after F. G. Donnan, who worked out the mathematics of this relationship in 1911). If the concentration of A^+ and B^- ions outside the cell were equal at the start, and if equal numbers of A^+ and B^- ions enter the cell, then it follows that at equilibrium the concentration of A^+ ions outside the cell must equal the concentration of B^- ions outside the cell. At the start there were already some A^+ ions present inside the cell but no B^- ions inside the cell. Thus if equal numbers of A^+ and B^- ions enter, it must be concluded that at equilibrium the concentration of A^+ ions inside the cell will be greater than the concentration of B^- ions inside the cell. In summary, then, we could state that at equilibrium: the concentration of A^+ ions outside the cell equals the concentration of B^- ions outside the cell; the concentration of A^+ ions inside the cell is greater than the concentration of B^- ions inside the cell; and the product $[A^+] \cdot [B^-]$ inside the cell equals the product $[A^+] \cdot [B^-]$ outside the cell. The only way these products could be equal is if the concentration of A^+ ions inside the cell was somewhat *greater* than the concentration of A^+ outside the cell while the concentration of B^- ions inside the cell was somewhat *lesser* than the concentration of B^- ions outside the cell. A Donnan equilibrium is obviously not a concentration equilibrium. The significant point is that when a Donnan equilibrium is attained, the diffusible ion (A^+ in this example) having the charge which is opposite in sign to that of the nondiffusible ion (X^- in this example) will always *exist in a greater concentration inside the cell than outside the cell* in the ambient solution. Since, in this example, a greater concentration of A^+ ions exists inside the cell than outside it, we must conclude that in achieving a Donnan equilibrium, at least some of the A^+ ions had to move into the cell *against* a concentration gradient (*i.e.,* from an area of lesser concentration outside the cell to an area of greater concentration inside the cell). The greater the concentration of non-diffusible ions held within the cell, the greater the number of oppositely charged diffusible ions that will have to move against a concentration gradient if a Donnan equilibrium is to be reached.

The significance of the establishment of Donnan equilibria lies in the fact that this phenomenon can explain how certain ions might become more concentrated inside a cell than in the ambient solution. Secondly, it is important to note that this phenomenon basically involves passive absorption, that is, the process itself does not require metabolic energy.

Transpiration Associated Passive Uptake. There has been some question over the years as to the role, if any, of transpiration in the uptake of mineral ions by the roots. Under certain conditions correlations are found between the transpiration rate and ion uptake, but under other conditions no such correlation is found. In general, where the ion concentration bathing the roots is high, the rate of ion uptake appears to be affected by the rate of transpiration. The transpiration rate seems to have little or no effect on the rate of ion uptake if the ion concentration around the roots is low.

These results might be explained by the fact that under conditions of high ion concentration, the entrance of ions into the root cells will be quite rapid.

Whether or not additional ions enter the root is dependent upon the maintenance of a steep concentration gradient between the root cells and the soil. The maintenance of such a steep concentration gradient is in part dependent upon the rate at which ions already present in the root cells can enter the transpiration stream and be transported away from the roots. The more rapid the rate of transpiration, the steeper the ionic gradients between the soil and the root cells. On the other hand, if the ionic concentrations around the roots are low, only a few ions will enter them per time increment. Approximately the same gradients will exist whether the rate of transpiration is slow, moderate, or fast, since each of these rates would be more than sufficient to move the relatively few entering ions away from the root cells and up through the transpiration stream. In the latter situation the limiting factor is not the rate of transpiration but instead is the rate of ion absorption by the roots. Thus in this situation, there is no reason to expect a correlation between ions absorbed and the rate of transpiration.

Active Uptake of Ions.   There is reason to believe that under some conditions metabolic energy generated in respiration is utilized in the uptake of certain ions. One line of evidence which supports this *active uptake* hypothesis is the work done on the rates of ion uptake with and without the inhibitors of respiration present. A typical curve for ion uptake plotted against time with no respiration inhibitor present is shown in Fig. 15.1. Two distinct phases of uptake can be

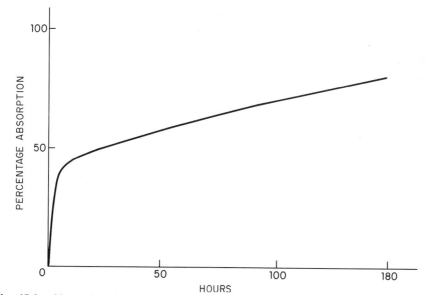

Fig. 15.1. Absorption of manganese ions by parsnip tissue over a period of time. Tissue first washed for 24 hours in aerated tap water. Data of Rees, *Ann. Botany,* n.s. **13,** 1949:31.

seen. There is usually a brief rapid uptake phase followed by a steady but much slower uptake phase. It has been found that the rate of ion uptake in roots treated with respiration inhibitors is very similar to that of noninhibited roots in terms of the rapid first phase ion uptake. With such inhibitors present, however, the slower steady secondary phase of the uptake is lacking. This indicates that the first phase does not require metabolic energy and is therefore a passive uptake while the second phase does require metabolic energy and is an active uptake. In addition to this evidence, it has been found that if noninhibited roots which have absorbed a specific type of ion over several hours time are then put in pure water, the given ion will diffuse back out of the root in an amount approximately equal to the quantity of ions absorbed passively. Ions equivalent in quantity to those absorbed actively do not diffuse back out, even though the roots are bathed in pure water (Fig. 15.2). It appears then that ions actively absorbed are somehow "captured" within the cell and held against a concentration gradient.

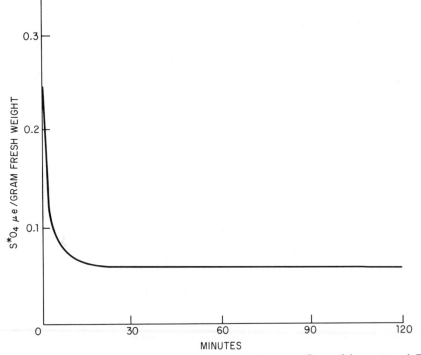

Fig. 15.2. Loss of labeled sulfate from excised barley roots. Data of Leggett and Epstein, *Plant Physiol.*, **31**, 1956:223.

Evidence of the type mentioned previously is taken by some physiologists to indicate that a given plant cell must contain an outer zone which can come to diffusion equilibrium with the external solution, and an internal core into which ions are actively transported and held against a concentration gradient. This outer zone is generally referred to as *outer space* or *free space* and the central core as *inner space*. The pectin and cellulose making up the middle lamella and the cell walls of plant cells are relatively permeable to both solute ions and water molecules. The lipid-protein plasmalemma membrane is more permeable to water molecules than to solute molecules, *i.e.,* differentially permeable. It would therefore seem probable that the portion of the cell involved in free diffusion (free space) would consist of intercellular spaces, the middle lamella, and the cell walls, while the inner space would consist of the cell volume from the plasmalemma inward. There are some differences of opinion among physiologists, however, in regard to the actual limit of the inner space.

The Carrier Hypothesis of Active Ion Uptake.   One of the more plausible hypotheses of active ion uptake involves the concept of a carrier molecule. The steps involved in the entrance of ions according to this hypothesis can be outlined as follows. An ion enters a cell and moves through outer space passively until it reaches the innermost limit of outer space where its inward progress is halted. A carrier molecule of opposite charge residing in inner space, perhaps within a membrane at the inner space perimeter, unites with the ion at the inner space/outer space interface. This union occurs possibly at the expense of metabolic energy. The ion when complexed to the carrier molecule could now enter inner space. The charge of the ion would be neutralized by its union with the carrier molecule, thus facilitating its entrance into inner space. After entering inner space, the ion-carrier complex could be broken, freeing the ion. Once in inner space, the ion is unable to diffuse back into outer space because it again has a charge and the ion is in a sense "captured" in inner space. The freed carrier molecule could then return to the inner space/outer space interface where the process could be repeated. Physiologists who favor the carrier hypothesis visualize the existence of a relatively small number of carrier molecules which are each repeatedly recycled (Fig. 15.3).

There are two main lines of evidence which suggest that carriers are involved in active uptake. The first is the so-called *saturation effect*. If the rate of active uptake measured in roots immersed in a low concentration of a given ion is compared to the rate when the roots are immersed in a slightly more concentrated solution, it is found that there is a marked increase in the active uptake per given increment of increase in the ion concentration at these low concentrations. At moderate concentrations the same increment of increase in ion concentration of the solution around the roots leads to only a small increase in active uptake rate. At high ion concentrations the same increment of ion concentration increase results in practically no change in active uptake rate. The explanation of these results in terms of the carrier hypothesis is that under low ion concentrations only a relatively small proportion of carrier molecules would be complexed to ions at any given moment. Thus additional

⊖ anions   ■ anion specific   --- innerspace/outerspace
⊕ cations   carrier molecule   interface

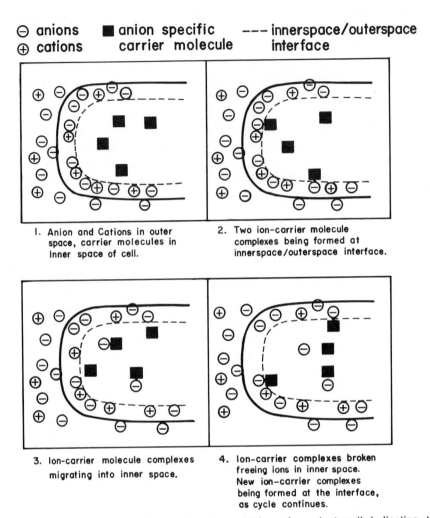

1. Anion and Cations in outer
   space, carrier molecules in
   inner space of cell.

2. Two ion-carrier molecule
   complexes being formed at
   innerspace/outerspace interface.

3. Ion-carrier molecule complexes
   migrating into inner space.

4. Ion-carrier complexes broken
   freeing ions in inner space.
   New ion-carrier complexes
   being formed at the interface,
   as cycle continues.

Fig. 15.3. Diagrammatic representation of a portion of a plant cell indicating the mechanism of ion absorption according to the carrier hypothesis.

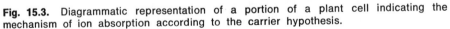

increments of ions could be complexed to the numerous free carrier molecules. Under high concentrations, a point would be reached at which nearly every carrier would be complexed to an ion. Adding more ions to the external solution at this point would have little effect on active uptake since there would be practically no free carriers available to unite with them. The recycling rate of the carrier would become the limiting factor under these conditions. One might say that at this point a carrier saturation point is reached.

The second type of evidence that lends credence to the carrier hypothesis are results which indicate *ion competition* in the active uptake process. If, for example, barley roots are immersed in a solution containing a given concentration of Cl^- ions, a rate of active uptake can be determined. If the experiment is repeated using the identical concentration of Cl^- ions, but in addition some Br^- ions are also added to the solution, the rate of active Cl^- uptake will be slower now that some Br^- uptake occurs than when Cl^- ions are present alone. This is taken to indicate that the carrier molecules which normally unite with Cl^- ions can also unite with Br^- ions. The two types of ions appear to compete for the same carrier molecule. The active uptake of Cl^- ions is thus inhibited to some extent because some of the carrier molecules are tied up in Br^- complexes. This type of competitive inhibition is analogous to the competitive inhibition of enzymes discussed in Chapter 9.

If the above experiment is modified such that the added anion is NO_3^- rather than Br^-, there is practically no effect of the NO_3^- ion uptake on the rate of active uptake of Cl^- ions. With these particular kinds of ions there are probably two different types of carrier molecules involved, one in Cl^- ion uptake, and another in NO_3^- ion uptake, and thus there is no competitive inhibition effect between these two ions.

Although the actual means by which metabolic energy is utilized in active ion uptake is not known, some hypotheses as to the possible use of this energy include: (1) the modification of the inner space limiting membrane so as to render it permeable to ions, or (2) the formation and/or breaking of the ion-carrier complexes, or (3) the transportation of the ions or ion-carrier complexes in a cyclosis-like movement.

Since the active uptake of ions requires respiratory energy, it is not surprising to find that ion uptake is markedly reduced under conditions of oxygen deficiency (Fig. 15.4). When plants are raised hydroponically (*i.e.,* in mineral salt solution rather than in soil), the bubbling of air through the liquid medium

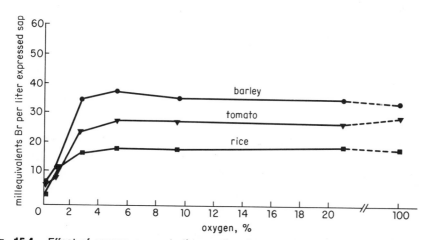

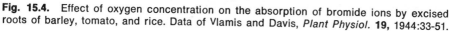

Fig. 15.4.   Effect of oxygen concentration on the absorption of bromide ions by excised roots of barley, tomato, and rice. Data of Vlamis and Davis, *Plant Physiol.* **19,** 1944:33-51.

is required if ion uptake is to be kept optimal. Likewise in waterlogged or otherwise poorly aerated soils, the absorption of mineral salts is greatly retarded.

The fact that the salt absorption of root cells is dependent upon respiration suggests that temperature may have a marked effect on the process, a supposition which has been confirmed experimentally. The curve of ion uptake versus temperature follows very closely the classical curve obtained when most plant tissue respiration rates are plotted against temperature (Fig. 15.5).

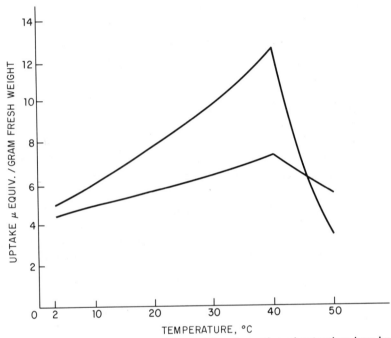

Fig. 15.5. Relation between temperature and the absorption of potassium ions by slices of washed carrot tissue over a period of 30 minutes (lower curve) and 2 hours (upper curve). Data of Sutcliffe, *Mineral Salts Absorption in Plants,* Pergamon Press, New York, 1962:48.

Aerobic respiration in the root cells is dependent for its substrate upon carbohydrates which are translocated to the roots from the leaves. Any conditions which markedly reduce the rate of photosynthesis in the leaves or the rate of the downward translocation of foods may therefore also bring about a diminution in the rate of salt accumulation. The salt-absorbing capacity of the roots of barley plants, for example, has been observed to differ appreciably from one season of the year to another. During the winter months when light intensities are low and days are relatively short, barley plants absorb much less salt from culture solutions in a greenhouse than they do during the summer months.

In some plants, the salt absorption by roots can be substantially increased if the roots are infected by fungi. These fungal associations are referred to as mycorrhizas. Mycorrhizas may also bring about an increase in the water uptake (Chapter 7).

General Aspects of Salt Accumulation.   The fact that ions may attain a higher concentration within living cells than in the ambient solution was first clearly demonstrated in the large cells of certain species of algae. Sap from the cells of the fresh-water alga *Nitella,* for example, can be obtained in sufficient quantities to permit its accurate analysis (Table 15.3). This analysis, and other similar ones, showed that anions and cations accumulate within the cells in concentrations which greatly exceed those in the bathing medium. Furthermore, the electrical conductivity of the sap was found to be approximately equal to that of a solution of electrolytes of the same concentration, indicating that the accumulated salts are present within the cell sap in the dissolved state. Both cations and anions are accumulated in cells by the operation of this mechanism, and often in approximately equivalent quantities. An increase in the concentration in the external solution can only be attained as a result of the entrance of those ions against a concentration gradient. Such an accumulation might be explained in part by the establishment of Donnan equilibria, but more probably it is the result of an active uptake process.

TABLE 15.3   ANALYSIS OF THE VACUOLAR SAP OF *NITELLA CLAVATA* AND OF THE POND WATER IN WHICH IT WAS GROWING[a]

	Sap concentration, milliequivalents per liter[b]	Pond water concentration, milliequivalents per liter[b]
Ca^{++}	13.0	1.3
Mg^{++}	10.8	3.0
Na^+	49.9	1.2
K^+	49.3	0.51
Sum of cations	123.0	
Cl^-	101.1	1.0
SO_4^{--}	13.0	0.67
H_2PO^-	1.7	0.008
Sum of anions	115.8	

[a]Data of Hoagland and Davis, *Protoplasma,* **6,** 1929: 611.
[b]A milliequivalent of an ion is one-thousandth its gram ionic weight divided by its valence.

In higher plants, the cells near the tips of roots up to and including those in the root hair zone also have the capacity of accumulating ions. If the initial salt content of the root cells is low and if other conditions are favorable, the concentration of ions in the absorbing cells may soon greatly exceed that of the same ions in the soil solution. The rapid accumulation of ions does not occur, however, if the cells already contain a relatively high concentration of the same ions. The rate of salt accumulation is often influenced, therefore, by the previous metabolic history of the absorbing cells.

The phenomenon of salt accumulation seems confined largely to cells which have retained the capacity for cell division and growth and are rapidly respiring. Meristematic cells and cells in the early stages of enlargement are particularly active in absorbing ions (Fig. 15.6). As cells lose their capacity for growth, they also lose their capacity for mineral salt accumulation. The parenchyma cells of an apple fruit, for example, are fully mature and are not

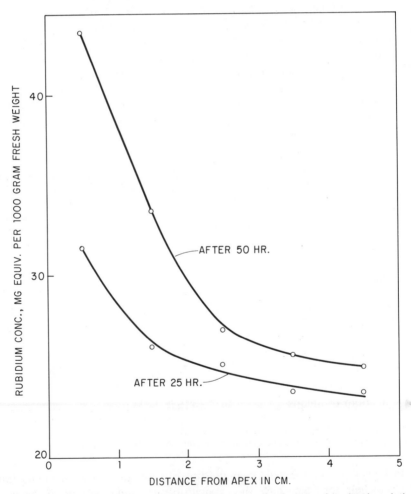

Fig. 15.6. Distribution of accumulated rubidium in excised roots of barley in relation to distance from the root apex. Data of Steward, *et al., Plant Physiol.,* **17,** 1942:414.

able to accumulate ions from dilute solutions, while in cells of potato tubers and some other storage organs which are capable of renewed growth, salt accumulation may occur under favorable conditions. It is probable that all or most plant cells are able to accumulate mineral salts when in a meristematic condition, but that this capacity decreases as the cells become mature.

Salt Accumulation and Organic Acid Metabolism. The number of cations and anions in the cell sap must be maintained in such a balance that the solution remains electrostatically neutral. It is common, however, to find that the cell sap contains a large excess of inorganic cations. The excess of inorganic cations is electrically balanced by organic anions synthesized in the cells. The high ratio of inorganic cations to inorganic anions in the cell sap may arise in at least two ways: (1) the cation may be accompanied into the cell by an OH^- ion or an HCO_3^- ion instead of its original inorganic partner in the external medium, or (2) certain inorganic anions may enter more rapidly into organic combinations and be removed from the cell sap by the metabolic activities of the protoplasm. Nitrate anions, for example, are often quickly reduced in the cells of the roots and synthesized into organic nitrogenous compounds (Chapter 16), leaving behind in the cell sap the inorganic cations with which they were paired upon entering the cell. There is, therefore, a close correlation between the excess absorption of inorganic cations and the increased organic acid content of plant tissues. Whenever surplus inorganic cations accumulate in the cells, organic acid anions also accumulate; and whenever the content of anions exceeds that of cations, the organic acid content of the cells is correspondingly decreased.

Absorption of Mineral Salts by Aerial Organs. Although we normally associate mineral absorption with the plant roots, there are certain situations in which absorption by stems and leaves is known to occur. For example, plants are sometimes "fertilized" by spraying the aerial organs with dilute solutions. This procedure has not been found very satisfactory for supplying significant amounts of most macrometabolic elements (see later) to plants. However, nitrogen in the form of urea can be successfully supplied to some crop plants, pineapple for example, in this way. This method has also been used as a satisfactory means of supplying micrometabolic elements (see later) to plants. An iron deficiency in pineapple plants, for example, may be rectified by spraying the plants with a dilute solution of ferrous sulfate. In similar fashion copper, manganese, and zinc are often supplied to citrus and other crops by spraying the foliage with a dilute solution of a salt of the metal which is to be supplied.

Not only can mineral salts be absorbed from the leaves, but they can also be lost from the leaves to at least a limited extent by leaching during rains. Potassium appears to be the most easily lost of the mineral elements as a result of leaching from foliage. Small quantities of amino acids and other organic compounds are also lost from leaves in this way.

General Roles of the Mineral Elements in Plants.   Strictly speaking mineral elements as such do not influence the physiological processes of plants. It is only when present in ionic form or as constituents of organic molecules that they assume important roles in plants. For the convenience of brevity, however, the term mineral elements is in common use to refer to these substances regardless of the exact form or combination in which they exert their effects in plants.

Considered as one group or class of substances found in plants, mineral elements function in a number of different ways. For example, a number of the mineral elements become constituents of structural molecules which are integral parts of the protoplasm and cell walls. Specific examples would be the sulfur in proteins, the phosphorus in nucleotides, the magnesium in chlorophyll, and the calcium in calcium pectate.

Certain effects of mineral elements in plants result from their participation in one way or another in catalytic systems. Iron, copper, and zinc are known to be prosthetic groups of certain enzymes, and this may also be true of some other mineral elements. Iron is also a constituent of the cytochromes. Other mineral elements such as magnesium, manganese, potassium, cobalt, molybdenum, and boron act as activators or inhibitors in one or more enzymatic systems.

In Chapter 5 it was shown that the osmotic potential of the cell sap of any plant cell results, in part, from the dissolved mineral salts which it contains. While in most plant cells the absolute concentration of mineral salts in the plant sap is so low that only a small proportion of the osmotic potential can be ascribed to their presence, there are some important exceptions to this statement as discussed previously.

The mineral salts absorbed from the soil often have an influence on the pH of the cell sap and other parts of plant cells, although usually not a very great one, as organic acids and other compounds resulting from the metabolic activities of plants ordinarily exert the predominant influence in determining pH values within cells. Two of the important buffer systems found in plants—the phosphate and the carbonate systems—have their origin in substances absorbed by the plant from its environment. The cation components of plant buffer systems, other than H^+, are mostly such mineral elements as potassium, calcium, sodium, and magnesium.

Many mineral elements in their ionic form have a marked toxic effect upon protoplasm, often resulting in its disorganization and death, even when present in very low concentrations. Among the elements which are known to be highly toxic to plants, at least under certain conditions, are aluminum, arsenic, boron, copper, fluorine, lead, magnesium, manganese, mercury, molybdenum, nickel, selenium, silver, and zinc. Included in this list are certain of the elements essential in plant metabolism which exert toxic effects when present within the tissues in concentrations exceeding physiological requirements. The toxic effects of some of these elements are discussed more fully later in the chapter.

Utilization of Mineral Salts.   A clear distinction should be drawn between the absorption of a salt and the subsequent utilization of it or its component ions. The term utilization is employed in a loose sense for the incorporation of mineral elements into the relatively permanent constituents of the cell walls and protoplasm or for their participation in fundamental metabolic reactions. The absorption of the ions or molecules of salts does not necessarily mean that they will be utilized. Many of the ions absorbed by a plant remain for more or less indefinite periods in the ionic state in the cells. Sooner or later many of these ions are usually incorporated either into the structure of more complex molecules synthesized by the plant such as storage proteins, glycosides, etc., or into the protoplasm or cell walls. There may, therefore, be a considerable time lag between the absorption of an ion and its utilization. Furthermore, some mineral elements may be utilized in one organ of a plant, subsequently released by the disintegration of cell constituents, translocated to other organs of the plant, and there reutilized. The redistribution of minerals which have accumulated in cells but are never actually utilized is also of common occurrence in plants.

Essential and Nonessential Elements.   Of the large number of elements that have been identified as occurring in plant tissues, only a limited number have been found to be indispensable. Beginning about 1860 a number of extensive investigations were undertaken to determine specifically which elements are essential for green plants and which are not. Some of the earliest workers on this problem were the German botanists, Sachs and Knop. Their investigations, conducted by the method of solution cultures (see later) and subsequently confirmed by a number of other workers, indicated that in addition to the elements carbon, hydrogen, and oxygen, obtained by plants from water or from atmospheric gases, the only essential elements were nitrogen, phosphorus, sulfur, calcium, magnesium, potassium, and iron, all of which enter the plant from the soil.

Although this precept of "ten essential elements" was strongly entrenched for some decades, in time it became more and more questioned. The likelihood of the presence of contaminating substances from impure chemicals, dust, or from other sources was not given sufficient consideration by the earlier investigators of the essential plant elements. The possibility existed that elements required only in minute quantities were present in their solution cultures as unknown contaminants, and hence the necessity of such elements for plants was overlooked. Refinements have been introduced into sand and solution culture techniques in more recent times which greatly reduce the possibility that unrecognized elements could be present in physiologically significant quantities.

As a result of critical investigations using much more refined methods, it is now generally recognized that six other elements—boron, manganese, chlorine, copper, zinc, and molybdenum—must be added to the group recognized to be essential in the metabolism of green plants. All of the elements just listed are required in only minute quantities. Hence they may be termed *micro-metabolic elements* in contrast with those necessary in relatively large amounts, which may be termed *macrometabolic elements*. All of the ten elements long recognized as being essential for plants, except iron, belong to the latter group. This brings to sixteen the number of elements recognized as necessary for green plants in general. In addition certain species, especially among the algae, appear to require other elements as discussed later in this chapter.

Specific Roles of the Mineral Elements in Plants. The specific known roles of the elements considered to be generally necessary for plants will be discussed under this heading. Other elements, which are apparently necessary only for certain species, and still others known to have marked effects on plants but not known to be essential, will also be discussed.

Nitrogen. Although N_2 gas makes up nearly 80 per cent of the atmosphere, nitrogen cannot be directly utilized by most plants in this form. Nitrogen in a relatively oxidized state as nitrate ions in the soil is readily utilized by the majority of plants. Once within the plant, the nitrogen of the nitrate ion is sequentially reduced in a series of reactions leading to the ultimate incorporation of nitrogen into the highly reduced amino groups of amino acids and other compounds (Chapter 16).

Because nitrate ions are easily leached from the soil as noted previously, supplementation with nitrate salts in the form of fertilizers is generally a necessary agricultural practice. With certain crops, the use of ammonium salts has been found to be a satisfactory method of providing nitrogen to the plants.

There are many phases of plant metabolism and development which are directly or indirectly affected when nitrogen is deficient. Nitrogen is a structural component of amino acids, proteins, enzymes, coenzymes, nucleic acids, chlorophyll, most plant membranes, many plant hormones, and a large number of other metabolically important organic compounds within the plant. The most typical nitrogen deficiency symptoms in plants are the marked reduction in growth rate and the development of a chlorotic condition. The latter condition is a general yellowing of the plant which is the result of normally masked carotenoid pigments becoming visible as chlorophyll synthesis declines. Because soluble forms of nitrogen can be fairly readily transported within the plant, young developing leaves sometimes remain green slightly longer than the older, more mature leaves of the plant.

Sulfur. As a rule this element is fairly well distributed throughout the tissues and organs of plants. Sulfur is a constituent of cystine, cysteine, and methionine, which are three of the amino acids from which enzymes and proteins in general are composed (Chapter 16). The vitamins thiamine and biotin are both important sulfur containing compounds. In addition, sulfur is also a component of coenzyme A, which is involved in many aspects of the Krebs cycle (Chapter 13) and fatty acid metabolism (Chapter 14). Sulfur is also a constituent of mustard oil glycosides such as sinigrin, which impart characteristic odors and flavors to such species as mustard, onions, and garlic.

Sulfur is usually absorbed by roots as the SO_4^{--} ion, but may also enter the leaves as SO_2 when that gas is present in the atmosphere. It should be noted, however, that SO_2 gas becomes toxic to plants at very low concentrations (Chapter 20). Although sulfur enters the plant in oxidized form, it is reduced, usually to the sulfhydryl (−SH) group, in the formation of amino acids or other organic sulfur-containing molecules. This process goes on largely in the leaves. The sulfur of organic molecules in living cells apparently may be reconverted into inorganic sulfur, usually the sulfate ion, in which form it may be redistributed within the plant and re-utilized in the formation of organic sulfur compounds in other tissues. Relatively large amounts of sulfur may be moved in this way from the leaves to ripening seeds and fruits.

Because sulfate is usually available in most soils, the addition of sulfur-containing salts to agricultural fertilizers is rarely required. The symptoms of sulfur deficiency in plants are similar in general to those characteristic of insufficient nitrogen. Amino acids and other nitrogen-containing compounds accumulate in the tissues of sulfur-deficient plants, probably because protein synthesis is not maintained at a rate comparable with that in plants receiving adequate sulfur.

Phosphorus.   Phosphorus is absorbed by plants principally as the $H_2PO_4^-$ ion. Unlike nitrogen and sulfur, however, phosphorus is not reduced in plant tissues, but is linked into organic combinations in highly oxidized form. Phosphorus enters into the composition of sugar phosphates, nucleotides, and nucleic acids. Phosphate carriers, phosphorylation, and the energy of phosphate bonds are of primary importance in fat and carbohydrate metabolism (Chapter 14), respiration (Chapter 13), photosynthesis (Chapter 10), and in many other metabolic processes.

A very large proportion of the phosphorus in a mature plant is located in the seeds and fruits, accumulating there during the period of their development. In growing plants, phosphorus is most abundant in meristematic tissues, which is consistent with the important roles played by phosphorylated compounds in metabolism.

The roles of phosphorus and nitrogen in plant metabolism appear to be interrelated in a number of ways. Inorganic nitrogen compounds are rapidly absorbed and accumulate in plant tissues when the available phosphates are low. When the available phosphates are abundant in the rooting medium, on the other hand, the absorption of inorganic nitrogen compounds is depressed. The application of phosphate fertilizers, therefore, may alter the nitrogen balance of the plant. Illustrations of this effect are the earlier maturation of plants that often occurs when available phosphorus is high, and the delay in reaching maturity occasioned by phosphorus deficiency. The synthesis of proteins (Chapter 16) apparently does not occur at the usual rates in phosphorus-deficient plants. Correlated with this decrease in protein synthesis there is often an accumulation of sugars in the vegetative organs of the plant. The purple coloration of leaves associated with phosphorus deficiency in certain varieties of maize, tomatoes, and other species reflects the relatively high concentrations of sugars in the leaf tissues, which often favors anthocyanin synthesis (Chapter 14). There is also good evidence that phosphates are more rapidly absorbed and accumulated in plants when nitrogen is supplied in organic form (urea) than when nitrate is being absorbed.

Phosphorus is readily redistributed in plants from one organ to another. During periods of phosphorus deficiency a large proportion of the phosphorus in the older leaves may move into other tissues, and developing fruits of tomatoes may obtain phosphorus from even the youngest leaves when phosphorus is unavailable in the rooting medium. Studies with radioactive phosphorus have contributed evidence of the high mobility of phosphorus and of the rapid rate with which it may move out of leaves and into growing tissues when the external supply is deficient.

Calcium.   A large part of the calcium in most plants is located in the leaves and, in contrast to phosphorus and potassium, more calcium is present in the older than in the younger leaves. An analysis of strawberry plants, for example, showed more than twice as much calcium in the leaves as in the crowns and roots together. Much of the calcium in plant tissues may be permanently fixed in the cell walls as a calcium salt of the pectic compounds of the middle lamella. Leaves of squash were found to have as much as 70 per cent of their calcium immobilized in the walls of the cells. In many plant species calcium is present in the form of insoluble crystals of calcium oxalate. Calcium also forms salts with other organic acids and probably enters into chemical combinations with protein molecules. Calcium ions are generally present as one constituent of the vacuolar sap.

Calcium is necessary for the continued growth of apical meristems. In the absence of calcium, mitotic divisions become aberrant or suppressed. Calcium is also known to have a role in the nitrogen metabolism of plants. In the absence of calcium, some species at least, are unable to absorb or assimilate nitrates. Organic forms of nitrogen such as urea, however, seem to serve as sources of nitrogen when calcium is absent. These observations are interpreted to mean that calcium is important in the reduction of nitrates in plant tissues (Chapter 16).

Calcium is known to serve as an activator of several enzymes in plants. It is known to be essential to amylase activity. In addition, the role of calcium in the stimulation of pollen germination and pollen tube growth may involve an enzyme activation effect.

Calcium is relatively immobile and is not readily redistributed in plant tissues when it becomes deficient in the rooting medium. Older leaves of a plant may be relatively high in calcium content at the same time that younger leaves on the same plant are deficient in calcium. However, crystals of calcium oxalate in the old leaves of peanut plants disappear at times of severe calcium deficiency and are reformed in very young leaves, indicating that some degree of redistribution is taking place. Similar observations have been reported for other species, but the redistribution of calcium does not appear to be sufficiently rapid or complete to meet the metabolic requirements of the younger tissues.

Magnesium. This element is a mineral constituent of the chlorophyll molecule. A large proportion of the magnesium present in plants is therefore in the chlorophyll-bearing organs. Seeds are also relatively rich in this element. Magnesium generally occurs in soils in sufficient abundance to meet the requirements of plants, although occasional exceptions to this statement are found. A deficiency of magnesium usually results in the development of a characteristic chlorosis and in some species in the appearance of a purple coloration in the foliage. The redistribution of magnesium from older to younger organs of plants occurs rapidly.

Magnesium plays a role in the phosphate metabolism of plants and indirectly therefore in the respiratory mechanism. Magnesium ions appear to be specific activators for a number of enzymes, including certain transphosphorylases, dehydrogenases, and carboxylases.

Magnesium also plays a role in maintaining the integrity of ribosomes, the organelles associated with protein synthesis (Chapter 16). When magnesium is present in very low concentration, the ribosomes fragment into a multitude of subunits which are no longer capable of protein synthesis.

Excess quantities of magnesium may prove toxic in solution cultures, an effect which may be offset by the presence of sufficient amounts of calcium. Magnesium toxicity in soils is not common, but may occasionally occur in alkali or serpentine soils.

Potassium. Unlike all of the other macrometabolic mineral elements required by plants, potassium is not definitely known to be built into any organic compounds essential for the continued existence of the plant. It occurs in plants principally as soluble inorganic salts, but potassium salts of organic acids also are found in plant cells. In spite of these facts potassium is an indispensable element and cannot be completely replaced even by such chemically similar elements as sodium or lithium. The young actively growing regions of plants, especially buds, young leaves, and root tips, are always rich in potassium, while as a rule the proportion of potassium is relatively low in seeds and mature tissues. The fundamental roles of this element in plant metabolism are largely regulatory or catalytic. However, as described in Chapter 6, potassium ions play a role, probably osmotic, in the opening and closing of the stomates.

A large number of experimental studies involving many species of plants have furnished considerable information about what happens to plants when potassium is deficient. Plants deficient in potassium usually contain a higher percentage of soluble nitrogen compounds (amino acids and amides) than plants supplied with adequate potassium. The protein content of the potassium-deficient plants, on the other hand, is relatively low. These facts suggest that potasssium is in some way involved in the synthesis of protein from amino acids. Further support for this hypothesis is furnished by the very different behavior of potassium-deficient plants when grown with ammonium compounds as compared with those grown with nitrates as sources of nitrogen. Plants supplied with ammonium nitrogen soon develop signs of serious injury, apparently resulting from the rapid accumulation of ammonia in the tissues. The reduced nitrogen compounds are not synthesized into proteins and so accumulate in toxic quantities in the leaves and stems.

The carbohydrate metabolism is also disturbed by inadequate supplies of potassium. There is evidence that photosynthesis is checked and that respiration is increased by severe potassium deficiency. The effects of low potassium are usually first apparent in the disturbed nitrogen metabolism which, because of the failure of protein synthesis, may lead to an initial increase of carbohydrates in the tissue. As potassium deficiency continues, carbohydrates rapidly decrease in quantity, probably as a result of decreased photosynthesis and increased respiration.

Potassium is highly mobile in plants. Internal redistributions of this element occur readily and more or less continuously during the life history of the plant. Older leaves and other organs frequently lose potassium, which is translocated in growing regions. Those tissues of the plant that are undergoing the most active growth appear to have the greatest capacity for accumulating potassium in contrast with cells that are physiologically less active.

The potassium ion is usually the most abundant monovalent cation in plant cells. Although it cannot be replaced entirely by any other element, symptoms of potassium deficiency may appear much sooner and are more severe in barley in the absence of sodium ions than when these are present in the culture solution. It is probable, therefore, that at least during the early stages of growth, potassium may be partially replaced by sodium, at least in some species of plants.

Iron. The deficiency of iron in soils is usually a consequence of its insolubility rather than its actual absence. In general, a larger proportion of the iron is in a soluble state in relatively acid soils than in approximately neutral or alkaline soils. One of the most common causes of iron deficiency is an excess of lime in the soil. Even in alkaline soils, however, some iron may be absorbed by plants as a result of the intimate contact between the root surface and the soil particles.

Iron is indispensable for the synthesis of chlorophyll in green plants. Deficiency of this element results in the development of a characteristic chlorosis. Iron does not, however, enter into the constitution of the chlorophyll molecule. The state of the iron in plant tissues is also often a factor determining its influence in chlorophyll synthesis. Chlorosis, as a result of iron deficiency, is sometimes found in leaves which contain as much iron as green leaves, the iron being present in an unavailable form in the chlorotic tissue. Iron is physiologically active in the ferrous state, and although often absorbed as the ferric ion, much of it is rapidly reduced within the cells. The rate at which iron is reduced in the living cells seems to be influenced by the quantity of manganese in the cells as discussed later.

Some of the enzymes and carriers which operate in the respiratory mechanism of living cells are iron compounds. Specific examples are catalase, peroxidase, ferredoxin, and the cytochromes. The participation of iron in the form of such compounds in the oxidative mechanism of cells is undoubtedly one of its more important roles in cellular metabolism.

The proportionate amount of iron in plant tissues is very low; much of that present is a constituent of organic compounds. Iron is one of the most immobile of all elements in plants, little redistribution occurring from one tissue to another. If plants which have been supplied with iron are transferred to a solution culture lacking this element, the subsequently developing leaves exhibit a marked iron chlorosis, while the older leaves retain their normal green color. This is a graphic demonstration of the fact that no appreciable transfer of iron occurs from older to younger leaves.

Manganese.   Only small quantities of this element are required by plants, manganese compounds being distinctly toxic to plants except in very low concentrations. Manganese, as a rule, seems to be most abundant in the physiologically active parts of plants, especially leaves. It is a relatively immobile element. little redistribution occurring from one part of a plant to another.

The roles of manganese in plants are undoubtedly those of a primary or accessory catalyst. This element probably plays a direct part in oxidation-reduction phenomena, especially in relation to iron compounds. Iron is commonly absorbed as the ferric ion and reduced in cells to the ferrous condition unless some oxidizing agent is present which prevents this reaction. According to some investigators, manganese plays the role of such an oxidizing agent, and an excess of manganese may therefore induce symptoms of iron deficiency by converting the available iron into the physiologically inactive ferric condition.

Manganese is an activator of a number of enzyme systems, including some of the same enzymes which are also activated by magnesium.

Manganese is also related in some way to chlorophyll synthesis, as the chloroplasts are soon affected by a deficiency of this element. Chlorosis resulting from manganese deficiency is distinctive in appearance as compared with that resulting from inadequate iron or magnesium.

Certain "deficiency diseases" of plants have been traced to an inadequate amount of manganese within the tissues. Some of the better known of these are gray speck of oats, the speckled yellows of sugar beet, and the frenching of tung. Manganese deficiency in soils is most likely to occur when the soil reaction is relatively alkaline. Manganese toxicity as a result of soil conditions is not common, but occurs occasionally, especially in acid soils.

Boron.   Although in general, this element is necessary in only small quantities, the boron requirements of different plants vary over a considerable range. Investigations by solution or sand culture techniques show that only a trace (less than 1 ppm) of boron is required for the best development of some species such as tomato and carrot, while others such as sugar beet and asparagus do not show maximum development unless boron is available in a concentration of 10–15 ppm. A boron deficiency in plants generally leads to serious injury or death of cells in the apical meristems of stems and roots. The injury to tissues of such plants has been traced to high concentrations of phenolic compounds. Boron, when present, tends to inhibit the synthesis of phenolic compounds. Some general aspects of this inhibition have recently been worked out. The synthesis of phenolic compounds involves erythrose-4-phosphate as a source molecule. Erythrose-4-phosphate is also an intermediate of the pentose phosphate pathway (Chapter 13). When boron is present at normal levels, both the pentose phosphate pathway and the phenolic compound synthesis pathway are generally inhibited. The actual metabolic block appears to occur when the 6-phosphogluconic acid of the pentose phosphate system complexes with boron. The 6-phosphogluconic acid thus complexed does not proceed further in the pentose phosphate pathway. Under boron deficiency conditions this complex cannot form and thus the unchecked pentose phosphate reactions lead to a large erythrose-4-phosphate pool, which in turn leads to an inordinately rapid rate of phenolic compound synthesis, which in turn results in damage to the tissues of the plant.

In most plants boron is not a readily mobile element. It reaches its greatest concentration in the leaves and seems to be fixed in the leaf cells in some manner which precludes its translocation to other tissues. The results of certain investigations have indicated that boron may be essential for the translocation of carbohydrates in plants. The boron concentration of meristems, roots, fruits, and storage tissues is much lower than that of leaves. Boron may be present in the lower leaves of a plant in such quantities as to cause them injury at the same time that cells of the apical meristems exhibit symptoms of boron deficiency.

Boron has been found to stimulate the growth of pollen tubes in some species. The role of boron in pollen tube growth, however, is not clearly understood.

A number of "physiological diseases" of plants have been found to be caused by a deficiency of boron within the tissues. Among these are heart rot of sugar beet, leaf roll of potato, brown heart of turnip, and browning of cauliflower. Injury as a result of boron toxicity is also fairly common occurrence in field-grown plants. The use of irrigation water containing compounds of boron in solution is a frequent cause of boron toxicity in plants.

Zinc.   This element is highly toxic to plants except in very dilute concentrations, but traces must be present if normal plant metabolism is to be maintained. The deficiency of zinc results in structural aberrancies in the root tips, the dwarfing of vegetative growth, and the failure of seed formation.

Zinc is known to be a constituent of the enzymes carbonic anhydrase and glutamic dehydrogenase, and is an activator of at least several other enzyme systems. Zinc is also necessary in the synthesis of indoleacetic acid, an important hormone in plants.

A number of disease conditions, especially of tree species, have been recognized to result from zinc deficiency. Among these are rosette of pecan, little leaf of various species of deciduous fruit trees, bronzing of tung, and white bud of maize. The abnormal growth behavior of fruit trees and other species when deficient in zinc may result at least in part from the necessity of zinc for the synthesis of indoleacetic acid.

Copper.   Like zinc this element is highly toxic to plants except in very dilute concentrations. Because of the fact that copper is a constituent of certain oxidizing-reducing enzymes, such as tyrosinase and ascorbic acid oxidase, as well as of plastocyanin (Chapter 10), it is certain that at least one of its roles in plant metabolism is participation in oxidation-reduction reactions.

The dieback disease of citrus and some other trees and the so-called "reclamation disease" have been found to have their origin in a deficiency of copper. The latter disease has been found in plants of northern Europe. Livestock grazing on such land is also subject to a copper-deficiency disease.

Molybdenum.   Of all the elements definitely considered to be essential, molybdenum is required in the smallest amounts. One part in 100,000,000 of a culture solution is sufficient to prevent the appearance of molybdenum deficiency symptoms in tomato plants. This element is usually supplied to plants as the molybdate ion. One of the roles of molybdenum in plants is that of playing a part in the enzyme system which catalyzes the reduction of nitrates to ammonium ions.

Molybdenum-deficient soils have been found in regions as far apart as Australia, California, New Jersey, and central Europe. Applications of MoO_3 at very low rates have resulted in marked increases in the yield of certain crops on such soils. Excess molybdenum is also harmful to animals. The teart disease of livestock, which occurs in certain parts of England, has been found to result from a excess of this element in the diet. Soils of areas in which this disease occurs are relatively rich in molybdenum, and usually large amounts of molybdenum compounds accumulate in many of the forage plants, from which it is obtained by the animals which browse on them.

Chlorine.   Chlorine seems to be of universal occurrence in plants and is apparently present almost wholly in the form of soluble inorganic chlorides. For many years the experimental results which were obtained upon supplying chlorides in considerable quantities to plants were very variable. In some species a definite beneficial effect was observed; in others the applications of chlorides resulted in a retardation of plant growth; and in still others no apparent influence could be detected. More recent investigations, however, indicate that chlorine is an essential micrometabolic element for plants. It appears to act as an enzyme activator in the water-splitting reactions of photosynthesis.

Plants indigenous to salt marshes and saline soils can endure the presence of relatively large quantities of chlorides, usually sodium chloride, in the soil.

Asparagus is an example of a crop plant which not only tolerates but actually requires treatment with sodium chloride for its best development.

Sodium. There is no convincing evidence that sodium is actually essential in the metabolism of higher plants except possibly in some halophytes. However, it is invariably present in the ash, sometimes in relatively large amounts. Furthermore, the addition of sodium compounds to the soil has been found to result in more vigorous development of many kinds of plants. This is true of some species only when potassium is deficient, but is true for some other species when potassium is present in a sufficiency. Examples of some plants in the first category are barley, carrot, cotton, flax, and tomato; of plants in the second category, celery, table and sugar beet, turnip, and Swiss chard.

Cobalt. Although not definitely known to be an essential element for higher plants, cobalt may exert marked effects on plant metabolism and development. Cobalt ions are known to activate some plant enzymes, such as certain carboxylases and peptidases. Cobalt also has an enhancing effect upon the enlargement or elongation phase of growth in etiolated leaves and certain other plant organs. Cobalt deficiency leads to the development of a diseased condition in cattle and sheep which has been called by a number of names (pining, bush sickness, Morton Mains disease) and which has been recognized in such widely separated parts of the world as Scotland, New Zealand, and Australia. This disease can be combated by soil treatments with a cobalt salt, resulting in an increased content of cobalt in the forage plants on which the animals browse. The fact that cobalt has been found to be a constituent of vitamin B_{12} is probably significant in relation to such deficiency diseases of animals. Cobalt does appear to be an essential micrometabolic element for certain species of blue-green algae.

Vanadium. Although not definitely known to be required by higher plants, vanadium is an essential micrometabolic element for at least some species of green algae.

Silicon. This element constitutes a very large proportion of the ash of some species, particularly of the aerial portions of members of the grass and horsetail families. It is also relatively abundant in the bark of trees. Earlier investigators believed, principally because of the large amounts present in the ash of many species, that silicon is essential for plants. Many years ago, however, it was shown that even those species in which silicon was most abundant could be grown to maturity in culture solutions to which no silicon was added. The results of certain investigators with sunflower, barley, and beet, on the other hand, strongly suggest that small amounts of this element may be essential, at least for some species of higher plants. Silicon is definitely an essential element for the normal development of the diatoms.

Aluminum. This element is one of the most abundant of those present in the soil, although it occurs chiefly in insoluble forms. A larger proportion of soluble aluminum is generally present in relatively acid soils (below *p*H 5.0) than in soils of higher *p*H.

Aluminum is probably universally present in plants, although in terms of percentage composition the amount in the ash of most species is very small. Aluminum is not usually considered to be one of the essential elements, but traces may be necessary, at least for some species.

The color of the flowers of *Hydrangea macrophylla* is related to the aluminum content of the floral tissues. Blue flowers always contain more aluminum than pink flowers. The addition of soluble aluminum compounds to the soil in which hydrangea plants are growing induces a shift in the flower color from pink to blue.

Selenium.   For many years a serious and often fatal disease of livestock, usually called "alkali disease," has been known to occur in certain regions of the great plains of western North America. The cause of this disease was long unknown, but it was finally recognized to be a result of selenium poisoning. The distribution of the disease was found to correspond closely with that of seleniferous soils. Such soils are of common occurrence in a belt extending from Alberta and Saskatchewan across the United States and into Mexico and also occur in other parts of the world. Some of the plants native to such seleniferous soils, notably certain species of *Astragalus,* a kind of vetch, accumulate selenium in relatively large quantities. Selenium poisoning is a common outcome of the browsing upon such plants by cattle, sheep, or horses. On the other hand, very little selenium accumulates in many kinds of plants, including most crop plants, even when growing on highly seleniferous soils. Plants in this latter category exhibit symptoms of selenium toxicity at relatively low concentrations in solution cultures. Selenium-accumulating species, on the contrary, may acquire a selenium content as high as 100 to 10,000 times that of nonaccumulating species without injury, and there are indications that selenium may even be a necessary element for such species.

Symptoms of Mineral Element Deficiency.   The absence or deficiency of any of the necessary mineral elements in the soil or other substratum in which a plant is rooted will sooner or later become apparent in the development of the plant by the appearance of growth aberrations of one kind or another. Symptoms of the deficiency of a mineral element may occur even when that element is present in the tissues of a plant, if for one reason or another it cannot be used in plant metabolism. Plants may exhibit symptoms of nitrogen deficiency, for example, even when the tissues contain an abundant supply of nitrates if the reduction of nitrates is for some reason prevented.

In a general way the symptoms of the deficiency of a given mineral element are similar in all species of plants. Certain deficiency symptoms assume, however, more or less distinctive aspects in some species. For example, manganese deficiency results in the development of a characteristic mottled chlorosis in the leaves of many species. In maize and other cereals, however, chlorosis resulting from manganese deficiency assumes the pattern of an alternate yellow and green striping running lengthwise of the leaves. Symptoms of the deficiency of a given element also frequently differ somewhat in woody plants from herbaceous plants. It is therefore important that the symptoms of mineral element deficiency be studied for each species of plant individually. Once such symptoms have been distinguished for a species, they are of assistance in diagnosing the abnormal development of plants of that species under natural or cultural conditions.

Deficiency symptoms of those elements which are relatively immobile in plants, such as iron, manganese, boron, and calcium, appear first in the younger leaves. Deficiency symptoms of the relatively mobile elements such as nitrogen, phosphorus, potassium, and magnesium, on the other hand, appear first in the older leaves. Various deficiency symptoms of tobacco are shown in Fig. 15.7.

Keys to the symptoms of various elemental deficiencies have been devised for a number of species of plants. An example of such a key is presented in Table 15.4.

Solution and Sand Cultures.   Much of our knowledge regarding the roles of mineral elements in plants has been obtained by means of solution culture experiments. The growing of plants which are rooted in solution and sand cultures

Fig. 15.7. Symptoms of mineral element deficiencies as shown by tobacco plants. The deficient elements are: (1) nitrogen, (2) phosphorus, (3) potassium, (4) calcium, (5) magnesium, (7) boron, (8) sulfur, (9) manganese, (10) iron. All of the elements were supplied to (6). Photograph, U.S. Department of Agriculture.

TABLE 15.4 KEY TO MINERAL ELEMENT DEFICIENCY
SYMPTOMS IN TOBACCO

A. Effects localized on older leaves or more or less general on the whole plant.

   B. Local, occurring as mottling or chlorosis with or without necrotic spotting of lower leaves, little or no drying of lower leaves.

      C. Lower leaves curved or cupped under with yellowish mottling at tips and margins. Necrotic spots at tips and margins.   *Potassium.*

      C. Lower leaves chlorotic between the principal veins at tips and margins of a light-green to white color. Typically no necrotic spots.
                                                      Magnesium.

   B. General; also yellowing and drying or "firing" of lower leaves.

      C. Plant light green, lower leaves yellow, drying to light-brown color. Plants dwarfed.   *Nitrogen*

      C. Plants dark green, leaves narrow in proportion to length; plants dwarfed.   *Phosphorus.*

A. Effects localized on terminal growth, consisting of upper and bud leaves.

   B. Dieback involving the terminal bud, which is preceded by peculiar distortions and necrosis at the tips or bases of young leaves making up the terminal growth.

      C. Young leaves making up the terminal bud first light green followed by a typical hooking downward at tips, followed by necrosis, so that if later growth takes place, tips and margins of the upper leaves are missing.   *Calcium.*

      C. Young leaves constricted and light green at base, followed by more or less decomposition at leaf base; if later growth takes place leaves show a twisted or distorted development; broken leaves show blackening of vascular tissue.   *Boron.*

   B. Terminal bud remains alive, chlorosis of upper or bud leaves, with or without necrotic spots, veins light or dark green.

      C. Young leaves with necrotic spots scattered over chlorotic leaf, smallest veins tend to remain green, producing a checkered effect.
                                                      Manganese.

      C. Young leaves without necrotic spots, chlorosis does or does not involve veins so as to make them dark or light green in color.

         D. Young leaves with veins of a light-green color or of the same shade as interveinal tissue. Color light green, never white or yellow. Lower leaves do not dry up.   *Sulfur.*

         D. Young leaves chlorotic, principal veins characteristically darker green than tissue between the veins. When veins lose their color, all the leaf tissue is white or yellow.   *Iron.*

has been one of the most widely used experimental techniques employed by plant physiologists. The necessary solutions, often rather inappropriately called "nutrient solutions," are prepared by dissolving the requisite salts in certain definite proportions in distilled water. Numerous formulae have been used for such solutions, but most species of plants grow almost equally well over a considerable range of formulations provided the solution is "complete," *i.e.*, contains all the mineral elements known to be essential, in concentrations which are not too far out of an optimum range for each element. One such solution which has been found satisfactory for the growth of a number of species in solution or sand culture contains four principal salts at the following concentrations: 0.003 M $Ca(NO_3)_2$, 0.002 M KNO_3, 0.002 M $MgSO_4$, and 0.002 M KH_2PO_4. Ions of the six macrometabolic elements are supplied by these four salts. A sufficient quantity of an iron ethylene-diamine-tetraacetic acid solution is added to bring the concentration of iron in the solution up to 5 ppm. Micrometabolic elements are introduced as a supplementary solution at the rate of 1 milliliter per liter of culture solution. This micrometabolic element solution has the following composition per liter: H_3BO_3, 2.5 g; $MnCl_2 \cdot H_2O$, 1.5 g; $ZnCl_2$, 0.10 g; $CuCl_2 \cdot 2H_2O$, 0.5 g; MoO_3, 0.05 g.

Large-scale applications of sand and solution culture (hydroponic) techniques have been shown to be entirely feasible for crops. Except in certain special situations, however, such as the raising of high unit value greenhouse crops or the raising of vegetable crops on isolated and barren islands, the commercial employment of sand and solution culture techniques does not appear to be economically feasible at the present time.

SUGGESTED FOR COLLATERAL READING

Epstein, E. *Mineral Nutrition in Plants.* John Wiley & Sons, Inc., New York, 1972.

Fried, M., and H. Broeshart. *The Soil-Plant System in Relation to Inorganic Nutrition.* Academic Press, Inc., New York, 1967.

Gauch, H. C. *Inorganic Plant Nutrition.* Dowden, Hutchinson & Ross, Inc., Stroudsburg, Pa. 1972.

Hambridge, G., Editor. *Hunger Signs in Crops, 2nd Edition.* Judd and Detweiler, Washington, 1949.

Jennings, D. H. *The Absorption of Solutes by Plant Cells.* Iowa State University Press, Ames, Iowa, 1963.

Kramer, P. J. *Plant and Soil Water Relationships. A Modern Synthesis.* McGraw-Hill Book Company, New York, 1969.

Lamb, C. A., O. G. Bentley, and J. M. Beattie, Editors. *Trace Elements.* Academic Press, Inc., New York, 1958.

Price, C. A. *Molecular Approaches to Plant Physiology.* McGraw-Hill Book Company, New York, 1970.

Stiles, W. *Trace Elements in Plants.* University Press, Cambridge, 1961

Sutcliffe, J. I. *Mineral Salts Absorption in Plants.* The Macmillan Company, New York, 1962.

Wallace, T. *The Diagnosis of Mineral Deficiencies in Plants, 3rd Edition.* Chemical Publishing Company, New York, 1961.

16 NITROGEN METABOLISM

Nitrogen-containing compounds occupy the heartland of metabolism in all kinds of living organisms. Discussion in this book will naturally be centered on plants. Preeminent among the nitrogen-containing compounds in plants are the proteins. One reason for the preeminence is that all enzymes are basically proteins, although not all of the proteins in plants function in an enzymatic capacity.

Important as they are, and central as they are to the operation of metabolic processes, the proteins are not the only metabolically significant kinds of nitrogen-containing compounds occurring in plants. Among the other important ones are amino acids, nucleosides, nucleotides, nucleic acids, pigments, coenzymes, alkaloids, acid amides, auxins, and cytokinins. The physiological roles of many of the compounds in these categories are discussed either later in this chapter or in other chapters in the book.

The Proteins.   The molecules of proteins are relatively huge in size and enormously complex in structure. Together with water they are the principal constituents of the cytoplasm and cell organelles. Proteins also occur in plants cells in the form of stored foods, especially in the seeds of many species. Such "reserve" or storage proteins differ in some of their properties from the protoplasmic proteins. In basic structural organization, however, the two kinds of proteins appear to be essentially similar.

All proteins contain carbon (50–54 per cent), hydrogen (about 7 per cent), nitrogen (16–18 per cent) and oxygen (20–25 per cent). Although some animal proteins do not contain sulfur, this element is apparently present in all plant proteins, but never exceeds 2 per cent.

Their percentage composition, however, gives no idea of the structure of protein molecules, nor of their tremendous size relative to most molecules. Even the smallest protein molecules have molecular weights of at least 10,000, whereas some of the most complex ones have molecular weights of millions. The structure of these gargantuan molecules is discussed more fully on the following pages.

The Amino Acids.   When proteins are hydrolyzed by treatment with acids, alkalies, or suitable enzymes the end products of the hydrolysis is always a mixture of amino acids. It is clear, therefore, that the amino acids are the structural units out of which the protein molecules are constructed.

Amino acids are, as the name suggests, compounds with the properties of both acids and amines. Every amino acid contains at least one carboxyl (−COOH) group and one or more amino (−NH₂) groups. Although other amino acids occur in plants, we will confine our attention to twenty known to be structural components of proteins (Table 16.1).

Two of the amino acids listed — cysteine and methionine — are sulfur-containing. Another sulfur-containing amino acid, cystine, is formed by the union of two cysteine molecules through a -S-S- linkage with the elimination of 2 H; this compound is formed only after the initial stages of protein synthesis have been completed. Two of the compounds listed in Table 16.1, asparagine and glutamine, are simultaneously both amino acids and acid amides. These two compounds are discussed more fully later in this chapter. Proline is an imino acid, *i.e.,* it contains a −NH group instead of a−NH₂ group.

In all of the amino acids listed, the amino group, or one amino group if several occur in the molecule, is always attached to the α carbon atom, which is the one next to the −COOH group. This is not true, however, of some of the nonprotein amino acids which occur in plants. Furthermore, in all of the amino acids listed except glycine the α carbon atom is asymmetric and theoretically these molecules can all exist in both D and L forms. All of the amino acids which become constituents of proteins have the L-configuration.

Of the twenty amino acids listed in Table 16.1, it has been found that eight—lysine, tryptophan, phenylalanine, threonine, valine, methionine, leucine, and isoleucine—must be supplied in the diet of man and higher animals. The animal body cannot synthesize these amino acids, at least not in the amounts required for its normal metabolism. Plants are the ultimate source of these eight amino acids in human and higher animal nutrition.

Origin of the Nitrogenous Compounds Used by Plants.   Higher green plants cannot utilize directly the gaseous nitrogen of the atmosphere in the synthesis of nitrogen-contaning organic compounds, and there is no nitrogen in the rocks from which the great bulk of the soil particles are derived. Despite the depletion of nitrogen compounds from the soil by leaching and, in agricultural areas, as a result of the removal of plant cover by harvesting or grazing, few if

TABLE 16.1 AMINO ACIDS WHICH ARE CONSTITUTENTS OF PROTEINS

Glycine	$CH_2(NH_2) \cdot COOH$
Alanine	$CH_3 \cdot CH(NH_2) \cdot COOH$
Valine	$CH_3 \cdot CH(CH_3) \cdot CH(NH_2) \cdot COOH$
Leucine	$CH_3 \cdot CH(CH_3) \cdot CH_2 \cdot CH(NH_2) \cdot COOH$
Isoleucine	$CH_3 \cdot CH_2 \cdot CH(CH_3) \cdot CH(NH_2) \cdot COOH$
Serine	$CH_2(OH) \cdot CH(NH_2) \cdot COOH$
Threonine	$CH_3 \cdot CH(OH) \cdot CH(NH_2) \cdot COOH$
Aspartic acid	$HOOC \cdot CH_2 \cdot CH(NH_2) \cdot COOH$
Glutamic acid	$HOOC \cdot CH_2 \cdot CH_2 \cdot CH(NH_2) \cdot COOH$
Asparagine	$H_2NOC \cdot CH_2 \cdot CH(NH_2) \cdot COOH$
Glutamine	$H_2NOC \cdot CH_2 \cdot CH_2 \cdot CH(NH_2) \cdot COOH$
Lysine	$CH_2(NH_2) \cdot CH_2 \cdot CH_2 \cdot CH_2 \cdot CH(NH_2) \cdot COOH$

Arginine

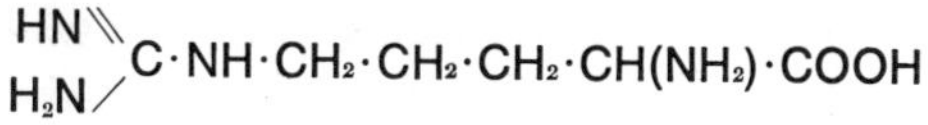

$$\begin{matrix} HN\diagdown \\ \qquad C \cdot NH \cdot CH_2 \cdot CH_2 \cdot CH_2 \cdot CH(NH_2) \cdot COOH \\ H_2N\diagup \end{matrix}$$

Histidine

$$\begin{matrix} N\text{——}CH \\ \| \qquad \| \\ HC\diagdown \quad \diagup C \cdot CH_2CH(NH_2) \cdot COOH \\ N \\ H \end{matrix}$$

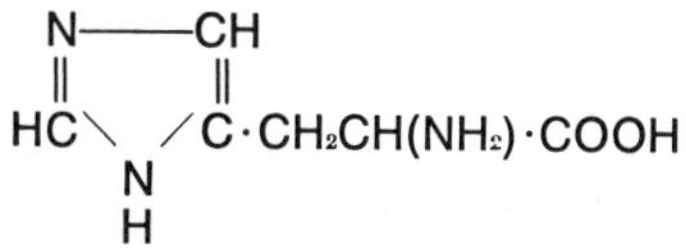

Cysteine	$HS \cdot CH_2 \cdot CH(NH_2) \cdot COOH$
Methionine	$CH_3 \cdot S \cdot CH_2 \cdot CH_2 \cdot CH(NH_2) \cdot COOH$

Phenylalanine   ⬡—$CH_2 \cdot CH(NH_2) \cdot COOH$

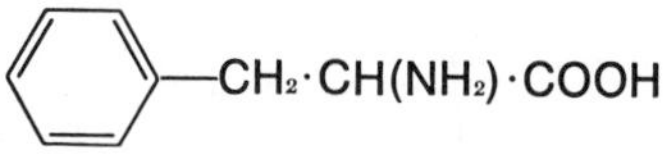

Tyrosine   HO—⬡—$CH_2 \cdot CH(NH_2) \cdot COOH$

Tryptophan

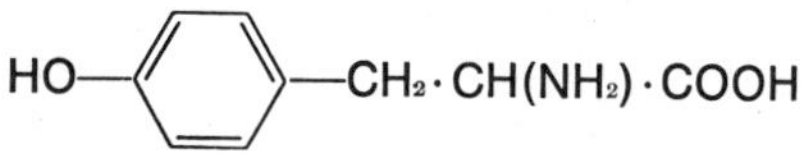

$CH_2 \cdot CH(NH_2) \cdot COOH$ (attached to indole ring with N—H)

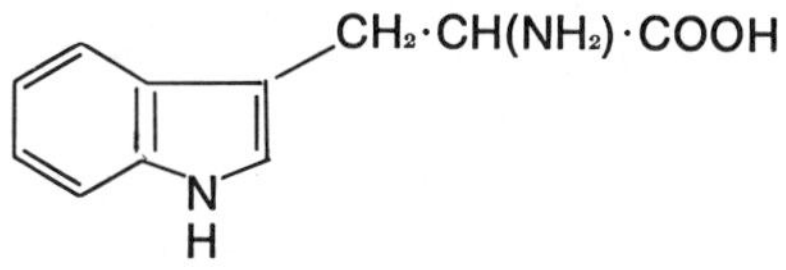

Proline

$$\begin{matrix} H_2C\text{————}CH_2 \\ | \qquad\qquad | \\ H_2C\diagdown \quad \diagup CH \cdot COOH \\ N \\ H \end{matrix}$$

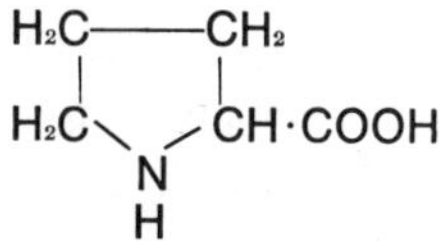

any soils, ever become completely deficient in nitrogen. It is obvious that processes exist whereby the supply of nitrogen in the soil is constantly being replenished.

Small amounts of oxides of nitrogen (formed during electrical storms) and ammonia are washed out of the atmosphere into the soil by rain. Large amounts of nitrogenous fertilizers are added every year to agricultural soils in many parts of the world. On a world-wide basis, however, the biological process of *nitrogen fixation* is by far the major source of the nitrogen used by vascular plants. This process and several other pertinent ones are discussed in the following paragraphs.

Nitrogen Fixation.   This process supplies the great bulk of the nitrogen used by plants in the natural vegetation but also, to a substantial degree, by agricultural crops also. Nitrogen fixation is the process whereby atmospheric nitrogen is converted into organic nitrogen-containing compounds by living organisms. The process is not known to occur in any vascular plant as such but does take place in many species of bacteria and blue-green algae, and in at least some kinds of fungi and actinomycetes.

The process of nitrogen fixation has been studied most extensively in the nitrogen-fixing bacteria. These organisms fall into two main groups: (1) saprophytic bacteria which obtain their energy from dead organic matter in the soil, and (2) symbiotic bacteria which live in the roots of plants, especially leguminous species.

At least fifteen genera of free-living saprophytic bacteria which fix atmospheric nitrogen are known to exist and doubtless more remain to be discovered. The best known of these are species of the genera *Azotobacter* and *Clostridium,* which have been studied largely in agricultural soils. The former consists of coccus-like aerobic organisms and the latter of rod-shaped anaerobic bacteria. Species of both genera are common in well-aerated soils. The aerobic ones occur around the surface of the soil particles; the anaerobic ones are found within aggregations of soil particles or in localized regions of the soil in which the oxygen content has been depleted. These bacteria fix atmospheric nitrogen in organic form using carbohydrates and other substrate materials resulting from organic decay as a source of energy and substance.

Symbiotic nitrogen fixation is largely the result of the activities of species and varieties of the genus *Rhizobium* which are rod-shaped bacteria. These organisms enter the tips of the root hairs of the leguminous host plant and penetrate to the cortical region of the root, engendering the development of root nodules to be described later. There is considerable specificity between the strain of bacterium and the specific host plants which it will invade. The bacteria obtain the carbohydrates and perhaps other kinds of organic compounds necessary in their metabolism from the host plants. They synthesize organic nitrogen-containing compounds from such carbohydrates and from atmospheric nitrogen.

The process of symbiotic nitrogen fixation has been studied most extensively in certain leguminous crop plants such as alfalfa, clover and soybean. Many other members of the legume family, largely wild species, have symbiotic bacteria associated with them. In most if not all of the species in which nitrogen-fixing bacteria become associated, nodules form on the roots. The invading bacteria multiply within the nodules. Most of the organic nitrogenous compounds synthesized by the bacteria remain within the host. Those which are not utilized in the metabolism of the bacteria are available for use by the host. Soybean plants, for example, are estimated to use about 90 per cent of the nitrogen fixed by the bacteria in the nodules of their roots. Under some conditions some of the nitrogen-containing compounds in the nodules move out and into the surrounding soil where they can be absorbed by other species of plants. Nitrogenous compounds can also be supplied to the soil by the sloughing off or decay of roots of plants in which nitrogen-fixing bacteria have been present.

The metabolic processes whereby nitrogen is reduced by bacteria are not entirely understood. Apparently molecular nitrogen is reduced to ammonia by a complex of enzymes called *nitrogenase,* using reductants derived, at least indirectly, from the host plant. The ammonia thus formed is incorporated as amino groups into glutamic acid and similar compounds probably by reactions like those described later in this chapter. The presence of three micrometabolic elements, iron, molybdenum, and cobalt is usually necessary for the operation of the nitrogen-fixing process in bacteria.

Symbiotic nitrogen fixation and root nodule formation have also been found to occur in more than two hundred species of nonleguminous plants, mostly woody perennials. Not surprisingly many such species are characteristically plants which pioneer into areas in which the soil is highly deficient in nitrogenous compounds. For example, a nitrogen-fixing species of alder is one of the first plants to invade recently deglaciated areas in Alaska. The symbiotic microorganisms in such nitrogen-fixing plants seem most commonly to be actinomycetes.

More than forty species of autotrophic blue-green algae are known to possess the capacity of nitrogen-fixation and more will probably be discovered. Some of the genera including nitrogen-fixing species are *Anabaena, Calothrix, Cylindrospermum, Nostoc,* and *Stigonema.* Most such algae grow independently in aquatic or at least moist habitats. Nitrogen fixation in rice fields and paddies, for example, is accomplished largely by blue-green algae. About four-fifths of the earth's surface is occupied by oceans. In these extensive marine habitats, as well as in fresh-water ones, blue-green algae are probably the most important nitrogen-fixing organisms.

Ammonification. In the process of decay the complex nitrogenous compounds present in plant and animal tissues are broken down into a number of simpler compounds, most of the nitrogen being released in the form of ammonia. This process is called ammonification and is the result of the activities of a number of kinds of microorganisms including bacteria, fungi, and actinomycetes.

Nitrification. The ammonia formed in the decomposition of proteins and other organic nitrogenous compounds may be acted upon by the aerobic nitrifying bacteria and transformed in two steps to nitrates. The first step is the oxidation of ammonia to nitrates. This is accomplished by organisms belonging to the two genera *Nitrosomonas* and *Nitrosococcus*. Neither of these organisms can oxidize the nitrate which they produce, but this compound is oxidized to nitrates by a different organism, *Nitrobacter*. All of these organisms differ from each other morphologically, but they are similar physiologically in that they use the energy obtained from the oxidation of ammonia or nitrates in the synthesis of carbohydrate compounds from carbon dioxide and water. All are therefore chemosynthetic autophytes.

Denitrification. A large number of organisms are capable of reducing nitrates to nitrites and ammonia. This occurs commonly in the tissues of the higher plants, as discussed later. Certain soil organisms, however, can reduce nitrates all the way to molecular nitrogen. These organisms are known as deni- trifying bacteria and include a number of species of which *Bacterium denitrifi- cants* is probably the best known. Denitrification occurs only in the absence of atmospheric oxygen, as in water-logged soils, and most effectively when an abundant supply of carbohydrates is present in the soil. It does not normally occur in well-cultivated soils.

Absorption of Nitrogenous Compounds From the Soil. As the preceding discussion has shown, much of the nitrogen available to vascular plants orig- inates as a result of the nitrogen-fixing activities of free-living or symbiotic microorganisms. The nitrogen made available by symbiotic root nodule micro- organisms can be used directly by the higher plant without passing through the soil medium. The nitrogen made available by the free-living, nitrogen-fixing or- ganisms, must, however, be absorbed from the soil as must also available nitro- gen from all other sources. In general plants can utilize four kinds of compounds as sources of nitrogen: (1) nitrates, (2) nitrites, (3) ammonium compounds, and (4) organic nitrogenous compounds. The mechanism of the absorption of ionic forms of nitrogen is believed to be essentially similar to that involved in the intake of other ions (Chapter 15).

Many plants absorb most of their nitrogen in the form of nitrates. Nor- mally metabolizing plants usually contain only relatively small quantities of nitrates, because the nitrogen of nitrate ions is reduced to other forms almost as rapidly as it enters the plant. Under certain conditions, however, plants accu- mulate relatively large quantities of nitrates in their tissues without any toxic effects. Subsequently such accumulated nitrates may be utilized in the nitrogen metabolism of the plant. Plants sometimes exhibit acute symptoms of nitrogen deficiency while they still contain considerable quantities of nitrates. Although such plants have been able to absorb nitrates, metabolic conditions within the plant have been such that they have been unable to utilize them in the forma- tion of nitrogenous organic compounds.

As shown in the next section, the first step in the utilization of nitrates by plants is their reduction to nitrites. It seems probable, therefore, that plants can utilize nitrates as a source of nitrogen, and this supposition has been confirmed by solution culture experiments. However, nitrites are rarely if ever an important source of nitrogen for plants in nature.

Many species of plants, when grown in sand or solution cultures under suitable conditions, develop as well or better when supplied with ammonium salts as when supplied with nitrates. This is not surprising since the nitrogen in ammonium compounds is in a highly reduced form similar to that found in amino acids and related comopnds. In certain types of soils it is probable that ammonium compounds are the chief form in which nitrogen is available to plants. This is apparently true of the acid podsolic soils of northern latitudes and of many uncultivated soils in the southern United States. Such soils contain little nitrate, but considerable quantities of ammonium compounds, and plants growing in them apparently obtain their nitrogen in the latter form. Unlike nitrate ions, plants seldom accumulate appreciable concentrations of ammonium ions. Even when ammonium fertilizers are applied to agricultural soils, much, if not most, of the absorption of nitrogen by plants occurs in the form of nitrates. In such soils nitrification often occurs very effectively as a result of the activity of nitrifying bacteria, resulting in the rapid conversion of ammonium compounds to nitrates.

As a result of the decay of organic remains, there are present in most soils at least small quantities of amino acids and other organic nitrogenous compounds. There is considerable evidence that plants can absorb and utilize such compounds in the synthesis of proteins. This is most likely to occur from soils which are relatively rich in organic matter.

Reduction of Nitrates and Sulfates.   Since in nitrates the nitrogen is a highly oxidized state ($-NO_3$), while in amino acids and other organic compounds common in plants, it is in a highly reduced state (commonly $-NH_2$), it is evident that the reduction of nitrogen is one of the steps in the synthesis of amino acids and other organic nitrogenous compounds whenever nitrates are the source of nitrogen. A similar statement holds for sulfur, which is absorbed by plants in the highly oxidized sulfate ($-SO_4$) state but is incorporated into amino acids and other organic compounds in a reduced state (usually $-S^-$ or $-SH$). The process involved in the reduction of nitrates are closely interrelated with the synthesis of amino acids as described in the next section.

There is clear evidence that the first step in the reduction of nitrate is its conversion to nitrite. The inducible enzyme *nitrate reductase* (Chapter 9) which catalyzes this reaction has been isolated from the tissues of higher plants as well as from bacteria and some fungi. The reaction may be summarized as follows:

$$XNO_3 + NADH + H^+ \xrightarrow{\text{nitrate reductase}} XNO_2 + NAD^+ + H_2O$$

In addition to the $NADH + H^+$, which supplies the energy for the reduction, molybdenum and FAD must be present as cofactors. The nitrites are further reduced to NH_3, perhaps through the intermediate steps of hyponitrous acid and hydroxylamine. Other enzymes are required for any such steps, as well as reduced pyridine nucleotides as a source of the necessary energy. The NH_3 formed, however, normally does not accumulate in appreciable amounts in the cells because it is almost immediately metabolized in the synthesis of amino acids or related compounds.

Temperature exerts a marked effect on the nitrate-reducing capacity of plants. In the tomato plant, for example, although nitrates are quickly absorbed, their reduction and the synthesis of organic nitrogen compounds occur very slowly at 13°C. At 21°C, on the other hand, both the absorption and reduction of nitrate ions occur very rapidly.

The metabolic sequence of reactions whereby sulfate ions are reduced in plants is even less clearly understood than the one involved in nitrate reduction; although there is evidence that the first step is conversion of sulfate to sulfite. Sulfate reduction occurs in plants on a much smaller scale than nitrate reduction and like the latter process is closely interrelated with amino acid synthesis.

Synthesis of Amino Acids.   On the average about 85 per cent of the weight of amino acid molecules represents non-nitrogenous components, chiefly carbon, hydrogen, and oxygen. It is evident, therefore, that amino acid synthesis cannot occur without an adequate supply of carbon compounds in addition to nitrogen and, for some of them also sulfur, in suitable chemical combination.

Two principal metabolic systems seem to be operative in the synthesis of amino acids. Radiocarbon tracer experiments have shown that amino acids appear within a few seconds after the start of photosynthesis (Chapter 10). Such amino acid syntheses must be accompanied by nitrate reduction and sulfate reduction. These processes occur in the chloroplasts in the light and are undoubtedly powered by photophosphorylation. The non-nitrogenous molecular frameworks which go into the synthesis of the amino acids are derived from intermediates of the carbon reduction cycle. The synthesis of amino acids in leaves or other photosynthesizing organs occurs by this mechanism.

Amino acid synthesis and the cognate processes of nitrate and sulfate reduction also occur in the nongreen organs of plants. In some plants the smaller roots appear to be important centers of these processes. In such plant organs they take place in the mitochondria, and the necessary energy is supplied by oxidative phosphorylation. The non-nitrogenous molecular frameworks which go into amino acids synthesized by this mechanism come from intermediates of the Krebs cycle (Chapter 13).

There is substantial evidence that much of the amino nitrogen which becomes incorporated into amino acids enters by way of the synthesis of glutamic acid. When, for example, tomato plants were supplied with $(NH_4)_2 SO_4$ containing the ^{15}N isotope, much larger amounts of this isotope were found after 12 hours in glutamic acid than in any of the other amino acids. Glutamic acid is synthesized in plant cells by a reaction between ammonia and α-ketoglutaric acid as follows:

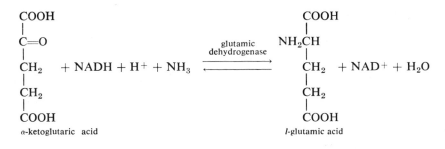

α-ketoglutaric acid   l-glutamic acid

This reaction is catalyzed by the zinc-containing enzyme *glutamic dehydrogenase*. This enzyme is of widespread occurrence in plants in which it is largely or entirely localized in the mitochondria. This type of reaction is called *a reductive amination*. It is reversible and when it proceeds from right to left is called an *oxidative deamination*. This latter reaction probably occurs quite commonly in senescent plant organs such as aging leaves.

Transamination reactions, which occur freely in the metabolic system, are the key to the synthesis of most, if not all, amino acids in plants. In such reactions an amino group is transferred from one amino acid to the carbon framework of another. The following are examples of such reactions:

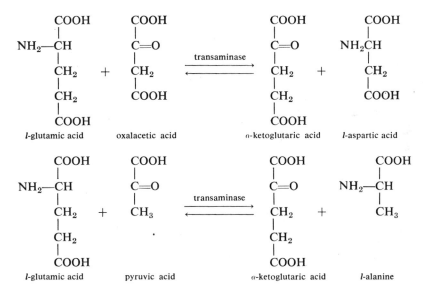

Transaminases, the enzymes which catalyze such reactions, are widely distributed in the higher plants and occur largely in the mitochondria. A derivative of vitamin B_6, either pyridoxamine phosphate or pyridoxal phosphate (Chapter 9), is a necessary coenzyme for transaminase activity.

The overall and generalized pattern of amino acid synthesis seems to be that glutamic acid, which is highly reactive in transaminations, serves as the principal portal of entry of NH_3 into the metabolic web of reactions, and that other amino acids are derived from this compound, directly or indirectly, as a result of transamination reactions.

Virtually nothing is known regarding the synthesis of the sulfur-containing amino acids except that the reduction of sulfate, as discussed previously, must be a closely interrelated process. There are some indications that cysteine may be the key compound in this phase of amino acid synthesis.

Some of the amino acids present in plant cells are of secondary origin, resulting from the chemical transformation of acid amides (see next section) or other nitrogenous compounds, or from the hydrolysis of proteins. A number of protein-digesting enzymes have been found to be present in plant cells. Such enzymes hydrolyze proteins to amino acids or to intermediate products of protein hydrolysis. Two of the best known of the plant proteases are *papain* from the latex of the papaya (*Carica payaya*) and *bromelin* from fruit of the pineapple (*Ananas sativus*) and other members of the Bromeliaceae. The effectiveness of papain as a protein-digesting enzyme, in particular, has long been recognized. Natives of Central and South America, where the papaya grows wild, have long known that its leaves have a digestive effect on meat. It is said that simply wrapping meat in crushed papaya leaves will increase its tenderness. Commercial preparations of papain are made from the latex of the papaya.

Asparagine and Glutamine.   These two compounds have been mentioned previously as constituents of proteins. They are simultaneously both amino acids and acid amides. They also commonly occur in the free state in plant tissues, although not in equivalent quantities. In beets, tomato plants, potato tubers, and rye grass, for example, more glutamine accumulates than asparagine. On the other hand, asparagine predominates in many legume seedlings, in Sudan grass, and in asparagus in which it was first discovered. Some species, such as maize, synthesize appreciable amounts of both amides when metabolic conditions are favorable for their formation, although asparagine is usually more abundant than glutamine.

Glutamine synthesis occurs by a direct reaction of glutamic acid with ammonia:

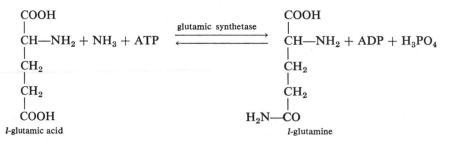

A certain amount of energy is necessary for the establishment of the amide bond, which accounts for the role of ATP in the reaction. Asparagine is apparently synthesized from aspartic acid by an analogous reaction. The synthesis of these acid amides represents an important route of entry of ammonia into organic molecules.

The appearance of asparagine and glutamine in plant tissues is closely linked with the presence of ammonia. The amides usually appear, for example, whenever large amounts of ammonium ions are being absorbed by the root system. Asparagine and glutamine are also formed in young seedlings when the proteins of the food storage tissues of the seed are being utilized in growth. At times of carbohydrate starvation, amino acids and related compounds may be oxidized in respiration, releasing ammonia within the cells. Under such conditions the ammonia set free is rapidly used in amide synthesis, thus obviating any toxic effect it would otherwise exert. If carbohydrate deficiency becomes very severe, asparagine and glutamine may also be oxidized, resulting in the liberation of ammonia within the tissues. When this occurs the tissues are usually injured.

It has long been assumed that one reason the amides are important in nitrogen metabolism is because, as a result of their synthesis, the accumulation of free ammonia in plant tissues is prevented. The correlation known to exist between amide formation and metabolic or environmental conditions which favor the accumulation of ammonia (the rapid absorption of ammonium ions or carbohydrate starvation leading to the oxidation of amino acids) is considered to be a consequence of the utilization of ammonia in amide synthesis.

TABLE 16.2 COMMON NUCLEOSIDES AND NUCLEOTIDES

Base	Nucleoside	Nucleotide[a]
Ribonucleotides		
Adenine	Adenosine	Adenosine monophosphate
Guanine	Guanosine	Guanosine monophosphate
Uracil	Uridine	Uridine monophosphate
Cytosine	Cytidine	Cytidine monophosphate
Hypoxanthine	Inosine	Inosine monophosphate
Deoxyribonucleotides		
Adenine	Deoxyadenosine	Deoxyadenosine monophosphate
Guanine	Deoxyguanosine	Deoxyguanosine monophosphate
Thymine	Thymidine	Thymidine monophosphate
Cytosine	Deoxycytidine	Deoxycytidine monophosphate

[a]In ribonucleotides the phosphate linkage may be at the 2', 3' or 5' position; in deoxyribonucleotides the phosphate linkage may be only at the 3' and 5' positions. Di-and tri-phosphate nucleotides can also form when the mononucleotide linkage is at the 5' position.

Nucleosides, Nucleotides, and Nucleic Acids. Substituted purines or pyrimidines, which collectively are referred to as *nitrogenous bases,* play a central role in the metabolism of both plants and animals. They are constituents of the nucleic acids (DNA and RNA), the nucleosides (Table 16.2), the nucleotides (ATP, ADP, AMP and similar compounds—Table 16.2), the nicotinamide nucleotides (NAD^+ and $NADP^+$), thiamin, flavin adenine dinucleotide (FAD), coenzyme A and the cytokinins, the latter being an important group of plant hormones.

The following are the formulas for pyrimidine and three of the most important substituted derivatives of this compound:

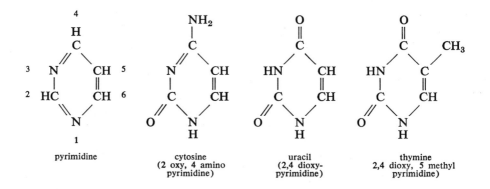

| pyrimidine | cytosine (2 oxy, 4 amino pyrimidine) | uracil (2,4 dioxy- pyrimidine) | thymine 2,4 dioxy, 5 methyl pyrimidine) |

The following are the formulas for purine and two of the most important substituted derivatives of this compound:

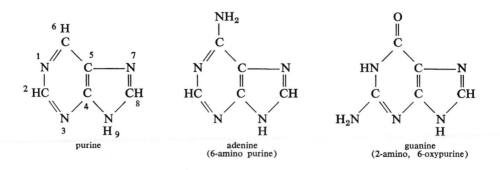

| purine | adenine (6-amino purine) | guanine (2-amino, 6-oxypurine) |

The *nucleosides* are compounds in which purines and pyrimidines are linked to sugars. In the naturally occurring nucleosides the sugar is a pentose, either D-ribose or deoxy-D-ribose. The nature of the pentose-nitrogenous base linkage is shown in the following examples:

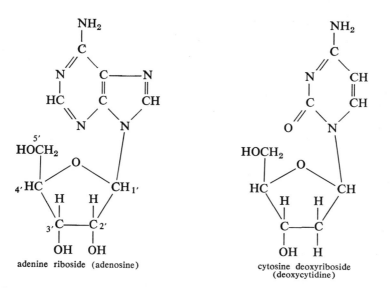

adenine riboside (adenosine)

cytosine deoxyriboside
(deoxycytidine)

The names of the more important naturally occurring nucleosides are listed in Table 16.2.

The phosphate ester of a nucleoside is called a *nucleotide*. Some of the commoner nucleotides are listed in Table 16.2. In the ribosides, linkage of phosphate can occur at the 2', 3', and 5' hydroxyl positions of the nucleoside moiety; in the deoxyribosides it can occur at only the 3' and 5' hydroxyl positions. The 5' monophosphates can be further modified into diphosphates such as ADP and triphosphates such as ATP. The formula for adenosine-5'-monophosphate illustrates the structure of a nucleotide:

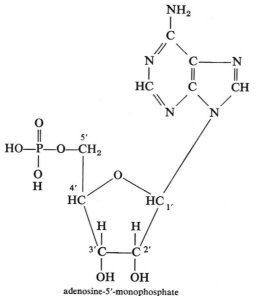

adenosine-5'-monophosphate

The *nucleic acids*—deoxyribonucleic acid (DNA) and ribonucleic acid (RNA)—are polymers of high molecular weight constructed of long chains of mononucleotide units. Because of their roles in the genetic mechanism of the cell and in protein synthesis these compounds have attracted an enormous amount of attention. DNA is found principally in the nucleus but also occurs in relatively small amounts in other parts of the cell, such as the chloroplasts and mitochondria. RNA is found principally in the ground cytoplasm and in the ribosomes but also occurs in small quantities in other cell organelles. The postulation of Watson and Crick in 1953 that a molecule of DNA consists of two polynucleotide chains arranged in a double helix has since been confirmed by much experimental evidence (Fig. 16.1).

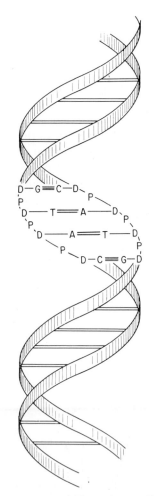

Fig. 16.1. Diagrammatic representation of a portion of a DNA double helix.

Four deoxynucleotides constitute the units out of which each DNA molecule is constructed. Each adenine nucleotide of one chain is paired with a thymine nucleotide of the other. Likewise each guanine nucleotide of one chain is paired with a cytosine nucleotide of the other. Thus one chain always has a structure which is complementary to the other. Some of the deoxycytosine nucleotides present are commonly present in the modified form of methyl deoxycytosine. This compound operates in the same fashion as the cytosine nucleotides in nucleic acid metabolism.

The base portions of the nucleotides project towards the center, where they are held together by hydrogen bonds (Fig. 16.2). Guanine and cytosine each have the potentiality of three such bonds; adenine and thymine have the potentiality of only two each. This difference accounts for the specificity of the base pairing in the structure of DNA.

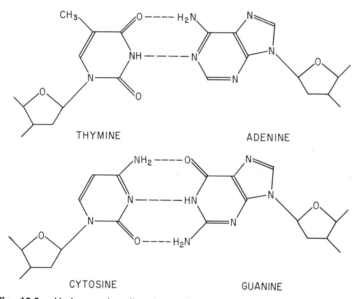

THYMINE   ADENINE

CYTOSINE   GUANINE

Fig. 16.2.   Hydrogen bonding between base pairs in a DNA molecule.

The successive base pairs are held together along the long axis of the molecule by phosphate diester bonds between the 3′ hydroxyl of one nucleoside and the 5′ hydroxyl of the adjacent nucleoside (Fig. 16.3). This ladder-like structure is twisted along its long axis into the form of the previously mentioned double helix (Fig. 16.4). The order of the adenine, thymine, guanine, and cytosine nucleotides lengthwise along one strand may follow a variety of sequences. It is the exact sequence of these bases along the length of a DNA molecule which ultimately determines the structure of a particular kind of protein by a mechanism to be described shortly.

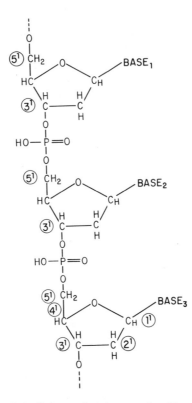

Fig. 16.3. Phosphate linkages between nucleosides in a DNA strand.

RNA molecules are single-stranded instead of double-stranded as in DNA. This kind of nucleic acid exists in several forms in the cell. The constituent bases are the same as those for DNA, except that uracil takes the place of thymine. The sugar component in RNA is ribose instead of deoxyribose as in DNA. Three principal kinds of RNA occur in cells: messenger RNA (m-RNA), transfer RNA (t-RNA), and ribosomal RNA (r-RNA). All of them play roles in protein synthesis.

Protein Synthesis.   It has long been conjectured that the nucleus is the controlling center of cellular metabolism and that this control is exerted through enzyme activity. It has been only in recent years, however, that an understanding of the mechanism of such control has begun to emerge. The bridge between genetic information and cellular metabolism lies in the control exerted by DNA on the synthesis of proteins since all enzymes are proteins.

Like other proteins, enzymes are built up as chains of amino acid residues, joined together by peptide linkages. A peptide linkage is established when the hydroxyl group of one amino acid unites with a hydrogen of the amino group

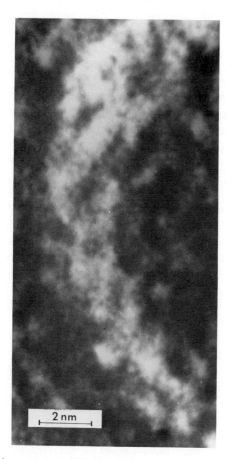

Fig. 16.4. Electron micrograph of DNA from a pea plant. Micrograph by Jack Griffith, California Institute of Technology.

of another amino acid, water being eliminated in the reaction. As shown later, the metabolic machinery whereby such linkages are established is very complex, but the essence of the process is shown in the following diagram, using two glycine molecules in the example;

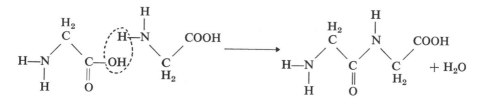

The dipeptide resulting from the linkage together of two amino acid molecules possesses an amino group and a carboxyl group to which other amino acids can be linked. The addition of further amino acid molecules by peptide linkages to these groups still leaves amino and carboxyl groups present in the resulting molecule. Polypeptides, peptoses, proteoses, and finally proteins, each a category of greater molecular weight than the preceding, are formed by the linkage together of more and more amino acids in this way. A variety of amino acids are linked together in the formation of a single protein molecule, but it is unlikely that any one protein contains all the amino acids listed in Table 16.1. Furthermore, as shown in the later discussion, amino acids are not coupled together in random fashion in the construction of a protein molecule, but in a precise order for each kind of protein. The number of amino acids which becomes incorporated into a protein molecule ranges from about 100 to many times this figure.

Amino acids are thus the "alphabet" in terms of which the structure of protein is spelled. There are almost as many amino acids involved in protein synthesis as there are letters in the alphabet. The number of words and thoughts which can be expressed by suitable arrangement of the alphabetical symbols is potentially infinite. The number of kinds of protein molecules which could be constructed from amino acid molecules is likewise potentially infinite. Not all of them exist, however, any more than all combinations of letters exist as meaningful words.

Ribosomes are the centers of protein synthesis. These approximately spherical organelles are mostly located in the cytoplasm and are closely associated with the endoplasmic reticulum. Ribosomes are also found in the nuclei, mitochondria, and the chloroplasts. Because of their very small size (about 25 nm in diameter) very little is known of their internal structure. Apart from water the ribosomes are composed principally of roughly equal parts of RNA and proteins.

DNA is a major constituent (30 to 40 per cent of the total mass) of the nuclear chromatin. During cell division the chromatin becomes organized into the chromosomes which duplicate themselves, one of each duplicate pair going to each of the daughter cells. Also during or immediately before cell division each double-stranded DNA molecule, many strands of which occur in a chromosome, duplicates itself. The two strands separate and each one generates the complement of itself from free deoxyribonucleoside triphosphates under the influence of enzymes called *DNA polymerases,* phosphate groups being split out in the process. Each adenine on one strand induces the formation of a partner thymine in the newly forming strand; each cytosine induces the formation of a partner guanine; each thymine induces the formation of a partner adenine; and each guanine induces the formation of a partner cytosine. As a result, each original strand plus its new complementary strand constitutes a double-stranded DNA which is identical with the original double-stranded DNA before duplication began (Fig. 16.5). One of the resulting double strands goes to one of the daughter cells in cell division, the other strand to the other cell. As a result of this process each of the two daughter cells will have the same complement of

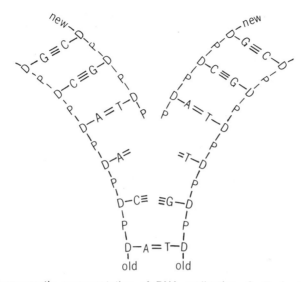

Fig. 16.5. Diagrammatic representation of DNA replication. G, T, A, and C, nitrogen bases; D, deoxyribose; P, phosphate.

DNA molecules as the cell from which they arose. A single DNA "molecule" may consist of thousands of linearly arranged base pairs, and have a molecular weight in the millions.

DNA meets the exact specifications needed for being the genetic material of the cell. It is constant in amount during the non-dividing periods of a cell's existence and it doubles precisely at the time of cell division. It has seemed reasonable, therefore, to assume that some profound interrelationship exists between DNA and the synthesis of enzymes and other proteins. Two apparent difficulties had to be surmounted, however, before such a relationship could be authenticated. One is that DNA is located in the nucleus whereas most protein synthesis takes place in the ribosomes. How was communication established between these two organelles? A second apparent difficulty was that only four distinctive kinds of units occur in the DNA molecule, whereas twenty different kinds of amino acids become incorporated into protein molecules. How could a meaningful relationship exist between two such disparate numbers?

Experimentally based answers to many aspects of these problems have been achieved, however, and by about 1965 a fairly definite picture had emerged regarding the relationship between genetic constitution as embodied in the chromatin of the nucleus and cellular metabolism as regulated by enzymes. This relationship lies at the heart of not only the synthesis of enzymes and other proteins, but at the very heart of life itself.

On the basis of initially theoretical considerations which were later supported by much experimental evidence, it was concluded that the sequence of bases in a DNA molecule constituted a code which controlled the synthesis of proteins. It is obvious that such a code could not be on a basis of one base to one amino acid since there are only four bases involved as against twenty amino acids. Neither would a code in which two bases corresponded to one amino acid be feasible because only sixteen (4 × 4) such combinations are possible. A triplet code, on the other hand, would provide sixty-four combinations which would be more than adequate. Modern interpretations of the relation between DNA and protein synthesis are based on this concept of a triplet code.

Each triplet is considered to consist of three adjacent nitrogenous base residues on one strand of the DNA molecule. The bases may all be the same, all different, or two alike and one different. By convention this code is expressed in terms of letters corresponding to the bases involved: A for adenine, G for guanine, C for cystosine, T (in DNA) for thymine, and U (in RNA) for uracil.

A major step in elucidating the operation of the genetic machinery was taken with the discovery that a form of RNA—called messenger RNA (m-RNA)—acted as an intermediary in transferring genetic information from the nucleus to the ribosome. A unit of messenger-RNA is generated as a single-stranded polynucleotide from free ribonucleoside triphosphates under the influence of one of the strands of DNA and an enzyme known as *RNA polymerase,* phosphate molecules being split off in the process. Although basically single-stranded, a m-RNA molecule may not always retain this configuration. The generation of a m-RNA molecule occurs in such a fashion that its structure is complementary to that of the DNA strand in conjunction with which it is formed. In other words each cytosine group in DNA results in the generation of a guanine group in m-RNA and vice-versa. Similarly, each thymine group in DNA results in the generation of an adenine group in m-RNA. Each adenine group in DNA results, however, in the generation of an uracil group in m-RNA, since this group is the fourth nitrogenous base present in RNA, instead of the thymine which is a constituent of DNA (Fig. 16.6). This coding of an RNA-molecule under the influence of DNA is referred to as *transcription.* It is estimated that, on average, an m-RNA strand consists of 600 linearly arranged base residues. This would correspond to a molecular weight in the neighborhood of 300,000. After its synthesis each m-RNA strand moves from the nucleus and makes contact with one or more ribosomes.

The arrangement of bases along the length of the m-RNA molecule constitutes its coding, triplet groups of sequentially arranged bases, each called a *codon,* being the code units. In general, each triplet is the code for a specific amino acid. As the result of various ingenious and imaginative experiments, it has been possible to spell out the genetic code for the synthesis of amino acids in considerable detail. Codons are stated in terms of the base order in the m-RNA molecule, and hence are complementary to the basic code of the DNA molecule from which they were transcribed.

As shown in Table 16.3 the number of codons for a given amino acid ranges from one to six. There appears to be only one triplet each which will code for the synthesis of methionine and tryptophan; at the other extreme lie

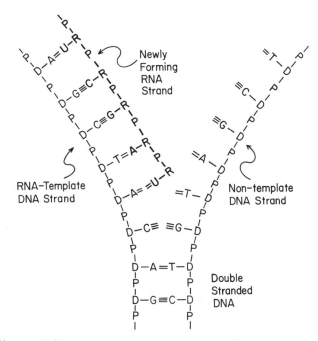

Fig. 16.6. Diagrammatic representation of the transcription of RNA from DNA. Note the substitution of U (uracil) for T (thymine) and R (ribose) for D (deoxyribose) in the newly forming RNA strand.

leucine, serine, and arginine with six different code triplets each. Other amino acids have two, three, or four codons each. Of the amino acids for which there is more than one codon, with only three exceptions (leucine, serine, and arginine), the first two members of the triplet are always identical. This suggests that the first two bases of the triplet are often of more significance in coding than the third one.

There are still many uncertainties and ambiguities in the genetic code. When more than one codon exists for a given amino acid, it is not certain, for example, that all of them are used equally. Neither is it certain that the code is unambiguous, that is, that some codons may not signal for the incorporation of more than one kind of amino acid into a protein molecule. Three triplets are unallocated. Two of them, UAA and UAG, are believed to be codons for the termination of the polypeptide chain which constitutes a protein. To date it has not been possible to assign a role to the third one, UGA.

Most of the information regarding the genetic code has been obtained from experiments upon the colon bacillus, *Escherichia coli,* but considerable confirmatory evidence has been obtained from work with other kinds of organisms including higher plants. It is highly probable that the genetic code is the same, or at least closely similar, for all types of living organisms. Undoubtedly such

TABLE 16.3 THE GENETIC CODE BASE SEQUENCES IN RNA CODONS

First base	Second base				Third base
	C	A	G		
U	UCU ⎫ UCC ⎬ Serine UCA ⎪ UCG ⎭	UAU ⎫ Tyrosine UAC ⎭ UAA ——— UAG ———	UGU ⎫ Cysteine UGC ⎭ UGA ——— UGG Tryptophan	UUU ⎫ Phenylala- UUC ⎭ nine UUA ⎫ Leucine UUG ⎭	U C A G
C	CCU ⎫ CCC ⎬ Proline CCA ⎪ CCG ⎭	CAU ⎫ Histidine CAC ⎭ CAA ⎫ Glutamine CAG ⎭	CGU ⎫ CGC ⎬ Arginine CGA ⎪ CGG ⎭	CUU ⎫ CUC ⎬ Leucine CUA ⎪ CUG ⎭	U C A G
A	ACU ⎫ ACC ⎬ Threonine ACA ⎪ ACG ⎭	AAU ⎫ Asparagine AAC ⎭ AAA ⎫ Lysine AAG ⎭	AGU ⎫ Serine AGC ⎭ AGA ⎫ Arginine AGG ⎭	AUU ⎫ Isoleucine AUC ⎭ AUA AUG Methionine	U C A G
G	GCU ⎫ GCC ⎬ Alanine GCA ⎪ GCG ⎭	GAU ⎫ Aspartic acid GAC ⎭ GAA ⎫ Glutamic acid GAG ⎭	GGU ⎫ GGC ⎬ Glycine GGA ⎪ GGG ⎭	GUU ⎫ GUC ⎬ Valine GUA ⎪ GUG ⎭	U C A G

a code was an early step in the evolution of living systems. Some biologists have speculated that it may have arisen as long as three billion years ago.

A second type of RNA which is functional in protein synthesis is transfer RNA (t-RNA). This kind of RNA becomes coded from DNA in the same manner as m-RNA, but as described later, only one of its triplets is active in controlling protein synthesis. There is at least one specific kind of t-RNA for each amino acid. The molecules of t-RNA are smaller than those of m-RNA. The exact structure of several kinds of t-RNA has been ascertained. The chain length of such molecules ranges from 75 to 87 nucleotide units. In yeast, for example, the t-RNA for alanine consists of a chain of 77 nucleotide units, that for tyrosine of 78 such units. Most t-RNA molecules have molecular weights in the neighborhood of 30,000. Transfer RNA molecules are basically single-stranded in structure, but a strand apparently can assume various configurations. The cloverleaf form depicted in Fig. 16.7 is considered to be a very likely one. A distinctive feature of t-RNA is that a large number of modified nucleosides—sometimes as high as 20 per cent of the total number present—occur in each molecule (Fig. 16.7). Most of these modified nucleosides represent relatively simple modifications of the four major nucleosides although some are much more complex. Such modifications in nucleoside structure appear to occur only after transcription of the t-RNA from the DNA.

A major step in protein synthesis is the conveyance of amino acids to the ribosomes by their association with transfer RNA molecules. Before the con-

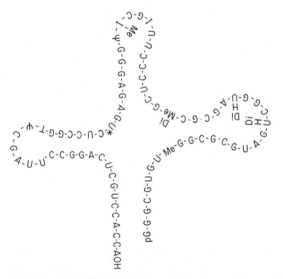

Fig. 16.7. The complete nucleotide sequence for alanine t-RNA isolated from yeast, and one of its possible configurations. From Holley, *et al., Science,* **147,** No. 19, March 1965:1462–1465. Copyright 1965 by the American Association for the Advancement of Science.

junction between a t-RNA molecule and its destined amino acid mate can take place the latter must be activated. This is accomplished by a reaction with ATP under the influence of an activating enzyme, phosphate groups being split off in the process. There is a separate activating enzyme for each amino acid. The entity which becomes associated with the t-RNA molecule appears to be an amino acid-AMP-enzyme complex. Each such complex is conveyed to the ribosome by a unit of t-RNA which is specific for it.

In the meanwhile the ribosome has become attached to a strand of m-RNA. As generally visualized the ribosome is considered to move along the strand reading off the codons in sequence as a basis for protein synthesis as described in more detail in the following paragraphs. A second ribosome may attach to the start position on the strand before the first has reached the terminal position. Others may follow in turn. The ribosomes of many cells are fastened together in clusters called polysomes (Fig. 16.8). The ribosomes in such a group are thought to be bound together by a strand of m-RNA.

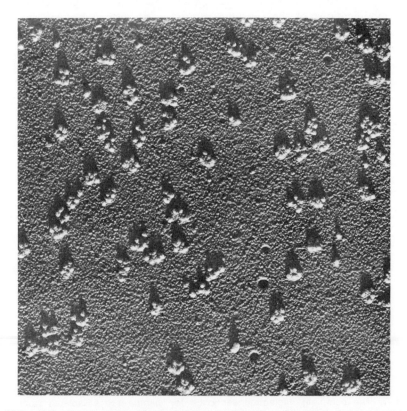

Fig. 16.8. Polysomes composed of clusters of ribosomes in rabbit tissue. × 104,000. Electron micrograph by A. Rich. From Jensen and Park, *Cell Ultrastructure*, Copyright 1967, Wadsworth Publishing Company, Inc.

Each t-RNA molecule carries a triplet code unit which is complementary to the codon on m-RNA which is the signal for the particular amino acid for which that kind of t-RNA is specific. Such a triplet group on a t-RNA molecule is called an *anticodon*. To cite an example, the anticodon for phenylalanine is GAA, while the codon for this amino acid on the m-RNA is UUC. The probable anticodon groups for three other kinds of t-RNA molecules are indicated in Fig. 16.9. By base pairing between the codon and anticodon the amino acid is brought into a position designated for it by the code on the m-RNA molecule. In such base-pairing the anticodon operates in reverse order from the manner in which it is usually written; in the example given above as AAG, instead of GAA (Fig. 16.9).

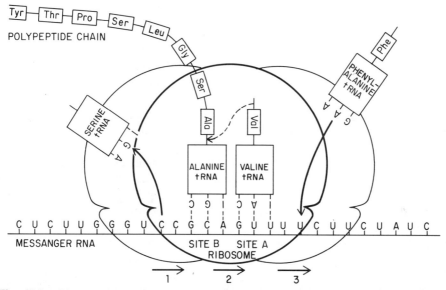

Fig. 16.9. Diagrammatic representation of the process of protein synthesis as discussed in the text. From "The Genetic Code III" by F. H. C. Crick, *Scientific American*, **215**, No. 4, 1966:55–62. Copyright 1966 by Scientific American, Inc. All rights reserved.

The synthesis of a protein molecule appears to be programmed from the $-NH_2$ end of the polypeptide chain to the $-COOH$ end. Successive amino acids are added to the chain in accordance with positions dictated by the coding of the m-RNA. Thus there is a *translation* of the codon sequence of the RNA chain into the amino acid sequence of a protein molecule. Peptide linkages are established from amino acid to amino acid throughout the length of the chain. A specific enzyme and certain cofactors are necessary for the establishment of each of these peptide linkages. A protein molecule thus consists of a chain of amino acid residues arranged in a precise order. Since the DNA molecules which control their synthesis are coded in many different sequences this is also true of the m-RNA molecules, and for each different kind of sequence a different kind of protein molecule is synthesized. Some of the protein molecules synthesized are structural; others are enzymes. The latter in turn exercise controlling influences on metabolic processes.

The molecules of t-RNA separate from amino acid molecules as the latter are incorporated into the polypeptide chain; they apparently can function repeatedly as escorts of amino acid molecules. The m-RNA molecules, on the other hand, appear to depolymerize after having served their coding function. This implies that new m-RNA molecules must constantly be in the process of formation as long as protein synthesis is to be maintained.

There are still many gaps in our knowledge regarding the mechanism of protein synthesis, and the picture as presented is necessarily a somewhat simplified one. Figure 16.9 is a diagrammatic representation of the process of protein synthesis.

Because of the mode of their synthesis, protein molecules are initially chainlike in structure, but subsequently can assume other configurations. Some proteins assume a generally fibrous structure in which individual polypeptide chains are linked together laterally largely by hydrogen bonding. In other proteins, back and forth folding of the polypeptide chain occurs, resulting in compact, closely-knit units which are more or less globular in shape. In proteins of this type other kinds of cross linkages occur besides hydrogen bonding, including disulfide (-S-S-) linkages.

The principal regions of protein synthesis in plants are the meristems and certain storage tissues such as the endosperm of seeds, although some protein synthesis probably continues in all living cells. The principal regions of protein synthesis do not necessarily correspond to the principal regions of amino acid synthesis. One major center of amino acid synthesis is in actively photosynthesizing leaves. Another, at least in many kinds of plants, is in the younger roots. Amino acids are commonly translocated from the tissues in which they originate to other, often distant, tissues before being converted into proteins. Little or no translocation of proteins as such occurs in plants. Much of the protein accumulated in plants, as in seeds, is ultimately digested back to amino acids which are translocated to other parts of the plant. In germinating seeds, for example, they are translocated largely to apical stem and root meristems where they are resynthesized into proteins. The reutilization of proteins which have become constituents of the cytoplasm may also occur. In senescent leaves, for example, the decomposition of cytoplasmic proteins may take place, and at least some of the resulting organic nitrogenous compounds may be translocated to meristems and resynthesized into proteins.

The Alkaloids. Alkaloids are a heterogenous group of complex cyclic compounds containing nitrogen that are synthesized only in certain species of plants. They are especially common in members of the Solanaceae, Papaveraceae,

Leguminosae, Ranunculaceae, Rubiaceae, and Apocynaceae. Species of plants which contain one alkaloid are very likely to contain others. More than twenty different alkaloids have been isolated from opium, which is the dried juice of the unripe fruits of certain species of poppies.

Some of the better known alkaloids are *nicotine* from tobacco *(Nicotiana tabacum), quinine* from cinchona *(Cinchona officinalis* or *C. pubescens), morphine* from poppy *(Papaver somiferum), strychnine* and *brucine* from *Strychnos nuxvomica, atropine* from the deadly nightshade *(Atropa belladonna),* and *colchicine* from meadow saffron *(Colchicum autumnale).* The formula for nicotine, which is structurally one of the simplest alkaloids, gives some idea of the chemical nature of these substances:

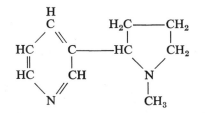

Most of the alkaloids are white solids, but nicotine is a liquid at ordinary temperatures. They are all basic in reaction as the name indicates and only slightly soluble in water.

It is common for the alkaloids synthesized in a given plant species to accumulate in certain organs of a plant. About 85 per cent of the nicotine in a tobacco plant, for example, is located in the leaves. However, when a tobacco scion is grafted on the root system of a tomato plant, no nicotine is found in the leaves; when the reverse graft is made, the leaves of the tomato contain nicotine. It appears, therefore, that the nicotine is synthesized in the roots, whence it is translocated into the leaves and other aerial organs. Of special interest is the fact that this upward translocation appears to occur in the xylem. The alkaloids in *Atropa belladonna* and *Datura stramonium* also appear to be synthesized largely, if not entirely, in the roots. Anabasine, however, another alkaloid of tobacco, appears to be synthesized in both the root and the shoot.

The physiological role of the alkaloids in plants is unknown. Their restricted distribution indicates that they are not essential in any process of general importance. It is possible that many of them are physiologically unimportant by-products of the nitrogen metabolism of the species in which they are found. Extensive studies of the role of nicotine in tobacco suggest, however, that this alkaloid can be converted into other nitrogenous compounds which may play a useful role in the nitrogen metabolism of the plant. Much of the interest which attaches to the alkaloids derives from the pronounced therapeutic or other physiological effects which many of them exert on the animal body.

SUGGESTED FOR COLLATERAL READING

Carpenter, B. H. *Molecular and Cell Biology*. Dickenson Publishing Company, Inc., Belmont, California, 1967.

Davidson, J. N. *The Biochemistry of the Nucleic Acids*. John Wiley & Sons, Inc., New York, 1965.

Hewitt, E. J. and C. V. Cutting, Editors. *Recent Aspects of Nitrogen Metabolism in Plants*. Academic Press, Inc., New York, 1968.

McKee, H. S. *Nitrogen Metabolism in Plants*. Oxford University Press, Inc., London, 1962.

Stewart, W. D. T. *Nitrogen Fixation in Plants*. Athlone Press, University of London, 1966.

Watson, G. D. *The Molecular Biology of the Gene*. W. A. Benjamin, Inc., New York, 1965.

Webster, G. C. *Nitrogen Metabolism in Plants*. Row, Peterson and Company, Evanston, Illinois, 1959.

17 | TRANSLOCATION OF SOLUTES

In most plants a large proportion of the living cells do not contain chloroplasts. All such nongreen cells are dependent for essential carbohydrates upon the chlorophyllous cells of the plant. Many of the nongreen cells are remote from the photosynthesizing cells. The cells in the root tips of trees, for example, may be a hundred or more meters distant from the nearest leaves. They, as well as all other nongreen cells in the body of the plant, are dependent for their existence upon carbohydrates which move to them, by either direct or indirect routes, from the chlorenchyma. Such a movement of soluble carbohydrates is only one example of various important kinds of solute transfer which occur in plants.

Movement of organic and inorganic solutes from one part of the plant to another is designated as the *translocation* of solutes. This term is generally restricted to movements of solutes in the tissues of the xylem and phloem in which the distance through which they are transported is usually very great in proportion to the size of the individual cells. It is not ordinarily used to refer to the cell-to-cell movement of solutes which may occur in any part of the plant.

Anatomy of Phloem Tissues.   Of the various stem tissues, only the xylem and phloem possess such a structure as to suggest that a relatively rapid longitudinal movement of solutes can occur through them. Both of these tissues are characterized by the presence of elongated cells and elements which are joined in such a way as to form essentially continuous ducts. Furthermore, it has been shown experimentally that the rate of the movement of solutes through other stem tissues, such as the pith and cortex, is totally inadequate to account for known rates of translocation through stems.

The structure of the xylem tissues has already been discussed in Chapter 8. Discussion of the anatomy of the conductive tissues will now be completed by a consideration of the structure of the phloem. Like the xylem, the phloem is continuous from the top to the bottom of the plant, the ultimate terminations of the phloem system being in the tissues of the stem tips or leaves and other lateral organs, and in the root tips (Chapter 19). Although in most species solutes move for the greatest distance through stems, it is important to realize that the conductive tissues of plants constitute a complicated but unit system which ramifies to all parts of the plant body.

The general arrangement of the xylem and phloem tissues in representative types of stems has already been considered (Figs. 8.2, 8.3). In the majority of dicot stems the phloem usually occurs as a continuous cylinder of tissues just external to the cambium layer. In a number of kinds of plants, strands of internal phloem are present inside of the xylem (Fig. 17.1). In roots the primary phloem and xylem are present, as seen in cross section, in a radial arrangement, but the secondary conductive tissues are oriented in essentially the pattern found in stems (Fig. 7.5).

Five principal types of cells are found in the phloem: (1) *sieve-tube elements,* (2) *companion cells,* (3) *phloem parenchyma,* (4) *phloem ray cells,* and (5) *phloem fibers.* The proportions of these types of cells are different in every species, and in some species not all are present.

Sieve elements are of two main types. In most angiosperms *sieve tubes* are present, each consisting of a linear series of elongated cells joined together end to end (Fig. 17.2). The individual cells are called *sieve-tube elements.* Groups of pores are present in the end walls and sometimes in the side walls of sieve-tube elements. These portions of the walls are called *sieve plates* or *sieve areas.* When the end wall of a sieve-tube element is transverse, it usually has only a single sieve plate; but if the wall is oblique, it commonly bears several sieve plates. Sieve-tube elements naturally differ in size with the plant and their location within the plant, but mostly they would fall within a range of 20–30 μ in diameter and 100–500 μ long.

In gymnosperms true sieve tubes are not present. The elongated cells in which conduction occurs are called *sieve cells* and resemble the sieve-tube elements in many ways. However, they are not regularly arranged end to end and do not possess such elaborate sieve plates as sieve-tube elements.

Sieve-tube elements usually develop by the longitudinal division of mother cells which have arisen as a result of the division of a cambium cell. The division of such a mother cell results in a sieve-tube element and a companion cell (Fig. 17.2). Sometimes the mother cell may divide longitudinally more than once, forming one sieve-tube element and two or more companion cells.

In general a young sieve-tube element is structurally and organizationally

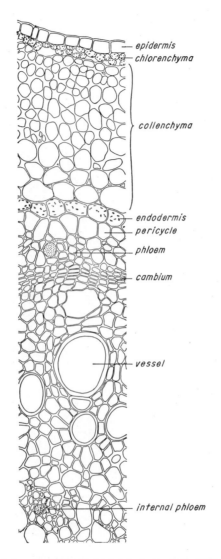

epidermis
chlorenchyma

collenchyma

endodermis
pericycle

phloem

cambium

vessel

internal phloem

Fig. 17.1. Cross section of a small portion of a tomato stem showing internal phloem.

similar to other newly formed plant cells. It contains a nucleus and most, if not all, of the other kinds of organelles. As such an element becomes older, but still remains alive and functional, it undergoes some marked structural reorganizations. A major change is the disintegration of the nucleus or at least its shrinkage into an apparently degenerate state. In some species the nucleolus is released from the disintegrating nucleus and remains intact within the cell. There is substantial evidence that the tonoplast disintegrates as the element ages. If this occurs the vacuolar sap and cytoplasm presumably become intermixed,

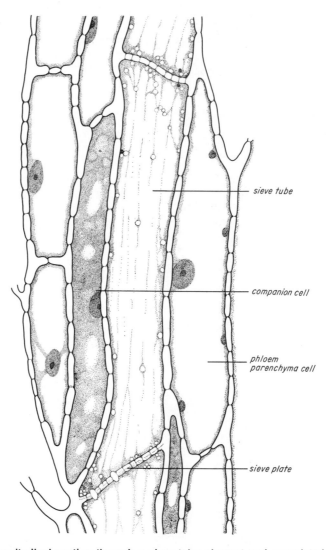

Fig. 17.2. Longitudinal section through a sieve-tube element and associated cells from the stem of tobacco. After Holman and Robbins, *Elements of Botany, 3rd Edition*, John Wiley & Sons, Inc., New York, 1940:58.

constituting a much more highly hydrated medium than the cytoplasm proper. The plasmalemma, however, retains its structural integrity and location as shown by the fact that mature sieve-tube elements can be plasmolyzed. Of the other principal organelles, the endoplasmic reticulum and mitochondria are retained but undergo considerable modification as the sieve-tube element matures. During this cellular reorganization most of the ribosomes and all of the dictyosomes which were present disappear. These structural changes suggest that the sieve-tube elements sink to a low state of metabolic activity while still functional in translocation, but this is not a certainty.

The pores in the sieve areas, which are mostly in the range of 2–7 μ in diameter, retain through them a continuum of cytoplasm or cytoplasm-vacuolar sap mixture (depending upon whether or not the tonoplast has disintegrated) as long as they are functional. The wall of a pore is bordered by plasmalemma. The structure of a sieve tube is thus such that a continuous conduit exists along its length from element to element.

One of the least understood features of a sieve-tube element is the presence of a material called "slime" which is generally considered to be proteinaceous in nature. During the early stages of sieve tube development the slime is present in the cytoplasm of many species as organized bodies which may assume various shapes. The subsequent fate of these slime bodies during the maturation of the sieve tube is a matter of some dispute but most investigators of phloem anatomy believe that they become reorganized into fine strands which are less than 0.5 μ in diameter. Some coarser strands appear to consist of several of these finer strands bound together lengthwise. Some of these strands are intracellular but others traverse pores in the sieve plate from one sieve-tube element to the next. Several strands may pass through a single pore, but in aggregate they do not occupy more than a relatively small proportion of the cross-sectional area of the pore.

The functional life of most sieve tubes is relatively short. In a typical woody species the zone of phloem in which translocation takes place is only a few tenths of a millimeter thick. In most such species all cytoplasm disappears from the sieve-tube elements by the end of the growing season in which they were formed. At about the time that the cytoplasm disintegrates, the pores in the sieve areas become plugged with callose, a β, 1–3 polymer of glucose (Chapter 14). Injury of any kind to a sieve-tube element also results in a rapid sealing off of the sieve plate pores by a deposition of callose.

In a few woody species such as grape, rose, linden, and tulip tree, the cytoplasm of most sieve-tube elements does not disintegrate at the end of the first season. In such species temporary callose plugs are formed at the approach of the dormant season which are dissolved at the beginning of the next growing season. The sieve tubes may remain functional in such species for two or even more years.

The *companion cells* (Fig. 17.2) are much smaller in cross section than the sieve-tube elements. They may have the same length as the sieve-tube elements or they may be substantially shorter. A companion cell retains such basic organelles as a nucleus, mitochondria, dictyosomes, and endoplasmic reticulum until maturity. Plasmodesmata interconnect companion cells with adjacent sieve-tube elements. Companion cells die at the same time that adjacent sieve-tube elements do, suggesting that an especially intimate physiological relationship exists between these two kinds of cells. In some species some sieve-tube elements do not have companion cells adjacent to them, whereas in other species every sieve-tube element is accompanied by one or more companion cells. Gymnosperms and pteridophytes do not have companion cells.

Phloem parenchyma (Fig. 17.2) is composed of cells which are usually somewhat elongate parallel to the long axis of the stem. Such cells do not occur in monocots and are absent from the phloem of some species of dicots. Phloem parenchyma cells remain densely packed with cytoplasm long after the cytoplasm of sieve-tube elements has shown signs of deterioration. They appear to remain quite active metabolically. The proportion of phloem parenchyma cells in the phloem varies widely according to species, and their arrangement within the phloem differs greatly from one species to another. Pits interconnect phloem parenchyma cells and also connect phloem parenchyma cells with sieve-tube elements, companion cells, and phloem ray cells.

Phloem ray cells are parenchymatous cells out of which the *phloem rays* are constituted. The phloem rays are those segments of the vascular rays (Chapter 8) which are located in the phloem and are a characteristic tissue of woody plants. Both the phloem ray cells and the xylem ray cells are generated from the vascular cambium (Chapter 19). As a rule a phloem ray consists of a band-like bundle of transversely oriented living cells varying from one to many cells in width and from several to many cells in height. Pit interconnections commonly exist between phloem ray cells and phloem parenchyma cells.

Phloem fibers, found in the phloem of some plants, are elongated cells with thick, often lignified, walls. They are more common in woody than in herbaceous species and, like the companion cells, do not occur in the gymnosperms or the pteridophytes. Phloem fiber cells have long tapering ends which overlap, forming strong fibrous strands.

Longitudinal translocation undoubtedly occurs principally through the sieve-tube elements or the sieve cells (in gymnosperms) since they are so constructed as to offer less resistance to the movement of materials through them than the other kinds of phloem cells. Companion cells probably play accessory roles in phloem translocation and phloem parenchyma cells almost certainly do, as evidence discussed later in this chapter indicates. Phloem ray cells are considered to serve as channels of lateral translocation. Phloem fiber cells are probably not involved in translocation processes.

As a result of cambial growth there is a continued formation of new xylem and phloem cells during part or all of the growing season. This results in most species in crushing the older phloem tissues, including the sieve tubes. Other changes which occur as phloem grows older include the lignification of fibers and sometimes also ray and parenchymatal cells into hard, thick-walled, dead cells known as *stone cells.* In woody stems and roots further profound modifications occur in aging phloem tissues as the result of the activity of secondary meristems called *cork cambiums* (Chapter 19).

General Aspects of the Translocation of Solutes in Plants. From the time that a plant starts to grow until its death, a more or less continuous movement of solutes is in progress through the conducting elements of every organ of the plant. In a very young seedling, foods are usually translocated upward in the growing stems and downward in the developing roots from the storage tissues of the seed. As soon as the rate of photosynthesis in the developing seedling becomes sufficiently high, at least part of the photosynthate moves in a downward direction from the leaves toward the roots while some of it is moving upward toward apical meristems. Furthermore, as soon as the developing roots make effective contact with the substrate, the absorption of mineral salts begins, followed by the translocation of a large part of them in a generally upward direction through the plant. Some such upward translocation of mineral elements takes place in organic combinations, as discussed later.

Mineral elements absorbed by the roots are mostly translocated to young leaves and other growing organs of the plant. All of them do not remain, however, in the organ into which they are first translocated. A considerable proportion of the mineral elements which move into a leaf or flower petal, for example, may sooner or later move out of such a lateral organ back into the stem, and become redistributed to other, usually younger, parts of the plant.

The translocation patterns within a plant often exert significant effects upon its behavior. At the time that young fruits and seeds are developing, for example, there appears to be a general migration of organic compounds from all parts of the plant toward the enlarging fruits and seeds. This movement may so nearly monopolize the food resources of the plant as to check severely the growth of vegetative organs. In cotton plants, for example, there is a conspicuous decrease in vegetative growth when the plant is fruiting heavily. A large part of the photosynthate and of the nitrogenous compounds in the above-ground organs of the plant moves into the enlarging fruits. The supply of carbohydrates reaching the root system becomes insufficient to maintain adequate respiration and root elongation. As a result there is a decrease in the rate of the absorption of mineral salts, which in turn results in checking or even stopping the vegetative growth of the plant.

As the preceding discussion indicates, the patterns of translocation in plants are complex and may be different at different stages in the life history of the plant. Nevertheless certain predominant translocation routes can be recognized as follows: (1) the downward translocation of organic solutes from leaves to other parts of the plant, (2) the upward translocation of organic solutes to growing or storage regions, (3) the upward translocation of mineral elements from roots to aerial organs, (4) the outward translocation of mineral elements from leaves and other lateral organs into stems, and (5) the lateral or cross transfer of solutes within stems.

Downward Translocation of Organic Solutes.   The downward translocation of organic compounds occurs through the phloem tissues. Much of the evidence indicating that organic solutes move toward the basal portions of the plant in the phloem has been obtained by ringing experiments. "Ringing," when the term is employed without qualification, as mentioned previously, refers to the removal of a narrow continuous band of tissues external to the xylem. Since ringing entirely encircles the stem, all tissues external to the xylem are completely intercepted. This operation is also called "girdling."

In a girdled tree carbohydrates and other organic compounds slowly accumulate in the tissues above the ring and slowly decrease in quantity in the tissues below the ring as they are utilized in respiration and assimilation. Unless special conditions intervene, such, for example, as the development of sprouts on the tree trunk below the ring, a girdled tree ultimately dies because of the starvation of the roots, showing that no appreciable amounts of foods are conducted downward through the xylem. The accumulation of organic compounds above a ring has also been demonstrated in certain herbaceous plants such as cotton.

The evidence from ringing experiments that the downward translocation of organic solutes occurs in the phloem has been reinforced by the results of experiments in which radioactive tracers have been used. If one or more leaves of a plant are allowed to photosynthesize in an atmosphere containing CO_2 made with the radioactive ^{14}C isotope, the photosynthetic products become "labeled" by having the ^{14}C atoms incorporated into them. That this artificially made radioactive photosynthate travels downward in the phloem has been shown by the autoradiography of the stem tissues.

The analysis of phloem exudates from cut stems (see later) shows that they are in effect a relatively highly concentrated sugar solution. Their concentration may approach 1 molar in value, but is usually somewhat less. Sucrose is invariably present and in some species is the only sugar in the sieve-tube sap. Other sugars commonly found in the phloem sap of many species are raffinose, stachyose, and verbascose. The sugar alcohols mannitol and sorbitol are also found in the phloem sap of some species (Chapter 14). Hexose sugars are either absent from phloem exudates or occur in very low concentrations. A number of different kinds of amino acids have also been found in phloem sap but in low total concentration—seldom exceeding 0.5 per cent.

Upward Translocation of Organic Solutes.   Under many conditions an upward translocation of organic solutes takes place in plants. This occurs, for example, in the stems of woody species when the buds resume growth in the spring. The tissues of the new shoots are constructed largely out of foods which move in an upward direction from the storage tissues of the stems, since during the early stages in their expansion the leaves do not photosynthesize at a rate sufficient to supply all the carbohydrates used in the growth of the shoot which bears them. The upward translocation of foods from the older leaves on a given shoot to developing leaves situated closer to its apex also occurs. As the leaves mature there is a reversal in the direction of translocation of carbohydrates; they are then translocated from the leaves into the stems, from which they move to other organs of the plant.

A number of other examples of the upward transport of foods in plants can be cited. The opening flowers and developing fruits are often attached to stems in such a position that some or all of the organic compounds translocated into them move through the stems in an upward direction. In the early stages of the development of seedlings, upward translocation occurs from the endosperm or cotyledons toward the apical portions of the plant in which rapid growth is taking place. Likewise the upward transport of foods invariably occurs during the earlier stages of shoot growth from bulbs, tubers, rhizomes, and other types of underground organs.

An older view that the upward translocation of soluble carbohydrates occurred in the xylem along with water and mineral elements (see later) is no longer entertained. It is true that sugars are present in the xylem sap of many woody species at least during some seasons. However, the concentrations present are very small (usually not more than a few tenths of a milligram per liter), and such soluble carbohydrates are nearly or entirely absent from the xylem sap during the late spring and early summer months when most such upward translocation occurs. Such upward translocation of sugars as occurs in the xylem must be regarded as an incidental process rather than a primary one.

On the other hand, as discussed later, some of the mineral elements absorbed from the soil are translocated upwards through the xylem in the form of organic compounds. The xylem appears to be the main pathway of translocation for such compounds when they originate in the roots.

Various experiments indicate that the upward translocation of solutes, principally soluble carbohydrates, occurs in the phloem. In one type of experiment the contrasting effects of intercepting the xylem and intercepting the phloem of woody tissues was investigated (Fig. 17.3). In these experiments a number of growing shoots were first defoliated. Some received no further treatment, thus serving as checks. In others a ring of the tissues external to the xylem was removed, and in still others a segment of the xylem was excised, leaving the phloem and cortical tissues intact. Every stem which was cut into was enclosed in a glass cylinder as shown in the figure. This cylinder was filled with water to the top of the stems in which the xylem was cut. The outcome of experiments in which the water jacket was rinsed once each day with distilled water was essentially the same in those experiments in which this was not done, indicating that translocation of solutes did not occur through the water.

The results of some of these experiments are presented in Table 17.1. Invariably the stems in which the xylem was cut showed greater elongation than those in which the phloem was cut, suggesting greater upward translocation of foods in the phloem than in the xylem.

Shoot elongation is a somewhat indirect measure of translocation, but the conclusions drawn from such observation were supported in a number of experiments by dry weight determinations and analyses for sugar. As shown in Table 17.1, the dry weight and total sugar content per stem were invariably

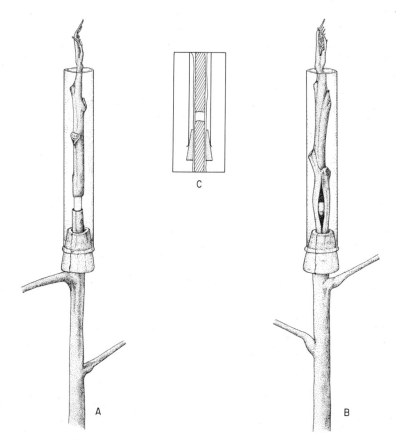

Fig. 17.3. Diagrams to show: (A) stem with phloem removed; (B) stem with xylem removed, the cut portion of each stem being enclosed in a water jacket; (C) sectional view of (B). After Curtis, *Translocation of Solutes in Plants,* McGraw-Hill Book Company, New York, 1935:37.

least in the ringed stems. This was also true for the percentage of sugar in terms of fresh weight.

Experimental results obtained using radioactively tagged molecules supports the older evidence obtained by surgical procedures that upward translocation of organic solutes occurs in the phloem. Such experiments are described later in this chapter.

In general, the phloem appears to be the principal tissue in which the upward translocation of organic solutes occurs. Mineral element-containing organic compounds, which move upward in the xylem from the roots, are an important exception to this statement.

Upward Translocation of Mineral Elements.   Strictly speaking, mineral elements as such are not translocated in plants. As in Chapter 15 this term is

TABLE 17.1 COMPARATIVE EFFECTS OF CUTTING THE XYLEM OR PHLOEM ON GROWTH, DRY WEIGHT, AND SUGAR CONTENT OF DEFOLIATED SHOOTS[a]

Species	Treatment	Average total growth, mm	Dry weight, per cent of fresh growth	Total sugar per stem, mg	Sugar, per cent, fresh weight	Sugar, per cent, dry weight
Mock Orange (Philadelphus pubescens) June 13–June 19	Check	63.6	10.8	3.08	0.12	1.12
	Phloem cut	7.8	9.0	0.08	0.03	0.35
	Xylem cut	49.2	10.8	5.32	0.22	2.03
Mock Orange (Philadelphus pubescens) June 25–July 1	Check	105.3	13.0	2.10	0.094	0.72
	Phloem cut	19.7	9.4	1.63	0.087	0.93
	Xylem cut	47.4	11.8	4.83	0.231	2.08
Sumac (Rhus typhina) June 26–July 1	Check	63.0	22.3	4.17	0.33	1.48
	Phloem cut	15.8	17.2	3.05	0.67	3.89
	Xylem cut	49.5	20.5	3.90	0.42	2.05

[a]Data of Curtis, Ann. Bot., **39**, 1925: 579, 589, and 581.

used as a convenient designation for compounds of which one or more mineral elements are constituents. The principal categories of such compounds are; firstly, inorganic salts, and secondly, organic compounds, the molecules of which are partly constructed from one or more mineral elements. As discussed in Chapter 16, the synthesis of amino acids occurs in the roots of many kinds of plants and the upward translocation of nitrogen and sulfur occurs in the form of such compounds as well as in inorganic combination. Phosphorus probably is translocated in an upward direction both as inorganic phosphates and as phosphorylated organic compounds. Most of the other physiologically important mineral elements are translocated upward in the plant as constituents of inorganic compounds.

Studies of the sap from xylem vessels shows that it usually contains at least traces of both organic and inorganic solutes. In proportion to the total quantities utilized, however, the concentration of mineral element-containing compounds in the xylem sap is usually much greater than the proportion of soluble carbohydrates and related compounds. Furthermore, appreciable concentrations of mineral elements are commonly present in the sap of vessels at seasons when the upward flow of water is occurring at rapid rates. At such times the xylem sap contains little or no soluble carbohydrate. The presence of mineral elements in the form of soluble compounds at such seasons is presumptive evidence that at least some of them are translocated upward in the plant through the xylem.

It has long been known that basally ringing stems of various species does not prevent the upward movement of mineral elements through the plant. The results of such experiments show that the upward movement of mineral elements can occur in the xylem, but they do not prove that some such movement can not also occur in the phloem.

More definite information regarding the path of the upward movement of the mineral elements in plants has been obtained by means of radioactive tracer experiments. In one investigation of this type small plants of cotton, geranium, and willow, rooted in sand or solution cultures, were used. Certain branches of each plant were stripped by cutting longitudinal slits about 20 cm long on opposite sides of the stem, and then carefully pulling the bark away from the wood, but leaving it attached at the ends. A sheet of paraffined paper was then introduced between the phloem and the xylem (Fig. 17.4). This treatment resulted in no visible signs of injury to the plants during the course of an experiment. Radioactive ions of potassium, sodium, phosphate, or bromide were introduced into the rooting medium. After a period of no more than a few hours, under conditions favorable to transpiration, distribution of the tracer ions in the stem was ascertained by measuring the quantity of radioactive elements in ashed segments of the xylem and phloem from the stem above, below, and in the region where the xylem and phloem were separated with paraffined paper.

As shown by the results from a representative one of these experiments

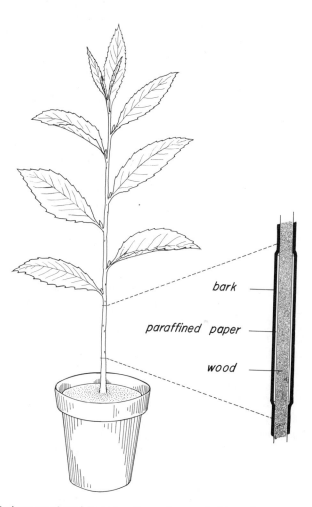

bark

paraffined paper

wood

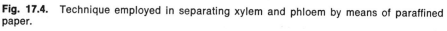

Fig. 17.4. Technique employed in separating xylem and phloem by means of paraffined paper.

(Table 17.2), radioactive potassium (^{42}K) was found to be relatively abundant in both the bark and the wood above and below the section of the willow stem in which the xylem and phloem had been separated by paraffined paper. Within this latter segment of the stem, however, almost all of the tracer element was located in the xylem. In the intact branch there was no such marked difference in the distribution of the radioactive potassium.

Similar results were obtained with the other plants used, and with each of the radioactive elements employed as a tracer.

In these experiments it is obvious that mineral elements absorbed by the roots were transported upward at a relatively rapid rate through the xylem.

TABLE 17.2   DISTRIBUTION OF RADIOACTIVE POTASSIUM IN THE STEM OF WILLOW AFTER AN ABSORPTION PERIOD OF FIVE HOURS[a]

Stem segment		Branch stripped 1–½ hrs before absorption period		Intact branch	
		ppm in Bark	ppm in Wood	ppm in Bark	ppm in Wood
Above strip	SA	53	47	64	56
	S6	11.6	119		
	S5	0.9	122		
Stripped section	S4	0.7	112	87	69
(see Fig. 17.4)	S3	<0.3	98		
	S2	<0.3	108		
	S1	20	113		
Below strip	SB	84	58	74	67

[a]After Stout and Hoagland, *Am. J. Botany*, **26**, 1939: 322.

During their upward passage some of the mineral elements moved laterally from the xylem into the phloem, thus accounting for the relatively large amounts of tracer elements in the phloem. When, however, their lateral movement from xylem to phloem was intercepted by interposing an impermeable barrier between these two tissues, practically all of the tracer element was found in the xylem.

These results demonstrate that the upward movement of mineral elements occurs in the xylem, but the possibility that some such movement occurs in the phloem also is not excluded. The data in Table 17.2 furnish some evidence of upward and downward translocation in the phloem. Bark sections S1 and S6, adjacent to the unstripped portions of the stem, are both higher in radioactive potassium than sections S2 and S5. This can be accounted for only by assuming that a slow movement of radioactive potassium has occurred into these sections from adjacent unstripped portions of the bark. Somewhat similar experiments also indicate that a limited amount of upward translocation of radioactive phosphate may occur through the phloem.

There is no doubt that the xylem is the main pathway along which the upward movement of mineral elements occurs from roots to leaves. Some upward translocation of mineral elements also occurs in the phloem under certain circumstances. As described in the next section, the upward movement of mineral elements in the phloem of younger stems appears to be a usual occurrence when they are "exported" from older, lower leaves into the stem.

Outward Movement of Mineral Elements from Leaves and Other Lateral Organs.   Not all of the mineral elements which are translocated into a given leaf remain in that leaf; some of them are "exported" back into the stem, whence they are translocated to other parts of the plant. This is shown by the results of the chemical analyses of leaves at different times of the day and also by the results of the periodic chemical analyses of leaf and stem tissues during the period just prior to leaf abscission. Nitrogen, phosphorus, potassium, sulfur, magnesium, and chlorine may all be exported from leaves in one form or another, but calcium, boron, iron, and manganese appear to be virtually immobile. Nitrogen, phosphorus, and sulfur move out of leaves at least partly in organic combination, while potassium, magnesium, and chlorine probably move out mostly in the form of inorganic ions. The translocation of mineral elements out of flower petals just prior to their abscission also occurs. In general, those mineral elements which are readily redistributed in plants (Chapter 15) are the ones which are the most readily exported from the leaves.

The earlier evidence of the export of substances other than carbohydrates from leaves and other lateral organs has been augmented and strengthened by more recent studies in which radioactive elements have been used. To study this phenomenon by means of tracer elements it is necessary to apply the tagged ions directly to the leaf tissues rather than to let them enter the plant in the usual way through the root system.

Radioactive phosphorus as phosphate has been introduced into the leaves of cotton plants in such a way as to insure that the tagged ions would move into the phloem tissue of the leaf. The xylem and phloem of the stem were separated by a membrane of waxed paper in a segment extending for about 10 cm immediately below the point of leaf attachment. The phosphate ions were allowed to migrate through the issues of the leaf and stem for them 1 to 3 hr before their distribution in the stem was determined. An examination of the treated portion of the stem revealed that almost all of the radioactive phosphate was located in the bark. The phosphate had moved both up and down after reaching the stem, but only traces of the tagged element were found in the xylem where it was separated from the phloem by the waxed paper. It is evident from these results that the phosphate was translocated out of the leaf and longitudinally within the stem almost entirely in the tissues of the phloem.

When the radioactive phosphate was applied to leaves of plants with intact stems, the tagged ions were found in both the xylem and the phloem of the stem and both above and below the point of leaf attachment. Again, much more of the radioactive element was present in the bark than in the wood, which suggests that the tracer ions reached the wood by radial movement from the phloem.

The general picture of the movements of mineral elements in plants seems to be about as follows. After entering the roots from the soil, most of them move across the cells of the young roots; some of them, after entering into organic combination, move into the xylem ducts. In the xylem conduits they are carried upward into the leaves and apical growing regions in the moving columns of water. Some of the molecules of which these elements are a part or of which they become a part as a result of chemical transformations in the leaf do not become permanently immobilized in the leaves, but sooner or later move out of them by way of the phloem. Such outward movements are often pronounced during a period just prior to leaf abscission. Once in the stem, such solutes move both upward and downward in the phloem, entering other, usually younger, leaves and often also reaching the apical growing regions of roots and stems. Some of the mineral-containing compounds in the phloem also migrate radially and enter the transpiration stream in which they are transported upward at a frequently rapid rate. In a sense, therefore, and to a certain degree, there is a circulation of at least some of the mineral elements in plants. Phosphorous, nitrogen, and potassium in one kind of chemical combination or another, seem to be the most freely circulating of the mineral elements.

Lateral Translocation of Solutes.   As was shown in Chapter 8, the lateral translocation of water in a tangential direction readily occurs in woody stems. This does not appear to be true of many solutes. In straight-grained trees the sugars from the leaves on one side of the tree are translocated to the roots directly below them; and if nitrates are added to the soil on one side of a tree, an increase in nitrogen occurs principally in the leaves and branches above the roots on that side. A similar lack of lateral translocation of solutes has been shown in certain species of herbaceous plants. On the other hand, the radial transfer of solutes from xylem to phloem or vice versa appears to occur readily. Some of the radial movement of solutes probably occurs along the vascular rays.

In perennial woody plants there is evidence that the reorientation of conductive tissues may occur in such a way as to offset the effects of a lack of lateral translocation of solutes. If all the branches on one side of a tree are removed or destroyed, for example, the effect of diminishing the growth of the trunk and roots on that side does not persist for more than one year. Subsequently developed conductive tissues are usually oriented in such a fashion as to permit translocation to the trunk and roots on the side of the tree which no longer bears branches.

Mechanism of Translocation of Solutes in the Xylem.   The probable mechanism whereby mineral salts are absorbed by the roots and delivered to the xylem ducts, has been described in Chapter 15. Dissolved mineral salts and organic solutes, when present, are carried along with the ascending streams of water in the xylem ducts which are pulled up through the plant according to the mechanism discussed in Chapter 8. During the upward translocation some of the solutes are lost by radial movement into living cells of the stem adjacent to the xylem conduits. In the leaves, the solutes of the xylem vessels migrate into the living cells of the mesophyll.

Basic Aspects of the Physiology of Translocation in the Phloem.   As a background for considering the principal theories which have been proposed to account for the mechanism of translocation in the phloem, some of the more pertinent facts regarding this process will be summarized.

Metabolic Activity of the Phloem Cells in Relation to Translocation. Even though they undergo a substantial amount of structural degeneration as they mature, the sieve-tube elements remain alive and metabolically active, at least at a low level, as long as translocation occurs through them. It is considered that the sieve-tube elements must at least maintain a high enough level of metabolic activity to preserve their structural integrity. The intimate anatomical relationship between sieve tubes and phloem parenchyma cells suggests that important physiological relationships exist between these two components of the phloem. Phloem parenchyma cells, and especially those associated with the terminal regions of the phloem, typically contain a dense cytoplasm and numerous mitochondria suggesting that they are capable of a high level of metabolic activity.

The exposure of stems or petioles to temperatures of 15°C or lower has been found to retard the rate of translocation of soluble carbohydrates as compared with their rate of movement at 20°C to 30°C. Treatment of stems or petioles with metabolic inhibitors such as cyanides, fluorides, azides, and 2–4 dinitrophenol has also been shown to retard the rate of the translocation of solutes through them.

The retarding effects upon the translocation of solutes of relatively low temperatures or metabolic inhibitors both suggest some close relationship between the metabolic activity of the phloem cells in the stem and the process of translocation. While this is a valid conjecture, it is weakened by the fact that alternative explanations of these results are plausible. Although the treatments described previously are applied to localized segments of stems, their effects are not necessarily restricted to such zones. Inhibitors can be readily translocated to other parts of the plant and, for reasons not clearly understood, localized application of a low temperature also may have effects in parts of a plant remote from the region to which they are applied. The effects may be largely or entirely on phloem parenchyma cells in certain regions of the phloem system which, as shown in the further discussion, almost certainly play important roles in translocation.

In analyzing the processes of translocation, and especially the movement of sugars, it is important to recognize the regions in which the translocate originates and the regions in which it is utilized or immobilized. The former are called "sources"; the latter are called "sinks." Important sources in sugar translocation are the photosynthesizing cells of the leaf and various tissues or organs in which carbohydrates have been stored. Important sinks in sugar translocation are rapidly growing regions, principally apical stem and root meristems, and developing storage regions such as fruits or tubers.

The numerous ultimate terminations of the phloem part of a leaf vascular bundle usually consist of sieve-tube elements and companion cells only. The smaller veins of the leaf are surrounded by a layer of usually parenchymatous cells called the bundle sheath. A square centimeter of a leaf of sugar beet has been shown to contain as many as seventy phloem endings. Considerable evidence supports the hypothesis that parenchyma cells associated with the phloem endings in the leaf accumulate solutes from the mesophyll cells and deliver them to the sieve tubes. Such a "loading" process requires the expenditure of metabolic energy. There is also evidence that solutes are removed from the sieve tubes in sink regions by processes requiring metabolic energy.

Adjacent parenchyma cells alongside the sieve tubes in any part of the phloem system may also serve at times as sources or at times as sinks for translocates in the sieve tubes, principally sugars. That such cells possess an accumulative capacity for solutes, which would of course require metabolic energy, has been demonstrated in a number of species.

Bidirectional Movement. There is no doubt that movement of solutes in the phloem can sometimes occur in one direction and sometimes in the opposite one. Experiments have shown that upward and downward movements of labelled sugars occur in the phloem tissue of the stem of a squash plant from the node at which a leaf was attached (Fig. 17.5). The sugars were labelled by allowing the leaf to photosynthesize radioactive carbon dioxide. Such experiments do not,

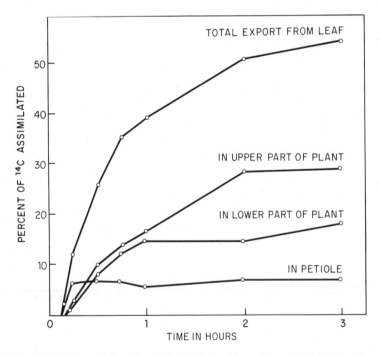

Fig. 17.5. Distribution of translocated ^{14}C from the primary leaf blade of a squash plant to other parts of the plant. Data of Webb and Gorham, *Plant Physiol.* **39**, 1964:667.

however, demonstrate the occurrence of a bidirectional movement through the same vertical segment of phloem tissue at the same time.

The simultaneous movement of phosphate and carbohydrate labelled with radioactive phosphorous (^{32}P) and radioactive carbon (^{14}C), respectively, has been shown to occur in opposite directions through the same vertical segment of phloem tissue of geranium stems at least over short distances and at relatively slow rates. Such movement occurred whether the carbohydrate was supplied to the stem in a position above or below the point at which the phosphate was supplied.

Translocation such as that just described could be accomplished by an upward movement in some sieve tubes and a concurrent downward movement in others. The other possibility is that both upward and downward movement can occur simultaneously in the same sieve tube. Several investigators claim to have demonstrated such an occurrence but others question such an interpretation of the experimental results on which such claims have been based. This very fundamental question regarding the mechanism of transport in the phloem still appears to be unanswered.

Turgor Pressure in the Sieve Tubes.   There is substantial evidence that turgor pressures of considerable magnitude often prevail in the sieve tubes. Values up to 30 bars have been reported but the usually prevailing pressures are undoubtedly lower. The physical condition of the sieve-tube sap of being usually under pressure contrasts with that of the xylem vessels in which the sap is commonly under tension.

As a result of the prevailing pressure, sap will exude from the stems of many plants if a cut is made into the phloem. A continued exudation of phloem sap has been shown to occur from the cut stems of squash plants for periods of at least 24 hours. Because the exudate from this plant rapidly coagulates in contact with the air, thus plugging the phloem, the flow is maintained only if a fresh cut is made at frequent intervals. Exudation occurs from the sieve-tube elements and takes place at rates ranging from 0.01 to 0.1 ml per minute.

It has been known for many years that exudation of phloem sap also occurs from woody plants. If a small puncture is made through the bark just to the inner phloem, but not into the xylem, in any of a number of different woody species, exudation of phloem sap ensues. In some species such an exudation may continue for hours or even days.

The feeding activities of certain species of aphids also furnish fascinating evidence of the existence of turgor pressure in the sieve tubes. These insects, some of which feed on herbaceous plants, and some on the younger stems of woody plants, insert their stylets through the outer stem tissues directly into the zone of phloem in which active translocation is occurring (Fig. 17.6). A stylet is sufficiently small in diameter that it can be inserted into a single sieve-tube element. The turgor pressure prevailing in the sieve tubes is sufficient to force phloem sap into and through the body of the aphid from which it is exuded at

Fig. 17.6 An aphid feeding on the underside of the branch of a linden tree. Photograph from M. H. Zimmerman. *Science,* **133,** No. 13, January 1961:Cover photo. Copyright 1961 by the American Association for the Advancement of Science.

the anal end as drops of sticky "honeydew" which is essentially a strong sugar solution. The composition of the honeydew is not exactly the same as that of the phloem sap because some of the constituents of the latter are removed by the digestive processes of the insect.

If an aphid is anesthetized and the stylet severed, the latter serves as a microconduit through which phloem sap will be forced, drop by drop, often for many days. The rate of exudation is such that an average-sized sieve-tube element must be refilled 3–10 times every second. This is evidence of a rapid flow of sugar solution through a long sequence of sieve-tube elements.

Amounts of Material Transported.   The amount of carbohydrate translocated through the phloem into such organs as fleshy roots, tubers, or fruits is almost incredibly great when considered in relation to the cross-sectional area of the conducting elements of the phloem.

For example, the gain in fresh weight of pumpkin fruits (Connecticut Field variety) has been calculated to be about 5500 g in a 33-day growing period. A large proportion of this increase represents water. However, calculations show that this corresponds to an average hourly increase of 0.61 g in *dry weight* throughout the growing season and that peak rates of about 1.70 g dry weight per hour were attained. All such material entered the fruit through the one slender stem connecting it with the vine. Similar rates of translocation have been demonstrated to occur into other fleshy fruits.

Velocity of Movement. Calculations of the velocities with which organic solutes must move through sieve tubes in order to account for a gain in dry weight of 0.61 g per hour (average) by a pumpkin fruit have been made. The required rate would be about 11 cm per hour if the organic compounds move in the dry state (an obviously hypothetical assumption), about 55 cm per hour if they move as a 20 per cent solution, and about 110 cm per hour if they move as a 10 per cent solution. These calculations assume that transport occurs in the entire lumen of the sieve tubes. It should be noted that these calculations are based on *average* and not on maximum rates of gain in dry weight.

Numerous measurements of the velocity with which translocates (usually sugars) move through the phloem have been made with radioactively tagged molecules. Results vary with the plant used and the conditions under which the measurements are made, but lie mostly in a range of 60 to 300 cm per hour.

Diurnal Periodicity of Translocation. Diurnal variations occur in the rate of movement of solutes through plants and do not follow the same pattern in all species. Most of the carbohydrates which enter growing cotton bolls, for example, appear to move into them during the daytime rather than at night. The rate of translocation of carbohydrates into date fruits in the preripe stage of development, on the other hand, appears to be more rapid at night than in the daytime.

Theories of the Mechanism of the Translocation in the Phloem. *The Mass Flow Hypothesis.* This hypothesis, also called the *pressure flow hypothesis,* was first proposed by the German plant physiologist Münch in 1926. The principles involved in this postulated mechanism can most easily be clarified by reference to a simple physical system (Fig. 17.7). As shown in the figure, two membranes permeable only to water, both dipping in water, are connected by a

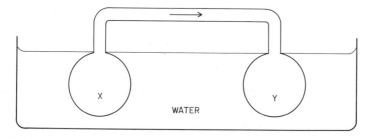

Fig. 17.7. Diagram of an osmotic system in which a mass flow of solution will occur.

tube to form a closed system. Membrane X is assumed to enclose a stronger solution of sucrose than membrane Y. Water will at first enter both membranes, but the greater turgor pressure developed in X will soon be transmitted through the system. This will result in a greater water potential in the water in Y than in the pure water in which the membranes are immersed. Water will therefore pass out of the membrane Y, even as it continues to move into X, and coincidentally there will be a flow of *solution* along the tube from X to Y. The mass movement of solution from X to Y will continue until the concentrations of the sugar solutions in both membranes are equal. At this point the flow of solution in the tube will stop and a dynamic equilibruim will be established between the solution in the closed system and the circumambient water.

If such an apparatus could be set up so that the sugar could be utilized or be converted into an insoluble form as fast as it was translocated into Y, and so that additional sugar could pass into solution in X as rapidly as it moved out of that membrane, flow of solution from X to Y would continue indefinitely.

The Münch pressure flow hypothesis assumes that a system analogous to that just described, accounts for translocation of solutes through the phloem. As originally postulated it was considered that chlorophyllous cells of the leaf corresponded to membrane X in Fig. 17.7 in activating the downward transport of solutes. Root cells in which sugars would be oxidized or converted into insoluble carbohydrates would correspond to membrane Y in Fig. 17.7. The phloem would be the continuous system connecting the mesophyll cells and the root cells, and water would move into the mesophyll cells through the xylem. The water component of the downward moving stream is supposed to rejoin the transpiration stream in the roots; smaller amounts are presumed to move laterally into the xylem along the stem; some may be used in the growth of the roots.

The osmotic potentials of the leaf cells are maintained at a relatively negative value as a result of photosynthesis, while the osmotic potentials of the root cells are usually less negative because the sugar translocated into them is used in metabolic activities or is converted into insoluble storage forms. Higher turgor pressures can therefore be generated in the leaf cells than in the root cells. Mass movement of solution from adjacent leaf cells into the sieve tubes and from sieve tubes into adjacent root cells is supposed to occur through the plasmodesmata.

According to the original postulation of this theory the higher turgor pressures which could be generated in the mesophyll cells as compared with the root cells cause a mass flow of solution through the phloem from the leaves to the roots. From the previous discussion of the active transport role of phloem parenchyma cells near the sieve tube termination in the leaves, it is evident that this theory is no longer tenable in its original form. As a result of the operation of these phloem parenchyma cells, solutes are accomulated in such cells and delivered to the sieve tubes. Both the phloem parenchyma cells and terminal sieve tube elements would have more negative osmotic potentials than the mesophyll cells. Water would move into these cells, directly or indirectly, from the xylem, resulting in an increase in their turgor pressure. However, the pressure flow theory can still be accepted as a valid possibility if its operation is visualized as confined to the sieve tube system. In other words, it could be considered that downward phloem sap movement is actuated by a gradient of turgor pressure, highest at the upper ends of the sieve tubes and lowest at their terminations in the roots or other sinks.

The pressure flow theory can account for transport at times in one direction and at times in the other. Flow would occur from the end in which the turgor pressure is highest toward the other. In the spring, for example, it can be postulated that if the turgor pressure in the sieve tubes of the root or stem is greater than that of the apical growing regions, upward flow will occur. Later in the season when the turgor pressure in the sieve tubes of the leaves increases, a reversal in the direction of flow would be expected. Daily variations in the rate of solute translocation can also be readily accounted for by this theory.

Translocation of certain viruses through the phloem is closely correlated with the translocation of sugars. Translocation of certain chemicals such as 2, 4-D (dichlorophenoxyacetic acid) which exert hormone-like effects on plants (Chapter 18) also seems to be closely correlated with the movement of sugars. This also appears to be true of certain vitamins (Chapter 18) and "florigen" (Chapter 22). Although these facts have been cited as supporting the concept that a mass flow of solution occurs through the sieve elements, they are actually amenable to alternative interpretations.

About the only tangible evidence which supports the pressure flow theory is that, as previously described, the phloem sap exhibits a turgor pressure and exudation of sap often occurs from the phloem tissue when an incision of any kind is made into it.

Although the pressure flow theory is a plausible one, at least in its modified form, a number of criticisms can be made of it. Only several of the major ones will be discussed. The theory does not account for the fact that when metabolic activity of phloem cells is reduced by treatment with inhibitors or lowered temperature, the rate of translocation slows down. As discussed previously, however, these effects may be primarily on the source and sink cells at the ends of the sieve tube system or possibly on the phloem parenchyma cells in intermediate portions of the phloem system and do not negate the possibility of mass flow through the sieve tubes.

The existence of consistent concentration gradients of sugars in the phloem of the stem in the direction in which translocation is occurring has been demonstrated in some kinds of plants. It is interesting, and perhaps significant to note that translocation (at least of sugars) follows a diffusion pattern, although the rate of movement is many times greater than can be accounted for by diffusion. Turgor pressure gradients from top to bottom in the sieve tube system of stems have also been demonstrated in some species in which downward translocation of sugars was occurring. It is not certain, however, that turgor pressure gradients are always so oriented as to account for the direction of movement of a translocate. It is even less certain that the turgor pressure differences which are established between sources and sinks are always sufficient in magnitude to account for known rates of flow.

The pressure flow theory obviously accounts for a translocation through a given sieve tube in only one direction at a time. It is conceivable that the upward movement of some solutes might be occurring in some sieve tubes while downward movement of other solutes was occurring simultaneously in other sieve tubes through the same zone of phloem tissue. This would imply that turgor pressure gradients are oriented in one direction in some sieve tubes and in the opposite direction in parallel sieve tubes. The pressure flow theory cannot be reconciled with the possibility that bidirectional movement of translocates occurs within individual sieve tubes. If such a situation is ever conclusively demonstrated, it would constitute a death warrant for the pressure flow theory, at least as currently conceived.

Cytoplasmic Streaming Theory.   De Vries and other nineteenth century investigators postulated that streaming of the cytoplasm in the cells of the phloem, primarily the sieve tube elements, might explain the relatively rapid rate of transport of solutes. The basic assumption is that rotational streaming of the cytoplasm occurs in the sieve tube elements, and that solute molecules, caught in the cytoplasmic matrix, are carried by its movement from one end of the element to the other. The molecules are assumed to pass from one sieve tube element to the next by diffusion, presumably largely through the sieve plate pores. Diffusion over short distances can occur very rapidly even if the molecules are moving along a gradient which is not very steep (Chapter 4). This theory could account for the simultaneous movement of solutes in both upward and downward directions in the same sieve tube at the same time.

About the only positive evidence which can be cited in favor of the cytoplasmic streaming theory is that solute translocation through the phloem is retarded, as mentioned previously, by conditions which lower the metabolic activity of phloem cells. A reduced level of metabolic activity would have a slower rate of cytoplasmic streaming as one of its consequences but there are other possible ways in which it could affect the rate of translocation, as discussed previously.

Streaming of the cytoplasm can be readily detected in young sieve tube elements but, as this phenomenon is usually visualized, there is no unequivocal evidence that it occurs in mature sieve tubes. Existence of the very mechanism upon which this theory is predicated thus seems to be doubtful.

Another objection commonly raised against the cytoplasmic streaming theory is that although the phenomenon is known to occur in many kinds of cells, known rates of streaming are inadequate to account for known rates of solute transfer through the phloem. Rates of streaming as high as 48 cm per hour have been observed in cells of the alga *Nitella,* but the rates in companion cells and phloem parenchyma cells of higher plants are much less. Even the highest rate observed in *Nitella* is substantially less than many known rates of translocation, which range up to 300 cm per hour.

While it seems ˊclear that cytoplasmic streaming, in the usual sense of the word, cannot be the mechanism which accounts for translocation of solutes through the phloem, variants of this hypothesis have been proposed which may have greater validity. Most of these modifications of the streaming of cytoplasm theory invoke the participation of special strands or fibrils in the cytoplasm. Although there is considerable evidence for the existence of such microstructures, the idea that they might in some manner actuate the translocation of solutes has not been generally accepted.

In brief, knowledge regarding the mechanism of the translocation of solutes is in an unsatisfactory state. It is probable that the viewpoints which have been adopted are oversimplified. It also seems probable that there are elements of truth in each of the principal theories which have been proposed.

SUGGESTED FOR COLLATERAL READING

Crafts, A. S. and C. E. Crisp. *Phloem Transport in Plants.* W. H. Freeman and Company, San Francisco, 1971.

Esau, Katherine. *Plant Anatomy, 2nd Edition.* John Wiley & Sons, Inc., New York, 1965.

Ledbetter, M. C., and K. R. Porter. *Introduction to the Fine Structure of Plant Cells.* Springer-Verlag, New York, 1970.

18 | PLANT HORMONES

The term *hormone* is applied to certain substances, made in one part of an organism, which, after translocation, induce pronounced physiological effects in other parts of that organism, being operative in very low concentrations. This concept originated in the realm of animal physiology and the term hormone is derived from the Greek root meaning "to excite." In the mammalian body most hormones are synthesized in the ductless glands; examples are thyroxin from the thyroid gland, insulin from the pancreas, and adrenalin from the adrenal gland. Analogous, although chemically very different, compounds exist in plants and the term hormone can properly be extended to them. Plant hormones, often called *phytohormones,* are mostly synthesized in meristematic, or at least young, tissues of one kind or another and often exert their effects after translocation to some relatively distant tissue from the one in which they originate.

Hormones share with enzymes, vitamins, and DNA the property of exerting effects, often of major physiological proportions, when present in low and often minute concentrations. In fact it is not always possible to distinguish clearly among these four categories of compounds. Some compounds classed as vitamins in animal nutrition behave as hormones in vascular plants. Examples are some of the B vitamins as discussed later.

The term hormone should be restricted to compounds naturally synthesized within an organism. A large number of substances, not known to occur in plants, exert effects similar to and sometimes apparently identical with the effects of one of the naturally occurring plant hormones. Many of these compounds resemble in chemical composition one of the endogenous plant hormones. Such compounds are sometimes also referred to as hormones, but this usage of the term should be discouraged. A preferred term for compounds which exert hormone-like effects on plants, but are not known to occur naturally in them, is *growth-regulators*.

The Auxins.   The action of compounds now classified as auxins was first clearly demonstrated in the leaf sheath or *coleoptile* of the oat plant (*Avena sativa*). This is a tubular, leaf-like structure, closed at the top, which is the first part of the plant to emerge from the soil (Fig. 18.1). Similar coleoptiles develop during early seedling growth of other members of the grass family. The coleoptile encloses the initial leaf and is eventually pierced at the tip as a result of growth of this leaf, soon after which all growth in length of the coleoptile ceases. Oat coleoptiles are approximately 1.5 mm in diameter, and when illuminated seldom attain heights of more than 2 cm. In the dark they may attain heights ranging up to 6 cm. Cell divisions cease relatively early in the life history of an oat coleoptile, and during approximately the last three-fourths of its growth period all increase in its length results from cell elongation.

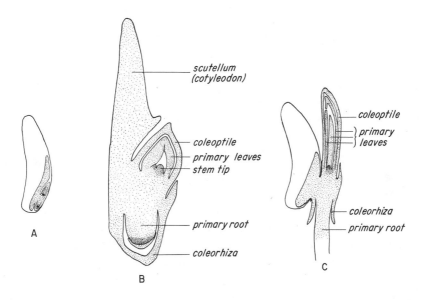

Fig. 18.1. Oats (*Avena sativa*): (A) Longitudinal section through grain, showing the location of the embryo. (B) Longitudinal section through the embryo. (C) Early stage in germination, showing the emergence of coleoptile. Redrawn with modifications from Avery, *Botan. Gaz.*, **89**, 1930:9.

If the tip of a coleoptile is removed by a clean cut made several millimeters below the apex, the rate of elongation of the stump is immediately retarded. If, however, the cutoff tip of the coleoptile or a similar tip from another coleoptile is affixed on the stump, its elongation will be resumed and may nearly regain

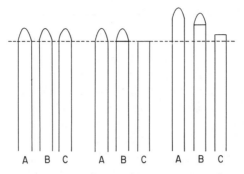

Fig. 18.2. Effect of the removal of tip on the elongation of oat coleoptile: (A) Check. (B) Tip severed and replaced. (C) Tip removed. The effects of the treatments are shown by the differences in increase in length of coleoptiles at the right.

the original rate (Fig. 18.2). Retipping the coleoptile with a short segment cut out of another coleoptile somewhat below the apex results in little or no increase in elongation rate. Such experiments indicate that the elongation of a coleoptile, which occurs in the more basal regions, is maintained only under the influence of some sort of "stimulus" originating in the tip, whence it is transmitted basipetally (apex to base) through the coleoptile.

That this stimulus is hormonal in nature is clearly shown by an experiment first performed by F. W. Went in 1928. Excised tips of oat coleoptiles were placed on a thin layer of 3 per cent agar gel from which they were removed after one hour, at which time the agar was sliced into a number of equal-sized small blocks (Fig. 18.3). If one of these blocks was placed upon the stump of a detipped coleoptile, the rate of elongation was accelerated just as if the stump had been capped with a fresh coleoptile tip. On the other hand, retipping a coleoptile with a block of pure agar had no appreciable accelerating effect on elongation. It seems evident from these results that some substance or substances (now known to be auxin) were transported out of the cutoff tip into the agar block and subsequently out of the block into the coleoptile stump, whence they were translocated downward to the elongating region of the coleoptile (Fig. 18.4).

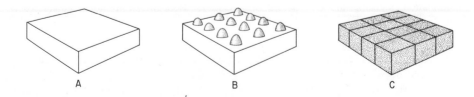

Fig. 18.3. Stages in the collection of auxin in agar from coleoptile tips.

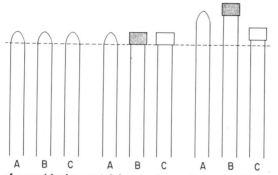

Fig. 18.4. Effect of agar blocks containing auxin on the elongation of oat coleoptiles: (A) Check. (B) Block containing auxin placed on decapitated tip. (C) Block of pure agar placed on the decapitated tip. The effects of the treatments are shown by the relative increases in the length of the coleoptiles at the right.

Chemical Nature of the Auxins.   Any compound which promotes growth along a longtitudinal axis when applied in low concentrations to coleoptiles which are initially relatively low in their content of such a growth promoting substance, but under conditions otherwise favorable for elongation growth, is called an auxin. The phrase low concentrations is a somewhat vague one but in general can be taken to refer to concentrations of less than 1×10^{-5} molar.

Only one compound, indole-3-acetic acid (IAA), has been positively identified as a naturally occurring auxin synthesized in plants:

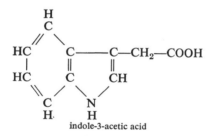

indole-3-acetic acid

There is also some evidence for the existence of non-indole auxins in plants but they have not been sufficiently characterized chemically to warrant their discussion in any detail. The presence of auxins has been demonstrated in a wide variety of species and they appear to be universal constituents of plants.

There is good evidence that indole-3-acetic acid is synthesized in plants from the amino acid tryptophan (Chapter 16) and that the immediate precursor of IAA is indole-3-acetaldehyde. Two main routes of synthesis of this latter compound from tryptophan have been recognized; one by way of indolepyruvic acid as an intermediate, the other by way of tryptamine as an intermediate. Different pathways of synthesis may be followed in different plants or even in different organs of the same plant.

The principal centers of auxin synthesis are apical meristematic tissue of

aerial tissues such as opening buds, young leaves, root tips, and flowers or inflorescences on growing flower stalks. The auxin synthesized in one tissue is
frequently translocated to other organs of the plant. The concentration of auxin
may vary greatly from one tissue to another; in general auxin is found in greatest concentrations in tissues in which it is synthesized or stored. Temperature is
a factor in auxin synthesis, but the optimum for this process is probably not the
same in all plants or tissues. In coleoptile tips the auxin is apparently synthesized from a precursor which is translocated acropetally (base to apex) through
the coleoptile from the grain.

A large number of compounds, not known to occur naturally in plants,
have been found to induce reactions in plants similar to those evoked by the
naturally occurring auxin. Some of the better known of these "synthetic" auxins
are α-naphthalene acetic acid, indole-3-butyric acid, 2,4-dichlorophenoxyacetic
acid, naphthoxyacetic acid and tri-iodo-benzoic acid (Fig. 18.5). Comparison
of the structural formulas of the many compounds which exhibit auxin activity
shows that they have three features in common: a ring-structured nucleus (benzene, indole, or naphthalene) with at least one double bond in the ring, a side
chain consisting of −COOH or bearing this group or one readily converted into
it, and a suitable steric relationship between the ring and the carboxyl or potential carboxyl group.

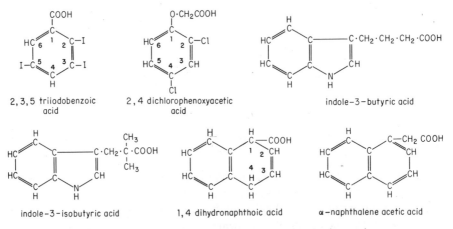

Fig. 18.5. Structural formulas for some synthetic auxins.

Auxins can be inactivated in plant cells by enzymatic activity. The nat-
urally-occurring indole acetic acid is broken down by such enzymes as oxidases,
peroxidases, and phenolases.

Auxin activity can also be offset or negated by the introduction into the
metabolic system of compounds known as *antiauxins* which, in general are quite
similar in chemical structure to the auxins proper. Antiauxins appear to exert

their offsetting effect on auxin activity through the mechanism of competitive inhibition (Chapter 9). The compound 2,6-dichlorophenoxyacetic acid, for example, is an antiauxin for both 2,4-dichlorophenoxyacetic acid and indole-3-acetic acid. Most antiauxins are not known to be naturally occurring compounds. *Trans*-cinnamic acid, however, is a naturally occurring antiauxin and is of special interest because its isomer *cis*-cinnamic acid is an auxin.

Roles of the Auxins. A basic characteristic of the auxins, which they share with most other kinds of plant hormones, is the multiplicity of the kinds of growth reactions which they influence. Only some of the more important of these effects will be covered in the following discussion. Auxins may exercise both enhancing and inhibiting effects on a given response, often depending upon the concentration of auxin present. The effectiveness of a given auxin depends not only on its concentration, but also the specific kind of growth response being influenced. A given kind of auxin which is highly effective in influencing one kind of growth reaction may have little or no effect on another. Some of the inhibitory effects of auxins, especially on the elongation of stem segments of some species and on bud growth in others, result from an auxin-induced generation of ethylene gas which, in itself, is a plant hormone (see later).

Elongation of Cells and Organs. This was the first effect of auxins to be recognized and its basic aspects have already been discussed with reference to the oat coleoptile. In fact, no compound is classed as an auxin unless it can be shown to promote elongation of the oat coleoptile. However, auxins play a role in the elongation phase of growth in many other plant organs similar to that described for the oat coleoptile. Cell elongation occurs only in the presence of auxins and usually with increase in auxin concentration over a wide range there is an increase in the rate of elongation if no other factors are limiting. The optimum range of concentration for cell elongation varies greatly with different tissues, and relatively high concentrations usually exert an inhibiting effect upon this phase of growth.

If the extreme tip of a maize or lupine root is cut off, its rate of elongation increases, although not greatly. Replacement of the root tip in a maize plant results in a retardation in elongation rate as compared with a detipped maize root. Furthermore, attachment of a coleoptile tip maize to a detipped root tip of the same plant results in a retardation in the elongation rate of the root tip. These results suggest that the same concentrations of auxin which accelerate elongation in coleoptile and other aerial organs retard elongation in roots.

This supposition has been confirmed by experiments in which the roots of oat seedlings were immersed in pure solutions of auxins. The growth of the roots was found to be retarded in proportion to the concentration of auxin used. However, when roots which contain either no auxin at all or virtually none are treated with auxin solutions of very low concentration, acceleration of growth as compared with similar but untreated roots often results.

The apparently contrasting effects of auxins upon elongation of roots and aerial organs may be explained by assuming that roots and stems react in a comparable way to auxins (Fig. 18.6), their growth being inhibited by relatively high and promoted by relatively low auxin concentrations. Elongation of roots is favored only at very low concentrations; at all higher concentrations their growth is checked. Stems and coleoptiles show a similar behavior except that the optimum range of concentrations for elongation is much higher than for roots. The same concentrations of auxins which favor stem elongation result in retardation of root elongation.

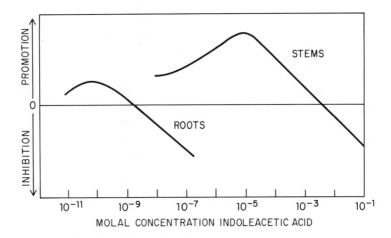

Fig. 18.6. Relation between indoleacetic acid concentration and its promoting or inhibiting effects on the development of stems and roots. After Leopold and Thimann, *Am. J. Botany,* **36,** 1949:344.

Root Formation.   It has been known for many years that the presence of buds on a cutting favors development of roots when the basal portion of the cutting is introduced into a suitable rooting medium. Developing buds are more effective in promoting root formation than quiescent buds. Leaves, especially if young, also often favor the production of roots on cuttings. These observations suggest that root initiation on cuttings is favored by hormones which are synthesized in the buds and young leaves and are subsequently translocated to the basal part of the cutting. Soon after the identification of indoleacetic acid as a naturally occurring auxin, it was found to be active in inducing root formation.

The effect of auxins on root formation should be clearly distinguished from their effect on root elongation. In general, the concentrations required for the former process are much greater than for the latter.

A number of the auxins not known to occur naturally in plants have also been found to be effective in promoting root formation in many species (Fig. 18.7). Among these are α-naphthalene acetic acid, naphthalene acetamide, and indolebutyric acid. Extensive experiments have been carried out on the suitablity

Fig. 18.7. Root formation on stem of tomato plant resulting from treatment with lanolin containing 2 per cent alpha napthalene acetic acid. Photograph from Zimmerman and Wilcox, *Contrib. Boyce Thompson Inst., 7,* 1935:212.

of treatments with various auxins as a practical method of aiding in the rooting of cuttings. Such treatments are not effective with all kinds of plants, but with cuttings of many species they lead to a speeding up of the process of root formation and to the development of a greater number of roots per cutting. Hormones do not induce root formation, however, on cuttings of species on which at least some roots develop without their application.

Phototropism.   Phototropic curvatures of plant organs are especially conspicuous in plants growing in situations in which they are subjected to unequal illumination on opposite sides. Under such circumstances, growing stems and other elongate organs usually bend toward the direction of most intense irradiation. Much of our knowledge of the mechanism of phototropic movements has been derived from studies of the behavior of the coleoptiles of grasses when subjected to one-sided illumination.

As a result of experiments on the coleoptile of the grass *Phalaris canariensis* Darwin showed as far back as 1880 that curvature of this structure will occur if only its extreme tip is exposed to unilateral illumination. This fact has been confirmed by various kinds of experiments mostly performed on oat coleoptiles. If the tip of the coleoptile is shaded by means of a tin-foil cap and the entire coleoptile is illuminated unilaterally, little or no curvature results. Likewise, detipped coleoptiles react only feebly to one-sided illumination but if coleoptile tips which have been subjected to one-sided illumination are placed upon unilluminated coleoptile stumps, marked phototropic curvature of the stump results. Experiments like these indicate clearly that the tip of the coleoptile profoundly influences the elongation phase of growth of cells below the tip of the coleoptile.

The positive phototropic curvature of oat coleoptiles is caused principally by greater elongation of the cells on the shaded side of the coleoptile than of the cells on the more illuminated side. Since cell elongation is known to be influenced by the quantity of auxin present, it is logical to seek an explanation of such phototropic reactions by studying the effect of light upon the distribution of auxin in the coletoptiles. Extensive investigations have shown that, in the oat coleoptile at least, unilateral exposure to light increases the quantity of the auxin reaching the shaded side of the coleoptile from the tip and decreases the quantity on the more illuminated side. The phototropic curvature of oat coleoptiles and presumably of many other plant structures results from the presence of unequal quantities of auxin on the two sides of the coleoptile.

Several types of experiments show quite clearly that light causes a lateral migration of auxin from the lighted to the shaded side of the coleoptile. When, for example, ^{14}C-labelled IAA was applied to the extreme tip of apical sections of unilaterally illuminated maize coleoptiles which had been grown in the dark, it was found that about three times as much would shortly be found in the half of the coleoptile away from the light as in the half towards the light. The total amount of radioactive auxin applied was recovered at the end of the experiment. Hence the unequal amounts of auxin present on the two sides were not brought about by its differential destruction but by displacement of a large proportion of it from the lighted to the shaded side of the coleoptile.

All wavelengths of the visible spectrum are not equally effective in inducing phototropic curvatures. The shorter wavelengths are most effective, maximum

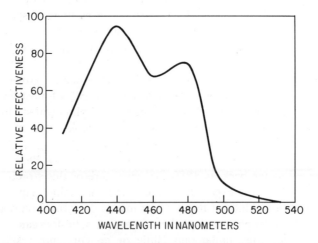

Fig. 18.8. Relation between wavelength of light and the phototropic curvature of oat coleoptiles. Data of Johnston, *Smithsonian Inst. Publ. Misc. Collections*, **92**, No. 11, 1934:13.

phototropic curvature occurring at 450 nm in the blue region of the spectrum (Fig. 18.8). The longer wavelengths at the red end of the spectrum are ineffective in causing a phototropic reaction in etiolated oat coleoptiles or seedlings. The absorption spectrum of β-carotene (Fig. 10.10) corresponds very closely to the action spectrum of phototropic curvature in etiolated coleoptiles and seedlings. This correlation suggests that β-carotene is responsible for the absorption of the light waves which are effective in inducing phototropic curvatures. It is also possible, however, that riboflavin, the absorption spectrum of which closely approaches that of β-carotene, may be the sensitizing pigment, since certain seedlings which, because of genetic mutations are free of carotenoid pigments, exhibit phototropic reactions.

Although extensive investigations have been conducted on the phototropism of coleoptiles, usually etiolated, only a few critical studies of this phenomenon have been made on leafy shoots. In experiments on sunflower seedlings it was shown that the cotyledonary leaves exerted a controlling influence on phototropism in this species. If one of the pair of cotyledons was shaded while the rest of the plant was well illuminated, the hypocotyl bent away from the side to which the shaded cotyledon was attached (Fig. 18.9). The shaded cotyledon simulates those leaves in nature which are receiving the less intense light, in other words, those leaves on the side of the plant "away from the sun." Growth in length of the hypocotyl was about twice as great on the side to which the shaded cotyledon was attached as on the side to which the lighted one was attached. No evidence of lateral transport of auxin was found, but substantially larger amounts of auxin were obtained from the basal ends of hypocotyls of seedlings kept in the dark than from those of seedlings kept in the light. In the sunflower seedling, therefore, phototropism appears to result from a differential in the amounts of auxin coming from the leaves (cotyledons) on the shaded and lighted sides.

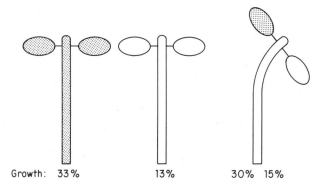

Growth: 33%   13%   30% 15%

Fig. 18.9. Effects of shading on growth and phototropism in sunflower seedlings. The percentage increase in growth of the one-centimeter portion of the hypocotyl just below the cotyledons. Where only one cotyledon was shaded (right) percentages are given for both sides of the hypocotyl. From Lam and Leopold, *Plant Physiol.,* **41,** 1966:848.

Geotropism. If a potted plant is placed in a horizontal position for a few days, the stem no longer lies prostrate but turns upward away from the direction of gravitational attraction. This change in position first appears in the region of elongation just back of the stem tip and with time may extend backward toward the older portions of the stem. If the primary root tips of the plant are examined they will also be found to have altered their position, but in exactly the opposite direction, by growing downward toward the center of the earth. The behavior of roots can be more easily observed in germinating seeds and, as in stems, the change in position first appears in the region of elongation (Fig. 18.10).

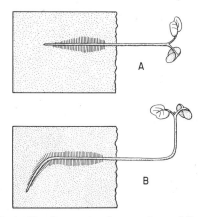

Fig. 18.10. Diagram illustrating the geotropic curvature of the root and hypocotyl of a mustard plant. (A) Plant just after having been placed horizontally. (B) Same plant one day later.

If horizontally placed potted plants are rotated slowly around the stem as an axis, so that every portion of the stem becomes successively the upper and then the lower side, no geotropic curvatures appear. No segment of the stem or root remains long enough in any one position for growth curvatures to occur. Similarly the roots fail to curve when germinating seeds are fastened to the rim of a wheel which is rotated rapidly in a horizontal plane. Growth is more strongly influenced by the centrifugal force generated by the rapid rotation than by the force of gravity. The roots react positively and grow toward the direction of the force (*i.e.*, outward) while the stems grow away from the direction of the force (*i.e.*, inward toward the hub of the wheel). If the rotation of the wheel is stopped or slowed down to a point where the centrifugal force is less than the pull of gravity, the usual geotropic curvatures of both root and stem soon appear.

Detipped primary roots usually fail to exhibit geotropic curvatures when placed in a horizontal position even though the excision of the root tip does not prevent the enlargement of cells below the tip. If tips from vertically oriented roots are placed upon such detipped horizontal roots, positive geotropic curvatures appear. Such experiments indicate that the tip of the root exerts a predominating influence upon its geotropic movements, a relation comparable to that of the coleoptile tip in phototropic reactions of the coleoptile.

In stems, however, geotropic reactions are not prevented by excision of the stem tip. In general, the stem tip is more "sensitive" to the force of gravity than zones at some distance below the tip, but this sensitivity is usually present throughout the growing region. In many grasses, geotropic reactions occur in mature nodes independently of the stem tip.

Tips of the horizontally placed coleoptiles generate the same amount of auxin as vertically oriented tips. The negative (upward) curvature of a coleoptile induced by gravity is not the result, therefore, of any total increase in the amount of hormones synthesized on the lower side of the coleoptile tip. When ^{14}C-labelled IAA was applied to the apical ends of horizontally placed maize or oat coleoptiles or sunflower hypocotyls it was found that substantially more auxin accumulated in an auxin block applied to the basal end of the lower half of the coleoptile than into a block applied to the basal end of its upper half. Apparently gravity, like light, influences the distribution of auxin in the coleoptile and the upward curvature obtained as a consequence of the force of gravity results from the greater concentrations of the hormone on the lower side of a horizontally placed coleoptile, which in turn results in greater elongation of the cells on that side.

The same concentrations of auxin that favor the elongation of stems and coleoptiles retard the elongation of roots. This fact suggests that the positive (downward) geotropic reaction of root tips may be caused by the same mechanism as invokes the negative (upward) curvature of stems or coleoptiles. A greater concentration of auxin in the lower half of horizontally placed roots would check elongation rather than increase it, and the growth of the upper side of the root would exceed that of the lower, resulting in its downward curvature. Experimental tests have confirmed this explanation. The lower halves of tips of horizontal roots have been found to contain higher concentrations of auxins than the upper halves. The growth rate of primary roots which are slowly rotated about their own vertical axis while in a horizontal position does not exceed that of vertically placed roots, i.e., auxin concentrations are equal in the root tips in both positions. The greater quantity of auxins in the lower half of the tips of horizontally oriented primary roots indicates that the auxin migrates to this region under the action of gravity. The downward curvature of the roots is therefore found to be correlated with the greater auxin concentration of the lower as compared with the upper side of the root.

Apical Dominance.   This term refers to the inhibiting or retarding effect of a terminal bud upon the growth of lateral buds, a phenomenon in which auxins play an important role (Chapter 23).

Parthenocarpy.   This term refers to the development of fruits without any associated development of viable seeds. Parthenocarpy occurs naturally in some species and can be induced in others by treatment with certain hormones, including some auxins (Chapter 21).

Abscission.   The separation of leaves, fruits and sometimes older structures from a plant occurs mostly by the process called abscission. Auxins are one of the hormones playing a role in this process (Chapter 23).

Activation of Cambial Cells.   As discussed in Chapter 23, auxins play a role in the resumption of cambial growth of woody plants in the spring and in other aspects of cambial activity. There is a considerable probability that they are involved in cell division generally.

Initiation of Flowers.   Auxins have been shown to enhance flowering of some long day plants (*Hyoscyamus niger* and *Silene armeria*) if applied while such plants are exposed to photoperiods which are almost long enough to be inductive. On the other hand, flowering of some short day plants, such as cocklebur, pigweed, and Biloxi soybean is inhibited if auxins are applied during the inductive dark period (Chapter 22).

It has been known for some years that certain synthetic auxins, of which naphthaleneacetic acid is the best example, stimulate flowering in the pineapple, an effect which has been employed in commercial production of this crop. There is evidence that this effect is actually mediated through ethylene which is generated as a result of the treatment with auxin (see later).

Toxic Effects of Auxins.   Reference has already been made to the inhibitory effect of relatively high concentrations of auxins on the process of cell elongation (Fig. 18.6). Applications of auxins in relatively high concentrations also result in growth malformations in plants such as distortions of leaves, stems, and roots, discolorations of leaves, inhibitions of stem or root elongation or flower opening, and the formation of tumors. It should be emphasized that the term "relatively high concentrations" as used in this discussion refers to concentrations which, in absolute terms, are very low, of the general order of magnitude of 0.1 per cent. In somewhat higher concentrations, but in absolute terms still very low, these compounds often result in death of the plant.

Realization that auxins, when applied in relatively high but actually very low concentrations, exert toxic or lethal effects on plants led to the suggestion that they could be employed for the purpose of killing weeds or other noxious plants. The phenoxyacetic acid auxins have proved especially effective as herbicides. The most widely used of these has been 2,4-dichlorophenoxyacetic acid ("2,4-D"). The effects on plants of this compound may be taken as an example of the action of an auxin used as an herbicide. This compound is readily absorbed from sprays or dusts when applied to leaves. It is quickly translocated to other parts of the plant and affects especially the meristems. The rapid distribution of this compound throughout the plant contributes greatly to its effectiveness as a toxicant. Death results from derangements of metabolism, especially in the meristems. Different kinds of plants differ markedly in their reactions to applications of 2,4-dichlorophenoxyacetic acid. Cereal grains and most other grasses are less susceptible, for example, than most broadleafed annuals, and most woody plants are less susceptible than most herbaceous species. This very selectivity of effect is one of the advantages of this compound, and of many similar ones, as herbicides. Broadleafed weeds can be eliminated from sugar cane fields or from lawns, for example, by application of a spray of "2,4-D" in the proper concentration and at the proper volume per unit area.

Translocation of Auxins.   If a block of agar containing auxin is affixed to the morphologically upper end of a segment of an oat coleoptile, and a block of pure agar to its lower end, auxin will move into and accumulate in the lower block. The final concentration of auxin in the basally attached block may greatly exceed that in the one affixed to the apex. If the agar block containing auxin is affixed to the morphologically basal end, no translocation of auxin will occur (Fig. 18.11). The results of such experiments show that transport of auxin in the oat coleoptile is *polar, i.e.,* occurs only basipetally, and that it can occur against a concentration gradient since the auxin accumulates in the lower block.

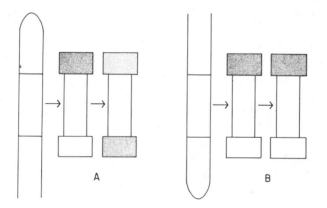

Fig. 18.11. Diagram to illustrate the basipetal movement of auxin: (A) Agar block containing auxin attached to the apical end of segment of oat coleoptile. (B) Agar block containing auxin attached to the basal end of segment of oat coleoptile. After Went, *Botan. Rev.,* **1,** 1935:164.

There is evidence that a similar basipetal translocation of auxin, either naturally occurring, or introduced, occurs in many other elongate plant structures. Among these are the coleoptiles of other species than oat, the veins and petioles of leaves, epicotyls, hypocotyls, and internodes. Such polarized downward movement of auxin occurs in parenchyma or phloem tissue.

In contrast with the translocation pattern of auxins in coleoptiles and other elongate above-ground structures is that in roots. Experiments on the direction of translocation in young root segments employing radioactively tagged IAA show the auxin transport to be highly polarized in the acropetal (base to apex) direction.

Even in structures in which the movement of auxin is predominantly basipetal, some acropetal movement occurs. In the internodes of coleus, for example, about three times as much basipetal movement takes place as acropetal movement.

If an auxin be applied to the roots of entire plants or to basal portions of cuttings it commonly moves up in the transpiration stream. After distribution to various tissues its further movement through the plant may occur along polarized pathways.

The Gibberellins.   The bakanae or "foolish seedling" disease of the rice plant has long been a serious problem in oriental countries. An early symptom of this disease is that infected plants grow substantially taller than those which have not been infected. In due time Japanese plant pathologists discovered that this disease resulted from infection with the fungus *Gibberella fujikuroi.* It was further discovered that organism-free filtrates from the medium on which this fungus had been growing would also induce symptoms of this disease. Eventually a crystalline material was isolated from such filtrates by Japanese investigators in 1938. Solutions of this material, which was named *gibberellin,* were shown to have the same effects on rice plants as the filtrates from the fungus cultures. The originally isolated material eventually turned out to be a mixture of four compounds now termed gibberellin A_1, giberellin A_2, gibberellin A_3 and gibberellin A_4. The best known of these is gibberellin A_3 which is also called gibberellic acid.

Partly because of poor communication during World War II, the discovery of the gibberellins aroused very little immediate interest outside of Japan. After a lapse of about 10 years, however, interest in these compounds awakened and it was soon discovered that they had a number of pronounced effects on the physiological behavior and morphogenic development of vascular plants when applied to them in very small quantities. This led to the inference that gibberellins probably were naturally occurring hormones in the higher plants. Before long this inference was confirmed by the actual isolation of gibberellins from such tissues. Thirty-five different gibberellins, all closely related chemically, have been found to be naturally occurring in higher plants or in the original fungus source or both. Others probably remain to be discovered. Gibberellins have been identified as occurring in a number of species and they are undoubtedly of ubiquitous occurrence, at least in the vascular plants. The concentrations of gibberellins present in plant tissues are, of course, extremely low. To cite one example, one hundred buds from sunflower seedlings yielded only 0.001 μg of a gibberellin. Two or even more gibberellins are often present in the same plant.

Chemistry of Gibberellins.   The gibberellins are structurally a distinctive and closely-knit group of plant hormones. The formula for gibberellic acid (A_3), one of the best known gibberellins, is as follows:

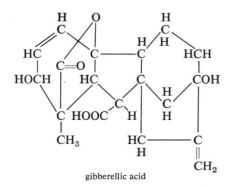

gibberellic acid

All of the known gibberellins, now totalling thirty-five, have this basic structure of four rings fused together, called the gibberellane ring. Each kind of gibberellin has 19 or 20 carbon atoms in the molecule, from 22 to 28 hydrogen atoms, and from 4 to 9 oxygen atoms. The gibberellins differ, one from the other, in the side chain substitutions which have been made in the basic gibberellane ring structure (Fig. 18.12). In addition to the gibberellane configuration, the only structural feature possessed by all the gibberellins in common is a −COOH group at the position shown in the formula for gibberellic acid, although most of them also possess a lactone ring. For convenience the gibberellins are designated A_1, A_2, A_3 . . . . A_n.

Roles of the Gibberellins.   The gibberellins rival the auxins in the multiplicity of growth reactions which they evoke in plants. Only the more thoroughly investigated of these responses will be discussed.

Cell Elongation.   One of the symptoms of the disease which led to discovery of the gibberellins was an undue elongation of the affected rice plants. Gibberellins have since been shown to promote elongation of stems in many other kinds of plants (Fig. 18.13). Treatment of dwarf varieties of peas, maize, and bean with gibberellins in appropriate amounts causes the plants to grow to a stature comparable to standard varieties. One of the symptoms of certain virus diseases of plants, such as aster yellows and corn stunt, is a dwarfing of the affected plants. Treating of such infected plants with a gibberellin causes them to grow to their usual height. The effects of gibberellins described above are all primarily a result of increased elongation of the internodes which in turn results principally from cell elongation, although some cell division may also be involved. Similar effects have been demonstrated on the hypocotyls of several species.

Both gibberellins and auxins play a role in the elongation of plant organs although their relative significance in terms of effective concentrations and degree of interaction with each other differs from one plant organ to another. Spraying intact plants of many kinds with a suitable concentration of gibberellin results in increased stem elongation as described previously. Spraying

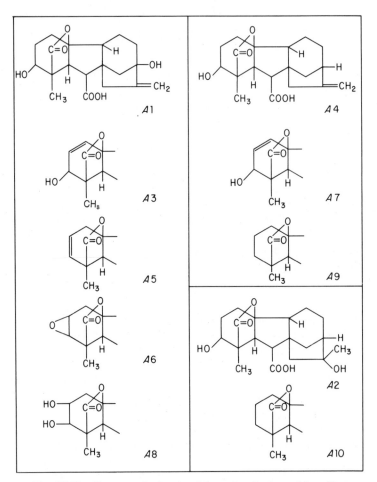

Fig. 18.12.   Framework structural formulas for ten gibberellins.

intact plants with an auxin, on the other hand, has little or no effect on stem elongation. However, in suitable concentrations, auxins cause an increase in the lengths of excised stem, coleoptile, hypocotyl and epicotyl segments when immersed in a suitable medium. Gibberellins, on the other hand, usually have little effect on the elongation of such organ segments.

However, when sections of dwarf pea stems are treated successively with GA_3 ($10^{-4}M$) and IAA ($5 \times 10^{-5}M$) they elongate much more than when treated with either of these two hormones individually. Such synergistic effects (see the following) probably also operate in the elongation of other plant organs.

Effects on Flowering.   Some of the most noteworthy effects of gibberellins are upon the flowering of plants. Some species of plants flower only under

Fig. 18.13. Comparative growth and flowering of bean 31 days after planting. Left, untreated plant. Right, growing tip of plant treated with 20 μg of gibberellins after the primary leaves unfolded. Photograph from Michigan State University.

long days (Chapter 22). Many, although not all, such species have been induced to flower when growing under short days by treatment with a gibberellin. Some of the long-day plants of which this is true are black henbane, petunia, spinach, radish, lettuce, endive, mustard, and dill. In such species a gibberellin substitutes for long-day conditions.

Short-day plants are species which require a minimum number of hours of darkness out of each 24 hour day in order for flowers to be initiated (Chapter 22). Treatment of short-day strawberry plants with gibberellic acid while the plants are exposed to short days inhibits their flowering. A similar effect has been found in some other short-day species although it is not certain how widely this is true of short-day species.

Some species of plants require exposure to a period of relatively low temperatures in order for flowering to be induced. This is true of temperate zone biennials which develop only vegetatively, usually as rosettes, during the first summer of their growth and then, after exposure to low winter temperatures, flower during the second and last summer of their existence. Such species can also be induced to flower by artificially exposing them to low temperatures toward the end of their first season's growth. A number of such biennials, among which are cabbage, collards, turnip, carrot, foxglove, stock, and pansy, have been induced to flower, once sufficient growth had been established, by treatment with a gibberellin, during the first season of their development. In other words, gibberellin treatment substitutes for a cold treatment in inducing blooming in such species.

Parthenocarpy.   Some gibberellins, have, like some auxins, been shown to induce parthenocarpy (Chapter 21) in a number of kinds of fruits, including those of tomato, apple, grape, cherry, and fig, but these two kinds of hormones are not equally effective in evoking this reaction in a given kind of plant. In some varieties of apple, for example, gibberellins (especially A_4) are effective in inducing parthenocarpy, but auxins are completely ineffective.

Root Elongation and Formation.   In suitable concentrations some gibberellins favor root *elongation* in at least some species. On the other hand, *formation* of roots on cuttings is inhibited by treatment with gibberellins and even auxin-enhanced root formation is counteracted by gibberellins.

Leaf Expansion.   When the seedlings of dicots develop in the dark the leaf blades expand only slightly. A short exposure of such seedlings to light brings about complete or nearly complete expansion of the leaf. Wavelengths in the short red region of the spectrum are most effective in evoking leaf expansion (Chapter 22). When small discs cut from leaves of dark-grown bean plants are floated on a dilute solution of gibberellic acid in the dark, considerable expansion occurs as compared with similar discs floated on distilled water. Simultaneous irradiation with red light results in a somewhat greater expansion of the leaf discs than treatment with gibberellin alone. In other words a gibberellin can substitute at least in part for red light induction of leaf expansion.

Seed Germination.   Seeds of a number of kinds of plants require exposure to light after they have absorbed water before germination will take place, red light being the most effective part of the spectrum (Chapter 24). When seeds of such a "light sensitive" variety of lettuce were treated with a dilute (100 mg/l) solution of gibberellic acid in the dark, increase in the germination percentage over the dark controls in water was as great as that obtained by exposure of water-soaked seeds to red light. Similar results have been obtained with some other kinds of light-sensitive seeds. In other words, a gibberellin substitutes for red light in breaking the dormancy of such seeds.

Breaking of Bud Dormancy.   The breaking of bud dormancy in some species of temperate zone woody plants is under photoperiodic control. In nature the buds of such species, of which beech is a good example, do not begin to proliferate in the spring until the photoperiods have attained sufficient length. However, if beech twigs be sprayed with a dilute solution of gibberellic acid while still under short days, bud development can be induced (Fig. 18.14). Similar results have been obtained with at least some other species of woody plants which require relatively long photoperiods for the breaking of bud dormancy.

Fig. 18.14. Effect of gibberellic acid on the opening of buds of beech *(Fagus sylvatica)* in early spring (March). Left, treated with gibberellic acid; right, untreated. Photographed 30 days after treatment; photograph from F. Lona.

The preceding discussion has dealt with gibberellins in general terms without always attempting to distinguish critically any differences in the effects of one gibberellin as compared with another. Much work was done before it was realized how many different kinds of gibberellins would be found to be constituents of plants. Investigators sometimes used mixtures of gibberellins and gibberellic acid (A_3), which because of its earlier availability, for a long while was the most widely used hormone in this group. Many differences exist, however, in the effectiveness of different gibberellins in promoting a given growth response (Table 18.1). For discussion of the effect tabulated in the extreme right column of this table see Chapter 24.

The effective concentrations of gibberellins when applied exogeneously to plants are, of course, very low. Because of the different methods of application used (solution, spray, dispersed in lanolin) and variations in the timing of the application (continuous or intermittent), it is difficult to designate an effective range of concentrations. For many plant responses concentrations of gibberellins in the range of 0.1 to 10 mg/1 are effective.

Synthesis and Translocation of the Gibberellins. The main sites of synthesis of the gibberellins in the higher plants are in meristematic leaves, root

TABLE 18.1 APPROXIMATE RELATIVE ACTIVITY OF GIBBERELLINS A_1–A_{15} AND A_{17}–A_{27}[a]

Gibberellin	Elongation dwarf pea epicotyl	Elongation dwarf maize 2nd leaf sheath	Elongation lettuce hypocotyl	Elongation cucumber hypocotyl	Elongation dwarf rice leaf sheath	α-amylase synthesis in barley aleurone layer
A_1	+++	+++	+++	++	+++	++++
A_2	++	+	++	++	+++	++++
A_3	+++	++++	+++	+++	+++	++++
A_4	+++	+++	+++	++	++	+++
A_5	++	++	+++	++	+++	++
A_6	+++	++	+++	+	+++	++
A_7	+	+++	+	+++	+	+
A_8	++	0	++	0	++	+
A_9	+	0	0	+++	++	+
A_{10}	0	0	0	++	+++	+
A_{11}	0	0	0	+	+	+
A_{12}	0	+	0	0	+	0
A_{13}	0	+	+	0	+	+
A_{14}	0	+	0	0	+	0
A_{15}	+	0	+	+	+	0
A_{17}	0	+	0	0	+	0
A_{18}	++	+	+++	0	+++	+
A_{19}	0	+	+	0	++	0
A_{20}	+	+	++	0	+	0
A_{21}	0	+	+	0	0	+
A_{22}	++	++	+	0	+++	+++
A_{23}	++	++	0	++	+++	++
A_{24}	+	+	0	+	+++	0
A_{25}	0	0	0	0	0	0
A_{26}	0	0	0	0	0	0
A_{27}	+	0	0	0	+	0

Relative activities: ++++ very high, +++ high, ++ intermediate, + low, 0 very low or inactive.
[a] Adapted from Crozier, et al. Can. J. Botany, **48**, 874, 1970.

tips, and developing seeds. One of the main facts known about the biosynthesis of the gibberellins is that an important intermediate in the synthesis of these compounds is the diterpene kaurene (Chapter 14).

Translocation of gibberellins occurs quite freely in both the xylem and the phloem. Interchange of gibberellins between the two conductive systems also occurs, probably by way of the phloem and xylem ray cells.

The Cytokinins. Hormones with a basically purine ring structure are called *cytokinins* (formerly *kinins*). The first such compound with hormonal properties to be isolated from natural sources was *kinetin,* the chemical name of which is 6-furfurylamino-purine (Fig. 18.15). This compound was isolated from degraded DNA obtained from herring sperm. Although it has many effects on plants it is not known to be a natural plant constituent and is probably not a true plant hormone. Eventually a naturally occurring cytokinin was isolated from *Zea mays* grains, and appropriately named *zeatin* (6-(4-hydroxy-3-methyl-2 *trans*-butenylamino) purine) (Fig. 18.15). Zeatin has since been found in other plants, including a puffball fungus *(Rhizopogon roseolus)*. It is the most active of the known cytokinins, exerting some of its promotive effects in concentrations at low as $5 \times 10^{-11}M$. Several other naturally occurring compounds, closely related to zeatin chemically, have been found to exhibit cytokinin activity (Fig. 18.15). Cytokinins are widely and probably ubiquitously distributed in higher plants.

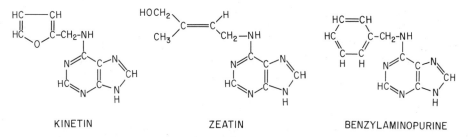

KINETIN   ZEATIN   BENZYLAMINOPURINE

Fig. 18.15. Structural formulas of cytokinins.

A large number of other compounds with an adenine nucleus, not known to occur in plants, have been found to influence plant development in much the same way that kinetin does. These synthetic cytokinins, of which benzylaminopurine (Fig. 18.15) is an example, occupy an analogous position to the synthetic auxins.

Roles of the Cytokinins. As is true of both the auxins and the gibberellins, the cytokinins evoke many growth responses in plants, only the more thoroughly investigated of which will be discussed.

Cytokinesis.   This effect of cytokinins, the one from which they derive their name, was the first to be recognized. Cytokinins induce cell divisions in many kinds of tissue cultures when present in extremely low concentrations and in the presence of IAA or a synthetic auxin (Fig. 18.16). Cytokinins also appear to be necessary for the correlated phenomena of mitosis and DNA synthesis.

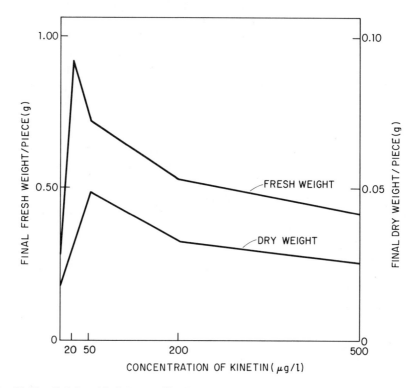

Fig. 18.16.   Relationship between kinetin concentration and the fresh and dry weight increase of tobacco callus tissue over 31 days as indices of cytokinesis. Data of Miller, *et al. J. Am. Chem. Soc.* **78,** 1956:1375.

Cell Enlargement and Elongation.   Kinetin promotes the expansion of discs cut from etiolated bean leaves in much the same manner that gibberellins or exposure to red light does. At least some of the other cytokinins have similar effects on leaves of a number of other species. In bean seedlings, treatment with kinetin promotes elongation of stems and petioles. On the other hand elongation of segments of pea stems or sunflower hypocotyls is inhibited by treatment with this compound. Most treatments of roots with cytokinins have resulted in a retardation in the rate of elongation, but enhancement in their rate of elongation has been observed in some kinds of roots when the concentration used was very low. In brief, cytokinins influence the enlargement or elongation phases of growth, but whether the effect is one of enhancement or retardation depends upon the organ involved, the specific kind of cytokinin, and its concentration.

Bud Initiation and Proliferation.   When small slabs of the internal tissues of a tobacco stem are cultured on a suitable medium under sterile conditions a callus tissue develops. Such callus tissue remains in the undifferentiated state indefinitely and through many subcultures if a proper ratio of kinetin to auxin is maintained. If, however, the proportion of kinetin to auxin is increased, buds and resultant leafy shoots are initiated. In sterile cultures developed from small fragments of lettuce leaves kinetin has a similar effect on bud development, but only in the presence of auxin (Fig. 18.17). Bud and leafy shoot formation have also been induced by treatment with kinetin of various plant organs such as isolated roots, petioles, and leaf cuttings.

Enhancement of shoot branching has also been induced by kinetin treatments in intact plants of a number of species. A probably related phenomenon

Fig. 18.17.   Small fragments of lettuce leaves in sterile culture: left to right: (1) basal medium only, (2) with 1 mg/l kinetin, (3) with 2 mg/l IAA, (4) with 1 mg/l kinetin plus 2 mg/l IAA. Photograph courtesy of C. O. Miller.

is the offsetting of auxin-induced inhibition of lateral bud development by treatment with cytokinins, an hormonal interrelation which has important implications for the phenomenon of apical dominance (Chapter 23).

Root Formation.   In cultures of tobacco callus tissue and in cultures developed from lettuce leaf fragments (Fig. 18.17), cytokinins in any appreciable concentration inhibit root formation and offset the tendency of any auxin present to promote the formation of roots. The effect of cytokinins on root formation parallels that of gibberellins. There is some evidence, however, that in extremely low concentrations cytokinins may favor root initiation.

Delay of Senescence.   Leaves of many species lose their chlorophyll and become yellow if floated upon water for a few days. If, however, kinetin at a relatively low concentration is present in the water, the leaves retain their chlorophyll and hence their green coloration much longer. In other words, kinetin delays the onset of senescence in the leaves, and cytokinins in general have a senescence-delaying effect on plant tissues.

Breaking Dormancy of Seeds.   In common with gibberellins and red light, kinetin in a proper dosage has the property of breaking the dormancy of light sensitive seeds of lettuce. Various other cytokinins have been found effective in breaking the dormancy of a number of kinds of seeds, not all of which are light sensitive.

Parthenocarpy.   A synthetic cytokinin (6-benzylamino-9-tetrahydropyran-2yl) purine) has been shown to be very effective in inducing parthenocarpy in the Calimyra variety of fig. It is noteworthy that hormones of both the auxin and the gibberellic groups also induce parthenocarpy in this species. It seems probable that parthenocarpic development of other kinds of fruits can be induced by treatment with cytokinins.

Effects on Flowering.   A few examples of the enhancement of flowering by treatment with cytokinins are known. Flower formation in short-day *Perilla* and *Chenopodium rubrum* has been evoked by treatment with kinetin under long days. Contrariwise, long-day mouse-ear cress *(Arabidopsis thaliana)* can be brought into bloom under short days by treatment with kinetin.

Synthesis of Cytokinins.   The natural cytokinins appear to be made principally in apical root meristems, inflorescences, and developing fruits. Certain cytokinins have been found to be constituents of certain transfer-RNA molecules in a number of different organisms, a provocative finding that may lead to some understanding of the basic role of the cytokinins.

When first discovered, cytokinins were believed to be relatively immobile in plant tissues but it is now clear that they are translocated in plants. There is evidence that cytokinins synthesized in root apices are translocated through the xylem to other parts of the plant.

Abscisic Acid.   Investigations of the abscission of immature cotton bolls led to the postulation that this phenomenon is caused by a hormone. In due time the effective compound, at first called *abscisin,* was isolated from such bolls and identified chemically. Investigations of an entirely different phenomenon, the induction and maintenance of dormancy in the buds of some woody species, led to the postulation that such effects were under the control of a hormone originating in the leaves. Eventually an effective compound, at first called *dormin,* was isolated from the leaves and identified chemically. Interestingly enough, abscisin and dormin turned out to be the identical chemical compound which is now called *abscisic acid* (3-methyl-5-(1'-hydroxy-4'-oxo-2',6',6'-trimethyl-2'-cyclohexen-1'-yl)-*cis, trans,*-2,4-pentadienoic acid). As with other plant hormones only minute quantities of this compound are present in plant tissues:

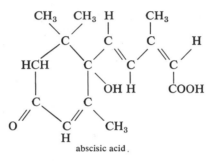

abscisic acid.

Abscisic acid (ABA) is of widespread and probably universal occurrance in the higher plants. It has been found in many different organs and tissues, including both young and old parts of plants. Several chemically related compounds, some naturally occurring and some synthetic, are known which have growth-regulating properties similar to this compound.

Roles of Abscisic Acid.   Like the other plant hormones previously discussed this compound plays roles in a number of different growth responses, some of which are probably as yet unrecognized. A characteristic of ABA is that many of its effects are relatively transient and it must be present in continuous supply in order to exert them. An inference from this is that this compound may be readily inactivated in plant cells.

Promotion of Abscission.   Investigation of the causes of abscission in young cotton fruits was one of the experimental routes which led to the discovery and isolation of abscisic acid. It has since been shown to induce abscission of the leaves of a wide variety of plants and of the fruits of a number of species. Other hormones, however, especially IAA and ethylene, interact with ABA in the control of abscission (Chapter 23).

Promotion of Bud Dormancy.   The buds of temperate zone woody plants pass into a state of dormancy during the late summer and early autumn. The main environmental factor inducing dormancy appears to be the shortening photoperiods during that season, but such a factor must of course work through an internal mechanism. When ABA was applied continuously to the leaves of young maple, birch, and currant under long photoperiods, cessation of extension growth and formation of resting buds with typical bud scales was induced. The ABA was obviously translocated from the leaves to the stem tips. Abscisic acid thus appears to be an internal factor inducing dormancy in the buds of at least some temperate zone woody plants.

Another example of this effect of ABA is shown by excised buds from potato tubers which remain dormant indefinitely if treated with as little as 0.006 μg of this compound per day.

Inhibition of Seed Germination. Abscisic acid prevents or delays the germination of many kinds of seeds, an effect which is equivalent to a prolongation of seed dormancy. It appears to be the germination inhibitor which is naturally present in some kinds of seeds, and may be the substance which, in some fruits, prevents germination of the seeds while they are still in the fruit. Abscisic acid also offsets the previously described promotive effect of gibberellins or cytokinins on lettuce seed germination.

Growth Retardation and Senescence. Abscisic acid retards the growth of a large variety of plant tissues and organs including leaves, coleoptiles, stems, hypocotyls, and roots. Retardation of stem growth involves development of shorter than normal internodes. Abscisic acid also inhibits or retards growth of plant tissues in culture.

A related effect is that of the senescence of plant organs (abscission can be considered to be one aspect of senescence). A representative example is the promoting effect of abscisic acid on degeneration of excised leaves. In isolated leaf disks ABA accelerates the decomposition of chlorophyll. Such a chlorophyll breakdown may occur even in the presence of a cytokinin which has a senescence-delaying effect.

Promotion of Root Formation. Abscisic acid has been found to promote root formation on cuttings of at least several species. Furthermore, treatment with ABA offsets, at least partially, the inhibitory effect on root formation of pre-treatment with gibberellic acid.

Effects on Flowering. Abscisic acid inhibits flower induction in long day rye grass *(Lolium temulentum)* (Chapter 22) when applied in amounts as small as 0.1 μg while the plants are under long days. The minute quantity required indicates that this is a true hormonal effect. It is not yet known whether or not ABA exerts such an effect on other long-day plants.

The reactions of short-day plants to treatment with abscisic acid have not been very consistent. Flowering of several such species, Japanese morning glory *(Pharbites nil),* pigweed *(Chenopodium rubrum),* black currant and strawberry, is induced by applications of ABA under long-day conditions. Attempts to induce flowering in some other short-day species such as Biloxi soybean, cocklebur, and the Maryland Mammoth variety of tobacco have not been successful. This promoting effect of ABA may not be as specific with short-day plants as its effect in inhibiting flower induction in long-day plants. The amount of hormone required to influence flowering seems to be much greater in short-day plants than in long-day plants. The promoting effect on the flowering of some short-day species may basically be one of retardation of growth which in itself often favors initiation of flowering.

Vitamins. Some of the substances classified as vitamins in animal physiology operate in a hormonal capacity in the higher plants. This is especially true of many of the so-called B group of vitamins. Although the higher green plants, unlike animals, synthesize all of the vitamins necessary for maintenance of their metabolic processes, not every tissue of a given plant necessarily synthesizes them in adequate quantities. This fact is clearly shown in sterile cultures of excised roots (Chapter 19) which can be grown indefinitely, with sub-culturing from time to time, in solutions containing only sucrose and mineral salts plus certain vitamins. All species of roots appear to require the presence of thiamin in the medium, and some require niacin and pyridoxine in addition. In intact plants these substances are synthesized in the leaves or other aerial organs, whence they are translocated to the roots.

The metabolic roles of the vitamins are better understood than those of other hormones. All vitamins which show a hormonal pattern of behavior in plants, as well as some other vitamins which do not appear to be hormonal in nature are constituents of important enzyme systems in plants (Chapter 9), and as such they are effective at very low concentrations.

Ethylene. It has long been known that ethylene ($CH_2{=}CH_2$) gas is synthesized in and released from plant tissues and that exposure to ethylene engenders certain growth reactions and physiological responses in plants. Only more recently has it been realized, largely because of improved methods of detecting the very small quantities of ethylene present in plant tissues, that this compound has all of the characteristic properties of a hormone. It is synthesized in and is present in many plant tissues. It exerts marked physiological effects at very low concentrations. Within limits, at least, ethylene is translocated from one part of a plant to another. The one feature that distinguishes ethylene from other known plant hormones is that it is a gas at physiological temperatures. As a result of its gaseous state ethylene exhibits certain properties not shown by other hormones. When present in excess it escapes readily from plant tissues. Also, ethylene from one plant source may move to and influence the development or physiological reactions of other nearby plants.

Effects of Ethylene. The following discussion of some of the principal effects of ethylene on plants includes some which are clearly hormonal and others which are not, at least in the present state of our knowledge. Only those physiological phenomena in which endogenously generated ethylene is known to participate can be regarded as truly hormonal.

Fruit Ripening. Many years ago it was observed that when a ship's cargo consisted partly of green bananas and partly of ripe oranges that the bananas ripened unusually fast. This led to the discovery that the active banana-ripening agent was ethylene which escaped from the oranges. In many, perhaps all, kinds of fleshy fruits, such as apples, oranges, bananas, mangoes, cantaloupes and squashes the ethylene content is very low while they are young. As the fruit enlarges, however, the synthesis of ethylene accelerates and measurable amounts of the gas are released. Associated with this, the rate of respiration continues to increase in some species until a peak value called the climacteric (Chapter 21) is reached. Ethylene in effect operates as a fruit-ripening hormone. Treatment of such fruits as oranges and bananas when mature, but not yet fully ripened, with ethylene hastens their ripening. Such treatments have found commercial applications.

Abscission.   One of the earliest noticed effects of ethylene on plants was that its presence in the atmosphere in relatively low concentrations promoted abscission. Endogenously generated ethylene plays a role in normal abscission processes which appears to involve interactions between ethylene and other hormones, especially auxins and abscisic acid (Chapter 23).

Epinasty.   The leaves of many species exhibit epinasty (downward bending) when exposed to very low concentrations of ethylene. The concentration required to induce epinasty is much less than that required to bring about leaf abscission. The leaves of tomato, for example, soon show epinasty after exposure to concentrations of ethylene as small as 0.1 part to one million of air (Fig. 18.18).

Fig. 18.18. Epinasty in tomato resulting from exposure to ethylene. Plant on the right was exposed to 0.1 part per million of ethylene for 48 hours; plant on the left was kept under the same conditions in an ethylene-free atmosphere. Photograph from Crocker *et al. Contrib. Boyce Thompson Inst.,* **4,** 1932:198.

Transverse Geotropism. When seedlings such as those of pea or bean are exposed to ethylene at atmospheric concentrations as low as 0.06 ppm the shoots which normally grow vertically (*i.e.* exhibit negative geotropism) assume a horizontal growth direction (*i.e.* become transversely geotropic). Accompanying morphogenic responses include reduced elongation rate of the stem, thickening of the stem, and inhibition of bud expansion. It is not known whether ethylene also plays a part in controlling transverse geotropism in such organs as stolons and rhizomes which normally are so oriented.

Flowering in Pineapple. Field observations that smoke from brush fires enhanced the flowering of pineapples led to the discovery by more critical experiments that ethylene was the active flower-inducing ingredient in the smoke. As far as is known, ethylene does not have this effect on any other kind of plant. As previously mentioned auxin-evoked flowering in this plant is mediated by ethylene synthesized as a result of the auxin treatment.

Other Hormone-like Substances in Plants. The hormones or hormone groups already discussed have all been well characterized as such. Many other substances are known to occur in plants which have promotive or inhibitory effects, or both, on various plant processes or responses. Some of these substances have been tested only in the form of crude extracts, others have been identified chemically. Of various examples of such substances which could be cited we will mention only coumarin which belongs to a group of compounds known as the unsaturated lactones:

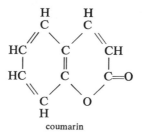

coumarin

The unsaturated lactones are characterized by the group $\overset{\displaystyle O}{\overset{\displaystyle \|}{C}}-O-\overset{\displaystyle O}{\overset{\displaystyle \|}{C}}-\overset{\displaystyle |}{C}=\overset{\displaystyle |}{C}-$.

A number of such compounds are known to occur in plants. Coumarin is of especially widespread occurrence in plants and exerts hormone-like effects when present in very low concentrations. It inhibits the germination of some kinds of seeds and retards the growth of roots in some species. On the other hand it is promotive of hypocotyl growth and leaf expansion in at least some kinds of plants. Other compounds with an unsaturated lactone structure have more or less similar effects on plants.

Mechanism of Hormone Action. Much investigative effort has been expended in attempting to fathom the mechanisms of operation of the plant hormones, especially the auxins. Although some leads have been obtained into this realm of knowledge, definitive information regarding the mechanism of plant hormone action is very sparse. Interpretations of this aspect of plant metabolism are complicated by interactions among the hormones as discussed in the next section. Only a few general comments on this topic are appropriate in an introductory treatment.

It is axiomatic that hormones must exert their effects by reacting at one or more points in the metabolic system, which would involve effects on either the synthesis or operation of enzymes. There are indications that some of them may operate at one stage or another in the sequence of metabolic events starting with DNA synthesis and leading to enzyme synthesis (Chapter 16). Some auxins, however, show an effect upon the elongation of coleoptiles or stem segments within a time span of 10 minutes or even less, and this may be true of some other hormones. This rapidity of action is not consistent with the concept that hormones act at a stage in the DNA-to-enzyme synthesis series of metabolic events which are believed to require a greater time interval than 10 minutes. The other possibility is that hormones are reactive at stages in metabolism which are somewhat remote from those which lead to enzyme synthesis.

In attempting to unravel the mechanism of hormone operation at the metabolic level, it is necessary to account for the fact that one kind of hormone often evokes a number of different end responses in plants. One possible interpretation of such a physiological pattern is that the hormone participates in some "master reaction" through which a number of subsequent reactions leading to different end responses are channeled. An alternative, but seemingly less probable interpretation, is that the hormone molecule is biochemically so versatile that it can participate in a number of different reactions within the metabolic system.

Interactions among Hormones. A number of examples of interactions among plant hormones have been described or mentioned on preceding pages and it is probable that no specific physiological response can be ascribed to the influence of only one kind of hormone. Such interactions may be synergistic, in which one hormone reinforces the effect of another, or antagonistic, in which one hormone offsets or negates the effect of another. Good examples of synergistic effects, previously mentioned, are the requirement of both an auxin and a cytokinin for cell division in tissue cultures and the requirement of both a gibberellin and an auxin for internode elongation. Good examples of antagonistic effects, previously mentioned and also discussed more fully in Chapter 23, are the counteracting of lateral bud inhibition by an auxin with a cytokinin, and the offsetting of the abscission-delaying effect of IAA with abscisic acid.

An especially clear-cut example of the interacting effects of hormones has been demonstrated in a species of duckweed *(Lemna minor),* a small, free-floating, aquatic angiosperm. Addition of ABA to make a very low concentration (1–2 mg/l) in the aqueous medium causes a drastic reduction in the growth rate of this plant. If, however, the synthetic cytokinin, benzyladenine, be added to the medium to make a concentration of about 0.1 mg/l, the growth-inhibiting action of the ABA is largely offset.

As indicated by some of the examples cited above, a given process or growth phenomenon is often influenced by one or more hormones which are inhibitory to its manifestation and one or more which are promotive. The actual outcome in terms of physiological behavior will be a resultant of the relative effectiveness of these two classes of hormones within the cells or tissues involved. It is a reasonable presumption that many of the physiological phenomena occurring in plants, not now known to be positively under hormonal control, are governed largely by the hormonal pattern prevailing within the metabolic systems involved. Such hormonal patterns must be regarded as being primarily under genetic control. However, the proportionate distribution of the various hormones in plant cells can be altered by shifts in environmental conditions, especially changes involving light and temperature. This is clearly one of the mechanisms whereby environmental factors influence the physiological behavior and morphogenic development of plants.

SUGGESTED FOR COLLATERAL READING

Audus, L. J., Editor. *The Physiology and Biochemistry of Herbicides*. Academic Press, Inc., New York, 1964.

Carr, D. J., Editor. *Plant Growth Substances 1970*. Springer-Verlag New York, Inc., 1972.

Galston, A. W., and P. J. Davies. *Control Mechanisms in Plant Development*. Prentice-Hall, Inc., Englewood Cliffs, New Jersey, 1970.

Leopold, A. C. *Plant Growth and Development*. McGraw-Hill Book Company, New York, 1964.

Phillips, I. D. J. *Introduction to the Biochemistry and Physiology of Plant Growth Hormones*. McGraw-Hill Book Company, New York, 1971.

Steward, F. C., and A. D. Krikorian. *Plants, Chemicals and Growth*. Academic Press, Inc., New York, 1971.

Vardar, Y., Editor. *The Transport of Plant Hormones*. North-Holland Publishing Company, Amsterdam, 1968.

Wareing, P. F., and I. D. J. Phillips. *The Control of Growth and Differentiation in Plants*. Pergamon Press, Elmsford, New York, 1970.

Wilkins, M. B., Editor. *Physiology of Plant Growth and Development*. McGraw-Hill Publishing Company, Ltd., Maidenhead, England, 1969.

19 | VEGETATIVE GROWTH

That plants more or less continuously increase in size and develop new organs at least intermittently throughout their life history is one of the most evident of natural phenomena. The term "growth" is popularly employed to designate this complex of processes and in a loose sense, at least, is so employed by botanists. Growth is the one plant process with which few persons are unfamiliar, even if they have never observed it on any larger scale than a potted plant on a window sill. For farmers, horticulturists, foresters, and all others who depend upon the productivity of plants for their livelihood, the phenomenon of plant growth holds the center of the stage of interest.

Assimilation. The dry matter which is incorporated into the structure of both protoplasm and cell walls during growth comes almost entirely from foods. The process whereby foods are utilized in the building of protoplasm and cell walls is called *assimilation*. In the synthesis of protoplasm the foods assimilated are largely nitrogenous, whereas those assimilated in the fabrication of cell walls are almost entirely carbohydrates. The chemical reactions involved in assimilation are principally of kinds in which relatively simple, soluble compounds are converted into complex, usually insoluble constituents of cell systems. The biosynthesis of many of these compounds has been discussed in previous chapters. As a result of assimilation a growing region invariably increases in dry weight during growth.

Apparent exceptions to the principle of increase in dry weight during growth are sometimes cited. A seedling developing in the dark, as described in Chapter 12, although obviously growing, continuously decreases in total dry weight. Even when seeds germinate in the light, the total dry weight of the plant decreases for a period prior to the initiation of photosynthesis in the developing seedling. Such examples, however, only obscure the crucial fact that the *growing region* invariably increases in dry weight during the process of growth. Such an increase in dry weight of growing regions is entirely possible, even when there is a simultaneous decrease in the total dry weight of the plant as a result of an excess of respiration over photosynthesis.

Growth as a Process. Growth is often referred to as a plant process, but more accurately should be thought of as a coordinated system of subprocesses. The most obvious aspect of growth is an increase in size which is accomplished by the usually overlapping processes of cell division and cell enlargement. Following and to some extent overlapping these two processes is the complicated phase of differentiation in which the distinctive kinds of cells characteristic of various tissues develop in coordination with each other. The obvious aspects of differentiation are structural and chemical but they must reflect more deep-seated modifications which occur in the operation of the metabolic mechanism. As discussed more fully later in this chapter and in the next chapter, operation of this mechanism is under the joint control of genetic and environmental factors.

Many authors restrict the term "growth" to the cell division and cell enlargement processes, and distinguish between growth thus defined, on one hand, and differentiation (*i.e.,* structural differentiation), on the other, as separate phenomena. Whenever the term "differentiation" is used without qualification in reference to growth, it can be assumed to refer to structural differentiation. Some term is needed, however, to designate the overall process whereby new organs and their constituent tissues develop. "Growth" is the most natural and suitable term for this process and will be used in this sense in this book. Separate phases of the growth process are designated by the terms "cell division", "cell enlargement" (or elongation) and "cell differentiation." Another term commonly used to designate the overall growth process is *morphogenesis.*

The discussion in this chapter is restricted to the vegetative growth of vascular plants, i.e., to the development of the roots, stems, and leaves. Reproductive growth, the formation of flowers, fruits, and seeds, is considered in Chapter 21.

Meristems. Growth does not occur indiscriminately in all parts of a plant but is initiated only in certain tissues of restricted distribution called *meristems.* A meristem is a tissue in which, under favorable conditions, new cells are more or less continually being formed as a result of repeated divisions of some or all of the cells. As a result of the subsequent further enlargement and structural differentiation of cells originating in a meristem, various tissues are developed according to a pattern which is more or less distinctive for each species.

The most important meristems in the body of the majority of vascular plants are the apical-root (Fig. 7.3) and apical-stem (Fig. 19.2, Fig. 19.3) meristems, and the vascular cambium (Fig. 19.6). An apical-stem meristem is present at every stem tip, including those of quiescent or dormant buds; an apical-root meristem is present at every root tip, and a vascular cambium is present in the stems and roots of gymnosperms and most dicotyledons. Differentiation of these three major meristems occurs very early in ontogeny of an individual plant.

Intercalary meristems also contribute significantly to the growth of some kinds of plants. Such meristems are found in the stems of some species and in the leaves of some species, especially monocots. In stems, intercalary meristems are most commonly located just above the nodes, as in maize and other monocots; but in some species they occur just below the nodes, as in some mints; and in still others they occur in the middle of the internode. Leaves of grasses and many other kinds of monocots have basal intercalary meristems. This fact can be attested to by anyone who has ever pushed a lawnmower. An intercalary stem meristem can be regarded as a portion of an apical-stem meristem which has become separated from the latter by differentiation of the tissues between it and the intercalary meristem. Such meristems ultimately disappear, sooner or later being entirely converted into non-meristematic tissues. Intercalary leaf meristems, however, often persist for an entire growing season.

Growth which is initiated in apical-stem and root meristems is called *primary growth*. Primary growth results in the construction of the *primary tissues* of a plant, accounts for all increase in length of the plant axis at both stem and root tips, and is responsible for the formation of lateral appendages such as branch stems, leaves, floral parts, and branch roots.

In many species the primary tissues constitute the entire plant. This is true of the pteridophytes and of many monocotyledons. In gymnosperms and most dicotyledons, however, stems and roots not only grow more or less continuously by a proliferation of the fundamental tissues of which these organs are composed, but also increase in diameter as a result of the activity of the vascular cambium.

Some increase in diameter may occur, however, even in stems and roots which do not possess a vascular cambium. Increase in the girth of a young stem as a result of primary growth may continue for some time after increase in length of that region of the stem has ceased, as a result of a slow continuance of cell division and enlargement in some of the tissues, particularly those near the periphery. In species in which the primary tissues constitute the entire body of the plant, this is the sole mode of growth in diameter.

Tissues which arise as a result of the growth activity of vascular or cork cambiums are called secondary tissues (see the following).

Dynamics of Primary Growth.   *Primary Growth of Roots.*   Examination of a median longitudinal section through a root tip (Fig. 7.3) shows that the region just back of the root cap is one of intense meristematic activity. The cells throughout the meristematic region are relatively small, thin-walled and approximately isodiametric. The vacuoles are minute, and the nuclei are relatively large, although in absolute size they are smaller than in mature cells. Intercellular spaces are lacking. Cell divisions are of frequent occurrence in this region and are accompanied by some enlargement of the cells.

All new cells formed in the growth process at a root tip arise from certain of the apical cells called *initials*, either directly, or after intervening cell divisions. One of the daughter cells resulting from each division of an initial retains its position and identity as an initial. From several to many generations of cell progeny may arise from the other, but ordinarily all of the ultimate offspring of this daughter cell become differentiated into cells characteristic of the tissue of which they become a part. The cells of the root cap are also formed from the initials. Cell divisions in an apical-root meristem occur both longitudinally and in variously oriented lateral planes, thus giving rise to the typically cylindrical configuration of a root.

Just back of the meristematic zone is a short region (seldom more than a millimeter in length) in which elongation of cells continues, but in which cell divisions have ceased. Enlargement or elongation of cells is an integral part of the growth process. Some enlargement occurs during cell division, but usually the increase in the size of cells occurring during this phase of growth is small compared with that taking place subsequently. Increase in volume of the cells as a result of the stretching of the cell wall caused by the turgor pressure developed within the cell. This increase in the volume of all cells does not occur equally, neither do they all enlarge symmetrically along all axes. Hence cells of very diverse sizes and configurations arise in plant tissues.

Since many of the cells originating at an apical-root meristem enlarge principally in a direction parallel to the long axis of the root, this phase of growth is often called *cell elongation*. The continued formation of new cells by division and their elongation result in projecting the root tip forward, which is one of the most obvious manifestations of its apical growth.

Differentiation of some of the kinds of cells in a growing root axis occurs much closer to the apex than others. Particular interest attaches to the stage in the growth of a root at which the vascular tissues develop, because of the part played by them in translocation processes. In roots both xylem and phloem differentiate only acropetally and as continuations of the older xylem and phloem in the more basal part of the root. In some root tips, developing sieve tubes are present within less than 200 μ of the apex of the meristem, which is well within the region in which cell divisions are occurring. First differentiation of xylem tissues typically occurs farther back from the apical initials than first differentiation of phloem tissues (Fig. 19.1).

The first phloem elements to form in a root tip are usually crushed or distorted by the continued division and elongation of surrounding cells. Likewise, the first vessels to form are usually stretched and often torn or destroyed as a result of the continued elongation of surrounding cells.

Primary Growth of Stems. The meristematic region at a stem tip is fun-

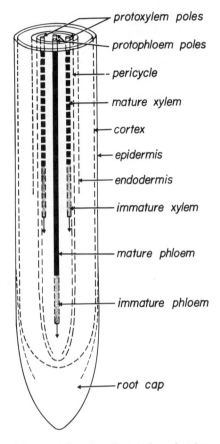

protoxylem poles

protophloem poles

pericycle

mature xylem

cortex

epidermis

endodermis

immature xylem

mature phloem

immature phloem

root cap

Fig. 19.1. Diagram of the root tip of a dicotyledon showing origin of the first vascular elements. After Esau, *Botan. Rev.,* **9,** 1943:165.

damentally similar to that at a root tip, but its behavior is more complicated since growth at a stem tip involves not only proliferation of the stem axis, but also the formation of leaves and other lateral organs which are structurally quite different from the stem (Fig. 19.2). The youngest and most apical portion of a stem tip is not more than a few hundred microns in length (Fig. 19.3).

The cells in a stem-tip meristem have thin, delicate walls and are approximately isodiametric. During active growth, cell divisions are of frequent occurrence through this region. The mass of cells in the meristematic region appears to be in a highly plastic condition, and its continuity is not interrupted even by minute intercellular spaces. The nuclei are relatively prominent, although actually they are smaller than in mature cells. The cells in the younger portions of apical-stem and other meristems contain only minute vacuoles. The cells in older portions of an apical-stem meristem show evidence of simultaneous occurrence of division and vacuolation.

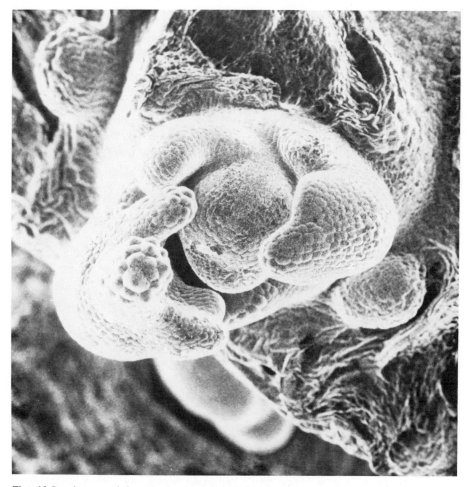

Fig. 19.2. Apex and four leaf primordia of the vegetative shoot tip of nasturtium, as shown under the scanning electron microscope. From Falk, Gifford, and Cutter, *Science,* **168,** No. 19, June 1970:1471–1472. Copyright 1970 by the American Association for the Advancement of Science.

As in roots there is present at or near the apex of a stem one or more initials. All new cells formed at any growing stem tip originate from these persistently meristematic cells. Cell divisions in apical-stem meristems occur both transversely and in various tangential planes, thus giving rise to the more or less cylindrical configuration of the typical stem.

As in root-tip meristems, some enlargement of cells occurs during the cell division phase of growth, but in the development of many kinds of cells most enlargement or elongation occurs after cell division. As in roots, cells of very different shapes and sizes arise as a result of differential enlargement. Many cells become distinctly elongate in shape during this phase of growth. During

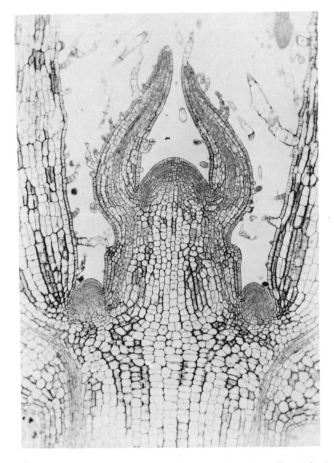

Fig. 19.3. Longitudinal section of the apical-stem tip of coleus. The apical meristem appears between the two first, only partly developed leaves. Just below, appearing as shoulders on the sides of the stem, are bases of the second pair of leaves, which are borne oppositely to the first pair. Farther below is the third pair of leaves, in the axils of each of which is a young lateral bud. The distance from the top of the apical meristem to the base of the section is 0.86 mm. Photomicrograph by Tillman Johnson.

enlargement of cells in an apical-stem meristem, intercellular spaces may be formed. Both the continued formation of new cells by division and their elongation contribute toward the lengthening of a stem, which is one of the most obvious aspects of apical stem growth.

Cell division in stems is generally restricted to the uppermost internodes, but the elongation of cells often extends over a long series of internodes. The rate of elongation becomes progressively slower, however, with increasing distance of the internode from the stem tip. The elongation region back of a stem tip is sometimes as much as 10 cm in length, and in twining plants even longer.

Investigators of the differentiation phase of growth in apical-stem meristems have focused most of their attention on the development of the vascular tissues. The first vascular strands to appear in stem apices originate in connection with the developing leaf primordia. Differentiation of procambium, in continuity with the procambium of the older bundles in the axis, and acropetally into the leaves, precedes differentiation of the vascular tissues. As in roots, phloem tissues usually differentiate earlier in the ontogeny of a stem than the xylary elements of the same strand (Fig. 19.4). The first xylem generally appears at the base of the leaf primordium or in the primordium itself, whence it differentiates both acropetally into the leaf primordium and basipetally into the stem axis until it unites with older strands of xylem. Differentiation of the phloem, on the other hand, appears to occur only acropetally into each leaf primordium as a continuation of the older phloem in the stem axis. Crushing or distortion or stretching of the first phloem or xylem elements formed in a stem tip may occur in much the same manner as previously described for a root tip.

Physiological Aspects of Cell Division.   The portion of a meristem in

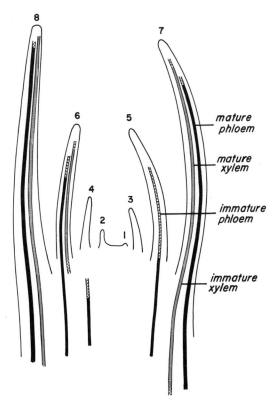

Fig. 19.4.   Diagram of an apical meristem of tobacco, showing the order of development of the leaves and the position of first sieve tube and first xylem vessel in each leaf. After Esau, *Botan. Rev.,* **9,** 1943:155.

which cell divisions are of frequent occurrence is a center of intense assimilatory activity. Since every cell formed as a result of cytokinesis contains its own complement of protoplasm, and since every cell division, with only a few exceptions, involves the formation of a cross wall between the two daughter cells as well as some extension of existing walls, both carbohydrates and nitrogenous compounds are assimilated in relatively large quantities during cell division.

Utilization of water in the hydration of newly formed protoplasm and cell walls, and to a limited degree in vacuolation, also occurs during cell division.

All of the essential mineral elements must be present, as consituents of one kind of compound or another, in meristematic cells if their physiological activity is to continue unimpaired. Some of these are translocated to meristematic cells as simple inorganic salts, others as components of more complex organic compounds.

During cytokinesis, therefore, a continuous translocation of water and various kinds of solutes is in progress toward the dividing cells. Significant in this connection is the fact that the sieve tubes, through which at least part of the solutes are presumably transported into a meristem, differentiate well into the zone of dividing cells. Translocation of solutes beyond the limits of the vascular elements must be by cell-to-cell movements. Since the rate of movement of solutes into such cells is usually too rapid to be accounted for by diffusion, it can be inferred they possess the capacity of accumulating solutes or of accelerating the rate at which they are translocated as a result of metabolic activity. Much cell-to-cell movement of water obviously also occurs in meristems. Such passage of water from one meristematic cell to another is undoubtedly accomplished largely if not entirely by osmotic and imbibitional mechanisms.

Regions of dividing cells are invariably centers of intense respiratory as well as assimilatory activity. A direct correlation has been demonstrated between the rate of cell division and rate of oxygen consumption in young leaves. Cell for cell the respiratory rate in meristematic regions is higher than in fully differentiated tissues. Dividing cells utilize energy in various ways, a number of which are mentioned in Chapter 12.

Maintenance of the process of cell division requires the presence in the cells of numerous growth-regulating substances. Many of these belong in the category of enzymes or coenzymes; others would ordinarily be classed as hormones; all of them participate in essential reactions of the metabolic mechanism.

Physiological Aspects of Cell Enlargement.   The increase in size of plant cells involves an increase, often manyfold, in the volume of the vacuoles and an areal extension of the cell walls. Some increase in the thickness of the cell walls also often occurs during this phase of growth. During the enlargement of a cell, additional protoplasm is usually synthesized, but the increase in quantity of cell-wall material, principally cellulose and pectic compounds, is proportionately greater than the increase in quantity of protoplasm. Hence, the proportion of carbohydrates to nitrogenous compounds assimilated during cell enlargement is greater than during cell division.

As cells increase in volume there is a movement of water into the enlarging vacuoles. Water is also used in the hydration of the protoplasm and cell walls which are contructed during cell enlargement. Relatively large quantities of water therefore become integral parts of each cell system during this stage of growth. In most kinds of plant cells, as they increase in size, the cytoplasm gradually becomes attenuated into a thin layer which lines the inside of the cell wall, against which it is held by the turgor pressure of the water in the vacuole.

Regions in which enlargement of most or all of the cells is proceeding rapidly are centers of relatively high respiratory activity. Cell for cell, the rate of respiration of enlarging cells is probably not much less than that of dividing cells.

Rapid translocation of both water and solutes is continuously in progress into any region of enlarging cells.

Despite the presence of differentiated vascular elements, considerable cell-to-cell movement of water and solutes must occur in meristmatic zones in which cell enlargement is prevalent. The same mechanisms of water and solute movement are doubtless operative in this part of a meristem as in the younger portion in which cell division is the predominant phase of growth.

As discussed in the preceding chapter, areal extension of cell walls occurs only in the presence of specific substances called auxins.

Two main views regarding the mechanism of cell enlargement have been advanced. One of these holds that the cell wall must first be subjected to elastic (reversible) or plastic (irreversible) stretching as a result of a turgor pressure developed by the cell sap. While in the stretched condition it is assumed that the material substance of the wall is increased, either by the intercalation of additional molecules in the wall (*intussusception*), by the deposition of additional molecules on the cell-wall layers already present (*apposition*), or by a combination of these two processes. Plastic extension of a wall alone would also result in an increase in its area, but if this occurs unaccompanied by the incorporation of new material the wall necessarily becomes thinner.

A second hypothesis holds that active growth of the cell wall is the primary step in cell enlargement. Growth of the wall is believed to result from the intercalation of additional molecules between those already present.

Physiological Aspects of Cell Differentiation. Size differentiation of a given type of cell is usually completed during the enlargement phase of its growth. The cells of various tissues differ, however, not only in spatial dimensions, but also in various structural features, most of which do not develop until enlargement of the cell is nearly or entirely complete. Such structural differentiation begins earlier in the ontogeny of some kinds of cells than others.

The cells which develop into pith, cortex, and certain other tissues do not elongate greatly along the axis of growth in length, although their elongation in this direction is usually greater than radially. Other kinds of cells, such as fibers and tracheids, elongate greatly parallel to the long axis of the stem or root and only slightly in other directions. Differentiation of cell walls ensues at about the time that cell enlargement ceases. During this phase of growth the walls of practically all cells become thicker, although usually not uniformly so. The walls of many kinds of cells and tissue elements become pitted while in others distinctive structural features are developed, the most striking of which are the spiral and other characteristic thickenings of the walls of the primary xylem vessels.

Chemical differentiation of cell walls accompanies their structural differentiation. The walls of some cells, such as those of the pith, the living cells of the phloem, and most of the cells in the cortex, retain their original cellulose-pectic composition indefinitely. The walls of other cells, such as those of most of the xylem tissues, become lignified. Similarly suberin lamellae develop in the walls of cork and some other kinds of cells.

In general the protoplasm soon disappears from those cells in which the walls become lignified, while those in which such a chemical modification of the wall does not occur retain their protoplasm in an unimpaired condition for a much longer period. In cells in which the protoplast dies, such as vessels, tracheids, and fibers, further structural differentiation of the walls occurs only by such purely physiochemical processes as continue in them, or under the influence of the activities of adjacent living cells. The changes occurring in the heartwood of trees, for example, are a result of such processes. Disintegration of certain parts of some kinds of cells may also occur while structural modifications are proceeding in other parts of the same cell. The most familiar example of this is the disappearance of the cross walls between the xylem elements in the formation of vessels.

Assimilation of carbohydrates occurs in all types of cell differentiation involving a thickening of the cell walls, but little or no net gain of protoplasmic proteins occurs during this phase of the growth process. The respiratory activity of even those fully differentiated cells which retain their protoplasm is generally less than that of dividing or enlarging cells.

Development of Lateral Organs.   Leaves originate from the *leaf primordia* which develop laterally from an apical-stem meristem close to its apex (Fig. 19.2 and Fig. 19.3). The histogenic development of a leaf from its primordium does not follow the same pattern in all species, although there are many points of similarity in the development of most kinds of leaves. As an example we may consider the development of the tobacco leaf. A single apical cell of the leaf primordium continues to form new cells until the leaf is 2–3 mm long; then its activity ceases. A midrib primordium develops as a result of subsequent divisions of these newly formed cells, no lamina being present until the leaf is about 0.6 mm long. The lamina originates from two rows of sub-epidermal meristematic cells, one on each side of the midrib primordium. Subsequent divisions, enlargements, and differentiation of these cells result in the development of all of the mesophyll tissues, including the lateral veins. The epidermis increases in area as a result of continued division and enlargement of the epidermal cells.

Cell divisions cease first in the epidermis, followed in order by the middle and lower mesophyll and the palisade layers. The tissues of the lateral veins may continue to develop long after cell division has stopped in other parts of the leaf. Although cessation of cell division occurs first in the epidermis, enlargement of the cells in this layer continues longer than in any other tissues of the leaf. Intercellular spaces do not develop markedly until the leaf has attained one-fourth to one-third its final size.

Evidence from various sources indicates that auxins play an essential role in the elongation of the petiole, midrib, and larger lateral veins of leaves, but growth substances other than auxin seem to be required for expansion of the leaf blade. When seedlings of dicotyledenous species develop in total darkness, little or no expansion of the leaf blade occurs. A very brief exposure to light of relatively low irradiance is sufficient to induce a considerable increase in area of the leaf blade. This expansion of the leaf blade is mediated by the red, far-red reversible photoreaction, discussed in the next chapter. Expansion of leaves in the dark can also be induced by treatment with a gibberellin or a cytokinin (Chapter 18) or with cobalt salts (Chapter 15). The mechanism of these promotive effects of certain compounds on leaf expansion and their relation to the light-activated enlargement of leaf blades is not clear.

In addition to the terminal bud present on most stem axes, lateral buds (which are essentially rudimentary side branches) develop in the axils of leaves. These first appear as moundlike meristems in the axils of the embryonic leaves (Fig. 19.3). In temperate zone woody plants, both terminal and lateral buds are encased in bud scales which are shed only when and if growth of the bud stem tip is resumed. The buds of most herbaceous plants are devoid of bud scales. A great many more axillary buds form on most plants than ever develop into lateral branches (Chapter 23).

The origin of lateral roots has already been discussed from an anatomical standpoint in Chapter 7. Auxins appear to be specifically required for root initiation to occur, whether on roots, stems or other organs, but other hormones also seem to be involved.

Lateral Growth of Stems and Roots. Growth in diameter of roots and stems results principally from the activities of lateral meristems called cambiums. Such meristems are responsible for the *secondary growth* of plants, *i.e.,* the formation of secondary tissues. The most important of these meristems is the vascular cambium (commonly referred to simply as "cambium") which is present in the stems and roots of gymnosperms and most dicotyledons. The vascular cambium consists typically of an uniseriate layer of cells that is located between the xylem and phloem. In most plants in which it occurs this cambium constitutes an almost continuous sheath of cells extending from just back of every root tip to just below every stem apex. The other important lateral meristem, especially in woody plants, is the *cork cambium,* the activity of which is considered later in the chapter.

Structurally, cambium cells are of two distinct types. The *vascular ray initials* from which the vascular rays develop are more or less isodiametric. The vertically elongated elements of the xylem and phloem develop from the more abundant *fusiform initials.* As viewed in the cross section the tangential width of cambium cells of this fusiform type is usually several times as great as their radial width. The length of such a cambium cell usually exceeds even the greatest of its cross-sectional dimensions by many times (Fig. 19.5). The cambium initials of a tulip tree, for example, are about 600 μ long, 25 μ in tangential width, and 8 μ in radial width. Much longer cambium cells have been reported in the stems of some conifers. In the white pine (*Pinus strobus*) they may attain lengths up to 4000 μ. Cambium cells are vacuolated and often prominently so, and often show protoplasmic streaming.

Fig. 19.5.   Perspective view of a typical fusiform cambium cell.

The cambium usually becomes active in the formation of new cells before primary growth has entirely ceased in all of the tissues at the corresponding level of the stem. Successive divisions of cambium cells along their radial axis result in the development of *secondary xylem* on the inner face of the cambium layer and *secondary phloem* on its outer face and cause an increase in the diameter of the axis. Increase in length of the horizontally oriented vascular rays also occurs as a result of cambial activity. Phloem ray cells originate by divisions of a ray initial on its outer face, xylem ray cells from divisions on its inner face. A vascular ray is thus continuous from the phloem into the xylem except for the interrupting layer of cambium. The *secondary xylem* and *secondary phloem* lie between the *primary xylem* and *primary phloem*. All increase in the diameter of stems resulting from the activity of the vascular cambium is caused by the formation of additional layers of phloem and xylem within the body of the stem. In annual plants, or in perennial species in which the stems die down to the ground at the end of each growing season, secondary growth never continues beyond the current season. In species with woody stems, new layers of both xylem and phloem are developed during each period of cambial activity, so that such species exhibit an annual increase in diameter.

Formation of a new cell in the xylem is initiated by division of one of the cells of the cambium layer, the new cell wall developing midway of the cell in the tangential plane. The outer one of the two daughter cells remains a cambium cell, but the inner one enlarges, usually in length as well as in cross-sectional area, and generally develops directly into one of the xylem elements. Often, however, the inner of the two cambium derivatives may divide one or more times before maturation of the developing xylem ensues. This usually happens in the formation of wood parenchyma cells during which the xylem mother cell is cut into a vertical series of cells by transverse divisions. Tracheids, fibers, vessels, wood parenchyma, and wood-ray cells are developed in the xylem from the cambial derivatives (Fig. 19.6; See also Chapter 8).

The enlargement of the xylem cells originating from the cambium initials results in outward movement of the cambium and all of the cells lying outside of this layer, necessarily resulting in an increase in the girth of the cambium cylinder. Enlargement of the immature phloem cells developed from the cambium, on the other hand, results in outward movement only of the phloem and tissues external to it. Generally several times as many new xylem elements as phloem elements are formed from the cambium during a period of growth activity.

In some woody perennials the vascular cambium continues to develop secondary tissues for many, often hundreds, of years, cambial activity being resumed periodically with the advent of each growing season. Since in such species the older phloem is first converted into bark and is eventually sloughed off, the bulk of the structure of all older stems and roots is composed of secondary xylem, commonly called wood. The seasonal periodity of cambial activity results in the presence in branches or trunks of trees, as seen in cross section, of the annual rings in the wood (Chapter 8).

Division of the cambium cells results not only in the formation of secondary phloem and xylem, but also, as the stem grows in diameter, in an in-

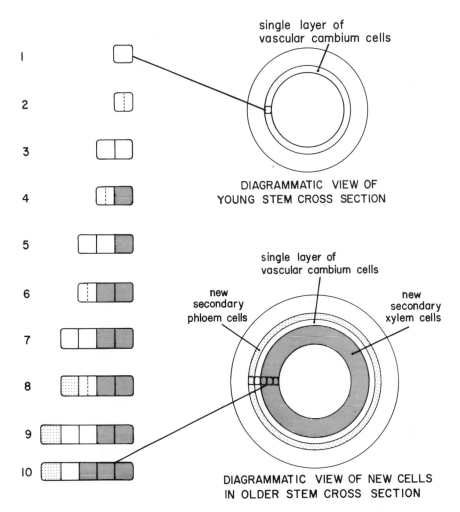

Fig. 19.6. Diagrammatic representation of the mode of origin of secondary xylem (shaded) and secondary phloem (stippled) from the cambium.

crease of the girth of the cambium layer. Some increase in the circumference of the cambium cylinder results from a lengthening of the cambium initials along their tangential axis as seen in cross section, but mostly this is brought about by an increase in the number of cells around the cylinder as the stem grows older.

Vascular cambiums, in the usual sense of the term, are not present in monocotyledons, although traces of cambial-like activity often occur in the vascular bundles. In some arborescent species of monocotyledons, such as palms, however, secondary thickening of the stem results from the operation of a kind of lateral meristem which is usually considered to be a type of cambium. This meristem occurs as a cylinder of tissue near the periphery of the stem. Unlike the cambium of dicotyledons and gymnosperms, it does not produce phloem on the outside and xylem ón the inside but forms complete vascular bundles and intervening parenchyma tissue toward the interior of the stem. These secondary vascular bundles are thus present in stems in a cylinder of tissues outside of the primary tissues in which the primary bundles are located. A limited amount of parenchyma tissue is also formed by such cambiums toward the exterior.

Still another kind of lateral meristem is the cork cambium or *phellogen,* present in most woody stems and roots. In young twigs the first cork cambium usually develops as a continuous uniseriate cylinder of tissue, either from the epidermis or more commonly from the layer of cells just beneath the epidermis. The operation of a cork cambium as a meristem is analogous to that of a vascular cambium. As a result of tangential divisions of cork cambium cells, layers of cork cells (*phellum*) are formed externally, and layers of *phelloderm* cells are formed internally (Fig. 19.7). In most species more phellum cells than phelloderm cells are formed as the result of cork cambium activity, but in a few the opposite is true, and in some no phelloderm cells are formed. Mature cells of the cork or phellum layer are nonliving and are characterized by the presence of suberin lamellae in the walls which render them highly impermeable to water. Phelloderm cells are living cells, more or less loosely arranged, and often resemble cortical cells. The entire system of phellum, phellogen, and phelloderm is often called the *periderm.*

In the majority of woody species the first cork cambium ceases activity

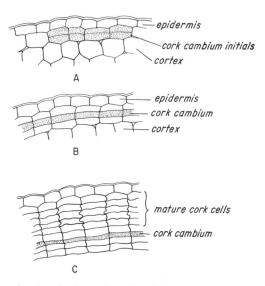

Fig. 19.7. Diagram showing the formation of cork in the stem of geranium *(Pelargonium).* (A) Division of the outer cortex cells, forming a cork cambium. (B) The first layer of cork cells has formed. (C) Several layers of mature cork cells have formed.

after a relatively brief time, and is replaced by other relatively short-lived cambiums which arise progressively more and more deeply in the tissues of the stem or root. Each such cork cambium is a uniseriate layer of cells, as a rule relatively small in area. Such secondary cork cambiums usually arise first in the cortex, later in the pericycle, and finally in the secondary phloem. They operate as meristems in essentially the same manner as the initial cork cambium. The various layers of cork cells formed overlap or join each other in one way or another so that a continuous corky layer, except for lenticels, completely encloses all woody parts of stems and roots. In older tree trunks and roots of many species all periderm layers have developed in the secondary phloem. The older bark of such trees thus consists of alternating layers of cork and dead phloem cells (Fig. 19.8).

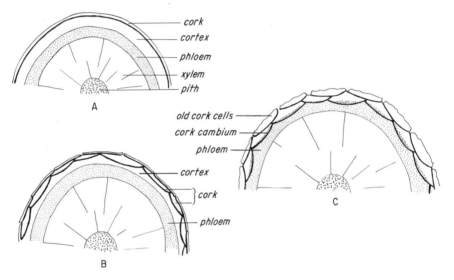

Fig. 19.8. Diagrams showing the development of cork in woody stems. (A) Cork cambium developing in the outer cortex of a young stem. (B) Later stage in which cork cambiums form deeper in the cortex and the epidermis is ruptured. (C) Still later stage in which cork cambiums form in the old phloem and the older cork is shed as the stem enlarges. After Eames and MacDaniels, *An Introduction to Plant Anatomy, 2nd Edition.* McGraw-Hill Book Company, New York, 1947:253.

The same, or at least very similar, physiological processes accompany the different stages in the growth of cells originating from the various kinds of cambial initials as accompany the corresponding stages in the growth of cells arising from apical initials.

Culture of Root Tips and Stem Tips on Synthetic Media.   The technique of culturing excised plant tissues or organs on synthetic media under sterile conditions has furnished significant information regarding many facets of the physiology of plants. The media used are either water or a dilute agar gel to which certain solutes, usually of known chemical composition, have been added.

Excised root tips of a number of species have been successfully cultured in this way. Such roots grow and branch in much the same way as roots attached to the plant. After a week or two, one of the root tips on an original explant can be cut off and transferred as the start of a new culture. Such a procedure of subculturing can be continued indefinitely. Roots of some species, such as tomato, have thus been maintained in sterile culture for many years. They are potentially immortal as long as mortal men continue to act as their caretakers.

The only necessary solutes which must be introduced into the medium for continued growth of the root tips are a carbohydrate (sucrose is generally used), essential mineral salts, and certain vitamins. All species of roots which have been successfully cultured appear to require the addition of thiamin to the medium, some also require niacin, some pyridoxine, and some both of the latter two compounds, if a good growth rate is to be maintained. An inference from these results is that roots do not synthesize, at least not in adequate quantities, those vitamins which must be in the medium if excised roots are to maintain their growth rate. In intact plants these substances are translocated from aerial organs to the roots, thus exhibiting a hormonal behavior (Chapter 18).

Stem tips of a number of species have also been cultured and some, asparagus for example, subcultured through a number of transfers on sterile media. Only sucrose and mineral salts need be present as solutes in the medium for the maintenance of such cultures. Apical-stem meristems thus appear to be more nearly self-sufficient in synthesizing essential growth substances than apical-root meristems. Nevertheless, formation of roots, which often happens on excised stem tips, usually favors an enhanced growth rate of the stem, and there is evidence that this effect is not principally a result of the presence of a better absorptive system. A reasonable inference is that one or more chemically unidentified hormones synthesized in the roots, are translocated into the stem, favoring the latter's rate of growth.

Totipotency.   It is generally considered that every living cell in a plant is *totipotent,* that is, it possesses the potentiality of generating an entire new plant. Many observations can be cited which support this concept, although they do not completely validate it. Fragmentary stem or root cuttings, or both, of many species can be so propagated that they will develop into complete plants. The same statement can be made regarding leaf and even petal cuttings of some species.

The most convincing evidence of the existence of totipotency comes from experiments using sterile cultures of single cells which have been isolated from the tissues of higher plants. Under proper environmental conditions such cells can be shown to act like a fertilized egg cell and to generate an entire plant. Some of the most noteworthy results are those which have been obtained by

Steward, working with cells obtained from carrot roots, although similar results have been obtained with cells obtained from other plant tissues by other investigators.

The single cells were obtained from the carrot tissue cultures by a washing technique. When such single cells were cultured individually many of them began to grow like zygotes into embryos and eventually developed into normal appearing carrot plants. In order that these already differentiated cells would exhibit this property of totipotency, two conditions had to be met. Firstly, each cell had to be entirely separated from its neighbors. Secondly, it was necessary to provide a proper environment, especially with regard to the substances present in the liquid culture medium. In addition to the usual mineral salts and a carbon source (usually sucrose) certain supplementary "nutrients" must be present. For carrot root tissues one such supplement, coconut milk, is adequate, as it is also in the sterile culture of immature embryos (Chapter 21). The fact that coconut milk is actually liquid endosperm doubtless accounts for its effective role as a supplement in some kinds of culture solutions.

Another aspect of totipotency is the phenomenon of *androgenesis*. This term refers to the production of haploid plants by the germination of young pollen grains or microspores within an anther. Under suitable conditions, such haploid plantlets can be raised for several months in culture tubes. After a sufficient degree of development has taken place, the plantlets can be planted in soil and raised to maturity. With certain species of tobacco, flowers have been produced, but seed from such flowers is generally non-viable.

Mutations can be fairly easily induced in haploid plants by a variety of mutagenic agents. A doubling of the chromosome number with the use of colchicine or by other means, leads to a homozygous diploid plant which is fertile. The coupling of these techniques; that is, the production of a haploid plant through androgenesis, the inducing of mutation in the resulting plant, and the subsequent doubling of chromosomes leading to a fertile plant, is a sequence being looked upon by geneticists as an important tool. The incorporation of mutations leading to better yields and disease resistance in food crop plants through these methods will undoubtedly receive much attention in the next few years. It has yet to be ascertained, however, whether androgenesis can be accomplished with pollen of a wide variety of plants, since only a relatively few kinds of plants have been tested thus far.

Genetic Basis of Morphogenic Development.   The discussion thus far of the various aspects of the growth process has been principally on a descriptive structural basis with some consideration of correlated metabolic processes. These more obvious facets of growth must, however, reflect even more fundamental changes which occur in the operation of the genetic system. All of the cells in a given plant contain (with rare exceptions) the same kinds of genes; in other words they possess an identical genetic code. Evidence for this statement comes from the property of totipotency, previously discussed, and also from the manner in which replication of DNA occurs during mitosis (Chapter 16). Doubling of the number of chromosomes in some of the cells of a plant sometimes occurs because of a failure of cytokinesis to accompany mitosis, and this is a characteristic feature of some plant tissues. Even such tetraploid cells, however, contain the same *kinds* of genes as the diploid cells, and differ only in the number of times each gene is present.

In spite of their endowment with a uniform genetic code, different cells develop differently during the ontogeny of the plant and, once developed, may function very differently. The only possible interpretation of this fact is that certain genes are operative in a cell at certain stages in its development and others at other stages. Obviously, if all genes were to function at the same time, this situation would lead to total chaos in cellular metabolism and development. Thus it is the turning off or repression of the action of certain genes which brings order and continuity to a given cell's development. It is also important to realize that genetic factors operate only in coordination with environmental factors as described more fully in the next chapter.

The mechanisms whereby the influence of genes can be turned on and off at different stages in the development of a cell or whereby different cells with the same genic constitution can be caused to function very differently is not known. It is virtually certain, however, that the genes must operate at least in part through the induction and repression of enzymatic activity (Chapter 9). Some imaginative hypotheses have been proposed to account for the controlling effect of genes on metabolic processes, but discussion of them is beyond the scope of this book.

Measures or Indices of Growth.   It is frequently desirable to give a quantitative expression to the amount of growth accomplished by a plant or group of plants during a given period of time. The principal indices which have been employed for this purpose are (1) increase in the length of the stem, root, or other organ of the plant, (2) increase in the area of the leaves, (3) increase in the diameter of the stem (or other organ), (4) increase in volume (especially of fruits), (5) dry-weight gain, and (6) fresh-weight gain.

All of these indices have at least a limited value as measures of growth, especially from various practical standpoints. Determination of the height and diameter growth of forest trees, for example, are standard forestry practices as indices of the productivity of forests and have considerable practical value for such purposes. Similarly, weight of dry hay or the fresh weight of cabbage or spinach produced per unit area of land would usually be an adequate measure of growth to the mind of the practical farmer.

Each of the indices listed above, however, measures only certain quantitative phases of growth, but growth phenomena generally involve not only such quantitative changes as expansion in length and girth and increase in weight but qualitative aspects as well. Qualitative differences in growth are often of scientific significance or practical importance as great as or greater than quantitative differences. The floriculturist is not primarily interested in the pounds of plant substance produced nor the height to which his plants grow. provided they bear flowers attractive to his customers. Likewise, the orchardist is much more interested in the development of fruit on his trees than in the increase in the height or weight of their vegetative organs. Evidence of this difficulty in giving adequate expression to the results of growth phenomena is seen in the common expedient of investigators in relying upon photographs as a means of recording the results of their experiments upon the growth of plants.

Rates of Growth. The absolute growth rates recorded in the botanical literature are mostly for increase in height of stems. A number of measurements have also been made of the fresh and dry weight increments of fruits, an example of which is given in Chapter 17.

The rate of height growth varies enormously with different species of plants and with the same species under different environmental conditions. Young bamboo shoots sometimes grow as rapidly as 60 cm in 24 hr and asparagus shoots as much as 30 cm in the same period of time. When a flower stalk develops on a century plant (*Agave spp.*) it often elongates as much as 15 cm in a single day. Under favorable growing conditions maize plants sometimes grow in height 8 to 12 cm during a 24-hour period.

The rates of elongation of the fastest growing stems are just a shade too slow for detection with the naked eye. By observing rapidly growing stem tips under a horizontally placed microscope, the externally visible aspects of growth can often be observed and measured directly.

Accumulation of Foods. The sum total of the food available to a green plant, except under special experimental conditions, in which carbohydrates or other foods are artificially supplied, can never exceed the quantity synthesized in photosynthesis. A large proportion of the photosynthate is normally used in the processes of assimilation and respiration. Any surplus which remains accumulates in one or more tissues or organs of the plant. Accumulation of foods, however, does not occur continuously. For considerable periods in the life cycle of most plants not only is no accumulation of food occurring, but a more or less rapid consumption of accumulated food is in progress. In woody plants during the dormant season, slow utilization of food in the processes of respiration and assimilation continues. When growth is resumed in the meristematic tissues of such species in the spring there is always a considerable drain on the foods stored in the plants, since much of this growth is accomplished before the photosynthetic rate is rapid enough to compensate for the necessarily speedy utilization of food which occurs. Similarly the sprouting of bulbs, corms, tubers, and rhizomes always occurs at the expense of accumulated foods in such organs. The same is true of seeds when germination occurs. Much of the food which accumulates in plants during periods when photosynthesis exceeds the food-consuming processes is utilized by the plant sooner or later in its life history.

The organs in which most accumulation of food occurs vary with different species. In annuals, food storage occurs predominantly in the seeds. Foods also accumulate in the seeds of most biennial and perennial species. During the process of germination the embryo uses food that was made by the preceding sporophyte generation.

Most of the accumulation of food in typical biennials such as beet, carrot, parsnip, and turnip occurs in fleshy rootlike structures. This accumulation of food by biennial species occurs mostly during their first season's growth.

In perennial species considerable storage of food often occurs in seeds and fruits, but the principal organs of food accumulation in many species which live for a number of years are the stems and roots. In woody species the pith, cortex, vascular rays, and wood parenchyma are the stem and root tissues in which most of the accumulation of surplus foods occurs. Rhizomes, (iris, many ferns, Solomon's seal), tubers (potato, Jerusalem artichoke), corms (crocus, gladiolus, jack-in-the-pulpit), and bulbs (onion, tulip, hyacinth) are almost invariably regions of food storage.

The principal storage carbohydrates are starch, sucrose, hemicelluloses, and inulin. Accumulation of oils (fats) in abundance occurs most commonly in seeds, although such compounds are stored in at least small quantities in the cells of many tissues. Proteins, like fats, accumulate principally in seeds.

Foods move into the cells in which they accumulate only in the soluble form; yet with few exceptions, sucrose being the most familiar example, the foods amassed in storage cells are converted into an insoluble form. All such conversions require enzymatic activity. Conversely, insoluble stored foods cannot be utilized by any part of a plant until they have first been digested into soluble forms as a result of enzymatic activity. Until such a transformation has occurred they cannot be translocated out of the cells in which they are situated into the cells in which they are utilized in assimilation or respiration.

SUGGESTED FOR COLLATERAL READING

Clowes, F. A. L. *Apical Meristems*. Blackwell Scientific Publications, Oxford, 1961.

Dormer, K. J. *Shoot Organization in Vascular Plants*. Syracuse University Press, Syracuse, 1972.

Esau, Katherine. *Plant Anatomy, 2nd Edition*. John Wiley and Sons, Inc., New York, 1965.

Kozlowski, T. T. *Growth and Development of Trees*. 2 vols. Academic Press, New York, 1972.

Ledbetter, M. C., and K. R. Porter. *Introduction to the Fine Structure of Plant Cells*. Springer-Verlag, New York, 1970.

Leopold, A. C. *Plant Growth and Development*. McGraw-Hill Book Company, New York, 1964.

Milthorpe, F. L. *The Growth of Leaves.* Thornton Butterworth Scientific Publications, Ltd., London, 1956.

O'Brien, T. P. and Margaret E. McCully. *Plant Structure and Development.* The Macmillan Company, New York, 1969.

Romberger, J. A. *Meristems, Growth, and Development in Woody Plants.* U.S. Dept. of Agriculture Technical Bulletin No. 1293, Washington, 1963.

Steward, F. C. *Growth and Organization in Plants.* Addison-Wesley Publishing Co., Reading, Massachusetts, 1968.

Torrey, J. G. *Development in Flowering Plants.* The Macmillan Company, New York, 1967.

Troughton, J., and Lesley A. Donaldson. *Probing Plant Structure.* McGraw-Hill Book Company, New York, 1972.

Wareing, P. F., and I. D. J. Phillips. *The Control of Growth and Differentiation in Plants.* Pergamon Press, Elmsford, New York, 1970.

ENVIRONMENTAL FACTORS INFLUENCING VEGETATIVE GROWTH

It is impossible to conceive of the existence of a living organism in which some kind of genetic mechanism does not exist. It is equally impossible to conceive of the existence of a living organism in the absence of an environment. Life is perpetuated by the interplay of environmental factors and genetic factors through their influence on the internal conditions, largely metabolic processes, within the organism. This interrelationship can be represented in a generalized way by the following diagram:

Despite the spectacular advances in knowledge of the genetic code in recent years, the principal aspects of which have been discussed in Chapter 16, the intricate interacting stages between the genetic system and environmental factors on the one hand and the development or reactions of an organism on the other, are far from being bridged in terms of present day knowledge.

The environment of any living organism can be analyzed as a complex array of factors, the effects of most of which are discussed later in the chapter. The dotted line in the above diagram indicates that environmental factors may sometimes have direct modifying effects on genetic factors, through the induction of mutations. By far the greater effect of environmental factors on plant development or reactions, however, is exerted through their integrated interaction with genetic factors.

As an example of the principle illustrated in the above diagram we can recall the process of chlorophyll synthesis as it occurs in maize. Most varieties of this species contain the genetic factors which ordinarily induce chlorophyll formation. Certain environmental conditions, including light, are also necessary for its synthesis in this species. A maize seedling developing in a dark room is devoid of chlorophyll, even if all the other environmental conditions necessary for chlorophyll formation are present. In a seedling growing in the light, however, interaction of the environmental factors with the hereditary mechanism will occur in such a way as to induce the process of chlorophyll synthesis in the leaf cells. So far, however, we have considered only one side of the story. Certain varieties of maize do not carry all of the genetic factors necessary for the development of chlorophyll. This trait is inherited in such strains as a Mendelian recessive and hence is apparent only in plants homozygous for this factor. Even if all the environmental factors necessary for chlorophyll synthesis are present, such seedlings cannot make chlorophyll, and they develop as "albinos." As soon as the accumulated food in the seed is exhausted, such albino seedlings die.

The genetic constitution of a given organism sets definite ultimate limits to the types of development and the reactions of which that organism is capable, beyond which no environmental conditions can carry it. Potato plants, for example, may vary greatly in their morphogenic development. Under some environmental conditions the plants may be large; under others, small. Under some they will flower; under others they will not. Under still others, tubers will develop; under others they will not. Nevertheless, all of these plants will remain unmistakably potato plants.

When a number of different species of plants develop under identical environmental conditions each one shows the characteristic development of its own kind, a manifestation of differences in genetic constitution. However, genetic differences, within limits, commonly exist even within a single species. Some of these become manifest as observable structural differences, but some of them are too subtle to be expressed in morphological language. Strains between which little or no detectable difference in structure can be discerned often differ markedly in physiological reaction. Such strains are often called "physiological races." Such varieties of a given species often show differences, one from the other, in such properties as cold resistance (see the following), drought resistance (Chapter 8), photoperiodic requirement for flowering (Chapter 22) and optimum temperature for growth (see below).

The specific environment to which a plant is subjected also sets limits upon its development. For example, under "short-day" conditions, discussed in Chapter 22, a radish plant continues to grow vegetatively for an indefinite period of time. Although radish plants possess the hereditary capacity for reproductive development, such an environment imposes a barrier to the expression of this particular potentiality. On the other hand, under long daylengths, if other environmental conditions are favorable, a radish plant will flower and fruit within the course of a few weeks. Innumerable other examples of the environmental limitation of the expression of hereditary factors can be cited.

The full gamut of the hereditary potentialities of a species can never be realized until individuals of that species have been observed growing in each of a wide range of environmental complexes. Since most observations of the behavior of plants are made while they are growing under natural or cultural conditions which represent only a rather narrow range of variations in the environmental complex, the many possible developmental reactions of a given species of plant are not always appreciated.

Environmental Factors Influencing Plant Growth.   The environment of living organisms is so complex as to defy any completely logical analysis or categorization. However, the principal factors of the environment which ordinarily exert a more or less *direct* effect upon the growth and development of terrestrial plants can be recognized and their effects are discussed later in this chapter.

The environment to which the roots are exposed is usually very different from that which the aerial organs of plants encounter. Because of reciprocal influences between the roots and tops of a plant, effects of any environmental factor upon the development or physiological process of the roots almost invariably will be reflected in the behavior of the aerial organs, and vice versa (Chapter 23).

Some important environmental factors, such as precipitation (rain, snow, and hail), are usually indirect in their influence on plants. Precipitation, for example, influences not only the soil-water content, but also soil aeration and atmospheric humidity.

Many of the environmental conditions to which plants growing under out-of-door conditions are subjected are in turn influenced by more remote factors. The intensity and quality of impinging sunlight, for example, are functions of the angle of the earth's inclination to the sun (which varies with the hour of the day, the latitude, and the season), the pitch of the slope upon which the light falls, and the direction toward which the slope faces. The soil-water content is controlled not only by the precipitation but by the surface runoff (which in turn is largely a function of the slope and the porosity of the soil and by factors which influence the rate of evaporation such as air temperature, humidity, wind and insolation. Similarly, with increase in altitude, differences in such physical factors as the irradiance, quality, and duration of radiant energy, soil and air temperature, and atmospheric pressure are encountered.

Furthermore, complex interrelationships exist among the medley of environmental factors which exert direct effects upon plants. Changes in the magnitude or duration of one factor seldom occur without inducing subsidiary changes in other factors. Increase in the intensity of radiant energy in any habitat results in an increase in soil and air temperature; increase in soil-water content diminishes soil aeration, and so forth.

In addition to the physical factors discussed above, plants are subject to the influence of another entirely different group of factors—the other living organisms in their environment. Among these are bacteria, fungi, green plants, and animals of many kinds, including man. The influences of such *biotic factors* upon the growth and development of plants are often as pronounced as the effects of physical factors. Biotic factors often operate as limiting factors in the survival or distribution of plants. The elimination of chestnut (*Castanea dentata*) from the forests of eastern North America, in which it was once a prominent tree species, by the chestnut blight disease, is an example of the profound effects sometimes wrought by biotic factors. From the standpoint of plants, man himself is the greatest of all biotic factors. He has exerted overriding influences on the distribution and prevalence of many kinds of plants and by hybridization, induced mutations and selection has modified the genetic constitution of a number of species.

In order to interpret the effect of changes in the magnitude of any one of the various factors influencing a process, such as growth, it is necessary to formulate certain guiding principles. In 1843 Liebig proposed his well-known "law of the minimum," which was the first attempt at such a formulation. Liebig was thinking primarily of the effect of fertilizers upon the yield of crop plants when he suggested this "law", which states in essence that the yield is limited by the factor which is present in relative minimum. Blackman's "principle of limiting factors" as applied to photosynthesis (Chapter 11) is essentially an extension of Liebig's principle.

In 1909 Mitscherlich proposed a somewhat different concept of the law of the minimum. His conception of the operation of a "limiting factor" may be stated as follows: "The increase in any crop produced by a unit increment of a deficient factor is proportional to the decrement of that factor from the maximum."

Both of these interpretations of the effect of minimal factors can be illustrated by a diagram (Fig. 20.1), in which it is supposed that five factors are affecting growth, but that each is present to a different relative degree as compared with its maximum effectiveness.

According to Liebig's law of the minimum, only an increase in factor *A* will cause an increase in the yield of a crop. According to Mitscherlich, increase in any one of these factors will cause an increase in yield. A unit increase in *A* will have the greatest effect, a unit increase in *B* the next greatest effect, and so forth. Factor *E* is so close to its maximum that a unit increase in it will have an almost negligible influence on yield. Mitscherlich's interpretation of the law of the minimum seems to be more nearly in accord with the results obtained in experiments on plants than Liebig's simpler formulation of this same principle.

The discussion in this chapter will be restricted to the effects of environ-

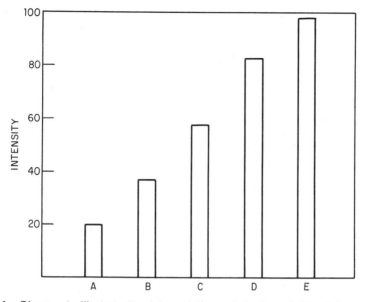

Fig. 20.1. Diagram to illustrate two interpretations of the law of the minimum, as discussed in the text.

mental factors on the vegetative growth of plants; their effects on reproductive growth are considered in Chapter 22.

Effects of Temperature upon the Rates and Incidence of Growth Processes.   Temperature is an all-pervading factor and its influence is inescapable. In all its effects on the rates of plant processes, including various aspects of vegetative growth, temperature follows the same general pattern. With increase in temperature from a relatively low value of usually a few degrees above 0°C the rate of the process increases to a peak rate, providing no other factor is limiting. Such a peak rate is usually attained somewhere in the range of 20–35°C, depending upon the kind of plant and the specific processes involved. With further increase in temperature above this optimum the rate of the process decreases, falling to zero at a temperature usually in the neighborhood of 45°C. This general pattern of effects of temperature upon the rates of plant processes results from the keying of all such processes to enzymatic activity which in itself follows such a temperature response (Chapter 9). The range of temperatures within which a physiological process will occur is often called the *physiological range*. In general, the range of temperatures within which growth will take place varies considerably with the species. Arctic and alpine species may grow at air temperatures at the freezing point or even slightly below, and their optimum growth temperature is often no higher than 10°C. Most species of temperate zone origin do not grow appreciably at temperatures below 5 or

10°C. Their optimum growth temperature is usually about 25 to 35°C, and their maximum about 35 to 40°C. The range of growth temperatures of most tropical and subtropical species is still higher. For maize, a crop plant of subtropical origin, the minimum growth temperature is about 10°C, the optimum about 30 to 35°C, and the maximum about 45°C.

In a comprehensive study of the effect of temperature upon the quantitative aspects of growth one investigator found that the rate of elongation of the roots of pea seedlings increased consistently with rise in temperature in the range of −2 to 29°C, and further that the rates within this range of temperatures, once established, showed little or no diminution with time. Above about 30°C, the higher the temperature, the lower the initial rate of growth and the more rapidly the rate decreased with time. Elongation ceased entirely at temperatures of 45°C or higher.

Different stages in the growth of a plant often have different optimal temperatures. For example, the optimum temperature for germination, initial elongation of the primary root, and initial elongation of the hypocotyl of cotton has been found to be about 33°C. After a few days, however, the optimum for root elongation falls to 27°C, whereas that for hypocotyl elongation rises to 36°C.

In a seemingly paradoxical fashion certain growth phenomena are *promoted,* by relatively low temperatures, usually in the range from a little below zero to about 10°C above zero. Among these are the breaking of dormancy in some kinds of seeds and buds (Chapter 23) and the phenomenon of vernalization (Chapter 22). Many such promotive temperature effects on plants are *inductive*; that is they evoke a delayed response.

Temperature Limitations upon Plant Survival.   A clear distinction should be drawn between the extremes of temperature at which the growth of a plant ceases and the extremes of temperature which that plant can endure without death resulting. The minimum temperature which a plant can tolerate without injury is almost invariably below that at which growth ceases; likewise plants can usually endure without lethal effects, at least temporarily, temperatures considerably in excess of the maximum at which growth occurs. A given plant, for example, may cease to grow when the temperature to which it is exposed rises to 40°C. Death, however, occurs only if the temperature of the plant is raised to some still higher temperature, perhaps 50 or 60°C. The most marked tolerance of temperature extremes by active plant cells is exhibited by certain species of algae. Some species of blue green algae survive temperatures up to about 75°C in hot springs. On the other hand a red colored species of green alga (*Chlamydomonas nivalis*) grows commonly on the summer snowbanks in high mountains.

The upper and lower extremes of temperature which plants or plant organs can endure vary greatly according to species and depend upon their capacity for heat resistance and cold resistance as discussed below.

Morphogenic Effects of Temperature.   No two of the many and varied physiological processes occurring in a plant are equally influenced by a change in temperature. Hence the morphogenic development of a plant, which is controlled by the pattern of its physiological activity, is often markedly different under one set of temperature conditions than under another. Such effects upon the structural development of plants are the most complex and most striking of the many influences of temperature upon plants.

Some of the ways in which temperature influences are undoubtedly mediated are through direct effects on operation of the genetic mechanism and involve differential repression and activation of genes. Other such effects, however, can be largely interpreted in terms of well-known physiological processes. Some are clearly related to the differential influences of temperature upon the processes of photosynthesis and respiration. The net daily gain in photosynthate by any plant is the difference between the quantity of carbohydrate synthesized during the photoperiod and the quantity of carbohydrate consumed per 24 hours in respiration. In a maize plant, for example, under favorable conditions, the average quantity of hexose synthesized per photoperiod during its life history is about 9 g, while the average quantity of hexose consumed in respiration is about 2 g per 24 hours. The average net daily gain in photosynthate by such a plant is therefore about 7 grams.

Only the net daily increment of photosynthate can be used by a plant in the assimilatory phases of growth or can accumulate as unused food in one or more organs of the plant. If its total daily respiration consistently exceeds its total daily photosynthesis, a plant can often survive for a while at the expense of previously accumulated foods, but eventually it will starve to death.

The rate of photosynthesis ordinarily does not change much in many species under favorable field conditions over a temperature range of 15 to 30°C, mainly, it appears, because of the limiting effect of carbon dioxide at atmospheric concentration on the process (Chapter 11). The temperature coefficient of respiration, on the other hand, over a similar temperature range, is approximately 2 (Chapter 13). With rise in temperature the rate of respiration therefore increases more rapidly than the rate of photosynthesis, and the net daily gain in photosynthate by the plant is curtailed. Respiration is not, of course, confined to the photosynthetic tissues, but occurs in all nongreen parts of the plant including the underground organs. Also significant is the fact that photosynthesis is restricted to the daylight hours while respiration goes on every hour of the twenty-four. Hence prevailing night temperatures markedly influence the daily net gain of photosynthate by a plant. Approximately twice as much food would disappear per hour as a result of respiration at 25°C, for example, as at 15°C. A daily alternation between relatively cool night temperatures and moderately high daytime temperatures will therefore usually result in a greater net gain of photosynthate by a plant than if night temperatures are also relatively high.

Translation of differential effects of temperature upon photosynthesis and respiration into morphogenic effects of growth can be illustrated by the process of tuberization in the potato. When tuberization occurs, foods are used in the construction of the cellular framework of the tubers and accumulate within the cells. The optimum rate of photosynthesis is apparently attained in the potato at about 20°C. The higher the daytime temperature above this value and the higher the night temperature, the greater the proportion of the photosynthate which will be consumed in respiration and the smaller the proportion which can be used in assimilation or which can accumulate as food. Higher temperatures also favor development of the aerial vegetative organs of potato plant, which further reduces the quantity of photosynthate which can be used in tuberization. Hence yields of potatoes are usually higher under cool growing conditions, other environmental factors being favorable, than under warm ones. Tuberization is a complex process however, and is influenced by a number of other factors in addition to the quantity of photosynthate available. One of these is the length of the photoperiod, as discussed later in this chapter.

Among the most pronounced of the morphogenic effects of temperature are those upon reproductive growth which are discussed in Chapter 22.

Thermoperiodicity. Temperature is one of the characteristically cyclical factors of the environment. Both rates of growth and morphogenic development of plants are markedly influenced by the pattern of the diurnal temperature cycle to which they are subjected. For example, when grown under a constant temperature the maximum rate of elongation of the stems of tomato plants was found to occur at 26.5°C. Elongation was still more rapid, however, if the plants were exposed to a 26.5°C daytime temperature alternating with a 17–20°C night temperature.

The term *thermoperiodicity* is used to designate the effects of a cyclical alternation of temperature between day and night periods upon the reactions of plants. Many other species, of which potato and pepper are examples, also show their best growth when moderate daytime temperatures alternate with cool nights. Some species are little affected in their growth performance by diurnal variations in temperature and a few, of which the African violet is an example, develop better if night temperatures exceed those in the daytime.

Cold Injury and Cold Resistance.   *Causes of Injury to Plants upon Exposure to Low Temperature.*   Several types of injury may occur in plants as a result of exposure to relatively low temperatures.

Desiccation.   Relatively high winter transpiration rates in evergreens during a period when absorption of water can proceed only a relatively slow rate often lead to a type of injury frequently called *winter-killing*. Injury under such circumstances results from desiccation of the tissues. A similar type of injury may result to some plants, especially herbaceous species, as a result of frost heaving of the soil. Such heaving often tears the roots loose from the soil or in extreme cases may even break them. If environmental conditions favoring high transpiration rates intervene before the root system can be securely reestablished in the soil, the plant may be so severely desiccated that death or marked injury results. This is often a serious source of injury to winter wheat during "open" winters. One of the advantages of mulching plants with straw, leaves, or other materials during the winter months is that it greatly reduces frost heaving of the soil.

Chilling Injury.   Many species of plants, particularly those native to tropical or subtropical regions, are killed or seriously injured by relatively low temperatures above their freezing point. This type of low temperature injury is often called *chilling injury.* For example, exposure to a temperature of 0.5 to 5.0°C for 24 to 36 hours is fatal or markedly injurious to rice, velvet beans, cotton, peanuts, or Sudan grass. Species which are only slightly injured include maize, sorghums, watermelons, and pumpkins, while soybeans, buckwheat, tomatoes, and flax show no evidence of injury from such a chilling. The cause of such pronounced effects of low but not freezing temperatures undoubtedly lies in disorders which are induced in the metabolic activities and physiological conditions within the cells. In general, the longer the exposure of such plants to chilling temperatures, the greater the resulting injury.

Freezing Injury.   Many plant tissues are killed or irreparably injured when exposed to temperatures low enough to cause ice formation within them. This is the most frequent and fundamental type of low-temperature injury in temperate climates.

When plant tissues are frozen very rapidly, a procedure which can easily be accomplished under laboratory conditions, crystallization of water takes place in the vacuoles and in the cytoplasmic layers. Such intracellular freezing is uncommon under natural conditions, but when it does occur, is invariably lethal. Under out-of-doors conditions, freezing of water in the tissues of higher plants is usually a gradual process. As a result ice-crystal formation almost invariably occurs in the intercellular spaces, presumably because of the lower solute content of the water impregnating the cell walls than of the water in the cytoplasm and vacuole. Because the water potential at the ice-crystal surfaces is more negative than that in the cells, water moves towards them and the crystals gradually enlarge. Water moves at first from cells bordering on the intercellular spaces, but also may move towards the proliferating ice crystals from more distant cells along water potential gradients. The lower the temperature, in general, the greater the proportion of water which freezes in the tissues and the larger the ice crystals become.

When ice formation occurs in the intercellular spaces of a plant tissue, death of the cells may or may not result, depending upon the hardiness (see the following) of the tissue. A probable explanation of freezing injury under these conditions is that the withdrawal of water from the cells results in a dehydration of the protoplasm which in turn induces various disorganizing effects in its structure. Freezing injury under some conditions appears to result, not at the time of ice formation in the intercellular spaces, but during the subsequent thawing. This is especially prone to occur when the thawing is rapid. Death under these conditions may result from mechanical distortions of the protoplasm attendant upon the too rapid re-entry of water into the cells from the intercellular spaces.

Cold Resistance and Hardening.   When liquid water gradually cools, freezing usually does not begin at 0°C but only after it has *undercooled* from a fraction of a degree to several degrees below its freezing point. Similarly the water in plant tissues does not usually freeze until the tissues have undercooled several degrees or more. Because of a considerable capacity for undercooling many plant tissues normally susceptible to low temperature injury, can survive short exposures to freezing temperatures without injury. This is true, for example, of certain species of cacti, the tissues of which can often undercool as much as 10–15°C. Unless a plant tissue possesses the property of cold resistance, as discussed later, it will be killed or severely injured once ice crystals form within it, regardless of the degree of preliminary undercooling.

Plant tissues which can survive exposure to below freezing temperatures are said to possess the property of *cold resistance* or hardiness. The degree of cold resistance is not a fixed quantity, but can be modified in many plants as a result of the natural or artificial conditions to which they are subjected. On the other hand there are many plants which are never cold resistant under any conditions and many, as noted earlier, can not even survive exposure of any length to low temperature just above the freezing point.

Increase in the cold resistance of a plant tissue is called *hardening;* decrease in its cold resistance, *dehardening.* Temperature is the principal, though not the only environmental factor inducing changes in the hardiness of plants. Exposure of many kinds of plants to temperatures just above the freezing point results in a marked increase in their cold resistance. Crop plants such as cabbage which are to be planted early in the spring are often hardened artificially before being set out in the field. This is usually done by transferring the cabbage seedlings from the greenhouse to a cold frame for a few days before they are transplanted into the field. Continuous exposure of hardened plant tissues to relatively warm temperatures, on the other hand, sooner or later results in dehardening.

Seasonal variations in hardiness are of normal occurrence in the organs of temperate-zone species which are exposed to freezing temperatures during the winter months. The leaves of temperate-zone evergreens and the buds and stems of temperate-zone deciduous trees and shrubs have little or no cold resistance during the summer, but pass into a hardened condition during the autumn, remain relatively cold resistant during the winter months, and undergo a dehardening process in the spring. During the colder months of the year such organs often survive exposure to temperatures of −20 to −30°C or even lower without injury.

Not all of the organs on a plant are equally cold resistant at a given time. In deciduous woody plants, mature leaves are commonly more hardy than young leaves, mature stems commonly more hardy than young stems, and stems in general more hardy than leaves. Floral organs may be more hardy or less hardy than leaves, depending upon the species. In temperate-zone woody species on which the flower buds open earlier in the season than leaf buds, such as elms, maples, and witch hazel, the floral parts are more cold resistant than the leaves. On the other hand, floral parts of apple and some other fruit trees at a certain stage in their development are notoriously more sensitive to low temperature injury than the young leaves which are present on the tree at the same time.

Complete immunity of any tissue to low temperature injury requires prevention of intercellular freezing or else a capacity of the cells to survive its effects. The basis for cold resistance is not the same in all organs or tissues. The flower buds of some species such as dogwood and some varities of azaleas apparently are cold resistant because freezing fails to take place in the tissues at any temperature normally encountered during the winter. The small size of the cells plus their low degree of hydration apparently make it virtually impossible for ice crystals to form in such tissues. The basis of the cold resistance of those tissues in which intercellular freezing normally takes place undoubtedly lies in certain properties of the protoplasm. The changes in protoplasmic properties which are known to be closely correlated with the hardening of plant tissues are an increased permeability of the protoplasmic membranes to water, a decreased structural viscosity of the cytoplasm, and a reduced liability to coagulation as a result of dehydration of certain layers of the cytoplasm. An increased hydrophily of the protoplasmic colloids—largely proteins—could account for all of these shifts in the properties of the protoplasm. The increased permeability of the protoplasmic membranes favors outward movement of water during freezing and reduces the likelihood of intracellular ice formation. The structural changes which occur in the protoplasm presumably make it better able to withstand mechanical stresses resulting from withdrawal of water when ice forms extracellularly.

Heat Injury and Heat Resistance.   *Causes of Injury to Plants at Relatively High Temperatures.*   Several types of injury result to plant cells either directly or indirectly from relatively high temperatures:

Desiccation Injury.   High leaf temperatures resulting from intense insolation or high air temperatures or both may result in excessive rates of transpiration. A relatively high rate of water loss, particularly at times when the rate of absorption of water is sluggish, often leads to death of some or all of the leaves or branches on a plant as a result of desiccation. The "leaf scorch" which occurs in many species as a result of exposure to high temperatures appears to be largely desiccation injury, although it is likely that direct thermal effects are sometimes also involved. In extreme cases entire plants are killed in this way.

Injury Resulting from Metabolic Disorders.   Relatively high temperatures often induce metabolic imbalances which are detrimental or even fatal to plants. With rise in temperature, increase in the daily photosynthesis in a plant usually fails to keep pace with increase in its daily respiration. Relatively high temperatures therefore frequently cause a stunting of plants because a disproportionate amount of the foods manufactured is consumed in respiration. Maintenance of such a condition for extended periods may result in the death of plants.

Direct Thermal Effects upon the Protoplasm.   The thermal death point of most active living plant cells is in the approximate range of 50 to 60°C. The exact temperature at which death of the protoplasm occurs depends upon the length of the period during which the cells are warming up to the lethal temperature. For example, if the leaf epidermal cells of *Rhoeo discolor* were heated at such a rate that death occurred in 4 minutes, the lethal temperature was 72.1°C. If the rate of warming was so slow that death did not take place until 150 minutes had elapsed, the thermal death point was only 52.0°C.

Air temperatures in temperate regions seldom exceed 40°C and although the temperature of an insolated plant often exceeds that of the atmosphere, lethal temperatures are seldom attained for reasons discussed in Chapter 6. The surface temperature of some soils, however, may attain values of 70°C or even higher when exposed to intense insolation. Attempts to reforest denuded areas in certain regions have sometimes failed because of such high soil-surface temperatures. The living cells of the stems of the young trees which had been set out were killed at the soil line by contact with soil at a temperature above their thermal death point, thus causing death of the entire transplant. On the other hand, many woody species are habitants of semi-desert regions in which high soil temperatures often prevail. The stems of such species are obviously more heat resistant than those which are injured or killed by contact with hot soils.

Another example of direct heat injury to plants is often evident after a "ground" forest fire sweeps through a woods. Such fires burn fallen leaves and branches on the forest floor and are often without any apparent immediate effects upon living trees and saplings. Subsequently the tops of many of the trees in an area burned over by such a fire die as a result of the killing of an encircling zone of cells at the base of the trunk by the high temperatures to which they have been exposed.

The most generally advocated theory of the cause of direct heat injury to plant cells is that it results largely, if not entirely, from a coagulation of enzymes and other protein components of the protoplasm.

Heat Resistance.   Certain types of tissues are more resistant to heat injury than others. Tissues low in water content generally can endure relatively high temperatures better than those of which the contrary is true. Dry seeds and spores of some species have endured exposure to temperatures of 125°C and even higher without loss of germinative capacity.

In some plants, otherwise susceptible tissues are protected against heat injury because they are enclosed within tissues which have a low thermal conductivity. The bark of many trees is so thick that it insulates the inner, living tissues against the destructive effects of forest fires. The bark of the well-known big tree (*Sequoiadendron giganteum*) of California, for example, is almost as resistant to fire as asbestos, has a very low thermal conductivity, and may be as much as 50 cm thick. Many other species of coniferous trees have a thick, incombustible bark of low thermal conductivity. Another well-known example of such a tree is the long-leaved pine (*Pinus palustris*) of the southern United States.

Effects of Radiant Energy on Vegetative Growth.   The spectrum of radiant energy (Fig. 10.1) ranges from the very long electric waves to the infinitesimally short cosmic waves. All kinds of radiant energy, including light, vary in several different ways, the most important of which are (1) irradiance ("intensity"), (2) quality, and (3) duration (Chapter 10). Light is essential to all green plants because of its primary role in photosynthesis. Numerous other effects of light upon physiological conditions and processes in plants have been discussed in previous chapters. Among these are (1) chlorophyll synthesis, (2) stomatal action, (3) anthocyanin formation, (4) temperature of aerial organs, (5) absorption of electrolytes, (6) phototropism, and (7) rate of transpiration. Several of the most important effects of light on plants remain to be discussed in this and the following chapters.

In this chapter the discussion will be restricted to a few of the more striking examples of the effects of differences in the irradiance, quality, and duration of light upon the vegetative development of plants.

Irradiance Effects on Vegetative Growth.   Variations in irradiance, especially of sunlight, are almost invariably accompanied by at least minor variations in the quality of light. This fact must frequently be considered in evaluating the effects of different irradiances upon growth or other plant processes. Generally speaking, however, under natural conditions, differences in irradiance have more significant effects upon the growth of plants than differences in light quality.

In general the results of a number of investigations of the effect of irradiance on plant development indicate that maximum height and leaf area are attained at irradiances which are considerably less than full summer sunlight. Relatively high irradiances result in most species in shorter internodes, plants of lower stature and smaller leaves; but the dry weight, size of the root system, and production of flowers and fruits is greater than in weaker irradiances. Many species show increased growth in terms of dry-weight increment with increased irradiance up to 100 per cent of summer sunlight, if no other factor is limiting. All phases of the growth of typical shade species are usually retarded, however, by high irradiances. Many tropical species, for example, are shade plants and attain their maximum development at irradiances considerably less than that of full sunlight.

These effects on growth actually represent the integrated influences of many processes occurring within the plant, a number of which are conditioned by light. Some of these effects of light are directly on one phase or another of growth; others are indirect effects on growth resulting from direct effects of light on other processes. For example, high irradiances may result in high rates of transpiration which are likely to lead to internal deficiencies of water within the plant and a consequent retardation or cessation of cell division or cell enlargement. Low irradiances, on the other hand, may lead to a retarded development of a plant because of the resulting low rate of photosynthesis.

Direct effects of light on different phases of the growth process have often been studied in comparison with the kind of growth exhibited by similar plants which have developed in complete absence of light. Growth of plants under such a condition can be maintained only if a supply of food is available to the growing parts from a storage organ such as a seed or tuber, or if the plant is artificially supplied with soluble carbohydrates.

Seedlings of dicotyledons which have developed in the absence of light have whitish or yellowish, elongate, spindling stems on which the leaves fail to expand (Fig. 20.2), and have relatively poorly developed root systems. When seedlings of monocotyledonous species develop in the absence of light, the chlorophyll-free leaves are relatively narrow and more attenuated than the leaves of similar plants which have developed in the light. Mesocotyls are also much more elongate than they are in similar plants which have developed in the light (Fig. 20.3). The distinctive development of plants in the complete absence of light is called *etiolation*.

Exposure of a plant to light of a very low irradiance is sufficient to prevent

Fig. 20.2. Seedlings of pea *(Pisum sativum)* grown in light (left) and total darkness (right). **Photograph by Alan Heilman.**

Fig. 20.3. Seedlings of maize *(Zea mais)* about two weeks old, grown in light (left) and in total darkness (right).

the development of any pronounced earmarks of etiolation. When seedlings develop in weak light, expansion of leaves and synthesis of chlorophyll occur much as in strong light, and the internodes do not elongate as much as in similar plants growing in the dark, although the plant will usually present a more attenuated appearance than in strong light. Even short temporary exposures to light result in the development of a much more nearly normal configuration in plants which are otherwise kept in the dark.

The attenuated structure of leaves and mesocotyls (monocotyledons) and of hypocotyls and stems (dicotyledons) of etiolated plants as compared with similar organs developed in the light appears to result chiefly from an increase in the length of their component cells, but is also in part a result of a greater number of cell divisions in the etiolated plants. Light thus has two distinctly retarding effects upon growth in length of plant organs—one on cell division and one on cell elongation. In the first internode of oats, inhibitory effects of light on cell division can be induced at much lower irradiances than inhibitory effects on cell enlargement. Within limits the magnitude of the retarding effect of light upon increase in length of plant organs becomes greater with increased irradiance or with increased exposure time at the same irradiance.

In contrast with its retarding effects on elongation and enlargement of many plant cells, exposure to light generally favors structural differentiation of cells.

Light Quality Effects on Vegetative Growth.   Different qualities of light are obtained in two principal ways for experimental work with plants. The most common method is to interpose a filter of colored glass, cellophane, or other transparent material between a light source and the plant or plant parts to be irradiated. The light source may be the sun but is more commonly an artificial one such as a tungsten filament, fluorescent, carbon arc, or mercury-vapor arc lamp. The wavelength composition ("quality"), of the light falling on the plant will be in part a function of the emission spectrum of the source and in part a function of the transmissive properties of the filter. By following such a technique it is possible to separate out from the visible spectrum a number of bands of wavelengths each corresponding approximately to a known color or narrow range of colors.

A second method which has been employed to obtain different qualities of light in plant experimentation is that of dispersion of a beam of "white light" through a system of prisms and exposing the plant material to different regions of the resulting spectrum. This method can be used only when the plant parts to be irradiated are relatively small, as even with the largest prisms available only narrow bands of light can be isolated by this method.

Comparisons between the effects of one quality of light and another on a given reaction in plants are valid only when their irradiances at the plant surface are equal.

All experiments upon the effect of limited ranges of wavelengths of light upon growth lead to the conclusion that overall development of a plant and increase in its dry weight take place more effectively in the full spectrum of visible light than in any spectral portion thereof. Overall growth of plants in the green is much less than in either the blue-violet or orange-red portions of the spectrum. Some more specific light quality effects are described in more detail in the next two sections.

The Phytochrome System. A number of species of plants are known in which the seeds require exposure to light before they will germinate. One such species which has been the subject of extensive investigations is the Grand Rapids variety of lettuce. If seeds of this plant are allowed to imbibe water in the dark they exhibit a very low percentage of germination. If, however, after imbibition, they are exposed briefly to a narrow band of light in the orange-red region of the spectrum with a peak value at about 660 nm and then returned to the dark, their germination percentage approaches 100 per cent. If exposure to light of this spectral composition is followed promptly by a brief exposure to a narrow band of far-red light with a peak value of about 730 nm, the effect of the 660 wave band is largely offset and a relatively small percentage of the seeds germinates (Fig. 20.4).

In the discussion of this and many related phenomena, the wave band with maximum effect at about 660 nm is generally referred to as red light (R) and the one with maximum effect at about 730 nm as far-red light (FR).

A further discovery regarding the effects of these two light bands on germination of lettuce seed is that they are repeatedly reversible. If the seeds are exposed last to red light, their germination percentage is high; if exposed

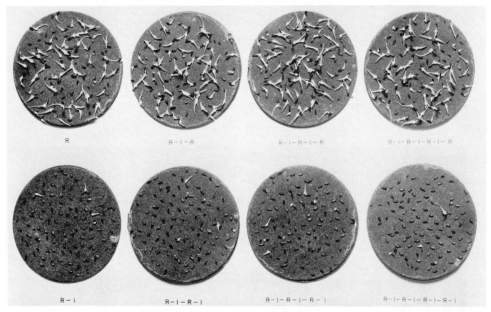

Fig. 20.4. Reversible effects of red and far red light on the germination of lettuce seed (Grand Rapids variety). Photograph from H. A. Borthwick.

TABLE 20.1   PERCENTAGE GERMINATION OF LETTUCE SEEDS (GRAND RAPIDS VARIETY) EXPOSED ALTERNATELY TO 1 MIN OF RED LIGHT AND 4 MIN OF FAR-RED LIGHT (200 SEEDS PER SAMPLE)[a]

Irradiation	per cent Germination
None (dark control)	8.5
R	98
R + FR	54
R + FR + R	100
R + FR + R + FR	43
R + FR + R + FR + R	99
R + FR + R + FR + R + FR	54
R + FR + R + FR + R + FR + R	98

[a]Data of Borthwick *et al.* Proc. Nat. Acad. Sci. **38**, 1952: 665.

last to far-red light, their germination percentage is much lower (Table 20.1, Fig. 20.4).

The seeds of a number of other species have been found to react in a similar fashion upon exposure to red or far-red light. Among these are pepper grass (*Lepidium virginicum*), henbit (*Lamium amplexicaule*), mouse-ear cress

(*Arabidopsis thaliana*), tobacco (*Nicotiana tabacum*), Virginia pine (*Pinus virginiana*), American elm (*Ulmus americana*), and long leaf pine (*Pinus palustris*).

This physiological reaction, generally called the red, far-red reversible photoreaction, plays a part in a wide variety of plant growth phenomena. Further examples of the operation of this photoreaction are illustrated in Fig. 20.5. As

Fig. 20.5. Effects of red and far-red light on the development of Red Kidney bean seedlings. All plants kept in the dark except during treatments as follows, left to right: dark control, 2 minutes of red light, 2 minutes of red light followed by 5 minutes of far-red light, 5 minutes of far red light. Photograph from R. J. Downs.

is shown in this figure, exposure briefly to red illumination induces these differences in the development of a bean seedling as compared with a dark-grown seedling: expanded leaf blades, shorter hypocotyl, longer epicotyl, a straightened plumular hook, and longer petioles. If the seedling is exposed only to far-red light its morphological development is not materially different from a dark-grown seedling. If exposure to red light is followed immediately by exposure to far-red light, the effects of the former are offset and the seedling resembles one which is entirely dark-grown. As with germinating light-sensitive seeds, these effects are repeatedly reversible, and this is shown for two of them in Table 20.2. Red light has an enhancing effect on the expansion of leaf blades and a retarding effect on the elongation of hypocotyls, while far-red light has exactly the opposite effects.

TABLE 20.2 AVERAGE HYPOCOTYL AND LEAF LENGTHS OF DARK-GROWN RED KIDNEY BEAN SEEDLINGS EXPOSED ALTERNATELY TO 2 MIN OF RED LIGHT AND 5 MIN OF FAR-RED LIGHT. (ENERGY VALUE OF RED LIGHT ABOUT 15×10^5 ERGS PER CM^2 PER MIN, ENERGY VALUE OF FAR-RED LIGHT ABOUT 2×10^5 ERGS PER CM^2 PER MIN.[a])

Treatment	Number of seedlings	Hypocotyl length mm	Leaf length mm
Dark control	30	310 ± 2	17 ± 0.8
R	29	230 ± 3	44 ± 0.8
R, FR	28	284 ± 2	21 ± 0.4
R, FR, R	28	222 ± 2	45 ± 0.2
R, FR, R, FR	31	287 ± 2	21 ± 0.4
R, FR, R, FR, R	23	212 ± 3	46 ± 0.3

[a]Data of Downs. *Plant Physiol.* **30,** 1955:2.

Other aspects of plant development in which this reaction is known to play a role are germination of fern spores, synthesis of anthocyanins, elongation of internodes, closing of leaflets in *Mimosa* (sensitive plant) and *Albizzia,* synthesis of a yellow pigment in tomato fruits, and elongation of the flower stalk in at least some long-day plants. In addition there are a number of other plant reactions in which this pigment system is thought to be involved, but regarding which incontrovertible evidence has not yet been obtained. Preeminent among the known roles are the induction of flowers in both long-day and short-day plants as discussed in Chapter 22. In some of these numerous phenomena, red light exerts an enhancing effect and far-red light a retarding effect; in a few others the opposite is true.

The red, far-red photoreaction is activated in both directions by light of very low energy values. For example, the energy values of both the red and far-red light, which were used to accomplish the results shown in Table 20.2 are far less than noonday summer sunlight which is about 5.9×10^7 ergs per cm^2 per minute (Chapter 10). Significant phytochrome mediated effects are accomplished in some kinds of reactions at substantially lower light intensities than those shown in Table 20.2.

Results of the experiments described above and many similar ones have led to the inference that a pigment capable of existing in two interconvertible forms exists in plants. One form of this pigment is considered to absorb red light which changes the pigment to a form of slightly different color. This second form of the pigment absorbs far-red light and can thereby be changed back to the red-absorbing form. Both forms are referred to as phytochrome although chemically they must differ slightly from each other.

Delicate spectrophotometric measurements have confirmed the postulation that such a pair of pigments actually exists in plant tissues. Relatively concentrated solutions of phytochromes have been obtained, starting with extracts from plant tissues. Such solutions of the red-absorbing form of phytochrome (Pr) are blue-green in color and of the far-red absorbing form (Pfr) light green in color. Phytochrome is similar in chemical composition to the phycocyanins. It is a pigment of widespread and probably universal occurrence in green plants. Phytochrome is present in plant tissues in such low concentrations as to be invisible and, generally speaking, is most abundant in meristematic tissues.

The main facts regarding the transformations which phytochromes undergo, as at present understood, can be summarized in the following diagram:

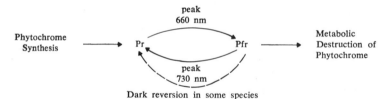

Pr, upon the absorption of red light, is converted into Pfr, which in turn is converted back to Pr upon the absorption of far-red light. As already indicated in the previous discussion, these are low energy reactions.

The lower dotted line indicates that a slow dark reversion of Pfr to Pr occurs in some plants. Such a reversion appears to be more common in dicots than in monocots, although there are also some dicots in which it does not occur. The diagram also indicates that the Pr form is the first to be synthesized; at least this is true in seedlings. Synthesis of phytochrome seems to be inhibited by light and presumably therefore goes on more readily in darkness. Metabolic destruction of Pfr occurs, usually at a relatively slow rate, in both light and dark. There is considerable evidence that the Pfr form of the pigment is the enzymatically active one in most physiological reactions in which it is involved.

In nature, plants are never exposed to narrow bands of red or far-red light, but to alternating periods of daylight and nighttime darkness. In those plants in which there is a slow reversion of Pfr to Pr, exposure to darkness may have the same effect on the development or reactions of a plant as a far-red radiation, but the pigment transformation occurs more slowly.

In sunlight, which includes both red and far-red radiations, an equilibrium between the two forms of phytochrome is attained. However, differences in energy values of the two wavebands, differences in the absorption coefficients of the pigments, and differences in their relative reactivities results in a greater proportion of Pfr present than Pr when plant tissue is exposed to sunlight.

Under an artificial source of mixed radiations, both forms of the pigment will be present, as in sunlight. The effect of the light source will depend upon the ratio of Pr to Pfr present at any given moment within the plant. Fluorescent lights, although in most ways a good artificial source of "white light", are relatively deficient in far-red radiations. In order to obtain well-balanced physiological responses of plants under artificial illumination therefore, it is usually necessary to provide a combination of fluorescent and incandescent lamps, light from the latter being relatively rich in far-red radiations.

It is important to realize that the values 660 nm and 730 nm, commonly quoted in connection with Pr and Pfr, respectively, represent only the wavelengths at which maximum absorption occurs. Actually both forms of phytochrome absorb some light over a wide range of wavelengths within the visible spectrum. This fact is believed to explain certain apparently anomalous results in plant reactions. For example, as described more fully in Chapter 22, if the long nights under which a short-day plant, such as *Chenopodium rubrum* (a pigweed), is growing are interrupted briefly with red irradiation, induction of flowering fails to take place, but if the exposure to red light is followed by a brief exposure to far-red light, the effect of the former is offset and induction of flowering occurs. However, prolonged exposure to far-red light also inhibits induction of flowering. A probable explanation of this result is that the red-absorbing form of the pigment, Pr, absorbs enough of the shorter far-red wavelengths to be partially transformed into Pfr (Fig. 20.6). Although the amount of Pfr thus generated is relatively small, if present for a relatively long time, it is inhibitory to induction of flowering in this short day plant.

The "High Energy" Reaction. Considerable evidence exists that another photomorphogenic reaction, in addition to the one known to be mediated by phytochrome, exists in plants. This reaction, commonly referred to as the "high energy reaction" (HER), has not been very clearly characterized. A number of examples of the apparent operation of such a reaction in evoking growth responses in plants can be cited. One of these is hypocotyl growth in the seedlings of the Grand Rapids variety of lettuce (Fig. 20.7). The blue-violet and far-red parts of the spectrum are most effective in this reaction which does not appear to be reversible. The energy requirement for this and other HER reactions is much higher than that for reactions known to be mediated by phytochrome, being of the order of 1×10^7 ergs/cm^2 or more.

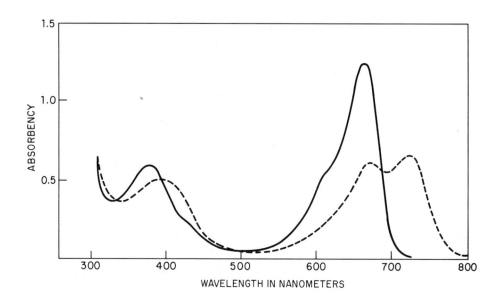

Fig. 20.6. Absorption spectrum of phytochrome from oats following irradiation with red light (dotted line) and with far-red light (solid line). Data of Siegelmen and Butler, *Ann. Rev. Plant Physiol.*, **16**, 1965:385.

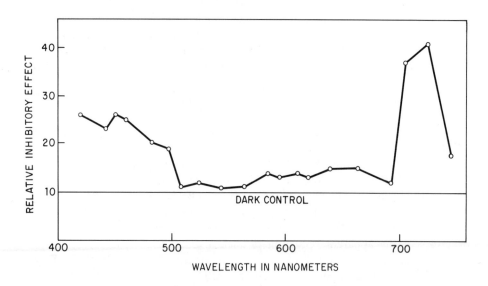

Fig. 20.7. Inhibiting effect of various wavelengths upon the elongation of the hypocotyl of lettuce *(Lactuca sativa)*, a manifestation of the high energy reaction. Data of Mohr and Wehrung, *Planta*, **55**, 1960:444.

The suggestion has been made that a special pigment may be involved in this reaction. However, as shown in Fig. 20.6, both forms of phytochrome absorb significant amounts of radiant energy in the blue-violet part of the spectrum and the Pfr form, of course, absorbs in the far-red. Hence it has been considered at least possible that this reaction is also mediated through the phytochrome system, operating at a relatively high energy level. Regardless of the precise mechanism involved, it seems clear that there is an interplay between the HER and the red, far-red reversible reaction in evoking some morphogenic responses.

Light Duration Effects on Vegetative Growth. In all parts of the world except the tropics and subtropics, marked seasonal variations occur in the lengths of the daylight period. At 40°N latitude (approximately that of Washington, D.C.; Madrid, Spain and Peiping, China) for example, on the shortest day of the year (December 21) the period from sunrise to sunset is a little more than 9 hours, on the longest day (June 21) about 15 hours. In the Southern Hemisphere the annual variation in daylength is the same as that at corresponding latitudes in the Northern Hemisphere, but the season of short days north of the equator is the season of long days south of the equator and vice-versa. At 40° S latitude (approximately that of northern New Zealand and southern South America) the longest day (December 21) is approximately the same length as June 21 at 40° N latitude, and the shortest day (June 21) is approximately the same length as December 21 at 40° N latitude. At higher latitudes the annual variation in daylength is greater; at lower latitudes less. At 30° N latitude (approximately that of Houston, Texas, Cairo, Egypt, and Delhi, India) and at 30° S latitude (approximately that of Santiago, Chili, Johannesburg, South Africa, and Perth, Australia) the shortest days are about 10 hours long and the longest ones about 14 hours. At 50° N latitude (approximately that of Vancouver, B.C., Paris, France, and mid-Russia) and at 50° S latitude, close to which there are only two small land areas, (southern New Zealand and southernmost South America) the short days are about 8 hours in length; the longest ones a little over 16 hours. During the summer months in the Arctic and Antarctic, daylight is continuous; during the winter months darkness is continuous. In tropical regions the daylengths approximate 12 hours the year around. Considering the earth's surface as a whole, therefore, the natural duration of light factor may range from a zero value up to a hundred per cent value —*i.e.,* continuous illumination.

The development of plants as conditioned by the relative lengths of the light and dark periods to which they are exposed, a phenomenon called *photoperiodism,* is one of the most notable of all reactions of plants to their environment. The foundations of our knowledge of this phenomenon were laid in 1920 when Garner and Allard observed the behavior of plants of the Maryland Mammoth variety of tobacco growing in a greenhouse during the winter months. This variety of tobacco grew to a height of 3–5 meters in the summer but did not ordinarily blossom during this season while growing in the vicinity of Washington, D.C. Plants grown in a greenhouse during the winter, on the other hand, did not exceed 1 meter in height, but blossomed profusely and bore excellent crops of seed (Fig. 20.8). These observations led to the hypothesis that the dissimilar development of the tobacco plants during the two seasons resulted from the difference in length of day; subsequently more critically performed experiments confirmed this hypothesis.

Fig. 20.8. Maryland Mammoth tobacco plants from Garner and Allard's experiments. Plant on the left was kept under natural winter daylight (short day); plant on the right was kept under natural winter daylight plus artificial light from sunset to midnight (long day). Photograph from U.S. Department of Agriculture.

Further experimentation on a variety of species soon resulted in the discovery that different kinds of plants react differently to a given length of photoperiod and that the most conspicuous effect of day length is on the reproductive growth of plants. Flowering in some kinds of plants, such as Maryland Mammoth tobacco, is favored by short days, in other kinds by long days, and in still other kinds by a wide range of daylengths (Chapter 22).

The length of the photoperiod also has important effects on many phases of vegetative growth. Discussion in this chapter will be restricted to several of the more important of such effects.

Photoperiodic effects upon reproductive growth may indirectly induce effects upon vegetative growth. Many short-day plants, exposed to long days, grow in height indefinitely, but if exposed to short days vegetative growth in height is soon checked as a result of the differentiation of a terminal inflorescence. Many long-day plants, on the other hand, develop only as leaf rosettes when exposed to short days, elongation of flower-bearing stems occurring only under long-day conditions.

Tuberization in a number of species is markedly influenced by the length of the photoperiod. In a number of varieties of potato, for example, few or no tubers form under long (16–18 hr) photoperiods, but there is a good yield of tubers when the plants are exposed to short (8–10 hr) photoperiods. Similarly, tubers form abundantly on the underground stolons of the Jerusalem artichoke (*Helianthus tuberosus*) when the plants are exposed to 9 hour photoperiods, but do not form when the plants are exposed to 18 hour photoperiods. Furthermore, exposure of only one leaf of a plant of this species to a 9 hour photoperiod while the rest of the plant is under an 18 hour photoperiod induces tuber formation just as if the entire plant were exposed to the short photoperiod, but exposure of the terminal bud to a 9 hour photoperiod while the rest of the plant is at an 18 hour photoperiod has no such effect. Obviously the leaves are the locus of a photoperiodic reaction, the effects of which are communicated in some manner or another to the underground organs of the plant and influence their development. Such an effect can be visualized most easily in terms of a hormonal mechanism.

The development of bulbs by certain species of plants is markedly influenced by the length of the photoperiod to which the plant is exposed. Bulb formation in most varieties of onions, for example, is favored by relatively long photoperiods, the minimum effective photoperiod varying from about 12 to 16 hours according to the variety.

As described more fully in Chapter 24, the length of the photoperiod has important effects on the dormancy of many temperate zone woody plants. The shortening photoperiods of late summer and early fall are a factor in inducing dormancy in many such species. Contrariwise, the lengthening photoperiods of late winter and early spring are a factor in breaking the dormancy of many such species.

Effects of the Internal Water Status on Vegetative Growth.   The dynamic condition of the water in a plant is largely controlled by the opposing effects of the processes of transpiration and water absorption (Chapter 8). Whenever the rate of transpiration exceeds that of water absorption for any appreciable period of time, the volume of water within the plant shrinks. This results in a diminution of cell turgidity, more negative water potentials in the cells, and a decrease in the hydration of the protoplasm and cell walls. A decrease in the hydration of the protoplasm in the cells of any meristematic tissue usually results in a cessation or checking of cell division or of cell enlargement or of both. Contrariwise, a shift in environmental conditions which brings about, directly or indirectly, an increase in the hydration of the protoplasm of a meristem usually results in an increase in the rate of these two phases of growth if no other factors are limiting.

Surprisingly small reductions in the water potential of cells have marked effects on growth in at least some tissues. Enlargement growth in sunflower leaves, for example, ceases almost entirely when the water potential falls to −4 bars. Water potentials of this value or lower often prevail in sunflower leaves during the daylight hours. Whenever such a situation prevails most enlargement of such leaves occurs during the hours of darkness.

Not all phases of growth are equally affected by a diminution in the volume of water within a plant. Both cell division and cell enlargement are adversely influenced by even a moderate deficiency of water. It should be recalled, however, that when an internal deficiency of water exists within a plant, water often moves from other organs towards the meristematic tissues (Chapter 8). The cell enlargement phase of growth is seemingly checked more by an internal shortage of water than the cell division phase. Maximum enlargement of cells during growth apparently can be attained only when the water supply to them is not appreciably restricted. If a shortage of water prevails within the plant, the enlargement of cells terminates earlier than when water is present in abundance, and structural differentiation of the cells ensues sooner. The deficiency of water, in general, favors the structural differentiation of cells rather than the cell division and enlargement phases of growth.

Water, as it affects growth processes, operates primarily as an internal factor; but as such an internal factor, it is influenced by a galaxy of environmental conditions. Any external factor which affects the rate of transpiration or the rate of absorption of water will therefore have an influence on the rate of growth. While temporary periods of internal water deficiency may result from the effects of other environmental factors, principally those influencing transpiration rates, prolonged periods of internal dearth of water in plants mostly result from an inadequate supply of available soil water.

In many habitats, periods during which available water is present in the soil may alternate with periods during which there is virtually no available water present. Most kinds of plants can survive such intermittent periods of soil drought if none of them is too prolonged. In general, however, the more frequent and more extended such periods of drought are during a growing season, the less the overall growth of a plant.

For most kinds of plants, a soil-water content in the vicinity of the field capacity is the most favorable to continued growth. With a decrease in the soil-water content, marked effects on growth do not appear until the permanent wilting percentage is approached (*cf.* effects on absorption of water, Chapter 7). At the permanent wilting percentage all growth ceases.

As the soil-water content increases above the field capacity, the growth of most species will sooner or later be retarded as a result of the concomitant decreased aeration of the soil.

While the roots of plants are usually in contact with liquid water in the soil, the aerial parts are continuously bathed in a gaseous medium in which water molecules are present as vapor. The humidity of the atmosphere is thus another water factor which affects growth and other processes in plants. In general, small variations in the vapor pressure of the atmosphere have very little influence on the water content of plants, and hence no appreciable effect on growth rates. This is especially true whenever the soil-water supply is adequate. When the differences in vapor pressure are considerable, however, marked effects on growth result. This is exemplified by the growth of two groups of tomato plants, both at 21°C, and provided with adequate supplies of soil water, exposed to vapor pressures of 8.8 mbars and 22.7 mbars, respectively. The plants at the higher humidity grew much faster and developed thin-walled, succulent tissues. Those at the lower humidity, in which there was undoubtedly a more marked internal deficiency of water, grew relatively slowly, developed thick-walled cells, and in general lacked succulence. At the higher humidity, cell division and enlargement, resulting in the development of succulent tissues, were obviously favored more than structural differentiation. At the lower humidity, structural differentiation, principally in the form of a thickening of walls, predominated over cell division and enlargement.

In general, plants which have developed under adverse soil-water conditions are dwarfed or stunted. Excellent naturally occurring examples of the effect of soil drought on the development of plants are always evident during years of deficient precipitation in normally mesic habitats. Wheat, maize, and other cereal crops, for example, never attain their usual stature during a drought season. Similarly the annual shoots of woody species do not increase as much in length during droughts as during seasons when water is present in abundance. Diameter growth of woody stems is also often retarded during the drought years. On the basis of this fact, numerous attempts have been made to trace past climatic cycles, in regions in which trees that have attained a considerable age are found, by a study of the growth rings in the trunks. Caution must be exercised, however, in interpreting such data since a wet period in late summer following an earlier dry period may cause the appearance of a second ring in the same year.

An inadequate soil-water supply may check the growth of a plant more at certain stages in its development than others. In many species, vegetative growth is more likely to be retarded by soil-water deficiency than is the development of reproductive organs. In some species there are certain critical periods during which the water supply must be adequate or else pronounced modifications in the morphogenic development of the plant will occur. This is true of wheat and the other small grains in which elongation of the inflorescence-bearing stem does not occur until the internodes and inflorescence have been differentiated. Deficiency of water during this elongation or "shooting" period exerts a marked retarding effect upon the growth in height of such species, and also upon their yield.

Effects of Concentration of the Soil Solution on Vegetative Growth.
Although the concentration of solutes in the soil solutions of most soils is so low
that their osmotic potentials are usually only negative by a fraction of a bar,
certain noteworthy exceptions to this statement exist. The most important of
these are salt marsh, saline and alkali soils in which soil solution osmotic potentials may attain negative values of tens or even hundreds of bars. Soil-solu-
tion osmotic potentials ranging down to at least -10 bars may also exist under
certain cultural conditions. Examples are soils which have been irrigated with
water containing relatively high concentrations of salts, or soils which have been
injudiciously overfertilized. This latter situation sometimes occurs in greenhouses. The practice of salting icy highways also sometimes results in osmotic
potentials of considerable negativity in strips of soil bordering on such highways.

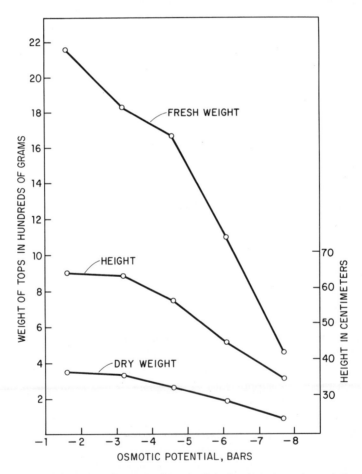

Fig. 20.9. Relation between the osmotic potential of substrate and vegetative growth of
tomato. Data of Hayward and Long, *Plant Physiol.*, **18,** 1943:559.

An increased negativity in the osmotic potential of the soil solution leads to a decrease in the rate of absorption of water and hence results in a retardation in growth. As an example, the relation between the osmotic potential of the soil solution and the growth of tomato plants is shown in Fig. 20.9. Although the primary cause of the retarding effect of an increased concentration of solutes in the soil solution is undoubtedly an osmotic one, some kinds of solutes, of which magnesium salts are an example, when present above a certain concentration, exert secondary toxicity effects which further retard growth.

The statements made in the above paragraph do not hold, at least not rigorously, for plants indigenous to saline or alkali soils. Mangroves, of which there are a number of species, grow only in coastal areas of tropical or semitropical regions with their roots in a sea-water saturated substratum. The osmotic potential of sea water is about -22 bars. Species which can survive and grow when rooted in alkali soils of high salt content include shadscale (*Atriplex*), picklewood (*Salicornia*) and iodine bush (*Allenrolfea*).

Effects of Concentration of Gases in the Soil Atmosphere on Vegetative Growth.   Except in very well-aerated soils, such as sands or sandy loams, the concentration of carbon dioxide in the soil atmosphere is usually higher and the concentration of oxygen usually lower than in the aerial atmosphere (Chapter 7). Marked deficiencies in soil aeration exist most commonly in overwet soils, but such a condition may also be present in fine-pored, tightly packed soils even when they are relatively dry.

In general, inadequate soil aeration results in a retardation in the growth of most kinds of plants. The principal exceptions to this statement are plants native to marshy or boggy terrains. As an example of such an effect we may consider the growth of sunflower and soybean plants which developed in unaerated sand and loam as contrasted with other plants of the same species grown in similar but aerated soils. Aerated plants were taller and heavier and had larger and more fibrous root systems and smaller shoot-root ratio (Chapter 23) than the unaerated plants. The ash, calcium, potassium, and phosphorous contents per plant were greater in terms of absolute weight in aerated than in unaerated plants. Similarly the total weights per plant of starch, total sugars, and nitrogen were greater in the aerated plants than in the unaerated plants.

The retarded growth of plants in poorly aerated soils undoubtedly results largely from the reduced absorption of water (Chapter 7) and from the reduced absorption of mineral salts (Chapter 15) which occur under such conditions. Carbon dioxide concentrations sufficiently high to exert a retarding effect on growth are not often attained even in poorly aerated soils. A carbon dioxide concentration of at least 30 per cent, for example, was necessary in the gas mixture saturating a solution culture to result in even a small inhibiting effect on the growth of cotton roots. Concentrations of carbon dioxide in the soil atmosphere apparently do not often exceed 15 per cent. On the other hand, concentrations of oxygen low enough to exert a retarding effect on growth are frequently attained in poorly aerated soils. Growth is retarded in the roots of many species when the oxygen concentration of the soil drops below 10 per cent. In poorly aerated soils, oxygen concentrations are frequently less than this and often approach a zero value. A deficiency of oxygen therefore appears to be a more important factor in causing diminished growth of both roots and plants as a whole in poorly aerated soils than an excess of carbon dioxide.

Effects of Mineral Elements on Growth.   Although it is conventional to speak of the effects of mineral elements on plant growth, strictly speaking they do not have effects as such, but only when present in the form of ions or as constituents of molecules. The many and complex relations of the various essential and some nonessential mineral elements to the growth and metabolism of plants have already been discussed in Chapter 15.

Effects of Nitrogen on Growth.   In the literal sense nitrogen as such has no effects on the metabolism or growth of the higher green plants. The word is conventionally used, however, for brevity, as a blanket designation for the various nitrogenous compounds which participate in the physiological processes of plants.

Both carbohydrate and nitrogenous compounds are necessary for the development of any plant. Deficiency of either soon results in the appearance of characteristic and recognizable peculiarities of growth. The supply of carbohydrate and nitrogenous compounds within a plant is influenced by many factors. Among these are the reciprocal relationships which exist between these two kinds of compounds. Primary synthesis of amino acids and similar compounds occurs only at the expense of carbohydrates or their derivatives, which serve both as building material (together with nitrates, or other nitrogen-containing compounds) and as a source of energy (Chapter 16). Rapid amino acid synthesis therefore results in a diminution in the proportion of carbohydrates in a plant, while plants in which amino acid synthesis occurs slowly will often be proportionately rich in carbohydrates.

As described in the preceding chapter, continued growth of any meristem requires that there be maintained to it a supply of both carbohydrate and amino acids. Both of these kinds of compounds are assimilated in relatively large quantities, especially during the cell division and enlargement phases of growth, and next to water, these two classes of substances are utilized in greatest quantities during growth. Considerable quantities of carbohydrates are also used up in any actively growing meristem in the process of respiration.

If the supply of amino acids to any actively growing vegetative meristem is abundant relative to the supply of carbohydrates, a large quantity of protoplasm will be formed relative to the amount of cell-wall material constructed. The resulting cells will usually be large and thin walled and will contain an abundance of protoplasm. Tissues composed mostly or entirely of such cells are usually soft and succulent. The proportion of mechanical tissue developing under such metabolic conditions is ordinarily small. If the opposite metabolic situation prevails within a plant, and carbohydrates are relatively more abundant than amino acids, proportionately more cell-wall structures and less protoplasm will be fabricated. The resulting cells will be small and thick walled and will contain comparatively little protoplasm. Tissues composed largely or entirely of such cells are usually compact and more or less woody. The development of fibers and of mechanical tissues generally is also favored by an excess of carbohydrates relative to nitrogenous compounds.

Effects of Atmospheric Gases and Pollutants on Growth.   At all lower altitudes the concentration of oxygen in the atmosphere is virtually constant and does not need to be considered as a variable which influences the development of plants under natural conditions. At high altitudes, however, the partial pressure of oxygen in the atmosphere is substantially less than at lower altitudes. Such a reduced oxygen concentration almost certainly has an effect on respiration and may indirectly affect the development of plants. Variations in the carbon dioxide concentration, although occurring over a much narrower range in the aerial than in the soil atmosphere, are often sufficient to have a considerable effect on photosynthesis (Chapter 11). The water vapor content of the air varies considerably and the pronounced influences of this factor upon transpiration, the internal water relations of plants, and indirectly upon growth have already been considered in preceeding chapters and in an earlier part of this one.

A non-gaseous aerial factor in the environment of many plants is salt-water spray, droplets of which may be carried a mile or two inland by only moderately strong winds, and under hurricane conditions much farther. Ocean-water spray is highly toxic to many land plants, young leaves and twigs being the most susceptible; but some species are much more liable to injury from such spray than others. The sheared-off configuration of many species of seaside trees and shrubs results at least in part from salt-spray injury. The distribution of plants in coastal areas is also controlled in part by their tolerance to ocean-water spray.

Atmospheric contaminants or "pollutants" from various industrial and domestic sources have become increasingly prevalent in recent years, especially in the more highly industrialized countries. Some of these substances are present as gases, others as finely particulate matter, often of colloidal dimensions. The principal atmospheric pollutants are sulfur dioxide, fluorides, ethylene, chlorine, ammonia, and the complex of variable composition commonly called "smog." Once a certain, and usually very low, concentration of any one of these substances or a combination of them is exceeded in the atmosphere the resulting effects on many kinds of plants are highly injurious and often lethal. Damage to crop plants, ornamentals, and even natural vegetation has been extensive in some areas, resulting in serious economic and aesthetic losses. Different species of plants, as might be expected, differ greatly in their susceptibility to injury from the various atmospheric pollutants.

The symptoms of injury from such causes vary from species to species and with the specific atmospheric constituent or constituents causing the damage. Identification of such symptoms constitutes a highly specialized field of study.

One of the first of the atmospheric pollutants to be recognized was sulfur dioxide. This gas is released into the atmosphere principally from copper, iron, zinc and lead smelters. Combustion of coal and oil also contributes to the content of this gas in the atmosphere. Sulfur dioxide, in any appreciable concentrations, is so highly toxic to plants that the countryside in the vicinity of smelters is often virtually denuded of vegetation. Most species of plants are injured by an exposure of only one hour to an atmosphere containing as little as one part of this gas in a million.

Under certain situations fluorides attain highly phytotoxic concentrations in the atmosphere. The term fluoride includes the gas hydrogen fluoride plus a number of its salts, which can be present in the atmosphere in a highly particulate form. Atmospheric fluorides originate from various industrial processes of which the refining of aluminum is a major one. Injury as a result of atmospheric fluorides ranges from chlorosis and necrosis of leaves to death of the entire plant, depending upon their concentration in the atmosphere and the sensitivity of the species, the latter being a highly variable factor. In this connection it should be recalled that fluorides act as enzyme inhibitors (Chapter 9). Another aspect of fluoride toxity is that livestock may be injured by browsing on forage plants in which relatively large quantities of fluorides have accumulated.

As discussed in Chapter 18, ethylene operates as a hormone in plants, but in somewhat higher concentrations than those of hormonal significance it often exerts inhibitory or toxic effects on plants.

Death or injury of house plants, plants in greenhouses, and even shade trees along a street have been traced to ethylene in the atmosphere. This gas is one of the products of the incomplete combustion of many kinds of organic compounds such as natural gas, coal, wood, and agricultural residues. Ethylene, along with other substances to be discussed later, is a constituent of automobile exhaust gases. This is probably the major source of atmospheric ethylene in metropolitan areas. Numerous crop losses in and near such areas can be largely ascribed to this gas.

Two other gases, ammonia and chlorine, sometimes escape into the atmosphere, largely as a result of industrial operations, in sufficient quantities to reach phytotoxic concentrations.

The word "smog" has no precise meaning but has been applied to a variety of smoky or hazy conditions prevailing in the atmosphere of industrialized regions with which various pollutants are associated. All of the previously mentioned atmospheric pollutants, plus others, may be found in one kind of smog or another.

Some of the most extensive investigations of a smog condition have been made in the Los Angeles area of California where, in addition to the irritations which it contributes to human existence, smog causes extensive damage to plants. Serious losses of agricultural crops as well as destructive damage to many kinds of ornamental plants have resulted.

The most phytotoxic constituents of the Los Angeles type of smog were found to be certain kinds of oxidants, specifically ozone and a group of compounds known as peroxyacyl nitrates, of which peroxyacetyl nitrate is the most injurious. All of these compounds originate as a result of reactions between hydrocarbons and nitrogen oxides, these latter compounds themselves being phytotoxic. The hydrocarbons and nitrogen oxides originate in the atmosphere as a result of various combustion processes and especially from the exhaust fumes of automobile engines.

SUGGESTED FOR COLLATERAL READING

Bickford, E. D., and S. Dunn. *Lighting for Plant Growth.* Kent State University Press, Kent, Ohio, 1972.

Evans, L. T., Editor. *Environmental Control of Plant Growth.* Academic Press, Inc., New York, 1963.

Jacobsen, J. S., and A. C. Hill, Editors. *Recognition of Air Pollution Injury to Vegetation: A Pictorial Atlas.* Air Pollution Control Association, Pittsburgh, 1970.

Kozlowski, T. T., Editor. *Water Deficits and Plant Growth,* 3 vols. Academic Press, New York, 1968, 1968, 1972.

Levitt, J. *Responses of Plants to Environmental Stresses.* Academic Press, Inc., New York, 1972.

Treshow, M. *Environment and Plant Response.* McGraw-Hill Book Company, New York, 1970.

Wilkins, M. B., Editor. *Physiology of Plant Growth and Development.* McGraw-Hill Publishing Company, Ltd., Maidenhead, England, 1969.

The term "reproductive growth" is used in this chapter to refer, in general, to the formation of flowers, fruits, and seeds. The discussion refers primarily to angiosperms since fruits, in the usual sense of the word, do not develop in gymnosperms, the other group of seed plants. Reproductive growth embraces a complex of closely interrelated processes and phenomena, some of which occur sequentially, while others overlap in time.

The principal events which usually occur during the reproductive growth of an angiosperm are initiation of flower primordia, maturation of floral parts, development of the pollen grains within the anthers, development of an embryo sac with an egg and a fusion nucleus in each ovule in the ovulary, pollination, formation of two sperms from the generative nucleus in the pollen grain or tube, growth of the pollen tube from the stigma into the ovule, fertilization (fusion of one sperm with the egg) and triple fusion (fusion of the other sperm with the fusion nucleus), development of the embryo from the fertilized egg, development of the endosperm from the endosperm nucleus, development of the seed from the ovule, and development of the fruit from the ovulary or from the ovulary and adjacent tissues.

Although all of the processes and phenomena listed previously are phases of reproductive growth, two main stages, the flowering stage and the fruiting stage, can be clearly distinguished. Not only are these two stages morphologically distinct, but they are physiologically even more distinct. Flowering, despite its morphological and physiological complexity, is a relatively transitory phase in growth. Vegetative growth and fruit growth, on the other hand, are usually processes of considerable duration and physiologically resemble each other more closely than either resembles flowering. These two kinds of growth are usually conditioned much more by the general nutritive conditions of the plant than is flowering, which appears to be predominantly under hormonal control.

Initiation and Development of Flowers.   Some vegetative apical stem meristems continue to grow as such indefinitely; but sooner or later in the life history of most plants, some of them become transformed into reproductive meristems. The phrase "sooner or later" may cover a span of time from a few days to many years. Many herbaceous plants complete a life cycle from the inception of vegetable growth to the maturation of fruits within a few weeks. Examples are many alpine species which have a short growing season, as well as many semidesert herbs which flourish only during a short rainy season. Other plants, of which bamboo and the century plant are examples, remain in the vegetative state for many years before any reproductive growth is initiated. Furthermore, the length of time which a meristem remains in the vegetative state before undergoing transformation to a reproductive meristem differs greatly from one meristem to another on a given plant. Even under the most favorable environmental conditions conversion of a vegetative into a reproductive meristem does not occur until the former has attained a metabolic status often called "ripeness to flower." This aspect of flower initiation is discussed more fully in the next chapter. Even when a meristem has attained the internal potentiality of initiating floral parts it may remain temporarily or indefinitely in the vegetative state because of unfavorable environmental conditions.

Some meristems start differentiating as reproductive meristems almost from the moment of their inception. This is true, for example, of the apical meristems of the lateral flower buds of many woody plants such as peach, redbud, elm, and some species of maples. Among temperate-zone woody plants, such floral meristems usually differentiate during one growing season but the flowers do not open until the next.

Examples of delayed transformation of a vegetative into a reproductive meristem can be seen in any herbaceous plant which bears a terminal inflorescence at the end of a leafy shoot which has first elongated for some weeks or months before the initiation of flower primordia begins. Examples include tobacco, mints, aster, goldenrods, phloxes, and many cereals. Similar growth behavior is exhibited by many woody plants such as dogwood, magnolias, rhododendrons, spiraeas, sumacs, and aralias. In many such woody species, conversion of apical vegetative meristems to reproductive meristems occurs toward the end of a period of shoot growth, resulting in the formation of terminal flower buds which do not open until the next growing season.

The first steps in the transformation of a vegetative to a reproductive meristem are invisible physiological changes resulting in metabolic conditions within the meristematic cells which completely alter the differentiation pattern of the meristem.

In some species of plants the transformed meristem becomes in effect an inflorescence primordium from which an influorescence bearing a number of flowers develops; whereas in the other species only a single flower becomes differentiated from the transformed meristem. Figure 21.1 shows the morphological stages in the transformation of a vegetative into a reproductive meristem as it occurs in green pepper. The first microscopically visible change in the transformation of a vegetative into a reproductive meristem is one of configuration. Growth of the central portion seemingly is inhibited, and the meristem becomes flattened on top instead of more or less conical. Small protuberances develop from this modified meristem in a regular spiral or whorled arrangement. These moundlike protuberances are the primordia from which flower parts develop in a manner analogous to that by which leaves develop from similar protuberances on a vegetative meristem. A marked difference in the development of the two meristems, however, is that there is no elongation of the axis between successive floral primordia such as usually occurs between successive leaf primordia. Although details of flower development differ considerably from one species to another, the fundamental pattern followed is similar in all species.

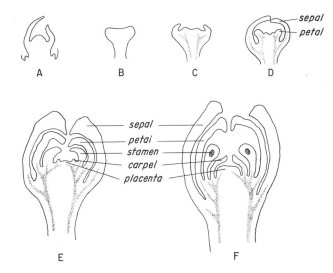

Fig. 21.1. Stages in the transformation of a vegetative to a floral meristem in pepper, as seen in longitudinal section: (A) Vegetative stem tip (*cf.* Fig. 19.3). (B) Apical meristem has become flattened. (C) Initiation of the first floral parts (sepals) has begun. (D) Later stage, showing the initiation of petals. (E) Still later stage, showing the initiation of stamens and carpels. (F) Well-differentiated flower bud, showing an early stage of carpel formation. After Cochran, *J. Agr. Research,* **56,** 1938:396, 397.

At first the floral parts of most flowers are tightly enclosed within a flower bud. Subsequently expansion of the flower bud into the opened flower occurs, a stage in flower development called *anthesis.* While sepals or both sepals and petals are present in the majority of flowers, the only essential parts are the stamens and the pistils. The majority of plants are bisporangiate species and have both of 'these structures present in each flower. In some plants (monoecious species), however, some flowers are staminate and others pistillate, but both kinds are borne on the same plant. Examples of monoecious species are maize, cucumber, squash, and many hardwood trees such as birch, beeches, and oaks. In other plants (dioecious species) some individuals bear only staminate flowers, others only pistillate flowers. Examples of dioecious species are date palm, holly, hemp, spinach, and willows.

The most pronounced physiological and developmental changes in the plant's existence occur in the brief period of floral differentiation. The whole structural pattern changes profoundly during the transformation of a vegetative to a reproductive meristem, reflecting equally deep-seated changes in metabolism. The rapid change from the male physiological state (stamens, etc.) to the female physiological state (carpels, etc.) over a short distance is another striking feature of the differentiation of most flowers. Physiologically these two states may differ from each other as much as either of them differs from the vegetative state.

Young developing flowers are centers of a myriad of metabolic changes. The metabolic complexity within the cells of developing flower parts is probably greater than within any other plant tissue. Very little is actually known, however, of the specific metabolic conditions which are associated with the reproductive state in general as contrasted with the vegetative state, or of the physiological differences between the male and female states. That hormones play a role in the initiation of flowers does, however, seem clearly indicated (Chapter 18, Chapter 22). Respiratory activity, an indirect index of metabolic activity, is always high in young floral meristems, but that is also true of other meristems. Assimilatory rates are also high, and there is a continuous translocation toward a developing flower of foods, water, compounds containing mineral elements, and hormones.

Pollen and Pollination.   Pollen grains differ greatly in size, ranging from about 5 μ to 200 μ in diameter, and also in configuration (Fig. 21.2) from one plant to another, but their physiological role is similar in all species. The quantity of pollen produced also differs greatly from one kind of plant to another. Some plants produce pollen only sparingly; others, with great prodigality. A single maize plant, for example, is estimated to release 50,000,000 pollen grains. Only about 1,000 pollen grains are required if fertilization of all the egg cells in all of the ovules of one maize plant is to be accomplished.

Ontogenetically the pollen grain and the pollen tube that develops from it represent the male gametophyte. When first formed, a pollen grain contains two nuclei, the tube nucleus and the generative nucleus. These nuclei are haploid, as a result of meiosis in the microporocyte, the forerunner of the pollen grain. The generative nucleus subsequently divides by mitosis, either

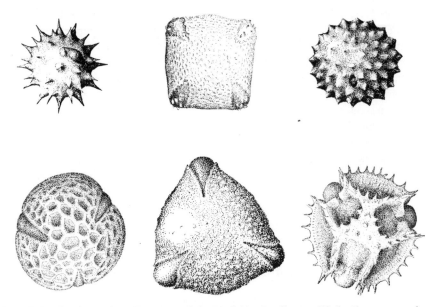

Fig. 21.2. Pollen grains. Top row, left to right: Sunflower *(Helianthus annuus)*, ash *(Fraxinus americana)*, ragweed *(Ambrosia trifida)*. Bottom row, left to right: privet *(Ligustrum ibota)*, white oak *(Quercus alba)*, dandelion *(Taraxacum officinale)*. From Wodehouse, *Pollen Grains*, McGraw-Hill Book Company, New York, 1935.

in the pollen grain or in the pollen tube, resulting in the formation of two identical haploid sperms which are the male gametes.

Pollination, or transfer of pollen from the anther to the stigma of the same or another flower, is effected principally by wind and insects. Some species are entirely wind-pollinated, some are entirely insect-pollinated, while in some both modes of pollen transfer may occur. Other less common agents of pollen dispersal are gravity, water, and certain animals, including man.

Self-pollination is the transfer of pollen to the stigma of the same flower, or of another on the same plant. Cross-pollination is the transfer of pollen from the anthers of one plant to the stigmas of another plant of the same kind. When the two plants involved in cross-pollination carry identical chromosome complements, however, the genetic effect of the "cross-pollination" is the same as self-pollination. Self-pollination is the usual occurrence in a number of species, such as wheat, oats, barley and tobacco, in which the pollen germinates before the flowers open. In some species both self- and cross-pollination may occur. Cross-pollination is obligatory in self-sterile plants, if normal fertilization is to occur.

Numerous pollen grains may alight on or be deposited on a single stigma. Under conditions of abundant pollination, from 600 to 900 pollen grains may be present, for example, on a single stigma of jimsonweed. Even "foreign" pollen grains sometimes germinate on a stigma, but growth of the pollen tube usually fails to occur or takes place very slowly. If a pollen grain from one plant falls on a stigma of the same plant, germination usually occurs, although not always. Pollen from a given plant often fails to germinate on the same individual plant; or if germination occurs, the pollen tube grows sluggishly. These are common causes of self-sterility in plants. On the other hand, pollen from a given plant usually germinates on the stigma of another plant of the same species, provided the stigma has reached an appropriate stage of development and environmental conditions are favorable.

Germination of a pollen grain, under favorable environmental conditions, commonly occurs within a few minutes of its making contact with the stigmatic surface. Usually only one pollen tube elongates from a pollen grain, but in some species more than one develops. When a plural number of tubes forms, however, ordinarily all except one soon cease growing. Even branching of pollen tubes may occur in some species.

The pollen tube serves as the mode of transport of the sperms from the stigma into the embryo sac. The distance which the pollen tube must grow if this transfer is to be accomplished is very short in some species, but may be as much as 30 cm in other kinds of plants, such as maize, with long styles (silks).

The time interval between germination of a pollen grain and fertilization differs greatly from one kind of plant to another, but for many lies within the range of 12 to 48 hours. In a few plants, however, such as barley, this time interval is less than an hour and in some, such as certain oaks and pines, it may be months, even exceeding a year in a few species. The absolute rate of elongation of pollen tubes ranges up to about 34 mm per hour.

The rate of growth of pollen tubes is markedly influenced by environmental conditions, especially temperature. In tomato, for example, their maximum rate of growth occurs at about 20°C, being less at higher or lower temperatures.

Apart from environmental conditions, the rate of pollen-tube growth is markedly influenced by the degree of physiological "compatibility" between the pollen tube and the tissues of the pistil. Incompatibility usually exists between the pollen and pistillary tissues in many species when self-pollination occurs. Although germination of the pollen grains may occur, the pollen tubes grow very slowly, and fertilization seldom results. When cross-pollination occurs in the same species, however, rapid growth of the pollen tubes is the usual occurrence. The exact physiological basis of such incompatibilities is unknown.

Even when a high degree of physiological compatibility exists between the pollen tube and pistillar tissues, some pollen tubes elongate through the tissues of the style and ovulary much more rapidly than others. The tube which enters a given embryo sac first is usually the one from which the sperms come which accomplish fertilization and triple fusion. If an ovulary contains but one ovule, only one pollen tube is required if fertilization of the egg in the enclosed embryo sac is to be accomplished; if it contains more than one ovule, one pollen tube is required for each if all egg cells in all the embryo sacs are to be fertilized.

The growing pollen tube is essentially parasitic upon the tissues of the pistil from which it undoubtedly obtains water, 'foods, mineral-containing compounds and probably also hormones. Pollen tubes are known to contain numerous kinds of enzymes some of which doubtless aid in their growth through the pistillary tissues. Enzymes or hormones released from growing pollen tubes apparently influence the development of the pistil and organs enclosed within it.

Many kinds of pollen grains germinate under artificial conditions if strewn on the surface of a sucrose-water or sucrose-agar (about 1.5 per cent)— water medium. The most favorable concentration of sucrose for germination differs considerably from one kind of pollen to another, but is usually in the range of 5–25 per cent. Germination of at least some kinds of pollen on such media occurs within a few minutes and growth of the pollen tube may be very rapid. Addition of certain substances, notably calcium and boron, to such media, has an enhancing effect on the germination percentage and growth rate of many kinds of pollen.

The length of time for which pollen grains of different species remain viable under ordinary air dry conditions ranges from a few hours to several months. The pollen of grasses, including maize, rye, barley, and wheat is notoriously short-lived, often retaining its germinative capacity for only a few hours and seldom for more than a few days even under the most favorable conditions.

The life duration of many kinds of pollen can be extended, in some species up to at least a year, by storage at low humidities in conjunction with relatively low temperatures. Prolongation of the life of pollen is often an important consideration in certain plant breeding problems—for example, when it is desired to cross two varieties which bloom at different seasons.

Pollen as the Causal Agent of Hayfever.   Wholly apart from its role in reproduction, pollen is of considerable medical interest because it is the causal agent in the well-known ailment of hayfever. This allergy is induced in many persons when the free-floating pollen of the atmosphere comes in contact with the mucous membranes of the eyes, nose, throat, bronchial tubes, and lungs. The number of kinds of pollen which cause hayfever is relatively small. Only pollens which are liberated in abundance and which are wind-borne are causal agents in this malady, and of them only those pollens which possess allergenic toxicity to at least some humans actually do cause hayfever. Pines, for example, are extravagant producers of pollen, but none is known to be a hayfever-causing plant. In much of temperate North America three main hayfever seasons can be distinguished. An early spring season, usually the least severe and of the shortest duration, results from the presence of certain tree pollens in the atmosphere, especially those from elms, maples, willows, poplars, birches, and oaks. Early summer hayfever is caused principally by the pollens of a number of species of grasses, especially June grass, orchard grass, timothy, and redtop. The incidence of early summer hayfever is much greater than that of spring hayfever, and the season is considerably longer. Late summer hayfever, which accounts for more victims of this affliction than the other two seasons combined, is caused principally by the pollen of the ragweeds, sagebrushes, cockleburs, false ragweeds, and salt bushes.

Initiation of Embryo and Endosperm Growth.   *Fertilization and Triple Fusion.*   The essential features of sexual reproduction is the union of an egg and a sperm, a process called *fertilization.* In the angiosperms the egg is an integral part of the embryo sac, which is enclosed within an ovule. In the majority of angiosperm species, the fully developed embryo sac is an eight-nucleate structure (Fig. 21.3). In some species, however, the embryo sac is four-nucleate; in others sixteen-nucleate. Ontogenetically the embryo sac represents the female gametophyte in the angiosperms. The nuclei of the embryo sac are normally all haploid as a result of meiosis in the megasporocyte nucleus, followed by subsequent mitotic divisions of resulting nuclei.

Fig. 21.3. Embryo sac (eight-cell stage) of *Epipactis.* Egg cell and two synergids at upper end: three antipodals at lower end. Subsequent union of the two nuclei nearest the center (polar nuclei) results in the formation of the fusion nucleus. After Brown and Sharp, *Botan. Gaz.,* **52,** 1911: Plate X.

Access of the pollen tube into the ovule is commonly, although not always, achieved through the micropyle. The tube then penetrates the nucellus and passes into the embryo sac between the egg and an adjacent nucleus. Fertilization of the egg by a sperm from the pollen tube usually occurs shortly after the tube enters the embryo sac.

In addition to the egg, especial significance attaches to the polar nuclei in the large central portion (Fig. 21.3) of the embryo sac. These two nuclei join, forming a fusion nucleus, with which the second sperm from the pollen tube unites, resulting in a triploid endosperm nucleus. This process is called *triple fusion*. In a few species only one polar nucleus is present, which unites with a male gamete in forming the endosperm nucleus, and in some species more than two polar nuclei are formed during embryo sac development, all of which fuse with a sperm in the formation of the endosperm nucleus. Ordinarily only one pollen tube penetrates an embryo sac. Occasionally more than one pollen tube may enter, and the sperm which unites with the egg may come from a different pollen tube than the sperm which unites with the fusion nucleus.

Apomixis. The reproductive apparatus does not operate identically in all plants, and many deviations from the most common mechanism as described above are known. Modes of asexual reproduction (*i.e.,* reproduction without fertilization) which outwardly appear to be sexual reproduction are of regular occurrence in some species and occasional in others. The generic term for this kind of reproduction is *apomixis,* of which three main types are recognized.

In one kind of apomixis the haploid egg develops into an embryo sporophyte without fertilization, either autonomously (*haploid parthenogenesis*), or under the influence of a promotive influence from the growing pollen tube or the process of triple fusion (*haploid pseudogamy*). Less commonly an embryo may develop without fertilization from another nucleus of the embryo sac (*haploid apogamety*). Plants resulting from this kind of apomixis have the haploid number of chromosomes and are usually sterile.

In a second type of apomixis the cells of the embryo sac, because of some deviation from the usual course in their development, all contain the diploid number of chromosomes. Development of the embryo may proceed according to methods analogous to those occurring in haploid apomixis, *i.e.,* by *diploid parthenogenesis, diploid pseudogamy,* or *diploid apogamety.*

In a third type of apomixis (*adventitious embryony*) the embryo develops directly from the ovular tissue of the parental sporophyte, usually from the integument or the nucellus. Such embryos are diploid and genetically identical with the parent sporophyte. This phenomenon can be taken as evidence in support of the concept of totipotency (Chapter 19).

Polyembryony. This term refers to the development of more than one embryo within a single ovule. Polyembryony is of frequent or regular occurrence in some species, and of sporadic occurrence in others. The causes of polyembryony are complex, and only the main mechanisms can be mentioned in this discussion. In many gymnosperms there are regularly two or more eggs in each female gametophyte, each of which, upon fertilization, develops into an embryo. An analogous situation occurs in some angiosperms, in which other nuclei of the embryo sac in addition to the egg may develop into super-numerary embryos, either with or without previous fertilization. Cleavage of the fertilized egg or young embryo into two or more units, and subsequent development of each into a genetically identical twin embryo, occurs in a number of species, especially among gymnosperms. Another cause of poly-embryony is the development of two embryo sacs within the same ovule, as in alder (*Alnus rugosa*), the egg in each of which gives rise to an embryo. Still another and common cause of plurality of embryos within an ovule is the apomictic development of one or more embryos from ovular tissue (adventitious embryony) as previously described. In some species one or more embryos of this origin may occur side by side in an ovule with an embryo derived from a fertilized egg as in citrus species; in other species only apomictic embryos are formed.

Development of the Seed.   The processes of fertilization and triple fusion not only set in motion the development of the embryo and the endosperm, respectively, but also exert a promotive influence on the development of the ovule into the seed and frequently also on the development of the fruit.

The fertilized egg cell does not usually divide at once, but only after a short delay. Once cell divisions start, however, they usually continue without interruption until a fully differentiated embryo has been developed. The foods used by the developing embryo come from the endosperm or other surrounding tissue such as the nucellus.

The endosperm, which in most species is a short-lived tissue, develops from the usually triploid endosperm nucleus. The endosperm ordinarily be-comes an actively growing tissue shortly after the occurrence of triple fusion. In most species considerable development of the endosperm has already taken place before cell division starts in the fertilized egg.

The extent to which the endosperm develops as an organized tissue differs greatly from one kind of plant to another. In some species, such as orchids, little or no growth of this tissue occurs. In the majority of species the endosperm grows rapidly during the early stages of seed development, but is later digested and used as a source of food by the growing embryo. The endosperm matrix appears to be an especially suitable medium for the growth of embryos, particularly in the early stages of their development. In such species the endosperm cells adjacent to the embryo disintegrate and disappear, and by the time the seed is mature, little if any of the endosperm remains. During the latter stages of embryo development in such species considerable quantities of foods usually accumulate in the cotyledons of the embryo. In a smaller number of species, such as the cereals, date, coconut, and castor bean, the endosperm persists as a storage tissue in the mature seed. In such species the foods in the endosperm are utilized by the developing seedling during the germination.

In the developing ovules of some plants the nucellus grows and becomes much enlarged, resulting in the mature seed of a tissue called the *perisperm*. Foods accumulate in this tissue much as they do in the endosperm. Representative species in which a perisperm is present include spinach, black pepper, beet, water lily, and coffee. In a few species both the endosperm and the perisperm persist as storage tissues in the mature seed.

Another major transformation which occurs during the morphogenesis of seeds is the development of the seed coats from the ovule coats (integuments). The latter are usually soft and fleshy, but because of loss of water and other changes during the maturation process, most seed coats are hard and dry.

Although the development of seeds is a usual outcome of the process of pollination it is by no means an invariable one. Failure of seeds to form within the fruit is especially common when an usually cross-pollinated species is self-pollinated, or when cross-pollination occurs between different species or different varieties of the same species. The principal causes of a failure of seeds to develop are: (1) Pollen fails to germinate after pollination. (2) Pollen tubes may grow too slowly for fertilization to be accomplished or the tubes may burst before reaching the embryo sac. (3) Pollen tubes grow but fertilization fails to take place. (4) Fertilization occurs, but abortion follows before more than a few divisions of the fertilized egg occur. (5) Fertilization occurs and embryo growth proceeds but is arrested at a later stage in its development. Failure of the embryo to develop to maturity appears to result from physiological conditions inherent in the young embryo or in the surrounding endosperm. The first four conditions listed above commonly lead to the development of "empty" seeds; the last one usually results in the formation of shrunken and usually nonviable seeds.

Development of Fruits.   The diversity in sizes, shapes, colors, textures, and arrangements of fruits is as great as that of the flowers which are their forerunners. Fundamentally, however, the morphogenic development of all fruits is similar. The simpler kinds of fruits develop solely from ovularies. Such gradual transformations of ovularies into fruits can easily be observed in many kinds of plants. Examples of simple fruits, which are in essence only modified ovularies, include the bean, pea, tomato, grape, avocado, orange, maple, elm, basswood, peach, cherry, and olive. In the development of some kinds of fruits, however, other parts ripen along with the ovulary and are incorporated into the final structure of the fruit. Examples are pomaceous fruits such as the apple in which the floral cup coalesces with the ovulary wall, the strawberry in which the fleshy receptacle becomes the prominent part of the fruit, and the fig in which the fruit is largely composed of a hollow fleshy peduncle.

In general, development of a fruit and its enclosed seed or seeds occurs concomitantly and in reciprocally coordinated fashion. During the transformation of some ovularies or ovularies plus adjacent parts into fruits, a considerable enlargement may occur, as much as a thousandfold or even more in some kinds. The period over which ripening occurs varies greatly from one kind of fruit to another, ranging from a week or two in some species to a year or even more in others. The resulting mature fruit tissues may be soft and fleshy as in tomato or grape, hard and dry as in nut fruits, or partly soft and partly hard as in peach and other drupes.

The processes of pollination and fertilization exert marked influences on the development of most kinds of fruits. Failure of pollination to occur commonly results in abscission of the pistil which of course means that no fruit will develop. Furthermore, most fruits do not develop normally unless pollination is followed by fertilization and a resulting development of the embryo or embryos. Some exceptions to the statements in this paragraph are discussed later.

Auxins are present in appreciable quantities in many kinds of developing fruits. Pollen tubes are at least a minor source of auxin and their elongation induces enchanced synthesis of auxins in the pistillary tissues through which they grow. Furthermore, young seeds are centers of auxin synthesis from which such hormones move out into the surrounding ovulary tisues. One role of the auxins which occur in a pistil is the prevention of its abscission (Chapter 23). Another is almost certainly a promotive effect on the development of the tissues of the enlarging fruit.

However, auxins are not the only kind of hormone which play a role in the development of fruits. Gibberellins and cytokinins are also involved. One evidence for this is that, as discussed later, these two kinds of hormones can be used to induce parthenocarpic (seedless) development of some kinds of fruits. In this connection it is worthy of note that young seeds are centers of gibberellin synthesis.

The patterns of hormonal control or influence on fruit development clearly differ from one kind of fruit to another. It resides in interactions among auxins, gibberellins, cytokinins, and probably other growth-promoting or growth-inhibiting substances not only with each other, but with components of the metabolic pathways within the cells.

Parthenocarpy. When mature, fruits usually contain seeds, but this is not invariably true. Seedless fruits are of regular occurrence in some kinds of plants such as certain varieties of banana, orange, grape, cucumber, and sunflower. Such fruits are also of sporadic occurrence in many other kinds of plants. This, condition of seedlessness is termed *parthenocarpy*. The designation "seedless" as applied in this connection refers to a lack of viable seeds; many parthenocarpic fruits contain ripened ovules which are empty, *i.e.,* lacking an embryo.

The sequence of events leading to the development of parthenocarpic fruits is different in different kinds of plants. In some species development of fruits proceeds in the absence of pollination and hence of fertilization, although the more usual consequence of a failure of pollination is for the pistil to abscise. In other plants the physiological effect of pollination not followed by fertilization is adequate to induce fruit development. Failure of fertilization to follow pollination is most commonly caused by such a slow rate of growth of the pollen tubes that they either never reach the egg or reach it after it is no longer viable. Even when pollination is followed by fertilization, development of the embryo from the fertilized egg may be arrested at an early stage, as has been mentioned, and seedless fruits or fruits with imperfect seeds is the usual result.

When both parthenocarpic and seeded fruits occur on the same plant, the former often fail to develop if the supply of foods within the plant is limited. Apparently the presence of developing seeds in a fruit aids in some way in promoting the translocation of soluble foods into the fruit. If parthenocarpy occurs in a given plant, parthenocarpic fruits are more likely to develop to maturity if none of the fruits on the plant are seeded.

Not only does pollination usually have a determinative effect on whether or not fruits will develop, but the genetic constitution of the pollen may have differential effects upon the growth of fruits. When, for example, date palms of the Deglet Noor variety are crossed with the Mosque variety as the pollen parent, the weights of the resulting fruits are considerably greater than when the Deglet Noor variety is crossed with the Ford No. 4 variety as the pollen parent. Such an effect of the source of the pollen upon the development of the parental tissue of the fruit is called *metaxenia*. Examples of this phenomenon are known in apple, cotton, oak, and other species. The phenomenon of metaxenia is almost certainly to be interpreted in terms of a hormonal mechanism.

As mentioned in Chapter 18, appropriate application of certain plant hormones will induce parthenocarpy in many fruits in the absence or failure of pollination. Auxins, gibberellins, and cytokinins have all been shown to have such effects with one species or another. Examples of plants in which at least one of the auxins is effective in inducing parthenocarpy are pepper, watermelon, tomato, eggplant, holly and pumpkin. Some of the species in which gibberellins are effective are apple, grape, peach, and tomato. In some species, apple for example, gibberellins are effective in inducing parthenocarpy, but auxins are not. Different gibberellins, however, do not induce parthenocarpy equally well. For example, GA_4 has been shown to be more effective than GA_3 in inducing parthenocarpy in some varieties of apples. In cherry both an auxin and a gibberellin must be applied if artificial parthenocarpy is to be induced. In some species, tomato for example, treatment either with certain auxins or with certain gibberellins induces parthenocarpy in default of pollination, but the latter type of hormone is far more effective in inducing such a growth reaction than the former. The Calymyra variety of fig represents an even more complex situation. Treatment of unpollinated flowers of this species with either an auxin (*p*-chlorophenoxyacetic acid), or gibberellin A_3, or a synthetic cytokinin [6-benzyl-amino)-9-(tetrahydropyran-2-yl)-9H purine], results in the development of parthenocarpic fruits.

Some Physiological Aspects of Fruit Development.   A developing fruit is a complex system of meristematic tissues. Simultaneously, each fertilized egg develops into an embryo, each endosperm nucleus into an endosperm, each nucellus (in some plants) into a perisperm, and the ovulary alone or ovulary and adjacent parts into a pericarp. Physiologically the growth of fruits, particularly those of the fleshy type, is closely akin to vegetative growth. Formation of the component parts of a fruit involves the same three morphological phases of growth as the development of vegetative organs, *i.e.,* cell division, cell enlargement, and cell differentiation. Development of the embryo within the seed, however, is largely limited to the cell division phase of growth. Water, carbohydrates, nitrogenous compounds, mineral-containing compounds, and hormones must be transported into the fruit from other parts of the plant. If the supply of any one of these substances to a growing fruit becomes deficient, the rate of growth is retarded. The proportions of these various kinds of substances which are required vary with the kind of fruit; being different, for example in a fleshy, succulent fruit from those in a hard, dry fruit. Within any given fruit the proportions of these various substances moving toward each of the several meristematic regions also differ considerably.

As pointed out in Chapter 17, developing fruits are one of the "sinks" toward which many translocates move in the plant. Such organs apparently have the capacity for inducing the movement of foods towards them from the vegetative parts of the plant to such an extent that vegetative growth is frequently curtailed. There is often also a drain on any accumulated foods within the plant. The effect of reproductive growth on vegetative growth is an example of a growth correlation and is discussed more fully in Chapter 23. The mechanism whereby developing fruits exert such an influence on other parts of the plant is not known but there is a strong presumption that it is hormonal in nature. It should be recalled that water as well as solutes moves preferentially towards developing fruits (Chapter 8).

The course of respiration in many kinds of developing fruits follows a characteristic cycle. The rate per unit fresh weight is relatively high in young fruits and gradually declines to a minimum value at about the time the fruits reach maturity. In some kinds of fruits, of which apple, avocado, banana, peach and tomato are examples, there is a secondary rise in the rate of respiration, beginning at about the termination of the maturation phase of growth, which is followed by a secondary decline in rate. This temporary rise in respiration rate is referred to as the *climacteric,* and occurs whether the fruits remain on the plant or are removed. Other kinds of fruits, of which the orange, grapefruit, lemon, cherry, and grape are examples, do not exhibit a climatteric rise in respiration rate (Fig. 21.4). The hastening effect of

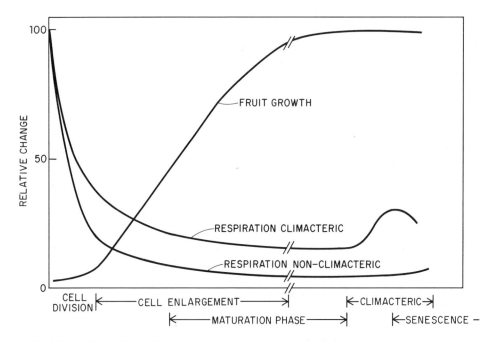

Fig. 21-4. Stages in fruit development and characteristic respiratory trends in fruits which exhibit the climacteric and those which do not. From Biale, *Science,* **146,** No. 13, November 1964:880–888. Copyright 1964 by the American Association for the Advancement of Science.

ethylene on fruit ripening, discussed in Chapter 18, may be causally related to the incidence of the climacteric. One reason for believing this to be true is that when fruits which do not normally exhibit a climacteric, such as orange, are exposed to adequate concentrations of ethylene, a temporary rise in their respiration rate is induced. The ultimate destiny of all fruits, whether of the climacteric or non-climacteric type, is senescence and death of all of the tissues of which they are composed except some of the tissues in the seed or seeds.

Embryo Culture.   Partly developed plant embryos can be cultured on sterile media according to techniques very similar to those used in culturing root tips and stem tips (Chapter 19). Mature or nearly mature embryos usually develop into seedlings on a medium which contains only inorganic salts and a sugar as solutes. Unlike mature embryos, immature embryos do not appear to be autotrophic with respect to hormones. Young embryos (about one-fourth mature size) of jimsonweed (*Datura stramonium*), for example, do not develop on a medium containing glucose, mineral salts and several physiologically active organic compounds, but do develop rapidly to many times their initial volume if a small amount of coconut milk, which is actually liquid endosperm, is added to the medium (Fig. 21.5). The similar effect of coconut

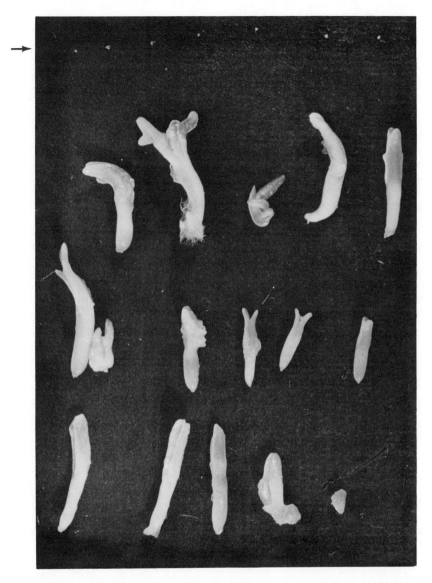

Fig. 21.5. Effect of coconut milk on the development of *Datura* embryos isolated 10 days after pollination. Top row (arrow): embryos after 8 days on a medium containing sugar, mineral salts, and vitamins. Almost no growth has occurred. Bottom three rows: embryos after 8 days on the same kind of medium plus coconut milk. From Van Overbeek *et al. Am. J. Botany*, **29**, 1942:474.

milk in inducing the development of single cells in culture into plantlets should be recalled (Chapter 19).

SUGGESTED FOR COLLATERAL READING

Kozlowski, T. T. *Growth and Development of Trees*. 2 vols. Academic Press, New York, 1972.

Ledbetter, M. C. and K. R. Porter. *Introduction to the Fine Structure of Plant Cells*. Springer-Verlag, New York, 1970.

Leopold, A. C. *Plant Growth and Development*. McGraw-Hill Book Company, New York, 1964.

Maheswari, P. *An Introduction to the Embryology of Angiosperms*. McGraw-Hill Book Company, 1950.

O'Brien, T. P., and Margaret E. McCully. *Plant Structure and Development*. The Macmillan Company, New York, 1969.

Salisbury, F. B. *The Biology of Flowering*. Natural History Press, New York, 1971.

Steward, F. C. *Growth and Organization in Plants*. Addison-Wesley Publishing Company, Reading, Massachusets, 1968.

Torrey, J. G. *Development in Flowering Plants*. The Macmillan Company, New York, 1967.

Wodehouse, R. P. *Hayfever Plants, 2nd Edition*. Hafner Publishing Company, New York, 1971.

22

ENVIRONMENTAL FACTORS AFFECTING REPRODUCTIVE GROWTH

Initiation of the internal physiological conditions which lead to the conversion of vegetative to reproductive meristems is brought about, as are all other physiological phenomena, by interactions between the genetic constitution of a plant and the factors which impinge upon it from its environment. In this chapter the effects of those environmental factors which exert major effects upon reproductive growth will be discussed.

Effects of Irradiance on Reproductive Growth. That the flowering and fruiting of many kinds of plants are less abundant in the shade than in full light is a common observation. The relation between irradiance and flowering for any given species is not a simple one, because it is complicated by the influence of the photoperiod as discussed later. The effect of an irradiance which is otherwise adequate for flower induction can be completely nullified, in some species, if the photoperiod is too long; in other species, if it is too short. In general, however, assuming daylight, temperature, and other conditions to be favorable for flowering, there is a minimum irradiance for each species below which no blooming occurs. The minimum varies considerably from one species to another, usually being lower, for example, in woodland shade species than in sun species. At irradiances only slightly greater than the minimum, flowering usually occurs only sparsely; at higher irradiances this factor ceases to be the limiting one in the initiation of flower primordia.

Effects of Light Quality on Reproductive Growth.   Although marked variations occur in the wavelength composition of daylight (Chapter 10), any effects which such variations may have upon reproductive growth are difficult to differentiate from other effects of light under natural conditions. Under laboratory conditions, on the other hand, some such effects can be recognized, the more important of which are discussed later in the chapter.

Effects of Duration of Light on Reproductive Growth.   The fundamental discovery of Garner and Allard that the length of the daily photoperiod has marked effects on the reproductive development of plants was mentioned in Chapter 20. In this chapter the term "photoperiodism" will be used solely to designate the effects of daylength on reproductive development.

As a result of the earlier investigations of photoperiodism a fourfold classification of plants based on the effect of the length of the natural daylight period upon their reproductive development has been used:

(1) "Short-day" plants flower readily only within a range of photoperiods shorter than a "critical photoperiod" (see later). Blooming may occur, at least in some species, although more slowly, or less profusely, within a range of photoperiods somewhat longer than the most favorable ones. Under still longer photoperiods or under continuous illumination, short-day plants do not bloom and remain in the vegetative state indefinitely (Fig. 22.1). Plants usually classed in the short-day category include coleus (*Coleus blumei*), cosmos, strawberry, aster, giant ragweed, cocklebur (*Xanthium strumarium*), poinsettia, chrysanthemum, sweet potato, violet, hemp (*Cannabis sativa*), Japanese morning glory (*Pharbitis nil*), pigweeds (*Chenopodium* spp.), kalanchoë (*Kalanchoë blossfeldiana*), some tobaccos, some soybeans, a duckweed (*Lemna perpusilla*), and most early spring or late summer blooming garden or wild flowers of the temperate zone.

(2) 'Long-day" plants bloom readily only under a range of photoperiods longer than a "critical photoperiod" (see later), up to and including continuous illumination. Blooming is usually induced most rapidly under continuous illumination or very long photoperiods, but may occur, although more slowly or less profusely, at least in some species, under shorter photoperiods. Under still shorter photoperiods the plants remain in the vegetative state indefinitely (Fig. 22.2). Plants usually classified in the long-day category include spinach, beet, radish, lettuce, English plaintain (*Plantago lanceolata*), black henbane (*Hyoscyamus niger*), most grains, timothy, clover, hibiscus, potato, and most late spring and early summer blooming garden or wild flowers of the temperate zone.

(3) "Indeterminate" ("day neutral") plants flower readily over a wide range of daylengths from relatively short photoperiods to continuous illumination (Fig. 22.3). Plants usually classified in the indeterminate group include zinnia, dandelion, chickweed, tomato, cotton, buckwheat, and some tobacco varieties.

Fig. 22.1. Effect of length of the photoperiod upon the flowering of salvia, a short-day species. Photograph from Arthur, *et al., Contrib. Boyce Thompson Inst.*, **2**, 1930:461.

Fig. 22.2. Effect of the length of the photoperiod upon the flowering of lettuce, a long-day species. Photograph from Arthur, *et al., Contrib. Boyce Thompson Inst.*, **2**, 1930:461.

(4) "Intermediate" plants bloom only under daylengths within a certain range and fail to flower under either longer or shorter photoperiods. Only a few examples of such species are known. Among them are some varieties of sugar cane, boneset (*Eupatorium hyssopifolium*), and climbing hempweed (*Mikania scandens*).

Fig. 22.3. Effect of the length of the photoperiod upon the flowering of buckwheat, an indeterminate species. Photograph from Arthur, *et al.*, *Contrib. Boyce Thompson Inst.*, **2**, 1930:461.

The listing of examples of species which fall into these four categories should not be accepted in too rigorous a sense. As indicated to some extent in the above listings, and as examples discussed later in this chapter also show, different varieties of the same species often differ in their photoperiodic reaction.

For at least some short-day plants the range of photoperiods under which initiation of flowers occurs is sharply delimited, and a definite *critical photoperiod* can be recognized. Such plants flower only in a range of photoperiods at or shorter than the critical. The critical photoperiod for cocklebur, for example, is about 15.5 hours; for Biloxi soybean, about 13.5 hours, for chrysanthemum about 10–11 hours, and for coleus, about 12 hours. A critical photoperiod can likewise be distinguished for long-day plants, such plants flowering only in a range of photoperiods at or longer than the critical. The critical photoperiod for black henbane, for example, is about 12 hours, for annual beet between 13–14 hours and for dill, between 11 and 14 hours. Indeterminate plants, like long-day plants, flower only in a range of daylengths longer than a critical, but the critical daylengths for such species are shorter than those for long-day plants. Intermediate species have two critical photoperiods, blooming only on photoperiods which are at or between them. One variety of sugar cane, for example, blooms only under photoperiods between about 11½ and 13 hours in length.

From the above discussion it should be clear that there are certain day-lengths which may induce flowering both in some long-day and in some short-day species. At a daylength of 13 hours, for example, both short-day plants with a critical photoperiod at or greater than this, and long-day plants with a critical photoperiod at or less than this will bloom.

The line of demarkation between the range of photoperiods favorable to flowering and the range of those which are not is much sharper in some kinds of plants than in others. As small a difference as 15 minutes in the length of the photoperiod determines whether or not some species will bloom. For most species, however, the length of the critical light period is more flexible and may vary with the strain of the plant as well as with certain environmental factors such as irradiance and temperature.

The fact that reproductive growth is a complex of integrated processes was emphasized in the preceding chapter. Various stages in reproductive growth may be affected differently by the length of the photoperiod. For example, in the Biloxi soybean, initiation of flower primordia occurs over a range of photoperiods from about 2 to about 13.5 hours. Subsequent development of the flower buds into macroscopic flowers, once primordia are present, occurs most rapidly under 8- to 13-hour photoperiods, more slowly under 14- and 15-hour photoperiods, and not at all under 16-hour photo-periods. No development of fruits takes place when photoperiods are longer than 13 hours.

Emphasis is placed in this discussion on the initiation of flowers rather than on subsequent stages of reproductive development, because this initiation is the pivotal morphogenic transformation and the one which has been in-vestigated most critically.

Seasonal and Geographic Aspects of Photoperiodism.   In temperate re-gions the season in which a plant blooms is largely controlled by its reaction to the length of the photoperiod. The natural blooming period of long-day plants is in the late spring and early summer. Short-day species which grow at rela-tively low temperatures bloom in the early spring. Differentiation of the flower buds of many such species occurs during the shortening days of the latter part of the preceding season. The majority of the members of this group do not develop flowers until the advent of the shortening days of late summer or early autumn. This type of behavior is found in most short-day annual species and is character-istic of many short-day herbaceous perennial species as well. Indeterminate species may flower at any season during which other environmental conditions are favorable.

Photoperiodic behavior is an important factor in determining the most favorable season for raising many crop plants in mid-latitudinal regions. Long-day lettuce and radish, for example, are best grown in the spring, because it is the vegetative parts of these plants which are of commercial importance. If sown later in the season, such plants rapidly "go to seed," rendering the crop unmarketable. If a series of plantings is made of short-day Biloxi soybean during the spring and early summer, none of the plants will flower until the daylength has decreased to the critical photoperiod in mid- or late summer. In other words the crops sown late in the season will bloom at about the same time as those sown early in the season. The late-sown plants, however, will have had very little time in which to develop vegetatively, hence they will bear fewer flowers and yield a smaller crop of fruits and seeds than the early-sown plants.

The geographic distribution of plants is also governed in part by their photoperiodic reactions. This is true both of crop plants and plants in the natural vegetation. Long-day species requiring photoperiods appreciably longer than 12 hours obviously cannot accomplish sexual reproduction in tropical regions where the daylengths approximate 12 hours the year round. Pronounced short-day species are for the most part excluded from extreme northern latitudes (there are almost no land areas suitable for occupancy by seed plants in extreme southern latitudes) unless they can be propagated vegetatively, because the growing season in such regions is restricted largely to the period during which very long photoperiods prevail. In mid-latitudinal temperate zones, plants of all photoperiodic categories flourish but bloom at different seasons as already described. Indeterminate species can bloom over such a wide range of daylengths that their geographic distribution is controlled by factors other than the length of the photoperiod.

Long-Day and Short-Day Plants as the Basic Categories.   The previously discussed early classification of plants into four groups according to their photoperiodic requirement for reproductive development under natural conditions has been widely employed, and is a useful one, but more recently the concept has emerged that there are basically only two broad catgories: long-day plants and short-day plants.

Long-day plants bloom not only under photoperiods longer than a critical but also under continuous illumination and indeterminate plants resemble them by doing likewise. No dark period is required within each 24 hour cycle for flowering to be induced. Indeterminates differ from long-day plants in having a much lower critical photoperiod which may be a photosynthetic rather than a photoperiodic requirement.

Short-day plants and intermediate plants are alike in that both of them require a dark period within each 24-hour cycle for flowering to be induced, *i.e.,* neither will bloom under continuous illumination. However, even pronounced short-day plants require a photoperiod of minimum length which is primarily a photosynthetic rather than a photoperiodic requirement. For some this is very short. Cocklebur, for example, requires only about a 1-hr photoperiod in a 24-hr cycle for induction of flowers, and Biloxi soybean only about 2 hours. Intermediates differ from short-day plants in requiring a longer minimum light period than the latter if flowering is to be induced.

Flowers are never induced in some kinds of short-day plants under unfavorable day lengths; such species are termed *obligative* short-day plants. In other kinds of short-day plants flowers may be induced ultimately, although perhaps only sparsely, even under unfavorable daylengths; such species are termed *facultative* short-day plants. The same two terms are applied, and with similar connotations, to long-day plants.

However, the broad long-day or short-day classification of plants on the basis of their photoperiodic behavior has proved to be not entirely adequate. Some species will not initiate flowers while exposed continuously to long-day cycles or to short-day cycles or to cycles of intermediate daylengths. Some such plants can be induced to flower by exposure to a sequence of long days followed by a sequence of short days. An example of such a "long-day-short-day" plant is the night-blooming jasmine (*Cestrum nocturnum*) which requires a minimum of five long days followed by a minimum of two short days for flowers to be induced. A number of other species are known which have similar requirements for blooming.

The reverse photoperiodic behavior pattern is shown by "short-day–long-day" plants. An example of such a plant is a variety of white clover (*Trifolium repens*) which requires a minimum of three short days followed by one long day for the induction of flowers. Other species are known to have similar photoperiodic requirements.

Photoperiodic Induction.   This phenomenon has been mentioned in the preceding section, but some aspects of it warrant further consideration.

Any photoperiodic cycle which induces initiation of flowers in plants is called a *photoinductive cycle;* one which does not is called a *nonphotoinductive cycle*. An 8-hour photoperiod alternating with a 16-hour dark period is one possible photoinductive cycle for short-day cocklebur plants having a critical photoperiod of 15.5 hours; a 16-hour photoperiod alternating with an 8-hour dark period is one possible nonphotoinductive cycle for this plant.

If a short-day plant which has been growing under long days is transferred temporarily to short days and then returned to a long-day environment, flowering will be initiated, even though the plant remains exposed to long photoperiods thereafter. The number of short-day cycles required to induce flowering differs from one short-day species to another. In some, of which cocklebur, pigweed, and Japanese morning glory are examples, one such cycle is adequate to induce initiation of flowering. In such one-cycle-inducible plants it is the preceding light period plus the dark period which constitutes the inductive cycle. Most short-day plants, however, require more than one inductive cycle for flowering to occur. Biloxi soybean requires a minimum of three consecutive short-day cycles for initiation of flower primordia, hemp requires about seven, and kalanchöe about twelve.

In some long-day plants one long-day cycle is adequate to induce initiation of flower primordia. Examples are darnel (*Lolium temulentum*), white mustard (*Sinapis alba*), and the arlo variety of rape (*Brassica campestris*).

Other long-day species require more than one long photoperiod in order for flower formation to be induced. Dill requires 2 to 4 such cycles, annual beet 15 to 20, English plaintain 15 to 20 and Sweet William (*Silene armeria*) about seven.

The number of inductive cycles required by any kind of plant varies somewhat depending upon the age of the plant and on environmental conditions, especially temperature, irradiance, and the length of the photoperiod.

One-cycle-inducible plants of both the long-day and short-day types are especially well adapted to use in certain kinds of critical experiments, some of which are described later in this chapter.

The Generalized Mechanism of Photoperiodism.   *Locus of the Photoperiodic Reaction.*   As was first shown in spinach, a long-day plant, initiation of flowers occurs only when the *leaves* are exposed to long photoperiods. Exposure of the apical meristem to long photoperiods while the leaves are kept under short photoperiods, results in maintenance of the vegetative condition. Similar results have been obtained with a number of other species of both the long-day and short-day types.

Not all the leaves on a plant are equally effective in serving as a locus of the photoperiodic reaction. Very young leaves and senescent leaves are often ineffective or at least less effective than mature leaves. In cocklebur, leaves with a mid-vein length shorter than about 2 cm are ineffective. Expanding leaves with a 7 to 9 cm mid-vein length are the most effective, while leaves larger or smaller than this are less so. In this species exposure of as little as 7–8 cm^2 of leaf surface to short photoperiods, while the rest of the leaves are under long photoperiods, induces the initiation of flowers.

Although leaves are undoubtedly the principal loci of the photoperiodic reaction, in some plants other organs may fulfill this role, at least in part. In Japanese morning glory the leaf-like cotyledons are effective loci of the photperiodic reaction, and in some herbaceous species the stems serve in such a role. In a short-day pigweed (*Chenopodium amaranticolor*), for example, defoliated plants react to photoinduction by initiating flowers almost as readily as intact plants.

Transmission of the Effect.   One of the most clear-cut aspects of photoperiodism is that some entity must be translocated from the leaves to the apical meristems. That this entity must be hormone-like seems obvious. Chailakhyan, one of the early Russian investigators of photoperiodism, proposed that it be called "florigen," a term which will be used in this book for convenience. However, as the later discussion indicates, it is becoming increasingly evident that the photoperiodic induction of flowering cannot be explained on the basis of the action of a single hormone.

Many fruitless attempts have been made over a period of several decades to extract from plants a substance or substances which would fulfill the role ascribed to florigen. The effective entity could conceivably be a hormone in the strict sense of the word, a mixture of several substances, or a hormone precursor which undergoes further chemical modification enroute or after arrival at the meristem.

One step towards characterizing florigen was taken in 1970 when it was discovered that acetone extracts from flowering cocklebur plants would induce floral initiation in a short-day duckweed (*Lemma perpusilla*) under noninductive conditions but similar extracts from non-flowering cocklebur plants would not. Similar results were obtained when cocklebur plants growing under noninductive conditions were treated with the extract from flowering cocklebur plants, but only if the extract was supplemented with gibberellic acid. No flowers were induced in cocklebur under noninductive conditions by extracts from non-flowering cocklebur plants even when the extracts were supplemented with gibberellic acid.

The distance over which florigen is transported in plants varies with the kind of plant and environmental conditions to which it is subjected. Translocation occurs in living cells and, in petioles and stems, through the phloem tissues.

There appears to be a close correlation between the translocation of soluble carbohydrates and that of florigen. If, for example, one branch of a two-branched cocklebur plant be exposed to short days and the other to long days, flowers occur on the former branch but few or no flowers are initiated on the latter branch. (Fig. 22.4). If, however, the long-day branch be defoliated or heavily shaded, flowers are initiated on it as well. As long as the long-day branch is photosynthesizing at a substantial rate, photosynthate is translocated both towards the apical region and also in a basipetal (downward) direction through that branch. Little or no translocate originating in the short-day branch will move up through the long-day branch. In the defoliated or shaded long-day branch, on the other hand, the rate of photosynthesis is greatly diminished and there is little basipetal translocation of photosynthate. When this condition prevails, photosynthate from the short-day branch moves into and is translocated upwards in the long-day branch which has become essentially a sink area. Florigen is thought to be carried along in this translocate stream and flowers are initiated on the long-day branch.

Another possible explanation of the results of such an experiment is that inhibitors are generated in those leaves exposed to the non-photoinductive long-day cycles which prevent florigen arriving from the short-day branch from exerting its usual effect in causing flowering of the long-day branch. Defoliating the long-day branch would remove the source of the inhibitor, thus facilitating the flowering of the long-day branch. While this possibility cannot be excluded as at least a partial explanation of the results obtained, considerable evidence indicates that a retardation in florigen translocation is the main factor involved.

Estimating the rate of movement of an unknown entity such as florigen presents special problems. One approach to this problem which has been used, applicable only to one-cycle-inducible plants such as cocklebur or Japanese

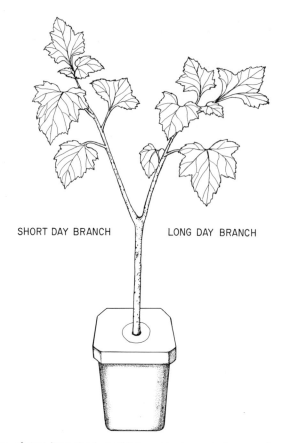

SHORT DAY BRANCH   LONG DAY BRANCH

Fig. 22.4. Type of two-branched cocklebur plant used in experiments described in the text.

morning glory, is that of defoliating plants at different time intervals after the termination of the single inductive cycle and examining the plants for the initiation of flowers after a suitable further time interval. If a plant is defoliated immediately after the termination of the inductive cycle, no flowers are initiated. If, however, defoliation is deferred for about 24 hours, flowering is induced, and often nearly as completely as on control plants indicating that florigen must have moved out of the leaves within the 24 hour period. Intermediate defoliation times result in intermediate degrees of flowering. Rates of translocation of florigen which have been inferred from most such experiments turn out to be surprisingly low, less than a centimeter per hour. However, strictly speaking, such experiments do not measure an absolute rate of translocation, but only the time required for threshold quantities of florigen to reach the meristems and induce initiation of flowers. Furthermore some movement of florigen occurs as a cell to cell process in leaf blade tissues in which the rates are undoubtedly lower than in the phloem.

In an improved version of this experiment, however, translocation of florigen through stems of Japanese morning glory 100 cm or more in length was estimated to be at least 51 cm per hour. This result is much more in accord with the concept that movement of florigen is associated with translocation of organic solutes, rates of translocation of which are discussed in Chapter 17.

Reactions at the Meristems.   Only apical stem meristems, which may be borne either terminally or laterally (usually in leaf axils) on stems can become transformed into floral meristems (Fig. 21.1). Several factors determine how many and which meristems will be transformed.

Before any reproductive development of a plant can occur it must reach a stage called "ripeness to flower." The period of growth required before this stage is reached differs greatly from species to species. Some plants such as certain pigweeds and Japanese morning glory can be induced to flower by exposure to short days while still in the seedling stage. Most herbaceous plants must have developed beyond this stage, however, before flowering can be induced. Trees stand at the other end of the scale; most of them will not flower until they have lived for a number of years. An extreme example is represented by certain species of bamboos which grow from 5 to 50 years, flower once, and then die.

Even when the plant as a whole has attained the condition of "ripeness to flower" the status of the apical meristems is still a factor. In general only active, and not dormant, buds can undergo conversion from the vegetative to the floral state. Cell division apparently must be in progress for this morphogenic change to occur.

The number of inductive cycles to which a plant is exposed is a factor in determining not only the number of meristems which become transformed from the vegetative to the reproductive state, but also the rapidity with which such transformations occur. For example, although one photoinductive cycle is sufficient to induce flowering in cocklebur, it requires a much longer period for the flowers to mature than if the plants are exposed continuously to photoinductive cycles, in which case flowers mature in about two weeks.

Role of Inhibitors in Photoperiodism.   The discussion of the mechanisms of photoperiodism up to this point has been largely in terms of a positive-acting hormone-like mechanism. Various lines of evidence indicate that inhibitors also play a role in this process, at least in some plants. Inhibitors may be generated in leaves when they are exposed to non-photoinductive cycles. As described previously, some results which could be explained by the presence of an inhibitor are seemingly a result of interference with florigen translocation rather than being the results of inhibition in a metabolic sense, although the latter possibility cannot be entirely excluded.

Experiments on fractional induction in various species of short-day plants, as described later in this chapter, furnish strong evidence that true metabolic inhibitors play a role in photoperiodism, at least in such species.

Defoliation experiments with certain species also suggest the presence of flowering inhibitors. Flowering in long-day black henbane and some other long-day species is promoted by defoliation while the plants are exposed to short-day conditions. Apparently a flowering inhibitor is synthesized in the leaves under short days and its synthesis stops or is counteracted under long-day inductive conditions. In plants which react in this fashion it is not necessary to postulate the existence of any positive-acting florigen at all in order to explain their photoperiodic behavior. In such species inhibitory substances appear to exert a predominant effect on flowering.

In general, the induction of flower primordia is probably the outcome of an interplay between promotive and inhibitory substances, all hormone-like in properties. The relative roles of these two types of substances almost certainly differ from one kind of plant to another. In some species a positive-acting mechanism, such as is envisioned in the florigen concept, appears to play a predominant role. In others inhibitory substances seem to exercise a major influence. In still others promotive and inhibitory effects are more nearly balanced, and small shifts in environmental conditions, such as temperature, will determine which will predominate.

Basic Differences in the Photoperiodic Reactions of Long-Day and Short-Day Plants.   No two kinds of plants have exactly the same photoperiodic requirements for flowering. This is true even of varieties or physiological races within a given species. Nevertheless, long-day plants as a group, and especially the obligative type, exhibit certain distinctive features in their photoperiodic behavior in common. The same general statement can be made for short-day plants. It should be emphasized, however, that the photoperiodic behavior of only a small number of either long-day or short-day plants has been studied comprehensively. It is virtually certain that a greater variety of patterns of photoperiodic behavior exist than have been revealed by investigations which have been made up to the present time.

Role of the Light and Dark Periods.   The essential feature in the photoperiodic reaction of long-day plants is the inhibitory or retarding effect, under 24 hour cycles, of long dark periods upon the initiation of flowers.

It should be emphasized that the above statement holds only for 24 hour cycles. Long-day plants will initiate flowers under *short* light periods if they alternate with *short* dark periods. Dill and zinnia, for example, flower on continuous cycles in which 1 hour of light alternates with 2 hours of dark. Furthermore, as discussed later under the topic of endogenous rhythms, extra long dark periods of certain lengths, in artificial cycles longer than 24 hours, are promotive of flowering in long-day plants.

In general, no kind of alternation of light and dark is superior to continuous light for floral initiation in long-day plants. Exposure to continuous illumination is not favorable, however, to vegetative development or subsequent floral and fruit development in many species.

The essential feature of the photoperiodic reaction in short-day plants is that a dark period of minimum length must be included in each 24 hour cycle for flowering to be induced. Under cycles longer than 24 hours, as discussed later, dark periods of some lengths are promotive to flowering, while dark periods of other lengths are inhibitory. Short-day plants do not flower under continuous illumination, which is probably the most definitive distinction between them and long-day plants (Fig. 22.5).

Fractional Induction.   In long-day plants, non-consecutive inductive cycles or sequences of such cycles are additive, a phenomenon called "fractional induction." In English plantain, flowering is induced by exposure to 15–25 long-day cycles. It is also induced if the plants are exposed, first to 10 long-day

Fig. 22.5.   Photoperiodic reactions of *Aster cordifolius,* an obligate short-day plant. Plants were transplanted from the field on December 16; photographed on March 28. Plant on left was kept under natural short photoperiods in a greenhouse. Plant on right was kept under continuous illumination of about 1000 foot-candles. Plant on the left is in bloom; plant on the right is not. Photograph from Mary Sigafoos.

cycles, then to 20 short-day cycles, and finally to 15 long-day cycles. Somewhat similar results have been obtained with black henbane in which alternating one long-day cycle and one short-day cycle six times results in the same amount of flowering as six consecutive long days. The inductive effect of one or more long days carries through intercalated short days and augments the effect of subsequent exposures to long days. Furthermore, the intercalated short days apparently do not exhibit any appreciable sort of an inhibitory effect on floral initiation.

Fractional induction is not known to occur in any kind of short-day plant. For example, 12 consecutive short days cause a high degree of floral induction in kalanchoë. If the 12 short days are alternated with long days, however, no floral induction occurs. Not only does the short-day effect fail to carry over from one short day to the next through an intercalated long day, but the long days appear to exert an inhibitory effect, since other experiments with this species have shown that each long day offsets the effects of from 1.5 to 2 short days. The long-day inhibitory effect is exerted on the following short day, not on the preceding one. A similar failure of partial induction to persist has been shown to be characteristic of several other short-day species.

Effects of a Light Break During the Dark Period. When a short-day plant is subjected to a usually photoinductive 16 hr dark period/8 hr photoperiod cycle, in which the dark period is interrupted more or less at its midpoint with a low intensity light break, floral initiation is inhibited. If, on the other hand, a long-day plant is subjected to a usually non-photoinductive short day cycle during which the dark period is interrupted with a low intensity light break, floral initiation is often promoted.

The duration of such a light break which is required to reverse the photoperiodic reaction differs from one kind of plant to another. It also differs for a given plant with the relative lengths of the light and dark periods in the main cycle, with the exact stage during the dark period at which the light interruption is applied, with the quality of the light, and with the temperature.

In order to be effective with some kinds of plants, light breaks must sometimes be several hours long, but much shorter light interruptions during the dark period are effective, especially in some short-day species in which this phenomenon manifests itself in a more dramatic fashion than in long-day species. In cocklebur, to cite one of the more extreme examples, exposure to one minute of irradiation in the approximate range of 10 to 100 foot candles during the long dark period is sufficient to inhibit initiation of flowers.

Examples of short-day plants in which flowering is inhibited by light breaks as just described include cocklebur, Biloxi soybean, *Perilla* (a mint), kalanchöe, strawberry, and chrysanthemum. Examples of long-day plants in which flowering is promoted by light breaks include black henbane, barley (Wintex variety), sugar beet, petunia, dill, and pimpernel (*Angallis arvensis*).

The effects described in this section are manifestations of the so-called "low intensity photoreaction" which is discussed later in this chapter.

Role of the Reversible Red, Far-Red and Related Photoreactions in Photoperiodism. Experiments have been performed to ascertain which regions of the spectrum are most effective in reversing photoperiodic behavior when employed as light breaks as described in the preceding section. Maximum effectiveness was found to be in the red region with a peak at about 660 nm. When applied as a light break during a long dark period, red light promotes floral initiation in many long day plants and inhibits it in many short-day plants.

As in other manifestations of this photoreaction (Chapter 20) the effect of light in the red region can be nullified, at least in some species, if the red light is followed promptly by exposure to low intensity far-red light with a peak value of about 730 nm. The effect is reversed both in short-day species (examples: cocklebur, pigweed, chrysanthemum) causing them to flower, and in long-day species (examples: Wintex barley, black henbane) in which flowering is inhibited.

In at least some species, chrysanthemum for example, this reaction is repeatedly reversible, at least up to a limit, as it is in other reversible photomorphogenic processes (Chapter 20).

It appears that conversion of phytochrome preponderantly into its Pfr form as a result of exposure of leaves to red light is inhibitory to the initiation of flowers in many short-day plants and favorable to flower initiation in many long-day plants. Virtually nothing is known, however, regarding the metabolic pathways through which phytochrome effects are mediated.

Blue light plays a role in the photoperiodic reactions of at least some kinds of plants, and at least some of its effects may operate through the phytochrome system. In long-day white mustard, for example, blue light is much more effective in initiating flowers when given as a break during a long dark period than is red light. Other photoperiodic roles of blue light are described later in this chapter.

Hormonal Effects. As discussed in Chapter 18, the photoperiodic reactions of both long-day and short-day plants can often be modified by treatment with hormones. Auxins, gibberellins, cytokinins, and abscisic acid have all been shown to have such effects. A hormone which promotes flowering in long-day plants under non-photoinductive cycles often inhibits flowering of short-day plants under photoinductive cycles, although there are exceptions to this statement. Gibberellins, for example, are promotive of flowering in many long day plants and inhibitory to flowering in at least some short-day plants.

Basic Similarities in the Photoperiodic Mechanism of Long-Day and Short-Day Plants.

In spite of the many dissimilarities in behavior of long-day as contrasted with short-day plants, several lines of evidence indicate that the photoperiodic mechanism in both is basically similar.

Apparent Identity or Similarity of the Florigen Synthesized in Short-Day and Long-Day Plants. A number of interspecies grafts have been made in which the scion was a long-day plant and the stock a short-day plant or vice-

TABLE 22.1 INDUCTION OF FLOWERS ON STOCKS OR SCIONS UNDER NON-PHOTOINDUCTIVE CYCLES[a]

Scion	Scion type	Stock	Stock type	Photoperiod	
Hyoscyanus niger (black henbane)	LD	Nicotiana tabacum (Maryland Mammoth var.)	SD	LD	Stock induced
Nicotiana sylvestris (tobacco)	LD	Nicotiana tabacum (Maryland Mammoth var.)	SD	LD	Stock induced
Xanthium strumarium (Cocklebur)	SD	Rudbeckia bicolor (Coneflower)	LD	LD	Scion induced
Kalanchöe blossfeldiana	SD	Sedum spectabile	LD	LD	Scion induced
Sedum ellacombianus	LD	Kalanchöe blossfeldiana	SD	SD	Scion induced
Sedum spectabile	LD	Kalanchöe blossfeldiana	SD	SD	Scion induced
Nicotiana tabacum (Maryland Mammoth var.)	SD	Hyoscyamus niger (black henbane)	LD	SD	Stock induced
Nicotiana tabacum (Maryland Mammoth var.)	SD	Nicotiana sylvestris (tobacco)	LD	SD	Stock induced

[a]Based on summary by Doorenbos and Wellensiek. Ann. Rev. Plant Physiol. **10**: 147–184. 1959.

versa (Table 22.1). When exposed to long days, the long-day portion of such a grafted plant flowered as well as the short-day portion. The same behavior obtained when such plants were exposed to short days. Under long days it appears that florigen in the long-day graft partner is translocated to the short-day partner and that both are induced to flower. Under short days it appears that florigen made in the short-day graft partner is translocated to the long-day partner and both are induced to flower. The conclusion seems inescapable either that an identical kind of florigen is made by both types of plants or else that the kind of florigen made by long-day plants is effective in short-day plants and vice versa.

Dodders are angiospermous parasites and hence flower-bearing. Various experiments have been performed regarding their photoperiodic behavior. When a species of dodder (*Cuscuta Gronovii*) is grown as a parasite on the long-day plant *Calendula officinalis,* both the host plant and the dodder flower only under long days. If, however, the dodder is grown as a parasite on short-day *Cosmos bipinnatus* both the host plant and the dodder flower only under short days. In addition to foods and mineral elements, dodder quite clearly obtains florigen from its host plants. Furthermore, the florigen from both long-day plants and short-day plants is either identical or sufficiently similar so that either will induce flowering in the dodder.

Varietal Differences in Photoperiodic Reaction Within a Species. The fact that varieties exhibiting differing photoperiodic reactions are found within a single species also suggests a basic similarity of mechanism. A good example of this is the side-oats grama grass (*Boutelua curtipendula*), an important range grass of wide latitudinal distribution on the Great Plains of North America. Strains of this species from Texas are short-day in photoperiodic behavior, strains from Oklahoma include both short-day and long-day plants, while strains from North Dakota consist almost entirely of long-day individuals.

Soybean is an example of a crop plant in which marked differences in photoperiodic reaction exist from one variety to another. Some varieties, including the often experimented with Biloxi variety, are short-day in photoperiodic reaction, others are essentially long-day. Many other examples of varietal differences in photoperiodic behavior within a species are known.

Reversal of Photoperiodic Behavior by Temperature. This effect is discussed further later in the chapter. Some plants are short-day in photoperiodic reaction within one range of temperatures but are long-day in reaction within another range. Such a mutability of photoperiodic reaction within a single kind of plant is virtually impossible to explain unless the mechanism of photoperiodism is fundamentally the same whether the plant reacts as a long-day or as a short-day type.

Requirements for Several Photoreactions. The photoperiodic process in both long-day and short-day plants involves at least two, and probably more, photoreactions. In nature these processes cannot be differentiated, but they can be separated by experimental procedures under laboratory conditions.

The two constituent reactions of photoperiodism that have been most clearly recognized are the so-called "high intensity" photoreaction and the so-called "low intensity" photoreaction. Energy requirements of these two reactions are very different and have been estimated for cocklebur to be about 30,000 foot-candle minutes for the high intensity reaction and about 10 to 100 foot-candle minutes for the low intensity reaction.

The high intensity reaction is often equated with the photosynthesis, considered broadly as a source of basic substrates. Support for this concept is furnished by the fact that, in at least some short-day plants, the high intensity reaction can be substituted for by feeding the leaves sugar in the dark or under very low light intensities. Basic substrates must be synthesized as a foundation for any kind of morphogenic development, including floral initiation. Also, as discussed previously, the high intensity reaction provides the carbohydrates with which florigen appears to be associated in the process of translocation from the leaves. The likelihood seems great, however, that the high intensity light process also plays other roles in photoperiodism besides the rather obvious possibilities which have been mentioned.

The low intensity light reaction is the most obviously, and perhaps only, truly photoperiodic reaction involved. This reaction operates at least in part, and perhaps wholly, through the phytochrome system. It accounts for the reversal of the photoperiodic response when either long-day or short-day plants under relatively short photoperiods are exposed to light breaks during a corresponding long dark period, as described previously.

Earlier in the chapter the enhancing effect of blue light on flowering in white mustard when employed as a light break during a long dark period was described. Blue light also appears to play a role in the main light period in this long-day species. If the 16-hr photoperiods to which such plants are exposed are composed entirely of blue light, induction of flowering is much more effective than if they are composed of red or green light. Blue light also plays a role in the photoperiodic behavior of some short-day species. In a duckweed (*Lemna perpusilla*), for example, flowering occurs under continuous blue light, but not under continuous white or red light. In effect the blue light changes the photoperiodic reaction of this species from the short-day to the long-day type. The effect of blue light may be mediated in part or wholly through the phytochrome system, in part or wholly through the HER reaction (Chapter 20), or in part or wholly through some yet uncharacterized reaction.

Rhythmic Behavior of Plant Processes.   A characteristic feature of plants is the rhythmic or periodic behavior of many of the processes which occur in them. A typical terrestrial plant occupies a fixed spot in space and is thus at the mercy of the environmental factors which impinge upon it. Many environmental factors fluctuate periodically in magnitude. Diurnal periodicities, seasonal periodicities, and even lunar periodicities in the environmental factors can be recognized. Because of their influence upon plant processes, it follows that periodic fluctuations in environmental factors are paralleled by periodic patterns in the rates of occurrence of many plant processes and phenomena. Some important seasonal periodicities, largely environmentally conditioned, are discussed in Chapter 23. Good examples of environmentally conditioned diurnal periodicities are transpiration (Fig. 6.7) and photosynthesis (Fig. 11.8).

Some approximately diurnal phenomena which take place in plants occur independently of the environment in which the plant is situated. The leaves of a bean plant for example, typically assume a nearly horizontal position during the daytime and a nearly vertical position at night. That this daily leaf movement is not environmentally controlled is shown by the fact that if the plants are exposed to a constant environment, this periodic pattern of movement is retained, often for many days. The results of an experiment which demonstrates the existence of such an endogenous rhythm in the up-and-down movement of bean leaves under constant light and temperature is shown in Fig.. 22.6. Many other herbaceous species exhibit a similar periodicity of leaf movement which is not environmentally conditioned.

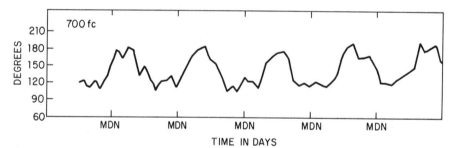

Fig. 22.6. Endogenous rhythm in the movement of the primary leaves of bean *(Phaseolus vulgaris)* under conditions of constant light and temperature. A horizontal leaf was assigned a value of 90° and a leaf pointing straight down 180°. Redrawn from Hoshizaki and Hamner, *Science,* **144,** 1964:1240. Copyright 1964 by the American Association for the Advancement of Science.

Such rhythms, each cycle of which approximates, but seldom exactly equals, 24 hours in length, are called *circadian rhythms* (circa=about; dia=day). Rhythmic responses or behavior of this circadian type are of widespread occurrence throughout the biological world. They appear to be operative in such diverse phenomena as, for example, petal movement of flowers, formation of conidia in the fungus *Neurospora crassa,* bioluminescence in *Gonyaulax polyedra* (a dinoflagellate), time of day at which adult insects emerge from pupae in *Drosophila,* and the activity cycles of various species of birds and rodents.

A distinctive feature of circadian rhythms is that their period lengths are relatively unaffected by temperature variations within the normal physiological range for the organism involved. Temperature may, however, affect the amplitude of the rhythmic oscillations, and temperature changes may operate in such a way as to reset the rhythm, as discussed later.

Circadian rhythms are manifestations of the *biological clock,* the existence of which has been detected in many types of living organisms. The metabolic basis for the operation of such rhythms is unknown.

So far we have considered environmentally conditioned diurnal rhythms and endogenous circadian rhythms as separate phenomena. However, an internal rhythm and an externally imposed rhythm can act simultaneously in affecting a plant process or response. The internal rhythm may operate in such a way as to reinforce the phenomenon affected or to diminish it depending upon the pattern of synchronization which exists between it and the environmentally-conditioned rhythm.

Endogenous Rhythms in Relation to Photoperiodism.   Beginning about 1936, Bünning, a German plant physiologist, and long-time investigator of circadian leaf movements, began to advance the view that there was an endogenous circadian rhythm component in photoperiodism. According to this concept an internal rhythm of approximately 24 hour duration operated in a plant which was sub-divided into two approximately 12-hour phases. Bünning originally proposed the terms *skotophil* and *photophil* for these two phases. The so-called photophil phase (meaning "light loving" phase) was supposed to promote flowering; the so-called skotophil phase (meaning "dark loving" phase) was supposed to hinder it.

For a while Bünning's hypothesis attracted very little credence, but more recently substantial evidence has accumulated that there is a circadian rhythm component in the photoperiodic behavior of at least some kinds of plants. The original Bünning interpretation of this phenomenon has, however, undergone substantial modifications.

Some of the most extensive investigations of this aspect of photoperiodism have been made by Hamner and his associates. Many of the implications of this work go beyond the scope of this book but the results of some of the most basic experiments will be considered.

In one experiment plants of short-day Biloxi soybean, which had previously been kept under long days, were exposed, in lots of ten plants each, to a series of different photoperiodic cycles. The photoperiod was 8 hours in all cycles, but the length of the dark periods was different in each cycle, ranging from 8 hours to 64 hours. Hence total cycle lengths ranged from 16 hours to 72 hours. Each lot of plants was exposed to its respective treatment for seven consecutive cycles. Maximum induction of flowering occurred under total cycle lengths of 24, 48, and 72 hours; while minimum induction of flowering occurred under cycle lengths of 16, 32 and 56 hours (Fig. 22.7). Thus flower bud initiation showed a rhythmical pattern of manifestation relative to total cycle length which suggests strongly that a circadian rhythm is involved.

A reasonable interpretation of these results is that at the 24, 48, and 72 hour total cycle lengths the imposed light-dark cycle is synchronized with an approximately 12 hour phase of the circadian rhythm which is favorable to, or

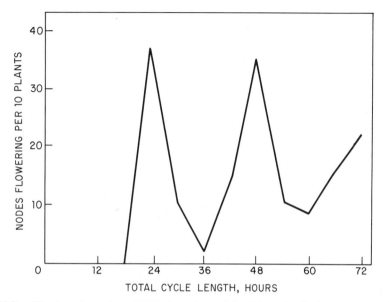

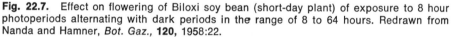

Fig. 22.7. Effect on flowering of Biloxi soy bean (short-day plant) of exposure to 8 hour photoperiods alternating with dark periods in the range of 8 to 64 hours. Redrawn from Nanda and Hamner, *Bot. Gaz.,* **120,** 1958:22.

at least not inhibitory to, flowering. On the other hand, at the 16, 32, and 56 hour total cycle lengths synchronization is with the other approximately 12 hour phase of the circadian rhythm which is suppressive of flowering.

It is noteworthy that this short-day plant does *not* initiate flowers under a "short day" cycle consisting of an 8 hour photoperiod, alternating with a 24 hour dark period. The term "short day" can meaningfully be applied to plants only in reference to their behavior under a 24 hour total cycle length.

Existence of a circadian rhythm component in their photoperiodic reactions has been demonstrated in several other short-day species, specifically cocklebur, coleus, Japanese morning glory, pigweed (*Chenopodium rubrum*) and a duckweed (*Lemna perpusilla*).

Evidence has also been obtained of the existence of an endogenous rhythm component in the photoperiodic reactions of the black henbane, a long day plant. In one experiment on this species, eleven groups of plants, previously kept under short days, were exposed to repeated cycles consisting of a 6 hour photoperiod followed by a dark period of a different length in each group. Dark periods ranged in length from 6 to 66 hours, hence total duration of each cycle ranged from 12 to 72 hours. The experiment lasted for a period of 42 days. In order to minimize physiological deterioration of the plants during the long dark periods, they were rooted in a culture medium containing both mineral salts and sucrose. The results are expressed in terms of the percentage of plants on which flower-stalk elongation became evident (Fig. 22.8).

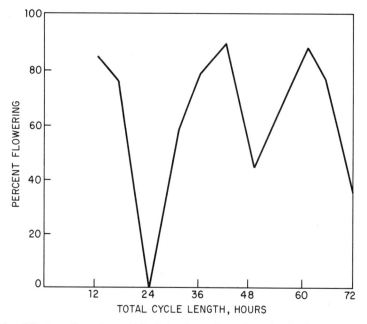

Fig. 22.8. Effect on flowering of black henbane (long-day plant) of exposure to 6 hour photoperiods alternating with dark periods in the range of 6 to 66 hours. Redrawn from Hsu and Hammer, *Plant Physiol.*, **42**, 1967:728.

Treatments with total cycle lengths of 24, 48, or 72 hours resulted in relatively little or no flowering, but intermediate cycle lengths were highly promotive to flowering. with maxima at 12, 42, and 60 hr cycles. Thus flowering showed a rhythmical pattern of response relative to the total cycle length which suggests very strongly that a circadian rhythm is involved. In brief, during one of the approximately 12 hour phases of the circadian rhythm, the plant is more receptive to a photoperiodic stimulus than during the other 12 hour period in each 24 hour span of time.

It is noteworthy that this "long-day" plant can flower under relatively short photoperiods even when they are associated with relatively long dark periods, if the latter are of an appropriate length and the total cycle length longer than 24 hours. The term "long day" can meaningfully be applied to plants only with reference to their behavior under a 24 hour total cycle length.

The peaks of flowering response in short-day Biloxi soybean (Fig. 22.7) are approximately 12 hours out of phase with those of the long-day black henbane (Figure 22.8). If the endogenous rhythms follow the same pattern in both species, the photoperiodic reactions in these two kinds of plants must be different in relation to the internal rhythm. The same phase of the rhythm which is suppressive of flowering in the short-day soybean is promotive of flowering in the long-day black henbane and vice-versa (*cf.* Fig. 22.7 and 22.8).

In some species in which the influence of an endogenous rhythm on the photoperiodic reaction has been sought, it has not been found. Whether this results from the absence of such a rhythm or from inappropriate experimental approaches is not certain.

A characteristic of circadian rhythms, including those involved in photoperiodism, is that they can be reset, relative to chronological time, under the influence of certain environmental factors. Resetting does not change the length of the individual cycle but, if the new environmental cycle is periodic, the circadian rhythm may again become synchronized with it. This phenomenon is known as *entrainment*. Light is the most common factor to operate in this manner. The cycle can be reset and entrained in some plants with the advent of light (often called the "dawn" or "light-on" signal). Circadian rhythms can also be reset in some species, such as Japanese morning glory, by the advent of darkness (often called the "dusk" or "light-off" signal). In other species, of which Biloxi soybean is an example, such rhythms can be reset by marked changes or differences in temperature.

Temperature Effects on Reproductive Growth.   Considered in general terms, reproductive growth, within the physiological range for plants, is favored in some species by relatively high temperatures, in others by relatively low ones, while in still others it occurs over a wide range of temperatures. Flowering and fruiting of plants in the first category are adversely affected in cool climates, flowering and fruiting of plants in the second category are adversely affected in warm climates.

The most favorable temperatures for vegetative growth of a given species are not always the same as the most favorable ones for reproductive growth of the same species. Lettuce (white Boston variety), for example, rapidly develops flower-stalks but no vegetative heads if grown at 21–27°C. At lower temperatures, however, vegetative growth of the heads is favored and flower-stalks develop only after a considerable delay, or not at all.

A complication in analyzing temperature effects on reproductive growth is that different stages in this complex of processes may be differently affected by temperature. Temperate zone species, especially those with a prolonged period of fruit maturation, are the most likely to exhibit different optimum temperature requirements between the flowering and fruiting stages of reproductive development. Such differences are much less marked in tropical and semi-tropical species.

The effects of temperature upon reproductive growth should be largely considered in conjunction with the effects of the length of the photoperiod, since these are the two major environmental factors which influence the development of flowers, fruits and seeds. Temperature, like photoperiod, is typically a cyclical factor of the environment. In all climates, night temperatures are regularly lower than day temperatures, although the range in alternation between maximum daytime and minimum nighttime temperatures varies greatly from one climate to another and from season to season and from day to day within a climate.

The reaction of plants to such temperature cycles has been termed *thermo-periodicity* (Chapter 20). The results of an experiment on tomato are an example of such a thermoperiodic effect on one phase of reproductive growth. With a daytime temperature of 26.5°C, fruit set was abundant in this plant only if nighttime temperatures were between 15 and 20°C and much less prevalent or absent at higher or lower night temperatures.

It appears that in many plants the flowering response is the outcome of the effects of both the temperature cycle and the light cycle operating simultaneously. In a number of species, however, the temperature of the dark period has more marked effects on the photoperiodic reaction than the temperature of the light period. In short day *Perilla,* for example, one experiment showed that flower buds were produced in 17 days, 37 days, and not at all, at respective temperatures of 20–22°C, 30–35°C, and 5–10°C during the 10 hr dark period.

There are several important ways in which the photoperiodic reactions of plants may be modified, depending upon the prevailing temperature regime. One such effect is on the basic pattern of the photoperiodic reaction itself. Strawberry, for example, is a short-day plant at moderate temperatures but operates as a long-day plant if grown at a temperature of 15°C or less. Sweet William (*Silene armeria*) shows an even more complex modifying effect of temperature on its photoperiodic response. A long-day plant at intermediate temperatures, it reacts as a short-day plant at either the lower temperature of 5°C, or at the higher temperature of 32°C.

The length of the critical period required for the induction of flowers is also affected by temperature. In cocklebur it was found that the length of the critical *dark* period at 21°C was about 8.5 hours but about 11 hours at 4°C for plants exposed to greenhouse temperatures during the photoperiod.

Still another effect of temperature on the photoperiodic reactions of some plants is that of a marked change in temperature in resetting the circadian rhythm.

Vernalization.   This term refers to the low temperature promotion of flowering. Vernalization is usually an inductive process; that is, the effect of the exposure to low temperatures does not become apparent until some time after the exposure. Vernalization, is in fact, one of the best examples of *thermal induction* in plants.

The phenomenon of vernalization has been studied extensively in cereal grains, such as winter rye and winter wheat, which are commonly planted in the fall, survive the winter as young seedlings and grow to maturity during the following summer. If, however, moist grains or very young seedlings of rye, for example, are exposed to low temperatures (0–10°C) for a few weeks, the time required for the developing plants to reach the stage of flower formation under moderate temperatures is reduced from about 14 weeks to about 7 weeks. Such vernalization of the grains of rye so speeds up the completion of its life cycle that it can be grown as a spring-sown crop instead of a fall-sown one. Winter varieties of wheat can be vernalized in a similar fashion.

If plants which have been exposed to vernalizing temperatures for an effective period of time are immediately thereafter exposed to relatively high temperatures (30°C or higher) the vernalized condition is nullified and induction of flowers does not occur. This phenomenon is called *devernalization*.

Another example of induced low temperature promotion of reproductive development is exhibited by many temperate zone biennials. Seeds of such plants germinate in the spring and develop only vegetatively, usually as rosettes, during the following summer. After passing the winter in this condition, flower-stalks elongate (a phenomenon often called *bolting*) during the second summer on which flowers, fruits, and seeds develop. Many well-known vegetable crop and floricultural plants follow the biennial pattern of growth: beets, cabbages, carrots, celery, stocks, foxglove, and pansies are examples. If, however, most biennials are not exposed to from a few days to several weeks of temperatures a little above the freezing point, as they usually are during the winter months, they remain vegetative during the second year. As discussed in Chapter 18, gibberellins can substitute for a cold period in evoking bolting and flower initiation in many kinds of biennials.

As discussed in Chapter 24, woody perennial plants of temperate zones usually must be subjected to a winter cold period in order for bud dormancy to be broken.

Various kinds of experiments show that the cold treatment is perceived in meristematic tissues, either in the embryo when seeds are vernalized, or in the buds when the low temperature is applied to seedling or more fully developed plants. This is in contrast with primary photoperiodic effects which are perceived in the leaves.

Carbohydrate and Nitrogen Metabolism. The flowering phase of reproductive growth is a relatively transitory one and is controlled predominantly by a hormonal mechanism rather than by nutritive conditions within the plant. Nevertheless, the construction of floral parts proceeds at the expense of foods translocated to the floral meristems, and some effects of the nutritional status of the plant are exerted even on this phase of growth, since no organ of a plant grows well unless adequately supplied with both carbohydrates and organic nitrogenous compounds.

Several phases of reproductive development appear to be linked to relative concentrations of carbohydrates and organic nitrogenous compounds. High nitrogen availability appears to favor development of carpellate rather than staminate flowers in a number of dioecious species. In the tomato, deficiency of carbohydrates induces microspore degeneration and pollen sterility, whereas nitrogen deficiency has no such effect and this seems to be true in at least some other species.

Development of fruits, particularly those of a fleshy type, is physiologically very similar to vegetative growth. The effect of different proportions of carbo-hydrates and nitrogenous compounds on the development of fleshy fruits is, in general, similar to their effects upon the development of vegetative organs (Chapter 20). In tomato, even when all other conditions are favorable for fruit development, deficiency of nitrogen results in the formation of small, tough, woody fruits. When nitrogenous compounds are present in adequate quantities and if other growth conditions are also favorable, large, juicy, succulent fruits develop. An adequate supply of water is also an obvious requisite for the maximum development of fruits of the fleshy type.

SUGGESTED FOR COLLATERAL READING

Bünning, E. *The Physiological Clock*. Springer-Verlag, Berlin, 1964.

Evans, L. T., Editor. *The Induction of Flowering*. The Macmillan Company of Australia, Melbourne, 1969.

Hillman, W. H. *The Physiology of Flowering*. Holt, Rinehart & Winston, Inc., New York, 1963.

Imamura, S., Editor. *Physiology of Flowering in Pharbitis nil*. Japanese Society of Plant Physiologists, Tokyo, 1967.

Salisbury, F. B. *The Flowering Process*. The Macmillan Company, New York, 1963.

Sweeney, Beatrice M. *Rhythmic Phenomena in Plants*. Academic Press, Inc., New York, 1969.

Wilkins, M. B., Editor. *Physiology of Plant Growth and Development*. McGraw-Hill Publishing Company, Maidenhead, England, 1969.

23 | GROWTH CORRELATIONS AND GROWTH PERIODICITY

Growth Correlations. The development of every organ of a growing plant is influenced to some degree by the physiological processes prevailing in some other organ or organs. Thus the vegetative growth of many plants is sharply checked during the period of fruiting because the presence of developing fruits strongly influences processes occurring in the root system. Similarly the size and vigor of root systems are influenced by photosynthetic activities of the leaves, and the formation of flower buds and flowers may be controlled by processes taking place in leaves. Such relationships, often reciprocal, existing among the organs of a plant are termed *growth correlations* or often simply *correlations*.

Growth correlations are not only exerted by one organ on another but also occur among tissues and even among cells. The harmonious development of the plant body as a whole is a result of correlative influences operating from organ to organ, tissue to tissue, and cell to cell. Hundreds of correlative influences are operating more or less continuously in the tissues of a growing plant. The discussion in this chapter will be restricted almost entirely, however, to some of the better known examples of the correlative influences of one plant organ on another.

Not all growth correlations result from operation of the same internal mechanism. Some result from the effect of one organ upon the supply and distribution of foods to other organs. One of the effects of leaves upon the root system is a correlation of this kind. Other correlations may be caused by the greater use of water or mineral salts in one organ than in another.

507

A large number of growth correlations apparently result from the influences of hormones or hormone-like substances. Examples of such correlations have been described in the preceding chapters.

Correlations Between Reproductive and Vegetative Development. The correlation between vegetative development and fruiting in the tomato plant is exemplified by the following experiment. When tomato plants were deflorated or the fruits were removed as rapidly as they set, the plants continued to grow vegetatively. If, however, the fruits were allowed to remain on the plant and enlarge, vegetative development and the formation of flowers gradually slowed down as more and more fruits began to develop. The steps in the inhibition of the development of such plants proceeded in approximately the following order: (1) loss of fecundity by the blossoms, (2) decrease in the size of the floral clusters, (3) abscission of the flower buds, (4) checking and later cessation of terminal growth of the stem, and (5) eventual death of all parts of the plant except the fruit.

The checking effect of the enlargement of fruits upon continued vegetative development and the development of flowers apparently resulted from the virtually complete monopolization of all of the nitrogenous compounds in the plants by the fruits. Carbohydrates, on the other hand, were accumulated in considerable quantities in both the fruits and vegetative organs. In general, the more nitrogenous compounds available, the more fruits that set and started to develop before inhibition of flowering and vegetative growth began. Removal of the fruits at any time before the vegetative parts died resulted in a renewal of vegetative growth and ultimately in another cycle of reproductive development.

The interrelationships between vegetative and reproductive growth have also been studied in cotton plants. The reduction in vegetative growth which was found to accompany the formation of bolls was attributed to the small quantity of carbohydrate reaching the root system. Most of the carbohydrate synthesized in the leaves was translocated into the developing fruits, with the result that the root systems received relatively small amounts of food. The effect of the low carbohydrate supply to the roots was to reduce markedly the absorption of mineral salts, which in turn restricted vegetative growth. Removal of the fruits resulted in tripling the sugar content of the root system and greatly increased the absorption of mineral salts.

Correlative effects between fruiting and flowering can be observed in most species which develop flower primordia over a considerable period of time, as is true of many summer-blooming species. If the blossoms of the sweet pea (*Lathyrus odoratus*) are allowed to develop, for example, flowering soon ceases; but if they are picked from time to time, flower primordia and blossoms develop continually throughout the growing season. All experienced flower gardeners know that if continued flowering is to be maintained in many species, especially annuals, the flowers must be cut regularly. Allowing fruit development to proceed soon results in a checking or even complete cessation of flowering.

The most satisfactory explanation of the growth correlations just described is that they result from modifications in the internal food relations of the plants. In general, they are believed to be caused by a diversion of such a large proportion of the available foods to developing flowers or fruits that other organs suffer a deficiency and hence are checked in growth. Both developing flowers and fruits are organs of high assimilatory and respiratory activity, and hence their maturation may result in a considerable drain on the available food supply. Some such correlative effects seem to result from a virtual monopolization of nitrogenous foods by the growing fruits; others appear to result mainly from the diversion of carbohydrate foods to the developing flowers or fruits.

The Shoot-Root Ratio.   A number of investigations have been made of the so-called shoot-root ratios in crop plants. Such ratios are usually calculated by dividing the dry weight of shoots formed by the dry weight of the roots formed during the growth period under consideration. The shoot-root ratio is influenced by reciprocal correlative influences between the aerial parts of a plant and its roots. The kind and magnitude of these correlative effects depend largely upon the environmental conditions to which the plant is exposed. For example, the nitrate concentration of the substratum has been shown to have a marked influence upon the shoot-root ratios of plants (Table 23.1).

TABLE 23.1   INFLUENCE OF NITRATE CONCENTRATION UPON THE SHOOT-ROOT RATIO OF BARLEY PLANTS[a]

Nitrate concentration in substratum	Dry weight of shoot, grams	Dry weight of roots, grams	S/R ratio
Low nitrate	9.64	1.81	5.33
Medium nitrate	11.81	1.43	8.28
High nitrate	10.55	1.17	9.08

[a]Duration of experiment, 49 days. (Data of Turner, *Am. J. Botany*, **9**, 1922: 427.)

The results of this experiment indicate a consistent increase in the shoot-root ratio with increase in the nitrate concentration of the solution culture. In this particular experiment there was also an absolute reduction in the dry weight of the roots developed with increase in nitrate concentration, but this was not found to be true in all the experiments performed by this investigator. Similar results have been obtained with a number of other species and by plants rooted in the soil as well as in solution cultures.

The effect of nitrates upon the shoot-root ratio can be interpreted in terms of their influences upon the internal food relations of plants. If the nitrate concentration of the substratum in which the plant is rooted is low, most of the nitrates absorbed are utilized in the synthesis of amino acids in the roots, the carbohydrates necessary for this process being translocated downward from the leaves. Most of these amino acids are used in the synthesis of protoplasmic proteins during the growth of the roots. Only a small proportion of the available nitrogenous compounds escapes utilization in the roots and is translocated, either as nitrates or as amino acids and related compounds, to the aerial portions of the plant. The tops are therefore relatively deficient in proteins. Hence the growth rate of the aerial portions of the plant will be relatively slow and the shoot-root ratio relatively low.

When the supply of nitrates is more abundant, however, a smaller proportion of the total quantity absorbed is utilized in the roots. A larger proportion of the nitrogen, as a constituent of one kind of compound or another, is translocated into the aerial portions of the plant, where much or all of it is usually utilized in the synthesis of protoplasmic proteins. The enhanced vegetative development of the aerial organs of the plant which is favored by such metabolic conditions results in the utilization of more carbohydrates as well as more proteinaceous foods by the aerial meristems. Because of the vigorous vegetative development of the shoot system, the proportion of the carbohydrate foods which are translocated to the roots may be relatively small. Hence, relative to the shoots, the roots are likely to be deficient in both carbohydrates and proteins, since synthesis of the latter requires carbohydrates as well as nitrates, and grow at a relatively slower rate than the tops. The net result is a higher shoot-root ratio than when the plants are grown in a soil which is deficient in nitrates.

Similarly, a decrease in the supply of carbohydrates within the plant as a result of a diminution in the rate of photosynthesis, or any other cause, influences the shoot-root ratio of plants. In general, diminution in the quantity of carbohydrate foods available in the tops results in an increased shoot-root ratio and vice versa. Plants grown in the shade, for example, have higher shoot-root ratios than other plants of the same species grown in full sunlight. Pruning commonly results in increasing the shoot-root ratio of woody plants, since the new growth following pruning is usually especially vigorous, resulting in monopolization of most of the available carbohydrates by the shoots. For similar reasons, defoliation from any cause, or cutting of leaves or tops (grasses, alfalfa, etc.) usually has the effect of increasing the shoot-root ratio. Removal of flowers, developing fruits, or developing buds, on the other hand, often favors root growth and may result in a decrease in the shoot-root ratio. The explanation of such effects follows a line of reasoning similar to that just presented in explanation of the relative influence of high and low nitrate supply on the shoot-root ratios of plants.

The shoot-root ratio is also influenced by the available soil-water content. In general, a relatively low soil-water content and adequate soil aeration favor relatively low shoot-root ratios, while the opposite conditions favor relatively high ones (Table 23.2). The shoot-root ratios as shown in this table are computed on a fresh weight basis but undoubtedly would show essentially the same relations if expressed on a dry weight basis. The results indicate clearly that

TABLE 23.2   SHOOT-ROOT RATIOS OF CORN SEEDLINGS GROWN
FOR 17 DAYS IN SAND AT VARIOUS WATER CONTENTS[a]

Per cent water in terms of dry weight of sand	Fresh weight of shoots, g.	Fresh weight of roots, g.	S/R ratio
38	3.63	4.95	0.90
30	3.54	4.21	0.84
20	3.36	5.18	0.65
15	2.35	4.90	0.48
11	1.56	4.30	0.36

[a]Data of Harris, *J. Am. Soc. Agron.*, **6**, 1914:68.

the shoot-root ratio increases with increase in the percentage of water in the
soil. The absolute weight of the shoots increases to a maximum at a soil-water
content of 20 per cent, after which it diminishes. The lesser development of
roots at the higher soil water contents is undoubtedly a result of the poorer
aeration of the wetter soils.

The relationship between photoperiodism and the shoot-root ratio has also
been studied. In general, long-day plants have higher shoot-root ratios under
long photoperiods, and short-day plants have higher shoot-root ratios under
short photoperiods. These generalizations are in agreement with the observation
that plants which are blossoming or those with young fruits have higher shoot-
root ratios than vegetative plants. The explanation probably lies in the monopolization of food materials by flowers and developing fruits. It is also possible
that the decreased formation of phloem tissues associated with flowering plays
a role in restricting the flow of foods into the root system.

TABLE 23.3   EFFECTS OF AERATION WITH DIFFERENT PROPORTIONS
OF OXYGEN AND NITROGEN ON SHOOT-ROOT RATIOS
OF YOUNG TOMATO PLANTS GROWN IN SOLUTION
CULTURES[a]

Concentration of oxygen, milliequivalents per liter	Dry weight, shoots, g.	Dry weight, roots, g.	S/R ratio
0.05	1.31	0.23	5.88
0.15	2.44	0.53	4.47
0.25	2.68	0.70	3.86
0.5	2.78	0.74	3.77
1.0[b]	3.11	0.78	4.05

[a]Data of Erickson, *Am. J. Botany*, **33**, 1946: 557.
[b]Approximate equilibrium with oxygen at atmospheric concentration.

Inadequate aeration results in a reduction in root growth in most species. This commonly leads to an increased shoot-root ratio (Table 23.3); increase in temperature within the physiological range for plants usually results in an increase in shoot-root ratio (Table 23.4).

TABLE 23.4   EFFECTS OF AIR TEMPERATURE ON THE SHOOT-ROOT RATIO OF CARROT[a]

Temperature, °F.	Fresh weight of shoots, g.	Fresh weight of roots, g.	S/R ratio
50–60	13.6	58.1	0.234
60–70	22.1	76.1	0.290
70–80	14.2	29.5	0.481

[a]Data of Barnes, *Cornell Univ. Agr. Ex. Sta. Mem.,* **186,** 1936: 9.

Apical Dominance.   In many herbaceous plants which produce aerial stems, growth in length takes place principally or entirely at the apex of the main axis of the plant. Although a lateral bud is present in the axil of every leaf, side branches do not often develop from these buds as long as the terminal bud retains its vigor and continues to grow. If, however, the terminal bud is destroyed or injured in any way, or is artificially removed, development of one or more of the lateral buds usually starts at once. This inhibiting effect of a terminal bud upon lateral bud development is called *apical dominance* and is much more pronounced in some species than in others.

If, for example, potato tubers begin to sprout, the apical buds grow rapidly but the lateral buds usually fail to elongate. If, however, the apical and lateral buds are separated by sectioning the tuber, they grow at similar rates (Fig. 23.1). Treatment of the tubers with ethylene chlorohydrin, which causes the destruction of auxin, results in the rapid growth of both lateral and apical buds.

The phenomenon of apical dominance is usually also in evidence in many woody plants on which true terminal buds form. The lateral buds on current

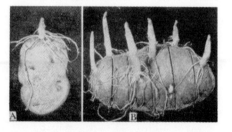

Fig. 23.1.   Growth of potato sprouts: (A) From the apical bud of an intact tuber, (B) From the apical and lateral buds of a sectioned tuber. Photograph from C. O. Appleman, *Univ. Maryland Agric. Expt. Sta. Bull.* No. 265, 1924.

shoots of such species usually do not develop unless the terminal bud is destroyed or injured. Development of the lateral buds on older shoot segments is of more frequent occurrence, indicating that the inhibitory effect of the apical bud diminishes with greater distance of the lateral buds from the apex of the stem.

Most coniferous trees and some broad-leaved trees as well have a single main stem which grows vertically upward. The lateral branches, however, assume an obliquely upright or almost horizontal position. If the apex of the main stem is destroyed or seriously injured, one or more (often all) of the lateral branches originating at the node or nodes immediately below the apex gradually turn upward as a result of greater growth on their lower than on their upper sides. Eventually these branches assume an approximately vertical position, often giving a candelabrum-shaped top to the tree. Subsequent vertical growth of the tree is accomplished by means of these reoriented branches. Maintenance of the more or less horizontal growth of the lateral branches in uninjured trees is obviously a result of some kind of control exerted by the apical growing region.

Mechanism of Apical Dominance.   The control of apical dominance apparently results mainly from auxin content. Auxin is synthesized in cells of the apical region of the stem. As noted previously, when these apical cells are removed from a stem, previously inhibited lateral buds on this stem will generally begin to grow. If auxin in agar or lanolin be applied to the cut end of the stem immediately after the apical cells are removed, inhibition of lateral bud growth continues as in intact plants. The maintenance of apical dominance by applied auxin as just described might suggest that auxin works alone in the inhibition of lateral bud growth. However, recent evidence indicates that other hormones are involved in the overall process.

For example, it has been shown that with some species of plants the application of cytokinins to inhibited lateral buds will release the inhibition of these buds, even though the stem apex is left intact. If only selected lateral buds on a given plant are treated with cytokinins, only the buds so treated will commence growth. It appears that the maintenance of inhibition of a lateral bud versus the release of the inhibition of this bud might be controlled by the relative abundance of auxin versus cytokinin within the bud (Fig. 23.2).

Although applied cytokinins can break apical dominance in some species, lateral shoots which develop on thus treated plants do not elongate as rapidly as lateral shoots on detipped controls. Auxin applied to these lateral shoots, after they are released by cytokinins, leads to growth rates which are comparable to those of detipped controls. Thus auxin, which has an inhibitory effect on the buds before they are released by cytokinins, has a promotive effect on the same buds after they are released. The latter effect of auxin on growth after release of the buds can in part be duplicated by applied gibberellins. Neither auxin nor gibberellins have any apparent releasing or growth promoting

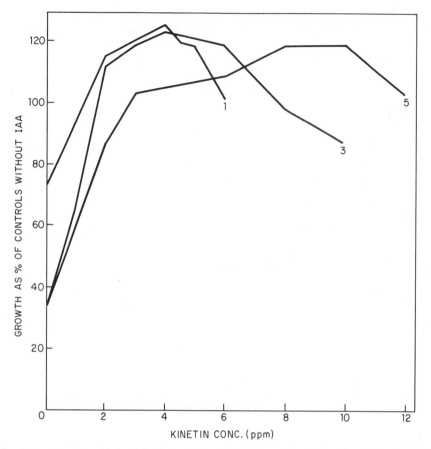

Fig. 23.2. Growth of lateral buds in different mixtures of indoleacetic acid and kinetin. Basal medium 1 per cent sucrose. Experiments carried out in 1, 3, and 5 ppm IAA. Data of Wickson and Thimann, *Physiol. Plantarum,* **11,** 1958:67.

effect on lateral buds if these buds are not first treated with applied cytokinins.

It appears likely that the actual release of lateral buds by cytokinins involves new cell divisions in the lateral bud, without much elongation of cells. The elongation phase which follows appears to be under the joint control of auxin and gibberellins.

Cambial Activity.   The first sign of growth in trees in the spring of the year is the swelling of the buds. This is quickly followed by the opening of the buds and the rapid elongation of the young stems. The resumption of growth by cambium cells in the older stems occurs much more slowly, however, and many days may elapse after the opening of the buds before cambium cells in the older stems begin to divide and enlarge. The division and enlargement of cambium cells first begins near the tips of the stems. A wave of cambial activity moves slowly from this region down the stems and branches and into the roots. Accompanying this basipetal migration of cambial activity is the differentiation of secondary xylem and phloem tissues. This progressive march of tissue differentiation from opening buds toward the roots has been known for many years, but until the discovery of hormones no satisfactory explanation of the phenomenon had been advanced.

There is considerable evidence that resumption of cambial activity is activated by auxins which move in a basipetal direction in stems from developing buds. In an experiment in which solutions of certain auxins were applied to the cut surface of detipped sunflower seedlings, activation of cambial growth was induced just as in intact plants, but no such activation occurred in detipped plants to which no auxin was applied. Similarly introduction of a crystal of indoleacetic acid into the cambium of willow and other woody species leads to rapid cambial growth below the point of insertion of the crystal. It has been shown that the auxin concentration of buds of apple and horse chestnut increases at the time of their enlargement and reaches a peak value just prior to the most rapid elongation of the current season's shoots. Movement of the hormone takes place from the current season's growth basipetally into older portions of the stem, paralleling the downward migration of cambial activity.

In herbaceous plants a growth correlation also appears to exist between cambial activity and flower production. Cambium cells divide rapidly throughout the length of the stem in plants that are strongly vegetative and growing vigorously. When flowering begins, the activity of the cambium cells is sharply checked. Cambial activity continues to decline as flowering progresses until, in profusely flowering plants, all of the cambium cells appear to have been differentiated into xylem and phloem tissues. The retardation of cambial activity first occurs in the region of the stem closest to the inflorescence and progresses basipetally from this region. The correlation between flower formation and retardation of cambial activity suggest the operation of a hormonal mechanism. Since transformation of a large proportion of vegetative to reproductive meristems would greatly reduce auxin output, this may be another auxin-controlled growth correlation.

Although the predominance of evidence indicates that auxin is the growth regulator involved in stimulating cambial activity, there are some plants in which gibberellic acid has also been found to be effective in stimulating cambial activity. This is true of some woody species including poplar, ash, and sycamore, as well as in several kinds of herbaceous plants.

Growth Correlations Between Leaves and Buds. Examination of a leafy shoot shows that a bud is present in the axil of every leaf. This same relationship between the position of leaves and the location of buds also holds at the shoot apex where minute buds develop above the center of each leaf primordium. When leaves are caused to form in unusual positions, buds also arise above the center of the point where the leaf joins the stem. The constancy of this relationship between leaves and buds suggest the existence of some controlling influences of leaves over bud development. Very soon after the leaf primordium begins its development, the first signs of bud formation also can be detected. However, if the young leaf primordium is carefully excised no bud will develop. Furthermore, partial separation of a very young leaf primordium from the stem apex by a vertical incision results in the formation of the bud upon the isolated portion of the leaf primordium—never upon the stem axis itself. Experiments of this kind demonstrate that the development of the buds is controlled by the leaf and not by factors in the stem axis. The mechanism of this control is unknown, but is probably of a hormonal type.

Polarity.   Many growth correlations are polar; that is, the two ends of a growing axis exhibit a marked dimorphism in development. The most familiar example of polarity in plants is that shown by cuttings, in which roots develop from the basal end and shoots from the apical end. Even if such cuttings are inverted and kept in a moist atmosphere roots will usually develop only from the morphologically basal end and shoots only from the morphologically apical end. It is not difficult, however, to induce the formation of roots at the upper end of a stem by the application of relatively high concentrations of growth regulators (Fig. 18.7).

While the obvious manifestations of polarity are morphological, basically all such phenomena depend upon a physiological mechanism. Many of the polar phenomena of plants probably result from the polar transport of auxins or other hormones. The polarity of cuttings, for example, can be explained largely if not entirely on a hormonal basis.

The movement of auxins in stems is usually polarized in the basipetal direction (Chapter 18). This polarity of movement appears to be associated with some fundamental organizational pattern of the protoplasm and cannot be changed easily. Segments of a stem with roots induced to form at the morphologically upper end of the stem axis can be inverted and grown for weeks in the inverted position without altering the original apex-to-base polarity of auxin transport. After three or four weeks in the inverted position a new polarity appears. The stem segments now transport auxin from the original base to the morphological apex as well as in the original direction. The inherent apex to base polarity of the stem persists, but a new polarity in the reverse direction is also present. Presumably the new polarity is limited to cells formed during growth in the inverted position.

Growth Periodicity.   The growth of a plant or plant organ never proceeds steadily hour after hour or day after day, but is subject to more or less regularly recurring, often rhythmical, daily and seasonal variations in rate. Seasonal variations in growth phenomena involve qualitative as well as quantitative differences in development during different stages of the growth cycle. Most plants, for example, produce flowers only at certain stages in their life history and either grow only vegetatively or not at all at other times. The more obvious examples of growth periodicity often correlate very closely with cyclical daily or seasonal variations in environmental conditions, but internal factors, including endogenous rhythms (Chapter 22), also play an important role in many periodic growth phenomena.

Daily Periodicity of Growth.   All actively growing plant organs characteristically exhibit a daily periodicity in growth rate. A number of studies have been made of daily variations in the rate of increase in the length of stems, the areal expansion or elongation of young leaves, and the diameters of growing fruits. Such measurements reveal that marked differences in the growth rates at different times of the day are of common occurrence in plants.

Cyclical variations in the rate of elongation of plant organs during the course of a day can be interpreted in terms of the principle of limiting factors. During the progress of the day, first one factor and then another is limiting. The rate of growth at any particular moment will be largely limited by the factor in relative minimum at that time. The three principal factors influencing the daily periodicity in the rate of elongation of plant organs are temperature, the internal water relations of the plant, and light.

As an example of the daily periodicity of growth, the growth (elongation) of the leaves plus stem axis of maize plants growing under field conditions will be considered (Fig. 23.3). In general, the rate of elongation was most closely correlated with temperature. Growth during the daylight hours was therefore usually greater than during the night hours, and growth during warm nights was greater than during cool nights whenever the internal water supply of the plants was not seriously deficient. Excessive temperatures (above about 35°C), however, apparently had a retarding effect on growth. Moderate internal water deficits reduced growth rates during the midday hours. Hence a double peak in growth rate, with one maximum in the early morning and the other in the evening, was common. Only when interval water deficits became relatively severe did the greatest growth occur at night. No evidence was found of a direct inhibiting effect of light on growth.

Numerous patterns of daily growth periodicity may occur in plants, depending upon the particular cyclical combination of environmental and internal factors which prevails. Maximum growth (elongation or enlargement) rates may take place at different times of the night, at midday, in the early morning, in the late afternoon, or in the early evening. Examples of conditions under which several of these different patterns of growth periodicity occur are given in the preceding paragraph. In some species, including maize, even transit of a cloud across the sun permits a temporary acceleration in growth rate. Under some conditions internal factors such as the supply of foods may play a determining role in shaping the daily periodicity of growth.

The preceding discussion has dealt with the influence on growth of varia-

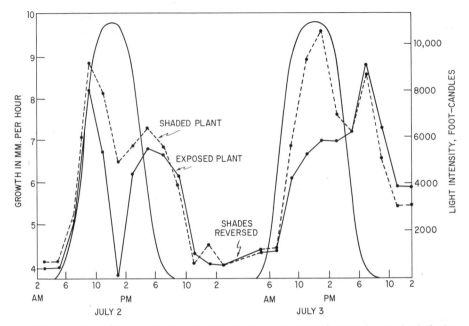

Fig. 23.3. Daily cycle of growth in height of two corn (maize) plants, one shaded, the other not, in relation to the daily cycle of light intensity. The plant which was shaded the first day was fully exposed to light the second, and vice versa. Data of Thut and Loomis, *Plant Physiol.,* **19,** 1944:120.

tions in temperature and internal water deficit within ordinary limits. In addition, it should be recognized that excessive internal water deficits, such as are engendered by drought conditions, or extremes of temperature, either high or low, may lead to a complete cessation of growth.

The factor of light must also be considered in relation to the phenomenon of growth periodicity. Although light is known to have direct effects on growth (Chapter 20), in plants growing under field conditions these appear to be relatively unimportant in comparison with its indirect effects. The influence of light on growth is exerted principally through its effects on the temperature of plant organs, rates of transpiration, and rates of photosynthesis. Plants exposed to intense light usually develop more severe internal water deficits than similar plants growing in the shade, because of the usually enhancing effect of light upon temperature and upon the rate of transpiration. Growth is slackened as a result of the greater water deficit within the plant. Low light intensities, on the other hand, may exert a retarding effect on growth because of a diminution in the total photosynthesis accomplished by the plant.

Measurements of the rate of growth (elongation or enlargement) of plant organs are sometimes complicated by the fact that there may be present in the organ mature cells which undergo reversible shifts in turgor and volume with changes in the intensity of the internal water deficit in the plant. The effect of shifts in the turgidity of cells on apparent rates of growth is well illustrated by the results of the following experiment. When growth in length of entire tomato stems was measured, elongation during the night apparently exceeded elongation during the daylight hours. Elongation of the meristematic stem tip (above the first node), however, was approximately the same for both day and night periods. Growth of such meristems continues even when a considerable internal water deficit exists in the plant (Chapter 8). During the day, smaller than actual rates for the entire stem were recorded because growth in length at the meristem was partially offset by shrinkage in volume of cells in older parts of the stem. During the night greater than actual growth rates for the entire stem were recorded, since in addition to growth in length at the meristem, there was an increase in the volume of cells in older parts of the stem. Similar effects of reversible changes in cell turgidity on the apparent growth of cotton bolls (Fig. 8.16) and other fruits have been recorded.

Seasonal Periodicity of Vegetative Growth.   All plants exhibit more or less clearly marked seasonal variations in the rate of vegetative growth. In temperate regions the periodic resumption of growth by woody perennials every spring is one of the most spectacular biological accompaniments of the march of the seasons. This topic will be discussed almost entirely in terms of such woody plants.

The seasonal periodicity of vegetative growth of any species, like the daily periodicity, is conditioned by both environmental and internal factors. Among the former, water, temperature and length of the photoperiod are especially important. Some of the internal conditions which are known to play a significant part in such phenomena are the genetic constitution of the species, hormonal relations, dormancy, correlative effects among organs, and the internal water relations. One of the most striking features of the seasonal growth of trees is its relative independence of fluctuations in the environmental factors. Trees growing in dry habitats may exhibit growth patterns very similar to trees of the same species growing nearby in locations where soil moisture is adequate. Although growth may be checked by low temperatures, many tree species initiate their growth in the spring before the close of the frost season, and elongation of shoots ceases long before the onset of frosts in the fall and winter. Similarly cambial activity of many coniferous and deciduous species decreases markedly long before environmental conditions become unfavorable. These aspects of growth periodicity may be under photoperiodic control.

The characteristic seasonal patterns of stem elongation in several tree species are illustrated in Fig. 23.4. Growth begins slowly with the appearance of the warmer days of early spring, accelerates rapidly during spring and early summer, and levels off by late summer or early fall. The shape of the growth curve cannot be correlated with environmental factors, nor is it appreciably modified by variations in rainfall unless these are exceptionally prolonged and severe.

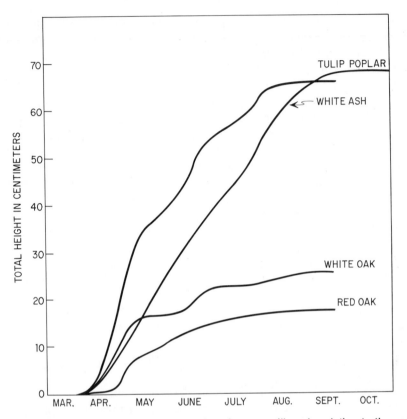

Fig. 23.4. Growth in height of several species of tree seedlings in relation to the season. Data of Kramer, *Plant Physiol.,* **18,** 1943:241.

In many species of woody plants buds do not ordinarily develop into shoots or flowers during the season in which they develop but remain in a dormant state. The length of time that buds remain dormant varies greatly according to the species (Chapter 24). Some lose their dormancy early in autumn, others retain it until late in the winter. The buds of temperate zone woody plants seldom open as soon as they lose their dormancy, but remain in a quiescent state until the favorable environmental conditions of spring. Low temperature is probably the principal factor preventing the development of quiescent buds in late winter and early spring, although a deficient water supply may also be involved, or growth inhibitors may be present (Chapter 24). Photoperiodic and thermoperiodic effects also play a part in the vernal resumption of vegetative growth in many plants. The breaking of dormancy of buds of some woody tree species such as red oak, sweetgum, birch, and beech can be shown to occur in response to long-day conditions. It appears that the buds themselves actually respond to the change in daylength. This is in contrast to induction of bud dormancy in many woody species which is thought to be brought about through a response by the *leaves* to the shortening photoperiods of autumn.

The development of new shoots from the buds on woody stems in the spring is not always a continuous process. In species such as cherries and willows which "leaf out" relatively early, the growth process is often intermittent. During this season, periods of warm weather often alternate with colder spells. Hence elongation of the developing shoots may take place in a series of short spurts, each terminated upon the advent of unfavorably cool weather. Apical growth is much more likely to proceed uninterruptedly in species such as beech and the hickories in which it is initiated later in the spring. Under favorable conditions practically all stem elongation and development of the new crop of leaves may occur in such species during a growth period lasting only two or three weeks. Termination of the spring burst of growth in such species is evidently a result of internal causes, since environmental conditions usually remain favorable for growth during much or all of the summer.

The stems of some ligneous species (sumac, dogwood, ailanthus, etc.) do not grow in the definite manner described above, but continue to elongate, producing leaf after leaf, for most or all of the summer, quiescent intermissions occurring only when environmental conditions are unfavorable. In some such species, growth is not terminated until the advent of frost.

Under some conditions the buds on woody stems open the same season they are formed. This is a commoner occurrence in certain species than in others and is more likely to happen on young trees or shrubs than on old ones. Defoliation of a tree relatively early in the season as a result of disease, insect ravages, drought, late frost, or any other cause usually results in a resumption of growth from buds developed during the current season. During wet summers, development of the current crop of buds into shoots occurs frequently in many species of woody plants. Among the oaks, especially when young, the development of two or even more successive shoots during a growing season is a common occurrence. When the terminal bud on the stem of an oak resumes growth during the season it is formed, lateral buds on that same segment also usually resume growth and develop into side branches. In the willow oak (*Quercus phellos*) three or even more prolongations of the same woody axis may take place during a single growing season. Almost always, however, when buds of the current crop on woody stems resume growth, there is a short dormant period between the time formation of the bud is completed and the time its active growth is resumed.

As noted previously, cambial growth in trees commonly begins near the opening buds and moves basipetally toward the roots. Weeks may elapse between the initiation of cambial activity in the young twigs and the resumption of cambium growth in the oldest region of the tree. Secondary thickening of the stems of most woody species generally continues later in the summer than elongation of the current shoots, although usually at a diminishing rate. Cambial activity usually ceases in the young twigs by midsummer but may continue until late summer or early fall in the older stems and sometimes until winter in the roots.

Less is known regarding the seasonal periodicity of the growth of roots than of the aerial organs of plants. The existence of an inherent dormancy in roots appears to be uncommon. Seasonal periodicity of root growth is probably largely controlled by environmental conditions. In colder climates little or no root growth occurs during the winter months. The results of one investigator indicate no growth of white pine roots, for example, took place in the relatively cold climate of New England between the middle of November and the first of April. In milder climates elongation of roots through the winter months occurs in at least some species of which apple and filbert are examples.

Cyclical Periodicity of Vegetative and Reproductive Growth.   The examples of seasonal periodicity which have already been described involve principally variations in growth rates. Growth periodicity is expressed not only in terms of seasonal variations in the quantitative aspects of growth, but also in the development of certain organs at one stage in the life cycle, and other organs at another stage. The most prominent periodicity in the qualitative aspects of plant growth is the cyclical development of vegetative and reproductive organs which is exhibited by most species of plants. It should be noted that flowering which occurs in response to seasonal changes in photoperiod is discussed in detail in Chapter 22.

The seasonal periodicity of all annual species is similar and involves in sequence: (1) seed germination, (2) vegetative development, (3) flowering and fruiting, usually accompanied, at least during the later stages, by slowly diminishing vegetative growth, (4) senescence, and (5) death of all organs except the seeds. All such species are perennial only by their seeds.

The seasonal periodicity of annual species is by no means immutable, however, but can be altered in various ways. As noted previously removal of flowers or fruits or both often leads to an acceleration or renewal of vegetative growth. Similarly a change in the length of the photoperiod at the onset of senescence often causes a rejuvenation of vegetative growth.

The cyclical development of vegetative and reproductive organs is similar in all biennial species. Plants of this type develop only vegetatively during their first growing season, forming underground organs which live over winter. In many biennials the leaves are cold resistant and survive the colder months of the year without injury. During their second growing season, vegetative development in often renewed, but before long is largely or entirely superseded by reproductive growth. Death of the plant follows closely after the formation of seeds and fruits. As with annuals the usual life cycle of biennials can be modified by various circumstances. For example, many biennials become annuals when growing in warmer or longer-season climates than in climates in which they normally behave as biennials. Many biennials can be converted into an annual pattern of development by treatment with a gibberellin (Chapter 18).

A greater diversity of cyclical patterns of reproductive and vegetative development is found in perennial species than in those which live for only one or two growing seasons. The following discussion refers primarily to plants of temperate regions.

In many woody perennials, flowers develop in the spring before vegetative growth is resumed or concurrently with the early stages in the development of the new leaf-bearing shoots. Examples of species which exhibit this type of periodicity include many fruit trees (peach, cherry, apple, etc.) and many forest trees species (elms, maples, oaks, chestnut, cottonwood, etc.). In some woody species such as the mulberry, in which flowers develop from axillary meristems on the current season's shoot, blooming occurs at the height of the season of vegetative growth. Flowers do not develop on many other woody perennials until after the season's vegetative growth is nearly or entirely completed. This is true of many species which bear terminal inflorescences at the end of the current season's shoots such as lilac, buckeye, and horse chestnut.

As in woody species, development of flowers in herbaceous perennials may take place before vegetative growth occurs during the same growing season, concurrently with the development of stems and leaves, or only toward the end of a period of vegetative development. The first of these types of growth periodicity, which is the least common, is found in certain spring-blooming species. The second is also characteristic of many spring-blooming herbaceous plants, but is by no means confined to such species. The third type of growth periodicity is especially common among summer- and fall-blooming plants and is characteristic of all species which produce terminal inflorescences on foliage-bearing stems.

Cyclical Periodicity of Vegetative Growth in Tropical Species.   Relatively little work has been done with tropical species with regard to cyclical growth patterns. The few studies which have been made reveal that at least some tropical species progress through growth periods (referred to as *flushes*) followed by dormant periods (referred to as *interflushes*). The flush period involves leaf expansion and internode elongation. In *Theobroma cacao* (the tree from which chocolate and cocoa are obtained), for example, a flushing period of about 22 days alternates with an interflushing period of about 43 days. This alternation of phases continues throughout the year even though seasonal variations in photoperiod and temperature are largely or entirely lacking in the areas where this plant normally grows.

When plants of *T. cacao* are raised under constant environmental conditions in the laboratory, the alternation of phases persists, although the interflush period is considerably shortened. Plants within a given experiment are not synchronous, that is, some plants were in the flushing period while adjacent plants were in the interflushing phase. The persistence of the growth pattern in a situation of constant environment is taken to be evidence that shoot-growth periodicity of this plant is under the control of an endogenous rhythm.

Seasonal Periodicity of Foliage Coloration.   A conspicuous and often spectacular aspect of growth periodicity in temperate regions is the autumnal leaf coloration of the leaves of deciduous woody species prior to leaf-fall. The "turning" of leaves in the autumn is not a result, as is commonly believed, of effects of frost. In fact, early frosts will greatly reduce the abundance of brilliance of the autumn leaf colors by killing or severely injuring the leaves before the pigments reach their maximum development. The sequence of events leading to the coloration of leaves in the autumn seems to be about as follows. In late summer or early autumn chlorophyll synthesis in the leaves ceases, while the destruction of the chlorophyll already present apparently proceeds at an accelerated rate. As the chlorophyll disappears, the residual yellow plastid pigments, carotene and xanthophylls (Chapter 10), become apparent. The yellow color of the leaves of many species at this season, as for example, the tulip tree, poplars, sycamore, and birch, is a result of the disappearance of the chlorophyll which has masked the presence of the yellow pigments during the summer season. The golden yellow effect produced in some leaves, such as those of beeches, results from the presence in the cells of a brownish pigment, probably a tannin, in addition to the yellow pigment.

The more prominent colors in most autumn landscapes, however, are the various shades of red and purplish red which develop in the leaves of such species as the red and sugar maple, many oaks, sumac, dogwood, and black gum. These result from the synthesis of anthocyanins (Chapter 14) in the leaf cells of such species. Autumnal development of anthocyanins is favored by periods of bright, clear, dry weather, during which cool but not freezing temperatures prevail.

Abscission.   Leaf-fall, particularly as it occurs from the stems of deciduous trees and shrubs in the autumn, is a distinctive phenomenon of periodic occurrence in plants of temperate regions. The *abscission* of leaves occurs at the point of petiole attachment to the stem. The phenomenon of leaf abscission is especially characteristic of woody dicotyledons, but also occurs in some herbaceous species such as coleus, begonia, and fuchsia. In most herbaceous species, however, the leaves are retained even after they die, and only disappear by decay or by mechanical disruption from the plant. In many herbaceous species most or all of the leaves are retained until after the death of the entire shoot system.

The abscission of leaves is associated with the so-called *abscission layer* which is made up of one or more layers of cells that undergo transverse divisions in a zone extending across the petiole near its base (Fig. 23.5). This zone of cells is often formed in the petiole before the leaf has reached its full size. The name "abscission layer" (*absciss layer, abscission zone*) has been applied to this region of the petiole because the separation of the cells of the petiole at the time of leaf-fall occurs between the cells of this zone. Separation of the cells of the abscission layer results from the dissolution of the middle lamella and sometimes also the cellulose wall of these parenchymatous cells. After separation of the cells of the abscission layer the petiole remains attached only by the vascular elements. These soon snap off under the pull of gravity or the pressure of the wind, and the leaf falls from the plant. The fractured elements of the vascular bundle usually become plugged with gums or tyloses.

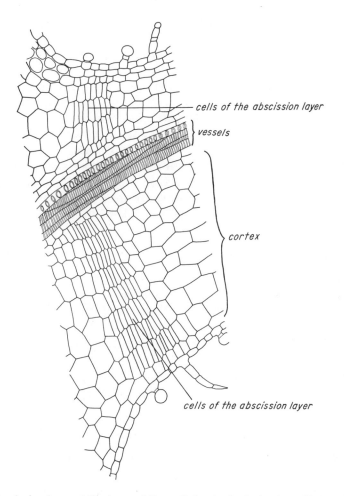

cells of the abscission layer

vessels

cortex

cells of the abscission layer

Fig. 23.5. Abscission layer at the base of the petiole of a leaf of coleus *(Coleus blumei)* as shown in vertical section.

It has been widely assumed that the abscission layer has a role of importance in the phenomenon of leaf-fall. There are at least three lines of evidence, however, which cast some doubt upon the validity of this assumption: (1) abscission occurs readily in species in which no abscission layer is present; (2) abscission may not occur in species in which abscision layers are present; (3) leaves of species in which an abscission layer usually develops can be induced to abscise before the abscission layer is formed. It is possible that the abscission layer may be of greater importance in the formation of the cork layer which covers the leaf scar after leaf-fall than in bringing about abscission itself. Whatever the role of the abscission layer may be, it is, when present, invariably the site of the separation of the leaf from the stem at the time of leaf-fall.

In species in which abscission occurs, this process may be brought about by changes in the environment of the plant. Such factors as soil-water content, temperature, light intensity, and photoperiod can play roles in abscission. A water deficit in plants brought about by drought conditions often leads to early leaf abscission. Cold temperatures can also trigger abscission in some plants. Reduced light intensity which might result from extreme shading is another factor which can lead to early leaf abscission. The abscission of lower and inner leaves of some crop plants raised in crowded conditions has been attributed to lack of light penetration between the plants. The length of the photoperiod is probably a major factor in the natural abscission process in many woody species. Short-day conditions stimulate abscission while long-day conditions can retard abscission. The late abscission of leaves of trees whose daylength is artificially lengthened by street lights is a good example of the abscission-retarding effect of long-day conditions.

Mechanism of Abscission.   When the blade portion of a coleus leaf is removed the petiole stump soon abscises, even if the leaf in question is relatively young at the time of de-blading. If auxin in lanolin paste is applied to the petiole stump of such de-bladed leaves, the abscission of the petiole is greatly retarded. This type of experimentation has been taken as evidence that IAA normally synthesized by young intact leaf blades inhibits abscission. When the blade is removed, the source of the auxin is thus removed, and abscission follows. This hypothesis is strengthened by the fact that the endogenous auxin level has been shown to fall off considerably just prior to normal abscission of the leaf (Fig. 23.6).

A modification of the above hypothesis has also been proposed which can be referred to as the *auxin-gradient hypothesis*. It has been suggested that it is not the absolute concentration of auxin in the blade *per se* which is the critical factor, but instead that the relative concentration of auxin on the blade side of the abscission zone versus the stem side of the abscission zone is critical in leaf abscission. When the auxin concentration is high in the blade relative to the auxin concentration on the stem side of the abscission zone, the leaf will not abscise. When the auxin level is nearly equal on both sides of the abscission zone, or higher on the stem side of the abscission zone, abscission will take place. Evidence for this hypothesis comes from tests which showed that auxin applied to the petiole side of the abscission zone of de-bladed leaves retards the rate of abscission but auxin applied to the stem side of the abscission zone of de-bladed leaves accelerates the rate of abscission. Under natural conditions the auxin concentration of the stem side of the abscission zone is probably maintained as a result of the basipetal movement of auxin from the stem tip or from young leaf blades near the top of the plant. The source of the auxin on the blade side of the abscission zone is the blade itself. Thus according to the auxin-gradient hypothesis, the lowering of the concentration of auxin in the blade as it matures (Figure 23.6) would result in a lower auxin concentration

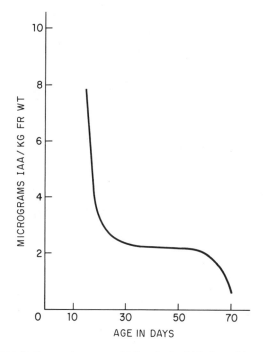

Fig. 23.6. Changes in the auxin concentration in leaf blades of bean with age. Approximate time required for complete leaf expansion is 30 days; first visual signs of senescence appear in approximately 60 days. Redrawn from Skoji, *et al., Plant Physiol.* **26,** 1951:190.

on the blade side of the abscission zone, and therefore the abscission process is initiated.

The discovery (Chapter 18) of the abscission-promoting growth substance, abscisic acid (ABA) has indicated that abscission may not be entirely under auxin control. Abscisic acid does not appear to be present in young expanding leaves but is present in leaves at the time of senescence. Applied ABA has been shown to be able to promote abscission of leaves, flowers, and fruits of various plants. The fact that usually fairly high concentrations of applied ABA are required to bring about abscission has led some workers to believe that this compound may not be as important in the natural abscission process as was generally thought when it was first discovered. Whether the increased concentration of ABA in leaves at the time of senescence is a cause of senescence or a result of this process is still not clear. Whether or not ABA interacts with the cytokinins, which are known to delay senescence, has also not yet been determined.

It has been known for some time that applied ethylene gas can cause rapid abscission of leaves. The mechanism by which ethylene plays a role in abscission is still not entirely clear, but at least a suggestion of its possible role is indicated in work relating to cellulase activity. It has been found that at the time of abscission, there is an inordinately large concentration of cellulase enzyme in the abscission zone as compared to tissue immediately on either side of this zone. It is assumed that the cellulase, as well as pectinases which have also been reported to occur in the abscission zone, degrade the cell wall constituents of cells in the abscission zone thereby weakening the mechanical strength of this tissue and facilitating the abscission process. It has been found that applied ethylene can cause an increase in the cellulase concentration in the abscission zone (Table 23.5). These results, however, do not prove that endogenous ethylene necessarily plays a role in bringing about a high concentration of cellulase in the abscission zone.

TABLE 23.5   CELLULASE ACTIVITY IN SEGMENTS CUT FROM BEAN SEEDLINGS *(Phaseolus vulgaris)*[a]

Segment	Relative cellulase activity
Pulvinus tissue (no treatment)	19.35 ($\pm$0.05)
Abscission zone (no treatment)	40.55 ($\pm$2.50)
Petiole tissue (no treatment)	20.10 ($\pm$4.05)
Abscission zone (No treatment with ethylene)	38.8 ($\pm$3.9)
Abscission zone (24 hrs. after treatment with ethylene)	62.6 ($\pm$1.8)

[a]Data of R. F. Horton and D. J. Osborne, *Nature* **214**: 1086–1088: 1967.

Abscission of Organs Other Than Leaves.   Leaves are not the only parts of plants which abscise. In compound leaves the individual leaflets usually drop one by one, leaving the petioles attached to the otherwise defoliated plant. Usually abscission of the petioles follows within a relatively short time. Similarly bud scales, inflorescences, petals, and fruits may be detached from the parent plant by abscission. Segments of the woody stems of some species also abscise. In many species of woody plants such as elm, cherry, birch, and linden, abscission of the leafy stem tips occurs at the termination of the spring growing period. In such species elongation of the stem continues the next season from the lateral bud just below the point of abscission, such lateral buds functioning essentially as terminal buds. Pines and some other species of conifers bear their needle-like leaves in fascicles, each fascicle being attached to a dwarf branch. In such species, leaves are shed in bundles by the abscission of the **dwarf**

branches rather than by detachment of the individual needles. In certain other woody species (oaks, cottonwood) segments of woody stems of considerable age and diameter are often lost by abscission. In some kinds of conifers, of which bald cypress (*Taxodium distichum*) and coast redwood (*Sequoia semper-virens*) are examples, individual leaves do not fall from the tree, but branches bearing numerous leaves abscise.

The phenomenon of the abscission of fruits presents certain practical problems to the horticulturist. If apple fruits, for example, drop from the tree before they can be picked, their quality is greatly impaired. The fact that the application of certain auxins and auxin-like compounds to leaves delays their abscission suggested that they would have a similar effect on fruits. Naphtha-leneacetic acid, when sprayed on apples or pears just before harvest, has proved especially effective in delaying the natural fall of the fruits. Postponement of fruit abscission as a result of such treatments is commonly as much as a week and sometimes longer. The length of time for which abscission is delayed differs considerably with the variety of fruit and with the prevailing environmental conditions. Postponement of fruit drop allows a longer period for picking and often permits the fruits to ripen to a higher quality on the tree before picking.

SUGGESTED FOR COLLATERAL READING

Leopold, A. C. *Plant Growth and Development.* McGraw-Hill Book Company, New York, 1964.

Wareing, P. I., and I. D. J. Phillips. *The Control of Growth and Differentiation in Plants.* Pergamon Press, Elmsford, New York, 1970.

Wilkins, M. B., Editor. *Physiology of Plant Growth and Development.* McGraw-Hill Publishing Company, Ltd., Maidenhead, England, 1969.

GERMINATION AND DORMANCY

The Structure of Seeds.   The development of seeds is discussed in some detail in Chapter 21. All seeds contain an embryo plant which is enclosed by one, or more commonly by two, seed coats. The seed coats originate from the integuments of the ovule and often exhibit external structural evidences of this origin even in the mature seed. Among these are the *hilum,* which represents the place where the seed was attached to the ovule stalk (*funiculus*), the *micropyle,* which frequently persists in the mature seed, and the *raphe,* a remnant of the ovule stalk which in certain kinds of seeds is adherent to the seed coats. When a single seed coat is present it is usually hard and woody, but when there are two seed coats the inner is almost invariably thin and membranous.

The embryos in seeds of different species of plants differ markedly in size and appearance, but all mature embryos are composed of one or more *cotyledons,* a *plumule,* and a *hypocotyl* (Fig. 24.1). The cotyledons vary in number from one in the monocotyledons, to two in the dictotyledons, to as many as fifteen in the embryos of some conifers. Cotyledons are generally thought to have a different anatomical origin than leaves and usually differ greatly in appearance from the foliage leaves of the same species. The cotyledons (or cotyledon) are attached near the upper end of the short thick stemlike axis of the embryo, the hypocotyl. The plumule or bud of the embryo is usually located just above the point at which the cotyledon or cotyledons are attached to the hypocotyl. The plumule consists of a meristem with several rudimentary foliage leaves. The primary root of the plant develops from the lower end of the hypocotyl. These structures can be seen in the stages of germination of the lima bean as shown in Fig. 24.2.

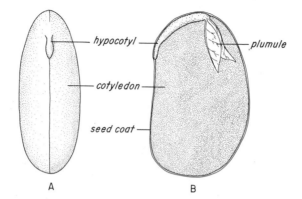

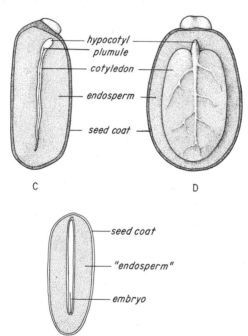

Fig. 24.1. Structure of seeds: (A) Seed of lima bean with seed coat removed. (B) Seed of lima bean with one cotyledon removed. (C) Seed of castor bean as seen in a longitudinal median section. (D) Seed of castor bean with endosperm removed down to the first cotyledon. (E) Seed of pinyon *(Pinus edulis)* as shown in a longitudinal median section.

An endosperm is also present in the seeds of many species (Fig. 24.1C,D). This tissue develops from the endosperm nucleus and usually contains considerable quantities of accumulated foods. In the seeds of those species which contain no endosperm, such as legumes, the cotyledons are usually enlarged and contain considerable quantities of reserve foods (Fig. 24.1A,B). Some

Fig. 24.2.   Stages in the germination of a seed of lima bean *(Phaseolus lunatus).*

seeds contain a *perisperm* which represents remnants of the nucellus. The so-called "endosperm" in the seeds of gymnosperms is not a true endosperm but represents the female gametophyte (Fig. 24.1E).

Insofar as the actual mechanics of seed germination are concerned, two principal groups of seeds may be recognized: (1) those in which the cotyledons emerge from the seed and (2) those in which the cotyledons remain permanently within the seed. Most seeds of dicotyledons and seeds of some monocotyledons such as onion belong to the first group, while the seeds of grasses and some dicotyledons such as peas and oaks belong in the second.

1. *Seeds in Which the Cotyledons Emerge.*   The sequence of events that takes place during the germination of the seed of the lima bean *(Phaseolus lunatus)* will be described as a type example in this group (Fig. 24.2). Germination is initiated by a marked swelling of the seed which usually ruptures the seed coat. This is followed by the emergence of the primary root which develops from the lower end of the hypocotyl and is the first structure of the embryo to make contact with the external environment. As the primary root grows downward in the soil, lateral roots and root hairs develop. The hypocotyl then elongates rapidly, pulling the cotyledons upward out of the soil into the air, where they separate into an approximately horizontal position on both sides of the

plumule. The plumule then begins active growth, giving rise to the stem and foliage leaves of the seedling. Since the bean is a seed without an endosperm, the food used during germination is largely derived from the accumulations in the thick cotyledons.

2. *Seeds in Which the Cotyledons Do Not Emerge.*   The seed of the pea is structurally very similar to that of the bean, but its germination behavior is very different. Elongation of the hypocotyl does not occur, and the cotyledons remain in the seed. The primary root elongates early in the process of germination much as in the bean. The plumule is elevated through the soil by rapid elongation of the epicotyl, which is the stem region between the cotyledons and the first true leaves—in other words, the first internode. This type of germination is also exhibited by oak acorns (Fig. 24.3).

Many monocotyledons also show this type of germination behavior. In the germination of the maize grain, for example, the primary root develops from the lower end of the axis of the embryo, growing through the coleorhiza and the wall of the grain (pericarp). As the primary root elongates, lateral roots soon appear and root hairs begin to develop just back of the elongating regions on all of the roots. The single cotyledon (*scutellum*) remains within the seed and acts as an absorbing organ through which soluble foods in the endosperm move into the tissues of the rapidly enlarging embryo. Soon after the appearance of the primary root, the plumule and the coleoptile, which completely encloses it, grow out through the walls of the grain and upward as a result of elongation of the region of the axis just below the plumule. About the time the coleoptile breaks through the surface of the soil or soon thereafter, the first foliage leaf grows through the tip of the coleoptile and emerges into the light and air.

Germination.   Prior to germination, many seeds pass through a phase in which they are dormant. Although the term *"dormant"* is often used in a loose sense to apply to seeds which fail to germinate, in the stricter sense the term is usually applied to seeds which fail to germinate for some internal reason even when all environmental conditions appear to be optimal for germination. The term *"quiescent"* is reserved for seeds which have not germinated because one or more environmental condition is not suitable for germination. For example, packets of seeds one might buy for spring planting in a garden would most likely be quiescent seeds which must simply be moistened to bring about germination.

We have briefly described the constituent parts of a seed from an anatomical point of view in the preceeding section. Before considering process of germination, it might be desirable to reevaluate the contents of a seed in a physiological sense.

In a seed prior to germination the seed coat is a dry, relatively hard tissue composed of generally nonliving cells. The cells of the seed coat, as well as the cells of the endosperm, if present, represent a barrier to the outward growth of the embryo. In some kinds of seeds the seed coat is also a barrier to the

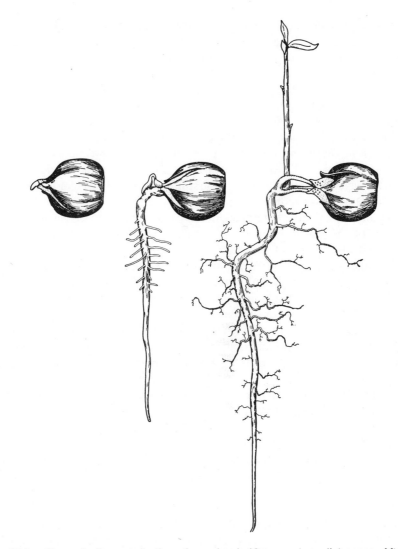

Fig. 24.3. Stages in the germination of a red oak *(Quercus borealis)* acorn. After Korstian, Yale Univ. *School Forestry Bull.* No. 19, 1927:18.

inward penetration of water and oxygen. Thus, if germination is to occur, the endosperm or seed coat or both must be rendered permeable to water and oxygen and also penetrable to the growing root tip of the embryo.

Aside from its cotyledonary portion, an embryo is largely composed of potentially meristematic cells. These embryonic cells are respiring at a very slow rate and probably not dividing in the dormant seed. Why aren't these meristematic cells dividing and enlarging? Answers to this question might include the following: (1) No water is reaching the cells of the embryo, thus growth of existing cells which is dependent upon the turgor created by an influx of water, cannot take place. (2) There is a relatively small amount of soluble respirable food in these cells. The large amount of accumulated foods which do exist in and around the embryo are in the form of polymers which must be hydrolyzed before they can be respired. (3) Aerobic respiration within the embryo cells may be limited because dry seed coat cells are generally much less permeable to oxygen than moistened seed coat cells. (4) Cells of the seed could contain inhibitors which have "turned off" some vital metabolic pathway which is prerequisite to growth of the embryo cells. (5) The concentration of active hydrolytic (digestive) enzymes is relatively low in the dry seed. (6) The hard seed coat cells present a physical barrier to growth of the embryo.

If a seed is to germinate, the previous six conditions limiting embryo growth must in some way be overcome, i.e., (1) water must reach the embryo cells, (2) non-respirable foods must be hydrolyzed, (3) an adequate amount of oxygen must reach the embryo cells, (4) inhibitors, if present, must be removed or counteracted, (5) the hydrolytic enzyme level must increase and (6) the seed coat must be rendered penetrable to the growing embryo. In the great majority of kinds of seeds, all six of these limiting conditions can be directly or indirectly overcome by simply placing the seed in a moist substrate. Some kinds of seeds to which water has been supplied still remain dormant, however. This indicates that one or more of the limiting factors in such seeds has not been overcome by the addition of water. With these types of seeds, the remaining limiting factors must be overcome by some additional modification in the environment of the seed as will be discussed in the more complete treatment of these factors that follow.

Water in Relation to Germination.   *Water uptake and penetration.*   Water will move from soil particles into seed cells providing that the water potential of the soil is less negative than the water potential within the cells of the seed (Chapter 5). Generally the water potential within a dry seed is very negative. For this reason water can generally move from the soil into the seeds, even though the soil might have a water-potential value somewhat below that of soil at field capacity. With some seeds, water vapor from very moist air is sufficient to result in imbibition and subsequently germination of the seeds. Water molecules entering dry seeds result in a considerable imbibitional force which is often sufficient to split the seed coat. Thus in some seeds, the physical restriction of embryo development by the seed coat is overcome by this imbibitional splitting of the seed coat. In other species, the seed coat does not split as a result of imbibitional force, but instead is ruptured by the internal pressure created by the growing primary root and/or by the enzymatic digestion of seed coat and other tissue surrounding the embryo, as will be discussed later.

Some intact seeds are so impervious to water that imbibition cannot take place. Such seeds will germinate only if *scarified*. This term is used to indicate any treatment, mechanical or chemical, which results in the weakening or rupturing of the seed coat. Scratching or cracking the seed coat is frequently used by agriculturalists to increase the germination percentage of such seeds. Soaking seeds in concentrated acid is another method used to weaken seed coats of some seeds. In nature, such seed coats are often degraded or weakened by the action of soil bacteria and fungi. Degrading of seed coats also often results when seeds pass through the digestive tracts of birds and other animals.

It is generally assumed that scarification leads to germination because water can penetrate a seed coat which previously it could not. However, it is entirely possible that with some types of seeds, the scarification results in germination because it facilitates the leaching of water-soluble germination inhibitors contained within the seed.

Seeds with seed coats relatively impervious to water include those of clover, alfalfa, black locust, honey locust, morning glory, and several types of berries.

Water and Enzyme Activation. Two of the factors which tend to keep a dry seed dormant are: 1) relatively few directly respirable food molecules are available to the embryo cells and 2) the relative concentration of activated hydrolytic enzymes is usually low. Let us consider how these two interrelated factors might be overcome by the simple penetration of water molecules into the seed. There are two possible reasons why the concentrations of active hydrolytic enzymes in a dormant seed might be low; (1) the enzyme molecules which are already present could be in an inactive form and (2) the metabolic pathway leading to synthesis of new enzyme molecules could be blocked.

It is known that hydration of certain dehydrated proteins can change the general configuration of the molecule. It is conceivable that enzyme molecules in the inactive dehydrated configuration might become activated (i.e. capable of catalyzing a given metabolic reaction) when they are in the hydrated form. If this were true then water would be directly activating the enzyme.

Water molecules may also play an indirect role in the synthesis of new enzyme molecules. One way in which synthesis of a given enzyme could be blocked would be as a result of the DNA code, which directs synthesis of a given enzyme, being repressed and therefore not transcribed to mRNA. Although at present we do not understand how derepression comes about, it is at least within the realm of possibility that the hydration of a specific type of molecule directly or indirectly triggers a chain of events leading to the derepression. There is reason to believe that intermediates in this chain of events leading to enzyme activation may be plant hormones, as will be discussed later.

Hydrolytic enzymes involved in germination are basically of two types: 1) those involved in the digestion of accumulated foods within the seed and 2) those which hydrolyze cell wall components.

The enzymes which act upon accumulated foods are important in that they convert non-respirable and non-assimilatable forms of foods to respirable and assimilatable forms. The conversion of polysaccharides to simple sugars which can then enter the glycolytic pathway is a typical example of such a hydrolysis. In addition, the hydrolysis of lipids can also lead to carbohydrates and other respirable organic molecules (Chapter 14). Some products resulting from hydrolysis of accumulated foods are not respired, but instead are translocated to newly forming cells of the growing embryo where they are used in the assimilation of new cell components. In some seeds the accumulated foods reside basically in the endosperm tissue, while in other types of seeds the accumulated food is present mainly in the cotyledons of the embryo itself.

The enzymes which act upon cell wall constituents are important in that, as a result of the reactions they catalyze, the seed coat or endosperm cells, or both, which constitute a physical barrier to the penetration of the embryonic root, become weakened. In addition, it is likely that some of the products resulting from these hydrolytic reactions enter the respiratory pathway while many others are simply translocated to new, growing cells of the embryo where they are re-assimilated.

Aside from any effect water molecules might have on the activation of the hydrolytic enzymes themselves, it should be remembered that water molecules also play a role in the hydrolytic processes as a chemical constituent of the reactions.

Light and Germination. A discussion of the phytochrome mediated photoreaction and its effects on the germination of light sensitive seeds such as those of lettuce has been taken up in Chapter 20. It will be recalled that red light promotes germination of such seeds while far-red light inhibits germination. With red light exposure, some of the phytochrome pigment within the light sensitive seeds is converted from the Pr to the Pfr absorbing form. However, it appears that it is not necessary for all of the phytochrome pigment to be converted to the Pfr form in order to trigger the next phase in the germination of such seeds. Instead, there is apparently a minimum critical proportion of the total phytochrome which must be in the Pfr form before the next phase of germination can be initiated. It is not clear, at present, the exact nature of the event triggered by the Pfr. There is, however, evidence that this event requires approximately twelve hours for completion. This supposition is based on the fact that the red light promotion of germination in lettuce seeds can be at least partially reversed by far-red light applied within twelve hours of the red irradiation. Far-red light applied twelve hours or more after red irradiation has no reversing effect on germination.

The response of light sensitive lettuce seeds to red and far-red light can be modified to some extent by temperature. For example, red light treatments which yield over 90 per cent germination at 25°C will yield under 10 per cent germination at 30°C. It is almost as though high temperature will substitute for far-red in reversing the effects of red light. In contrast to these results, it is found that lettuce seeds treated with far-red light can have at least some of the inhibiting effect of such light nullified by subsequent treatment at 2°C. In the latter case, the cold treatment is acting like a red light exposure in that it promotes germination. The temperature effects indicated above do not provide direct evidence that the phytochrome system itself is directly affected by temperature extremes. It is possible that temperature could be having some secondary effect on some aspect of the overall germination process.

In addition to low temperature promotion of light sensitive seed germination, certain chemical agents have also been found to have promotive effects on this process. Gibberellic acid, for example, can replace the red light requirement in lettuce seed germination. This fact is taken by some physiologists as an indication that the event triggered by the phytochrome conversion to Pfr discussed previously, may lead directly or indirectly to an increase in the level of natural gibberellic acid within the seed. Gibberellic acid has been shown to play a vital role in the germination of at least some nonlight-requiring seeds, as will be discussed later.

Temperature and Germination. Temperature effects on germination can be thought of in two distinct ways: 1) temperature as a factor in breaking dormancy of some seeds and 2) temperature as a general environmental factor affecting the germination process and the subsequent growth of the seedling. Temperature as a factor in breaking dormancy can itself be subdivided into at least two major areas; low temperature effects and alternating temperature effects.

Low Temperature Effects. Many seeds will not germinate immediately after removal from a fruit. These seeds are said to have an after-ripening requirement. The duration of the after-ripening requirement of some types of seeds can be lengthened or shortened by various environmental factors, one of which is temperature. For example, the after-ripening of certain seeds occurs more rapidly when these seeds are placed between layers of moist substrate at a low temperature (0 to 10°C), a treatment called *stratification,* than when these seeds are maintained in a dry condition at room temperature. Quite often two or three months of cold treatment are required to be effective. In nature, seeds of some species produced in a given year fail to germinate in that year. Such seeds must overwinter in the moist organic matter of the soil surface before they will germinate the following spring. Seeds of conifers as well as those of mountain ash, basswood, elder, bayberry, plum, cherry, and peach are examples of seeds in which germination is promoted by stratification.

Alternating Temperature Effects. In some seed-testing laboratories it is common practice to subject some kinds of seeds alternately to relatively low and high temperatures in order to break their dormancy. The temperature extremes of such treatments may not differ by more than 10 or 20 centigrade degrees and both temperatures are commonly well above the freezing point. The germination of seeds of Kentucky blue grass (*Poa pratensis*), for example, is greatly improved by subjecting the seeds alternately to temperatures of 20°C for 16–18 hours and 30°C for 6–8 hours; and the percentage germination of Johnson grass (*Holcus halpensis*) seeds is increased by alternate treatments at

30°C for 18–22 hours and 45°C for 2–8 hours. The dormancy of some kinds of seeds may be broken by alternate freezing and thawing, although this is decidedly harmful to other species. Alternating temperature treatments are entirely ineffective with seeds of some species. Seeds of carrot and timothy, for example, germinate just as well at constant temperatures as when temperatures are varied. In general, this method of treatment is used principally with seeds in which dormancy is inherent in the embryo.

It is not clear, as yet, exactly what specific effects stratification of alternating temperature treatments have on seeds. It is possible that these cold treatments affect certain seeds in a manner which parallels the effect of red light on light sensitive seeds. As with the red light effect, the cold treatments appear to trigger events which unblock formerly blocked metabolic pathways.

In addition to chemical changes, embryos of seeds of cherry have been shown to undergo anatomical changes during stratification treatments. Dormancy in this particular type of seed appears to be associated with lack of development of the embryo in nonstratified seeds, as will be discussed later.

Temperature Range as Overall Factor in Germination.   In the absence of other limiting factors, the seeds of any species will germinate within a specific range of temperature. At temperatures above or below this range, no germination will occur. As a rule, the seeds of species indigenous to temperate regions germinate within a lower range of temperatures than seeds whose native habitat is in tropical or subtropical regions. Wheat seeds, for example, germinate at temperatures only slightly above 0°C and at temperatures as high as 35°C, whereas the range of temperatures for germination of seeds of maize (a species of subtropical origin) lies between a lower value of 5 to 10°C and an upper limit of about 45°C. The optimum temperature is usually about midway between the two extremes of temperature at which germination will occur. It is not possible to designate any exact temperature as the optimum for germination, because this varies with the other prevailing environmental conditions and also with the exact criterion selected as an index of germination. The most favorable temperature for the elongation of the primary root, for example, does not always correspond to the most suitable temperature for the development of the plumule.

Embryo Development as a Factor in Germination and Dormancy.   Many species of plants have seeds in which the embryo does not develop as rapidly as the surrounding tissues, so that seeds which appear to be mature in a ripened fruit actually contain embryos which are not completely developed.

The range of incomplete embryo development in seeds of this type extends from embryos which are hardly more than beyond the fertilized egg stage to those which are very nearly completely developed at the time the seed appears outwardly to be mature. The germination of such seeds is necessarily delayed until formation of the embryo is complete. Thus undeveloped embryos are the cause of seed dormancy in some species. Further development of the embryo in such seeds often seems to be triggered by environmental factors, as exemplified by stratification effects on the seeds of cherry as discussed above, while in other species time seems to be the main requirement for further embryo development. Examples of other seeds which remain dormant as a result of incompletely developed embryos include those of ginkgo, European ash, holly, and many types of orchids.

Oxygen and Germination.   As mentioned previously, seed coats in the dry condition are generally less permeable to oxygen than seeds which have imbibed water. Although the change in permeability to oxygen is generally associated with the nonliving seed-coat cells, it is entirely possible that the permeability of membranes to oxygen in inner living cells is also directly or indirectly affected as a seed imbibes water.

Dormancy in seeds of cocklebur (*Xanthium*) seems in part, to involve the seed coat impermeability to oxygen. The two seeds in a fruit of cocklebur are not equally dormant. Under natural conditions the lower seed usually germinates in the spring following maturity while the upper seed remains dormant until the next year. If the seed coats are ruptured, or if the oxygen pressure is increased around intact seeds, both seeds germinate the first year. In addition to oxygen as a factor in cocklebur seed germination, other evidence indicates that the upper seed contains two leachable inhibitors of germination. Thus the rupture of the seed coat of the still dormant upper seed during the first year facilitates not only the entrance of oxygen, but also the removal by leaching of the germination inhibitors. In nature, the upper seed remains dormant a full year, during which weathering and the action of surface bacteria and fungi result in a partial degradation of the seed coat rendering it more permeable to oxygen and to the water soluble inhibitors.

In addition to cocklebur, seeds of a number of grasses and several members of the Compositae appear to have dormancy imposed in part as a result of seed coat impermeability to oxygen.

Oxygen is of course important to the germination process because of its relationship to aerobic respiration. The respiration of germinating seeds proceeds at a rapid rate, especially during the early stages of germination. The partial pressure of oxygen in the atmosphere can be reduced considerably, however, without greatly interfering with the rate of respiration (Chapter 13). In fact, the seeds of some water plants such as cattail (*Typha latifolia*) germinate better under low oxygen pressures than in air. Seeds of many terrestrial plants can germinate under water where the concentration of oxygen often corresponds to a partial pressure of oxygen very much less than that of the atmosphere.

During the early stages of germination of seeds of pea and some other species, respiration is largely or almost entirely of the anaerobic type because of the relative impermeability of even hydrated seed coats of such species to oxygen. As soon as the seed coats are ruptured, however, aerobic respiration replaces anaerobic oxidative processes even in seeds of this type.

The Germination Process in Barley Seeds. Unfortunately the germination of no particular type of seed has been studied so completely that all or even a majority of the various biochemical and biophysical steps in the germination process are completely understood. Perhaps the best inroads to the understanding of some of these steps have come from studies of the seed of barley. A summary of some of the basic facts thus far established for the sequence of events in barley seed germination is as follows: 1) as water is added to a dry barley seed, molecules of water reaching the embryo cells activate certain enzyme systems. 2) As a result of the action of now activated enzymes, gibberellic acid increases in concentration in the embryo cells. 3) Gibberellic acid moves from the embryo cells apparently through the endosperm tissue and eventually reaches the cells which form the outermost limit of the endosperm called the aleurone layer. 4) Once within the cells of the aleurone layer, the presence of gibberellic acid leads to the production of several hydrolytic enzymes (Table 18.1). 5) Hydrolytic enzymes such as α-amylase and protease begin digesting accumulated starch and protein in the cells of the endosperm. Possibly some hydrolytic enzymes also degrade cell wall constituents. 6) Molecules which result from the action of the hydrolytic enzymes are absorbed by the embryo cells and are utilized in respiration and assimilation processes within the now growing and dividing embryo cells.

It has been suggested that gibberellic acid brings about an increase in α-amylase enzyme concentration by making possible the synthesis of the specific m-RNA from DNA which codes for the synthesis of this enzyme. If this proves to be correct, then we can visualize the role of gibberellic acid as that of depressing formerly repressed genes. The recent finding that a second plant hormone, abscisic acid, can inhibit the gibberellic acid-induced synthesis of α-amylase in barley seeds suggests that abscisic acid may block the transcription of m-RNA from DNA. It is possible that abscisic acid could be acting here in the role of a gene repressor.

General Aspects of the Germination Process. As indicated previously, many types of seeds require rather special environmental conditions before they will germinate. This leads to the impression that every type of seed has its own specific requirements and proceeds through its own specific array of processes as it passes from the dormant phase to the germinated phase. Strictly speaking this may be true, but it is probably more valuable to visualize germination of any kind of seed as involving perhaps only two basic "themes" which are brought about in a variety of ways. These two basic themes, or fundamental requirements are that: 1) penetration of consumable requisites of germination (*i.e.,* water and oxygen) must take place and that sufficient quantities of these substances must reach the innermost cells of the seed, and 2) activation of certain enzyme systems must take place.

We have noted above that various seeds require pretreatments before penetration of oxygen or water or both will take place. These various pretreatments can be considered to be "variations" of the basic theme. Once the requisite molecules enter the seed, the processes they in turn engender are very similar, regardless of the type of seed involved.

There appears to be considerable variation in the trigger mechanisms which directly or indirectly initiate enzyme activation, *i.e.,* red light treatment, stratification treatment, etc. Where trigger mechanisms *indirectly* lead to enzyme activation, it is likely that the intermediate steps between trigger phase and activation phase will involve plant hormones, as was found in barley seed germination. In general, however, once the enzyme systems are activated, regardless of the trigger mechanism, the subsequent steps in germination appear to be fundamentally the same for all types of seeds.

Longevity of Seeds.   The life span of seeds varies from a few weeks to many years, depending upon the species and the environmental conditions to which the seeds are subjected. The silver maple (*Acer saccharinum*) may be cited as an example of a species which has short-lived seeds. When the seeds of this species mature in June, their water content is about 58 per cent. Once their moisture content drops below 30–34 per cent, the seeds die. Since this often happens within a few weeks in nature, seeds of this species soon perish. The seeds of the majority of crop plants are relatively short-lived under the usual storage conditions, generally remaining viable for only one to three years. The life span of such seeds can often be increased several fold by keeping them under suitable storage conditions. At the other extreme there are a few authentic records of seeds which have lived for more than a hundred years. Bequerel succeeded in germinating in 1934 seeds of *Cassia bicapsularis* which had been collected in 1819, and seeds of *Cassia multijuga* which had been collected in 1776. These are both South American species of legumes. Viable seeds of the Indian lotus (*Nelumbo nucifera*) have been found in Manchuria buried under layers of peat and soil to a depth that would indicate that they must have been at least 120 years old and may have been 200 to 400 years old. Although *Nelumbo* is an exception, generally seeds living for 75 years or longer have been found to be legumes.

At least some of the seeds of a number of wild plants will remain viable for 50 years or more. This is especially true of hard-coated species. As a general rule only seeds with a pronounced dormancy remain viable many years in nature. The seeds of many weed species are notoriously long-lived as compared with the seeds of most crop plants. This is illustrated by an experiment initiated by Beal at East Lansing, Michigan, in 1879. Seeds of twenty herbaceous species were mixed with sand and buried in bottles. Twenty such bottles were prepared. Once every five or ten years one of the bottles was disinterred and the enclosed seeds tested for their percentage of germination. Fifty years later seeds of five species remained alive and showed the following percentages of germination: yellow dock (*Rumex crispus*), 52 per cent; evening primrose (*Oenethera biennis*), 38 per cent; moth mullein (*Verbascum blattaria*), 62 per cent; black mustard (*Brassica nigra*), 8 per cent; and water smart-weed (*Polygonum hydro-*

piper), 4 per cent. At the end of seventy years some seeds of three species (*Oenethera biennis, Rumex crispus,* and *Verbascum blattaria*) were still viable. The germination percentage of the *Verbascum* seeds was still 72 per cent while the percentages for the seeds of *Rumex* and *Oenethera* were 8 and 14, respectively.

Dormancy of Buds.   Lateral and terminal buds ordinarily develop on the newly elongated shoots of temperate zone woody plants during the spring and early summer months. Such buds are not commonly in the dormant state during this period. Defoliation of the tree or shrub from any cause often results in the development of new shoots from buds formed during the current season. At or somewhat prior to leaf fall, however, such buds have passed into a state of dormancy.

Bud dormancy appears to be brought about in many woody species as a result of shortening of daylength. In some species a contributing factor in the initiation of the dormant condition in buds is the low soil moisture in late summer and autumn. Some workers feel, however, that "dormancy" brought about by water-stress conditions should actually be considered to be bud quiescence (see definition earlier) rather than bud dormancy. A third factor which results in bud dormancy of some species is low temperatures, but the effect of low temperatures appears to be enhanced when coupled with short-day conditions.

Bud dormancy which results from short-day conditions is a photoperiodic phenomenon. The photoreceptor in most woody plants is the leaves. Birch is an exception in that the buds of this tree can serve as the photoreceptive organ. For those species in which the leaves are the photoreceptor, it seems likely that there must be a transmission of some substance from the leaves to the buds in a manner which parallels the supposed transmission of a florigen in the flowering process. The plant hormone abscisic acid (originally named dormin) has been extracted from the leaves of the sycamore maple (*Acer pseudoplatanus*). This hormone induces dormancy in buds of this maple. It has not, however, been established that abscisic acid found in the maple leaves is there as a result of the photoperiodic treatment received by the leaf nor has it been proved that this hormone is transmitted from leaves to buds. Abscisic acid has been found to induce dormancy in buds of a number of woody species other than the sycamore maple.

Another parallel to photoperiodic induction of flowering exists in the induction of bud dormancy. When some types of woody plants under short photoperiods have the long night interrupted by light treatments, the buds do not go into a state of dormancy. The fact that red light is particularly effective in these night interruptions indicates that the phytochrome system may be involved, although this has not been definitely established.

Breaking of Bud Dormancy.   Buds of most woody species of the temperate zone will not break dormancy unless they receive a cold treatment near freezing for several weeks. The required duration of this cold treatment varies with the species of plant. Although the light stimulus which can photoperiodically induce bud dormancy is mediated via the leaves, the cold treatment which breaks dormancy appears to be mediated directly via the buds. In fact, there seems to be no transmission of a dormancy breaking stimulus, even between buds. This can be seen in tests of two-branched plants. If one branch receives a prolonged cold treatment but the other branch does not, only buds on the cold treated branch will break dormancy (Fig. 24.4).

Fig. 24.4. Breaking the dormancy of buds of blueberry by low temperatures. The branch on the right was exposed to low temperatures by allowing it to project through a small hole in a greenhouse during the winter. The branch on the left remained within the greenhouse. Photograph from Coville, *J. Agr. Research* **20**, 1920: Plate 22.

If dormancy is caused by the presence of abscisic acid in the bud, it remains unclear what effect the cold treatment is having on this compound which causes it to lose its effect. It would appear that as a result of the cold treatment, either the abscisic acid present in the autumn becomes physically or chemically destroyed, or else an abscisic acid antagonist such as gibberellic acid could be synthesized in the spring and thereby nullify the effects of the abscisic acid. It is possible that the trigger mechanisms which promote and break bud dormancy may eventually be found to be parallel to or even identical with those operating in seed dormancy and germination discussed previously.

Artificial Methods of Breaking Bud Dormancy.   Early in the twentieth century Johannsen discovered that dormant buds of many kinds of plants will begin to grow after being exposed to vapors of ether or chloroform for a day or two. The interval of time between the ether treatment and the beginning of growth differs widely with the time of the year at which the vapor is applied. In late summer or early fall, several weeks may elapse between the exposure to ether vapor and the active growth of the buds. Late in the winter, however, the interval is shortened to a day or two.

More recently a number of other chemical treatments have been discovered that are successful in breaking the dormancy of buds on potato tubers. Dormant tubers sprout freely when exposed to ethylene chlorohydrin vapor. Sodium and potassium thiocyanate, thiourea, dichloroethylene, carbon bisulfide, xylene, ethyl bromide, and a number of other compounds are also effective. In some tests, vapors of ethylene chlorohydrin so hastened the growth of tubers of the Irish cobbler variety of potato that vines more than half a meter high bearing young tubers 1 cm in diameter grew from the treated tubers before the sprouts of the untreated tubers appeared above the surface of the ground. Solutions of sodium and potassium thiocyanate gave almost equally striking results. Combinations of two or more compounds, each effective in overcoming the dormancy of buds in potato tubers, give better results than any single treatment. A mixture of ethylene chlorohydrin, ethylene, and carbon tetrachloride was more effective in breaking the dormancy of potato tubers than the sum of the effects of each compound used separately. Such synergistic effects are attributed to the increased penetration of the compounds as a result of changes in cellular permeability induced by one component of the mixture.

The vapors of ethylene chlorohydrin and ethylene dichloride also induce the growth of dormant buds of lilac (*Syringa vulgaris*), flowering almond (*Prunus triloba*), and some other species of woody plants (Fig. 24.5). The effect of the vapors was so restricted that when one of two paired buds of a lilac was exposed to the vapor, growth occurred only in the treated bud. The untreated bud remained as fully dormant as other buds more distant from the treated area. Apparently the factors causing the dormancy of the buds of these plants reside within the buds themselves and not in the adjacent tissues.

Methods of Prolonging Dormancy of Buds.

Often it becomes important to prolong, rather than to shorten, the dormancy of buds, especially of woody plants. In temperate zone regions the warm weather of an early spring may cause the buds to open before danger from frost is past. Under such conditions

Fig. 24.5. Effect of ethylene dichloride in breaking the dormancy of lilac buds. Plant on the right received no treatment. Plant on the left was exposed 48 hours to ethylene dichloride, 2.5 ml of liquid per .100 liters of space on December 10. Both plants were kept in a greenhouse and photographed in early January. Photograph from Denny and Stanton, *Am. J. Botany,* **15,** 1928: Plate XIX.

a severe frost can cause enormous damage. The prolongation of dormancy of buds of nursery stock held in storage or prepared for shipment is also very desirable. Dormancy of buds may be prolonged by spraying with certain chemical compounds. Some of the growth regulators are particularly effective in this respect. The results of experimental studies of orchard-grown fruit trees have not always been consistent, but greater success has been obtained in the treatment of nursery stock.

Probably the most complete and satisfactory control of dormancy through the use of chemical agents is that obtained over the buds of potato tubers. The sprouting of potato tubers in storage may be prevented almost completely by treatment with certain growth regulators. The methyl ester of naphthalene acetic acid is now widely used for this purpose. When shredded paper impregnated with small amounts of this ester is scattered among potato tubers in a storage bin, sprouting of the buds is inhibited. The tubers remain completely dormant and are firm and sound after overwinter storage. The dormancy induced in this way may be interrupted at any time with chemical treatments described earlier in this chapter.

SUGGESTED FOR COLLATERAL READING

Crocker, W., and Lela V. Barton. *Physiology of Seeds.* Chronica Botanica Company, Boston, 1957.

Forest Service, U. S. Dept. of Agriculture. *Woody-Plant Seed Manual.* U.S. Government Printing Office, Washington, 1948.

Kozlowski, T. T., Editor. *Seed Biology,* 3 vols. Academic Press, New York, 1972.

Leopold, A. C. *Plant Growth and Development.* McGraw-Hill Book Company, New York, 1964.

Mayer, A. M., and A. Poljakoff-Mayber. *The Germination of Seeds.* The Macmillan Company, New York, 1963.

Roberts, E. H. *Viability of Seeds.* Syracuse University Press, Syracuse, 1972.

Wareing, P. J., and I. D. J. Phillips. *The Control of Growth and Differentiation in Plants.* Pergamon Press, Elmsford, New York, 1970.

Wilkins, M. B., Editor. *Physiology of Plant Growth and Development.* McGraw-Hill Publishing Company, Ltd., Maidenhead, England, 1969.

INDEX

p/2
12 95